Lecture Notes in Computer Scie

Commenced Publication in 1973
Founding and Former Series Editors:
Gerhard Goos, Juris Hartmanis, and Jan van Leeuwen

Reinhard Koch   Fay Huang (Eds.)

Computer Vision – ACCV 2010 Workshops

ACCV 2010 International Workshops
Queenstown, New Zealand, November 8-9, 2010
Revised Selected Papers, Part I

 Springer

Volume Editors

Reinhard Koch
Christian-Albrechts-University Kiel
Computer Science Institute
Olshausenstr. 40
24098 Kiel, Germany
E-mail: rk@informatik.uni-kiel.de

Fay Huang
National Ilan University
Institute of Computer Science and Information Engineering
Shen-Lung Rd. 1
26047 Yi-Lan, Taiwan R.O.C.
E-mail: fay@niu.edu.tw

ISSN 0302-9743   e-ISSN 1611-3349
ISBN 978-3-642-22821-6   e-ISBN 978-3-642-22822-3
DOI 10.1007/978-3-642-22822-3
Springer Heidelberg Dordrecht London New York

Library of Congress Control Number: 2011936637

CR Subject Classification (1998): I.4, I.5, I.2.10, I.2.6, I.3.5, F.2.2, H.5, J.5

LNCS Sublibrary: SL 6 – Image Processing, Computer Vision, Pattern Recognition, and Graphics

Typesetting: Camera-ready by author, data conversion by Scientific Publishing Services, Chennai, India

Printed on acid-free paper

Springer is part of Springer Science+Business Media (www.springer.com)

Preface

During ACCV 2010 in Queenstown, New Zealand, a series of eight high-quality workshops were hold that reflect the full range of recent research topics in computer vision. The workshop themes ranged from established research areas like visual surveillance (the 10th edition) and subspace methods (third edition) to innovative vehicle technology (From Earth to Mars), from vision technology for world e-heritage preservation and mixed and augmented reality to aesthetic features in computational photograpy and human computer interaction.

From a total of 167 submissions, 89 presentations were selected by the individual workshop committees, yielding an overall acceptance rate of 53%. The reported attendence was quite attractive, between 40 and 60 participants in each of the workshops, sometimes over 70.

The two-volume proceedings contain a short introduction to each workshop, followed by all workshop contributions arranged according to the workshops.

We hope that you will enjoy reading the contributions which may inspire you to further research.

November 2010

Reinhard Koch
Fay Huang

Introduction to the 10th International Workshop on Visual Surveillance

Visual surveillance remains a challenging application area for computer vision. The large number of high-quality submissions is a testament to the continuing attention it attracts from research groups around the world. Within this area, the segmentation of the foreground (moving objects) from the background (residual scene) remains a core problem. Approximately half of the papers accepted for publication propose innovative segmentation processes. These include the modeling of photometric variations using local polynomials, the exploitation of geometric and temporal constraints, and the explicit modeling of foreground properties. The segmentation of foregrounds consisting of slowly moving objects is explored and there are two investigations into the improvements in segmentation that can be obtained using feedback from a subsequent tracking process.

Nonetheless, there is also an increasing interest in the detection of pedestrians, faces and vehicles using methods that do not rely on foreground–background segmentation. Several enhancements to the histogram of gradients method for pedestrian detection are proposed, leading to an improved efficiency and invariance of the results under rotations of the image. A method to improve the efficiency of the boosted cascade classifier is also proposed. A key problem for visual surveillance scene understanding is the tracking of pedestrians in arbitrarily crowded scenes across multiple cameras: there are several papers that offer contributions to the solution of this problem, including the modeling of pedestrian appearance as observed from multiple cameras in a network.

In the 12 years in which the Visual Surveillance workshops have been running, algorithms have become more sophisticated and more effective, more data sets have become available and experimental techniques and the reporting of results have improved. In spite of these advances, many of the classic problems in computer vision, such as optic flow estimation, object detection and object recognition, are still as relevant to the visual surveillance community as they have ever been.

The Workshop Chairs would like to thank the Program Committee for their valuable input into the reviewing process, and Reinhard Koch and Fay Huang for providing efficient liaison on behalf of the ACCV. The Chairs would also like to thank Graeme Jones, who dealt with many of the organizational aspects of this workshop.

November 2010

James Orwell
Steve Maybank
Tieniu Tan

Program Committee

Francois Bremond	INRIA Sophia-Antipolis Research Unit, France
Andrea Cavallaro	Queen Mary, University of London, UK
Patrick Courtney	PerkinElmer Life and Analytical Sciences, UK
Roy Davies	Royal Holloway, University of London, UK
Rogerio Feris	IBM Research, USA
Gustavo Fernandez Dominguez	AIT Austrian Institute of Technology Gmb, Austria
Gian Luca Foresti	University of Udine, Italy
Xiang Gao	Siemens Corporate Research, USA
Shaogang Gong	Queen Mary University London, UK
Riad Hammoud	Delphi Corporation, USA
R. Ismail Haritaoglu	Polar Rain Inc, USA
Janne Heikkila	Dept. of Electrical Engineering, Finland
Wei Ming Hu	NLPR, China
Kaiqi Huang	Institute of Automation CAS, China
Graeme Jones	DIRC, Kingston University, UK
Kyoung Mu Lee	Seoul National University, Korea
Peihua Li	Hei Long Jiang University, China
Stan Li	National Laboratory of Pattern Recognition, China
Xuelong Li	Birkbeck College, University of London, UK
Dimitrios Makris	Kingston University, UK
Steve Maybank	Birkbeck College, UK
James Orwell	Kingston University, UK
Vasudev Parameswaran	Siemens Corporate Research, USA
Federico Pernici	Università di Firenze, Italy
Justus Piater	Université de Liège, Belgium
Massimo Piccardi	University of Technology, Sydney, Australia
Ian Reid	University of Oxford, UK
Paolo Remagnino	Kingston University, UK
Gerhard Rigoll	Munich University of Technology, Germany
Neil Robertson	Heriot-Watt University, UK
Gerald Schaefer	Loughborough University, UK
Vinay Shet	Siemens Corporate Research, USA
Nils T Siebel	HTW University of Applied Sciences, Germany
Lauro Snidaro	Università degli Studi di Udine, Italy
Zoltan Szlavik	MTA SzTAKI, Hungary
Tieniu Tan	National Laboratory of Pattern Recognition, China
Dacheng Tao	Birkbeck College, University of London, UK

Introduction to the Second International Workshop on Video Event Categorization, Tagging and Retrieval (VECTaR)

One of the remarkable capabilities of the human visual perception system is to interpret and recognize thousands of events in videos, despite a high level of video object clutter, different types of scene context, variability of motion scales, appearance changes, occlusions and object interactions. As an ultimate goal of computer vision systems, the interpretation and recognition of visual events is one of the most challenging problems and has increasingly become very popular in the last few decades. This task remains exceedingly difficult because of several reasons:

1. There still remain large ambiguities in the definition of different levels of events.
2. A computer model should be capable of capturing a meaningful structure for a specific event. At the same time, the representation (or recognition process) must be robust under challenging video conditions.
3. A computer model should be able to understand the context of video scenes to have meaningful interpretation of a video event. Despite these difficulties, in recent years steady progress has been made toward better models for video event categorization and recognition, e.g., from modeling events with a bag of spatial temporal features to discovering event context, from detecting events using a single camera to inferring events through a distributed camera network, and from low-level event feature extraction and description to high-level semantic event classification and recognition.

This workshop served to provide a forum for recent research advances in the area of video event categorization, tagging and retrieval. A total of 11 papers were selected for publication, dealing with theories, applications and databases of visual event recognition.

November 2010

Ling Shao
Jianguo Zhang
Tieniu Tan
Thomas S. Huang

Program Committee

Faisal Bashir	Heartland Robotics, USA
Xu Chen	University of Michigan, USA
Ling-Yu Duan	Peking University, China
GianLuca Foresti	University of Udine, Italy
Kaiqi Huang	Chinese Academy of Sciences, China
Thomas S. Huang	University of Illinois at Urbana-Champaign, USA
Yu-Gang Jiang	City University of Hong Kong, China
Graeme A. Jones	Kingston University, UK
Ivan Laptev	INRIA, France
Jianmin Li	Tsinghua University, China
Xuelong Li	Chinese Academy of Sciences, China
Zhu Li	Hong Kong Polytechnic University, China
Xiang Ma	IntuVision, USA
Paul Miller	Queen's University Belfast, UK
Shin'ichi Satoh	National Institute of Informatics, Japan
Ling Shao	The University of Sheffield, UK
Peter Sturm	INRIA, France
Tieniu Tan	Chinese Academy of Sciences, China
Xin-Jing Wang	Microsoft Research Asia, China
Tao Xiang	Queen Mary University London, UK
Jian Zhang	Chinese Academy of Sciences, China
Jianguo Zhang	Queen's University Belfast, UK

Introduction to the Workshop on Gaze Sensing and Interactions

The goal of this workshop is to bring researchers from academia and industry in the field of computer vision and other closely related fields such as robotics and human – computer interaction together to share recent advances and discuss future research directions and opportunities for gaze sensing technologies and their applications to human – computer interactions and human – robot interactions. The workshop included two keynote speeches by Ian Reid at the University of Oxford, UK, and Chen Yu at Indiana University, USA, who are world-leading experts on gaze – sensing technologies and their applications for interactions, and seven oral presentations selected from submitted papers by blind review. This workshop was supported by the Japan Science and Technology Agency (JST) and CREST. We would like to thank Yusuke Sugano, Yoshihiko Mochizuki and Sakie Suzuki for their support in organizing this event.

November 2010

Yoichi Sato
Akihiro Sugimoto
Yoshihiro Kuno
Hideki Koike

Program Committee

Andrew T. Duchowski	Clemson University, USA
Shaogang Gong	Queen Mary, University of London, UK
Qiang Ji	Rensselaer Polytechnic Institute, USA
Kris Kitani	The University of Electro-Communications, Japan
Yoshinori Kobayashi	Saitama University, Japan
Yukie Nagai	Osaka University, Japan
Takahiro Okabe	University of Tokyo, Japan
Kazuhiro Otsuka	Nippon Telegraph and Telephone Corporation, Japan
Ian Reid	University of Oxford, UK
Yusuke Sugano	University of Tokyo, Japan
Yasuyuki Sumi	Kyoto University, Japan
Roel Vertegaal	Queen's University, Canada

Introduction to the Workshop on Application of Computer Vision for Mixed and Augmented Reality

The computer vision community has already provided numerous technical break-throughs in the field of mixed reality and augmented reality (MR/AR), particularly in camera tracking, human behavior understanding, object recognition, etc. The way of designing an MR/AR system based on computer vision research is still a difficult research and development issue. This workshop focuses on the recent trends in applications of computer vision to MR/AR systems.

We were proud to organize the exciting and stimulating technical program consisting of ten oral presentations and five poster presentations. We were very happy to have a distinguished invited speaker, Hideyuki Tamura, who has led the MR/AR research field since the 1990s. Finally, we would like to thank all of the authors who kindly submitted their research achievements to ACVMAR 2010 and all members of the Program Committee for their voluntarily efforts.

ACVMAR 2010 organized in collaboration with SIG-MR(VRSJ) and the GCOE Program at Keio University.

November 2010

Hideo Saito
Masayuki Kanbara
Itaru Kitahara
Yuko Uematsu

Program Committee

Toshiyuki Amano	NAIST, Japan
Jean-Yves Guillemaut	University of Surrey, UK
Ryosuke Ichikari	Ritsumeikan University, Japan
Sei Ikeda	NAIST, Japan
Daiske Iwai	Osaka University, Japan
Yoshinari Kameda	University of Tsukuba, Japan
Hansung Kim	University of Surrey, UK
Kiyoshi Kiyokawa	Osaka University, Japan
Takeshi Kurata	AIST, Japan
Vincent Lepetit	EPFL , Switzerland
Walterio Mayol-Cuevas	University of Bristol, UK
Jong-Il Park	Hanyang University, Korea
Gerhard Reitmayr	TU Graz, Austria
Chiristian Sandor	South Australia University, Australia
Tomokazu Sato	NAIST, Japan
Fumihisa Shibata	Ritsumeikan University, Japan
Ryuhei Tenmoku	AIST, Japan
Yuki Uranishi	NAIST, Japan
Daniel Wagner	TU Graz, Austria
Woontack Woo	GIST, Korea

Introduction to the Workshop on Computational Photography and Aesthetics

Computational photography is now well-established as a field of research that examines what lies beyond the conventional boundaries of digital photography. The newer field of computational aesthetics has seen much interest within the realm of computer graphics, art history and cultural studies. This workshop is intended to provide an opportunity for researchers working in both areas, photography as well as aesthetics, to meet and discuss their ideas in a collegial and interactive format.

The papers contained in these workshop proceedings make important contributions to our understanding of computational aspects of photography and aesthetics. The first paper, by Valente and Klette, describes a technique for blending artistic filters together. Their method allows users to define their own painting style, by choosing any point within the area of a triangle whose vertices represent pointillism, curved strokes, and glass patterns. The second paper, by Sachs, Kakarala, Castleman, and Rajan, describes a study of photographic skill whose purpose is to establish whether that skill can be identified in a double-blind manner. They show that human judges who are themselves expert photographers are able to identify up to four skill levels with statistical significance. The third paper, by Rigau, Feixas, and Sbert, applies the information theory of Shannon to model the channel between luminosity and composition. They show how changes in depth-of-field and exposure are reflected in the information channel, and formulate measures for saliency and "entanglement" in an image. The fourth paper, by Lo, Shih, Liu, and Hong, describes how computer vision may be applied to detect a classic error in photographic composition: objects which appear to protrude from a subject's head. Their method is able to reliably detect protruding objects in a variety of lighting conditions and backgrounds, with a detection rate of 87% and false alarm rate of 12%. The fifth paper, by Constable, shows how traditional drawing methods such as incomplete perimeters, lines that suggest colors, and lines that suggest form, can inform and improve non-photorealistic rendering (NPR). This paper provides a valuable artistic perspective to illustrate how engineering and art work collaboratively in NPR.

The workshop was fortunate to have a keynote presentation by Alfred Bruckstein. He described the problem of emulating classic engraving using non-photorealistic image rendering, and proposed to used level-set-based shape from shading techniques. The problem contains interesting mathematical challenges in connecting essential contours in natural, flowing ways, which Professor Bruckstein described.

November 2010

Ramakrishna Kakarala
Martin Constable

Program Committee

Introduction to the Workshop on Computer Vision in Vehicle Technology: From Earth to Mars

Vision-based autonomous navigation of vehicles has a long history which goes back to the success story of Dickmanns in Munich and the Mechanical Engineering Laboratory of MITI in Japan in the 1980th. At the time, DARPA had asked us to compete with autonomous land vehicles in their GRAND Challenges. Today, computer vision techniques provide methodologies to assist in long-distance exploration projects using visual sensing systems such those with the Mars rover project. Modern cars are now driven with the assistance of various sensor data. These assisted driving systems are developed as intelligent transportation systems. Among the various types of data used for driving assistance and navigation, we find visual information as the interface between human drivers and vehicles.

Today, data captured by visual sensors mounted on vehicles provide essential information used in intelligent driving systems. For applications of computer vision methodologies in exploration, evaluation, and quality-control techniques in the absence of ground truth information, it is essential to design robust and reliable algorithms.

In this workshop, we focus on exchanging new ideas on applications of computer vision theory to vehicle technology. In computer vision for driving assistance, tracking, reconstruction, and prediction become important concepts. Furthermore, real-time and on-board processes for these problems are required.

We received 21 papers and selected 11 papers for publication based on the reviews by the Program Committee and by the additional reviewer Ali Al-Sarraf.

November 2010

Steven Beauchemin
Atsushi Imiya
Tomas Pajdla

Program Committee

Introduction to the Workshop on e-Heritage

Digitally archived world heritage sites are broadening their value for preservation and access. Many valuable objects have been decayed by time due to weathering, natural disasters, even man-made disasters such as the Taliban destruction of the great Buddhas in Afghanistan, or the recent destruction by fire of a 600-year-old South Gate in Seoul. Cultural heritage also includes music, language, dance, and customs that are fast becoming extinct as the world moves toward a global village. Furthermore, most of the sites still face a problem of accessibility. Digital access projects are necessary to overcome those problems.

Computer vision research and practices have, and will continue, to play a central role in such cultural heritage preservation efforts. The proposed Workshop on e-Heritage and Digital Art Preservation aims to bring together computer vision researchers as well as interdisciplinary researchers that are related to computer vision, in particular computer graphics, image and audio research, image and haptic (touch) research, as well as presentation of visual content over the Web and education.

In this workshop, seven contributions to the field of e-heritage were presented, covering the areas of on-site augmented-reality applications, three-dimensional modeling and reconstruction, shape and image analysis, and interactive haptic systems. All submissions were double-blind reviewed by at least two experts. We thank all the authors who submitted their work. It was a special honor to have In So Kweon (KASIT, Korea), Hongbin Zha (Peking University, China) and Yasuyuki Matsushita (Microsoft Research Asia) as the invited speakers at the workshop. We are especially grateful to the members of the Program Committee for their remarkable efforts and the quality of the reviews.

November 2010

Katsushi Lkevchi
Takeshi Oishi
Rei Kawakami
Michael S. Brown
Moshe Ben-Ezra
Ryusuke Sagawa

Program Committee

Yasutaka Furukawa	Google, USA
Luc Van Gool	ETH Zurich, Switzerland
Yi Ping Hung	National Taiwan University, Taiwan
Asanobu Kitamoto	NII, Japan
In So Kweon	KAIST, Korea
Kok-Lim Low	NUS, Singapore
Yasuyuki Matsushita	MSRA, China
Daisuke Miyazaki	Hiroshima University, Japan
Tomokazu Sato	NAIST, Japan
David Suter	University of Adelaide, Australia
Jun Takamatsu	NAIST, Japan
Ping Tan	NUS, Singapore
Robby T. Tan	University of Utrecht, The Netherlands
Lior Wolf	Tel Aviv University, Israel
Toshihiko Yamasaki	University of Tokyo, Japan
Naokazu Yokoya	NAIST, Japan
Hongbin Zha	Peking University, China

Introduction to the Third International Workshop on Subspace Methods

We welcome you to the proceedings of the Third International Workshop of Subspace 2010 held in conjunction with ACCV 2010.

Subspace 2010 was held in Queenstown, New Zealand, on November 9, 2010. For the technical program of Subspace 2010, a total of 30 full-paper submissions underwent a rigorous review process. Each of these submissions was evaluated in a double-blind manner by a minimum of two reviewers. In the end, ten papers were accepted and included in this volume of proceedings.

The goal of the workshop is to share the potential of subspace-based methods, such as the subspace methods, with researchers working on various problems in computer vision; and to encourage interactions which could lead to further developments of the subspace-based methods. The fundamental theories of subspace-based methods and their applications in computer vision were discussed at the workshop.

Subspace-based methods are important for solving many theoretical problems in pattern recognition and computer vision. Also they have been widely used as a practical methodology in a large variety of real applications. During the last three decades, the area has become one of the most successful underpinnings of diverse applications such as classification, recognition, pose estimation, motion estimation. At the same time, there are many new and evolving research topics: nonlinear methods including kernel methods, manifold learning, subspace update and tracking. In addition to regular presentations, to overview these developments, we provided a historical survey talk of the subspace methods.

Prior to this workshop, we successfully organized two international workshops on subspace-based methods: Subspace 2007 in conjunction with ACCV 2007 and Subspace 2009 in conjunction with ICCV 2009. We believe that Subspace 2010 stimulated fruitful discussions among the participants and provided novel ideas for future research in computer vision.

November 2010

David Suter
Kazuhiro Fukui
Toru Tamaki

Program Committee

Toshiyuki Amano	NAIST, Japan
Horst Bischof	TU Graz, Austria
Seiji Hotta	Tokyo University of Agriculture and Technology, Japan
Masakazu Iwamura	Osaka Prefecture University, Japan
Tae-Kyun Kim	University of Cambridge, UK
Xi Li	Xi'an Jiaotong University, China
Yi Ma	University of Illinois at Urbana Champaign, USA
Atsuto Maki	Toshiba Cambridge Research Lab, UK
Shinichiro Omachi	Tohoku University, Japan
Bisser Raytchev	Hiroshima University, Japan
Peter Roth	TU Graz, Austria
Hitoshi Sakano	NTT CS Laboratories, Japan
Atsushi Sato	NEC, Japan
Yoichi Sato	The University of Tokyo, Japan
Shin'ichi Satoh	National Institute of Informatics, Japan
Terence Sim	National University of Singapore, Singapore
Bjorn Stenger	Toshiba Cambridge Research Lab, UK
Qi Tian	University of Texas at San Antonio, USA
Fernando De la Torre	Carnegie Mellon University, USA
Seiichi Uchida	Kyushu University, Japan
Osamu Yamaguchi	Toshiba, Japan
Jakob Verbeek	INRIA Rhône-Alpes, Grenoble, France
Jing-Hao Xue	University College London, UK

Table of Contents – Part I

Workshop on Visual Surveillance

Workshop on Video Event Categorization, Tagging and Retrieval (VECTaR)

Workshop on Gaze Sensing and Interactions

Table of Contents – Part II

Workshop on Application of Computer Vision for Mixed and Augmented Reality

Workshop on Computational Photography and Aesthetics

Workshop on Computer Vision in Vehicle Technology: From Earth to Mars

Workshop on e-Heritage

Workshop on Subspace Methods

Second-Order Polynomial Models for Background Subtraction

Alessandro Lanza, Federico Tombari, and Luigi Di Stefano

DEIS, University of Bologna
Bologna, Italy
{alessandro.lanza,federico.tombari,luigi.distefano}@unibo.it

Abstract. This paper is aimed at investigating background subtraction based on second-order polynomial models. Recently, preliminary results suggested that quadratic models hold the potential to yield superior performance in handling common disturbance factors, such as noise, sudden illumination changes and variations of camera parameters, with respect to state-of-the-art background subtraction methods. Therefore, based on the formalization of background subtraction as Bayesian regression of a second-order polynomial model, we propose here a thorough theoretical analysis aimed at identifying a family of suitable models and deriving the closed-form solutions of the associated regression problems. In addition, we present a detailed quantitative experimental evaluation aimed at comparing the different background subtraction algorithms resulting from theoretical analysis, so as to highlight those more favorable in terms of accuracy, speed and speed-accuracy tradeoff.

1   Introduction

Background subtraction is a crucial task in many video analysis applications, such as e.g. intelligent video surveillance. One of the main challenges consists in handling disturbance factors such as noise, gradual or sudden illumination changes, dynamic adjustments of camera parameters (e.g. exposure and gain), vacillating background, which are typical nuisances within video-surveillance scenarios. Many different algorithms for dealing with these issues have been proposed in literature (see [1] for a recent survey). Popular algorithms based on statistical per-pixel background models, such as e.g. Mixture of Gaussians (MoG) [2] or kernel-based non-parametric models [3], are effective in case of gradual illumination changes and vacillating background (e.g. waving trees). Unfortunately, though, they cannot deal with those nuisances causing sudden intensity changes (e.g. a light switch), yielding in such cases lots of false positives.

Instead, an effective approach to tackle the problem of sudden intensity changes due to disturbance factors is represented by a priori modeling over small image patches of the possible spurious changes that the scene can undergo. Following this idea, a pixel from the current frame is classified as *changed* if the intensity transformation between its local neighborhood and the corresponding neighborhood in the background can not be explained by the chosen a priori model.

R. Koch et al. (Eds.): ACCV 2010 Workshops, Part I, LNCS 6468, pp. 1–11, 2011.

Thanks to this approach, gradual as well as sudden photometric distortions do not yield false positives provided that they are explained by the model. Thus, the main issue concerns the choice of the a priori model: in principle, the more restrictive such a model, the higher is the ability to detect changes (sensitivity) but the lower is robustness to sources of disturbance (specificity). Some proposals assume disturbance factors to yield *linear* intensity transformations [4, 5]. Nevertheless, as discussed in [6], many non-linearities may arise in the image formation process, so that a more liberal model than linear is often required to achieve adequate robustness in practical applications. Hence, several other algorithms adopt *order-preserving* models, i.e. assume monotonic non-decreasing (i.e. non-linear) intensity transformations [6,7,8,9].

Very recently, preliminary results have been proposed in literature [10,11] that suggest how second-order polynomial models hold the potential to yield superior performance with respect to the classical previously mentioned approaches, being more liberal than linear proposals but still more restrictive than the order preserving ones. Motivated by these encouraging preliminary results, in this work we investigate on the use of second-order polynomial models within a Bayesian regression framework to achieve robust background subtraction. In particular, we first introduce a family of suitable second-order polynomial models and then derive closed-form solutions for the associated Bayesian regression problems. We also provide a thorough experimental evaluation of the algorithms resulting from theoretical analysis, so as to identify those providing the highest accuracy, the highest efficiency as well as the best tradeoff between the two.

2   Models and Solutions

For a generic pixel, let us denote as $\mathbf{x} = (x_1, \ldots, x_n)^T$ and $\mathbf{y} = (y_1, \ldots, y_n)^T$ the intensities of a surrounding neighborhood of pixels observed in the two images under comparison, i.e. background and current frame, respectively. We aim at detecting scene changes occurring in the pixel by evaluating the local intensity information contained in $\mathbf{x}$ and $\mathbf{y}$. In particular, classification of pixels as changed or unchanged is carried out by a priori assuming a model of the local photometric distortions that can be yielded by sources of disturbance and then testing, for each pixel, whether the model can explain the intensities $\mathbf{x}$ and $\mathbf{y}$ observed in the surrounding neighborhood. If this is the case, the pixel is likely sensing an effect of disturbs, so it is classified as unchanged; otherwise, it is marked as changed.

2.1   Modeling of Local Photometric Distortions

In this paper we assume that main photometric distortions are due to noise, gradual or sudden illumination changes, variations of camera parameters such as exposure and gain. We do not consider here the vacillating background problem (e.g. waving trees), for which the methods based on multi-modal and temporally adaptive background modeling, such as [2] and [3], are more suitable.

As for noise, first of all we assume that the background image is computed by means of a statistical estimation over an initialization sequence (e.g. temporal averaging of tens of frames) so that noise affecting the inferred background intensities can be neglected. Hence, $\mathbf{x}$ can be thought of as a deterministic vector of noiseless background intensities. As for the current frame, we assume that noise is additive, zero-mean, i.i.d. Gaussian with variance σ^2. Hence, noise affecting the vector $\mathbf{y}$ of current frame intensities can be expressed as follows:

$$p(\mathbf{y}|\tilde{\mathbf{y}}) = \prod_{i=1}^{n} p(y_i|\tilde{y}_i) = \prod_{i=1}^{n} \mathcal{N}(\tilde{y}, \sigma^2) = \left(\sqrt{2\pi}\sigma\right)^{-n} \exp\left(-\frac{1}{2\sigma^2}\sum_{i=1}^{n}(y_i - \tilde{y}_i)^2\right) \quad (1)$$

where $\tilde{\mathbf{y}} = (\tilde{y}_1, \ldots, \tilde{y}_n)^T$ denotes the (unobservable) vector of current frame noiseless intensities and $\mathcal{N}(\mu, \sigma^2)$ the normal pdf with mean μ and variance σ^2.

As far as remaining photometric distortions are concerned, we assume that noiseless intensities within a neighborhood of pixels can change due to variations of scene illumination and of camera parameters according to a second-order polynomial transformation $\phi(\cdot)$, i.e.:

$$\tilde{y}_i = \phi(x_i; \boldsymbol{\theta}) = (1, x_i, x_i^2)(\theta_0, \theta_1, \theta_2)^T = \theta_0 + \theta_1 x_i + \theta_2 x_i^2 \quad \forall i = 1, \ldots, n \quad (2)$$

It is worth pointing out that the assumed model (2) does not imply that the whole frame undergoes the same polynomial transformation but, more generally, that such a constraint holds locally. In other words, each neighborhood of intensities is allowed to undergo a different polynomial transformation, so that local illumination changes can be dealt with.

From (1) and (2) we can derive the expression of the likelihood $p(\mathbf{x}, \mathbf{y}|\boldsymbol{\theta})$, that is the probability of observing the neighborhood intensities $\mathbf{x}$ and $\mathbf{y}$ given a polynomial model $\boldsymbol{\theta}$:

$$p(\mathbf{x}, \mathbf{y}|\boldsymbol{\theta}) = p(\mathbf{y}|\boldsymbol{\theta}; \mathbf{x}) = p(\mathbf{y}|\tilde{\mathbf{y}} = \phi(\mathbf{x}; \boldsymbol{\theta})) = \left(\sqrt{2\pi}\sigma\right)^{-n} \exp\left(-\frac{1}{2\sigma^2}\sum_{i=1}^{n}(y_i - \phi(x_i; \boldsymbol{\theta}))^2\right)$$
$$(3)$$

where the first equality follows from the deterministic nature of the vector $\mathbf{x}$ that allows to treat it as a vector of parameters.

In practice, not all the polynomial transformations belonging to the linear space defined by the assumed model (2) are equally likely to occur. In Figure (1), on the left, we show examples of less (in red) and more (in azure) likely transformations. To summarize the differences we can say that the constant term of the polynomial has to be small and that the polynomial has to be monotonic non-decreasing. We formalize these constraints by imposing a prior probability on the parameters vector $\boldsymbol{\theta} = (\theta_0, \theta_1, \theta_2)^T$, as illustrated in Figure (1), on the right. In particular, we implement the constraint on the constant term by assuming a zero-mean Gaussian prior with variance σ_0^2 for the parameter θ_0:

$$p(\theta_0) = \mathcal{N}(0, \sigma_0^2) \quad (4)$$

The monotonicity constraint is addressed by assuming for (θ_1, θ_2) a uniform prior inside the subset Θ_{12} of $\mathbb{R}^2$ that renders $\phi'(x; \boldsymbol{\theta}) = \theta_1 + 2\theta_2 \cdot x \geq 0$ for all

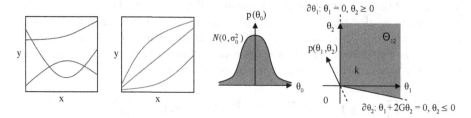

Fig. 1. Some polynomial transformations (left, in red) are less likely to occur in practice than others (left, in azure). To account for that, we assume a zero-mean normal prior for θ_0, a prior that is uniform inside Θ_{12}, zero outside for (θ_1, θ_2) (right).

$x \in [0 , G]$, with G denoting the highest measurable intensity ($G = 255$ for 8-bit images), zero probability outside Θ_{12}. Due to linearity of $\phi'(x)$, the monotonicity constraint $\phi'(x; \boldsymbol{\theta}) \geq 0$ over the entire x-domain $[0 , G]$ is equivalent to impose monotonicity at the domain extremes, i.e. $\phi'(0; \boldsymbol{\theta}) = \theta_1 \geq 0$ and $\phi'(G; \boldsymbol{\theta}) = \theta_1 + 2G \cdot \theta_2 \geq 0$. Hence, we can write the constraint as follows:

$$p(\theta_1, \theta_2) = \begin{cases} k & \text{if } (\theta_1, \theta_2) \in \Theta_{12} = \{(\theta_1, \theta_2) \in \mathbb{R}^2 : \theta_1 \geq 0 \ \wedge \ \theta_1 + 2G\theta_2 \geq 0\} \\ 0 & \text{otherwise} \end{cases} \quad (5)$$

We thus obtain the prior probability of the entire parameters vector as follows:

$$p(\boldsymbol{\theta}) = p(\theta_0) \cdot p(\theta_1, \theta_2) \quad (6)$$

In this paper we want to evaluate six different background subtraction algorithms, relying on as many models for photometric distortions obtained by combining the assumed noise and quadratic polynomial models in (1) and (2), that imply (3), with the two constraints in (4) and (5). In particular, the six considered algorithms (Q stands for quadratic) are:

Q_∞ : (1) $\wedge$ (2) $\Rightarrow$ (3) plus prior (4) for θ_0 with $\sigma_0^2 \to \infty$ (θ_0 free);

Q_f : (1) $\wedge$ (2) $\Rightarrow$ (3) plus prior (4) for θ_0 with σ_0^2 finite positive;

Q_0 : (1) $\wedge$ (2) $\Rightarrow$ (3) plus prior (4) for θ_0 with $\sigma_0^2 \to 0$ ($\theta_0 = 0$);

$Q_{\infty, M}$: same as Q_∞ plus prior (5) for (θ_1, θ_2) (monotonicity constraint);

$Q_{f, M}$: same as Q_f plus prior (5) for (θ_1, θ_2) (monotonicity constraint);

$Q_{0, M}$: same as Q_0 plus prior (5) for (θ_1, θ_2) (monotonicity constraint);

2.2 Bayesian Polynomial Fitting for Background Subtraction

Independently from the algorithm, scene changes are detected by computing a measure of the distance between the sensed neighborhood intensities $\mathbf{x}$, $\mathbf{y}$ and the space of models assumed for photometric distortions. In other words, if the intensities are not well-fitted by the models, the pixel is classified as changed. The minimum-distance intensity transformation within the model space is computed by a maximum a posteriori estimation of the parameters vector:

$$\boldsymbol{\theta}_{\text{MAP}} = \underset{\boldsymbol{\theta} \in \mathbb{R}^3}{argmax}\ p(\boldsymbol{\theta}|\mathbf{x}, \mathbf{y}) = \underset{\boldsymbol{\theta} \in \mathbb{R}^3}{argmax}\ [p(\mathbf{x}, \mathbf{y}|\boldsymbol{\theta})p(\boldsymbol{\theta})] \tag{7}$$

where the second equality follows from Bayes rule. To make the posterior in (7) explicit, an algorithm has to be chosen. We start from the more complex one, i.e. $Q_{f,M}$. By substituting (3) and (4)-(5), respectively, for the likelihood $p(\mathbf{x}, \mathbf{y}|\boldsymbol{\theta})$ and the prior $p(\boldsymbol{\theta})$ and transforming posterior maximization into minus log-posterior minimization, after eliminating the constant terms we obtain:

$$\boldsymbol{\theta}_{\text{MAP}} = \underset{\boldsymbol{\theta} \in \Theta}{argmin} \left[E(\boldsymbol{\theta}; \mathbf{x}, \mathbf{y}) = d(\boldsymbol{\theta}; \mathbf{x}, \mathbf{y}) + r(\boldsymbol{\theta}) = \sum_{i=1}^{n} (y_i - \phi(x_i; \boldsymbol{\theta}))^2 + \lambda \theta_0^2 \right] \tag{8}$$

The objective function to be minimized $E(\boldsymbol{\theta}; \mathbf{x}, \mathbf{y})$ is the weighted sum of a data-dependent term $d(\boldsymbol{\theta}; \mathbf{x}, \mathbf{y})$ and a regularization term $r(\boldsymbol{\theta})$ which derive, respectively, from the likelihood of the observed data $p(\mathbf{x}, \mathbf{y}|\boldsymbol{\theta})$ and the prior of the first parameter $p(\theta_0)$. The weight of the sum, i.e. the regularization coefficient, depends on both the likelihood and the prior and is given by $\lambda = \sigma^2 / \sigma_0^2$. The prior of the other two parameters $p(\theta_1, \theta_2)$ expressing the monotonicity constraint has translated into a restriction of the optimization domain from $\mathbb{R}^3$ to $\Theta = \mathbb{R} \times \Theta_{12}$. It is worth pointing out that the data dependent term represents the least-squares regression error, i.e. the sum over all the pixels in the neighborhood of the square differences between the frame intensities and the background intensities transformed by the model. By making $\phi(x_i; \boldsymbol{\theta})$ explicit and after simple algebraic manipulations, it is easy to observe that the objective function is quadratic, so that it can be compactly written as:

$$E(\boldsymbol{\theta}; \mathbf{x}, \mathbf{y}) = (1/2)\,\boldsymbol{\theta}^T H \boldsymbol{\theta} - \mathbf{b}^T \boldsymbol{\theta} + c \tag{9}$$

with the matrix H, the vector $\mathbf{b}$ and the scalar c given by:

$$H = 2 \begin{pmatrix} N & Sx & Sx^2 \\ Sx & Sx^2 & Sx^3 \\ Sx^2 & Sx^3 & Sx^4 \end{pmatrix} \qquad \mathbf{b} = 2 \begin{pmatrix} Sy \\ Sxy \\ Sx^2y \end{pmatrix} \qquad c = Sy^2 \tag{10}$$

and, for simplicity of notation:

$$Sx = \sum_{i=1}^{n} x_i \quad Sx^2 = \sum_{i=1}^{n} x_i^2 \quad Sx^3 = \sum_{i=1}^{n} x_i^3 \quad Sx^4 = \sum_{i=1}^{n} x_i^4 \qquad N = n + \lambda \tag{11}$$
$$Sy = \sum_{i=1}^{n} y_i \quad Sxy = \sum_{i=1}^{n} x_i y_i \quad Sx^2y = \sum_{i=1}^{n} x_i^2 y_i \quad Sy^2 = \sum_{i=1}^{n} y_i^2$$

As for the optimization domain $\Theta = \mathbb{R} \times \Theta_{12}$, with Θ_{12} defined in (5) and illustrated in Figure 1, it also can be compactly written in matrix form as follows:

$$\Theta = \{\boldsymbol{\theta} \in \mathbb{R}^3 : \ Z\boldsymbol{\theta} \geq 0\} \qquad \text{with} \qquad Z = \begin{pmatrix} 0 & 1 & 0 \\ 0 & 1 & 2G \end{pmatrix} \tag{12}$$

The estimation problem (8) can thus be written as a quadratic program:

$$\boldsymbol{\theta}_{\text{MAP}} = \underset{Z\boldsymbol{\theta} \geq 0}{argmin} \left[(1/2)\,\boldsymbol{\theta}^T H\,\boldsymbol{\theta} - \mathbf{b}^T\boldsymbol{\theta} + c \right] \tag{13}$$

If in the considered neighborhood there exist three pixels characterized by different background intensities, i.e. $\exists\, i, j, k : x_i \neq x_j \neq x_k$, it can be demonstrated that the matrix H is positive-definite. As a consequence, since H is the Hessian of the quadratic objective function, in this case the function is strictly convex. Hence, it admits a unique point of unconstrained global minimum $\boldsymbol{\theta}^{(u)}$ that can be easily calculated by searching for the unique zero-gradient point, i.e. by solving the linear system of normal equations:

$$\boldsymbol{\theta}^{(u)} = \boldsymbol{\theta} \in \mathbb{R}^3 : \quad \nabla E(\boldsymbol{\theta}) = 0 \quad \equiv \quad (H/2)\,\boldsymbol{\theta} = (\mathbf{b}/2) \tag{14}$$

for which a closed-form solution is obtained by computing the inverse of $H/2$:

$$\boldsymbol{\theta}^{(u)} = (H/2)^{-1}(\mathbf{b}/2) = \frac{1}{|H/2|} \begin{pmatrix} A & D & E \\ D & B & F \\ E & F & C \end{pmatrix} \begin{pmatrix} Sy \\ Sxy \\ Sx^2y \end{pmatrix} \tag{15}$$

where:

$$
\begin{array}{lll}
A = Sx^2 Sx^4 - \left(Sx^3\right)^2 & B = NSx^4 - \left(Sx^2\right)^2 & C = NSx^2 - \left(Sx\right)^2 \\
D = Sx^2 Sx^3 - Sx\, Sx^4 & E = Sx\, Sx^3 - \left(Sx^2\right)^2 & F = Sx\, Sx^2 - NSx^3
\end{array} \tag{16}
$$

and

$$|H/2| = N\,A + Sx\,D + Sx^2 E \tag{17}$$

If the computed point of unconstrained global minimum $\boldsymbol{\theta}^{(u)}$ belongs to the quadratic program feasible set Θ, i.e. satisfies the monotonicity constraint $Z\boldsymbol{\theta} \geq 0$, then the minimum distance between the observed neighborhood intensities and the model of photometric distortions is simply determined by substituting $\boldsymbol{\theta}^{(u)}$ for $\boldsymbol{\theta}$ in the objective function. A compact close-form expression for such a minimum distance can be obtained as follows:

$$E^{(u)} = E(\boldsymbol{\theta}^{(u)}) = \boldsymbol{\theta}^{(u)^T}(H/2)\,\boldsymbol{\theta}^{(u)} - \boldsymbol{\theta}^{(u)^T}\mathbf{b} + c = \boldsymbol{\theta}^{(u)^T}(\mathbf{b}/2) - \boldsymbol{\theta}^{(u)^T}\mathbf{b} + c$$

$$= c - \boldsymbol{\theta}^{(u)^T}(\mathbf{b}/2) = Sy^2 - |H/2|^{-1}\left(Sy\,\theta_0^{(u)} + Sxy\,\theta_1^{(u)} + Sx^2y\,\theta_2^{(u)}\right) \tag{18}$$

The two algorithms Q_f and Q_∞ rely on the computation of the point of unconstrained global minimum $\boldsymbol{\theta}^{(u)}$ by (15) and, subsequently, of the unconstrained minimum distance $E^{(u)} = E(\boldsymbol{\theta}^{(u)})$ by (18). The only difference between the two algorithms is the value of the pre-computed constant $N = n + \lambda$ that in Q_∞ tends to n due to $\sigma_0^2 \to \infty$ causing $\lambda \to 0$. Actually, Q_f corresponds to the method proposed in [10].

If the point $\boldsymbol{\theta}^{(u)}$ falls outside the feasible set Θ, the solution $\boldsymbol{\theta}^{(c)}$ of the constrained quadratic programming problem (13) must lie on the boundary of the feasible set, due to convexity of the objective function. However, since the monotonicity constraint $Z\,\boldsymbol{\theta} \geq 0$ does not concern θ_0, again the partial derivative of the objective function with respect to θ_0 must vanish in correspondence of the solution. Hence, first of all we impose this condition, thus obtaining:

$$\theta_0{}^{(c)} = (1/N)\left(Sy - \theta_1 Sx - \theta_2 Sx^2\right) \tag{19}$$

We thus substitute $\theta_0{}^{(c)}$ for θ_0 in the objective function, so that the original 3-d problem turns into a 2-d problem in the two unknowns θ_1, θ_2 with the feasible set Θ_{12} defined in (5) and illustrated in Figure 1, on the right. As previously mentioned, the solution of the problem must lie on the boundary of the feasible set $\partial\Theta_{12}$, that is on one of the two half-lines:

$$\partial\Theta_1:\ \theta_1 = 0 \wedge \theta_2 \geq 0 \qquad\qquad \partial\Theta_2:\ \theta_1 = -2\,G\,\theta_2 \wedge \theta_2 \leq 0 \tag{20}$$

The minimum of the 2-d objective function on each of the two half-lines can be determined by replacing the respective line equation into the objective function and then searching for the unique minimum of the obtained 1-d convex quadratic function in the unknown θ_2 restricted, respectively, to the positive $(\partial\Theta_1)$ and the negative $(\partial\Theta_2)$ axis. After some algebraic manipulations, we obtain that the two minimums $E_1^{(c)}$ and $E_2^{(c)}$ are given by:

$$E_1^{(c)} = Sy^2 - \frac{(Sy)^2}{N} - \begin{cases} \dfrac{T^2}{N\,B} & \text{if } T > 0 \\ 0 & \text{if } T \leq 0 \end{cases} \qquad E_2^{(c)} = Sy^2 - \frac{(Sy)^2}{N} - \begin{cases} \dfrac{V^2}{N\,U} & \text{if } V < 0 \\ 0 & \text{if } V \geq 0 \end{cases} \tag{21}$$

where:

$$\begin{aligned} T &= N\,Sx^2y - Sx^2 Sy & V &= T + 2\,G\,W \\ W &= Sx\,Sy - N\,Sxy & U &= B + 4G^2 C + 4G\,F \end{aligned} \tag{22}$$

The constrained global minimum $E^{(c)}$ is thus the minimum between $E_1^{(c)}$ and $E_2^{(c)}$. Hence, similarly to Q_f and Q_∞, the two algorithms $Q_{f,M}$ and $Q_{\infty,M}$ rely on the preliminary computation of the point of unconstrained global minimum $\boldsymbol{\theta}^{(u)}$ by (15). However, if the point does not satisfy the monotonicity constraint in (12), the minimum distance is computed by (21) instead of by (18). The two algorithms $Q_{f,M}$ and $Q_{\infty,M}$ differ in the exact same way as Q_f and Q_∞, i.e. only for the value of the pre-computed parameter N.

The two remaining algorithms, namely Q_0 and $Q_{0,M}$, rely on setting $\sigma_0^2 \to 0$. This implies that $\lambda \to \infty$ and, therefore, $N = (n + \lambda) \to \infty$. As a consequence, closed-form solutions for these algorithms can not be straightforwardly derived from the previously computed formulas by simply substituting the value of N. However, $\sigma_0^2 \to 0$ means that the parameter θ_0 is constrained to be zero, that is the quadratic polynomial model is constrained to pass through the origin. Hence, closed-form solutions for these algorithms can be obtained by means of the same procedure outlined above, the only difference being that in the model (2) θ_0 has to be eliminated. Details of the procedure and solutions can be found in [11].

By means of the proposed solutions, we have no need to resort to any iterative approach. In addition, it is worth pointing out that all terms involved in the calculations can be computed either off-line (i.e. those involving only background intensities) or by means of very fast incremental techniques such as Summed Area Table [12] (those involving also frame intensities). Overall, this allows the proposed solutions to exhibit a computational complexity of $O(1)$ with respect to the neighborhood size n.

3   Experimental Results

This Section proposes an experimental analysis aimed at comparing the 6 different approaches to background subtraction based on a quadratic polynomial model derived in previous Section. To distinguish between the methods we will use the notation described in the previous Section. Thus, e.g., the approach presented in [10] is referred to here as Q_f, and that proposed in [11] as $Q_{0,M}$. All algorithms were implemented in C using incremental techniques [12] to achieve $O(1)$ complexity. They also share the same code structure so to allow for a fair comparison in terms not only of accuracy, but also computational efficiency.

Evaluated approaches are compared on five test sequences, S_1–S_5, characterized by sudden and notable photometric changes that yield both linear and non-linear intensity transformations. We acquired sequences S_1–S_4 while S_5 is a synthetic benchmark sequence available on the web [13]. In particular, S_1, S_2 are two indoor sequences, while S_3, S_4 are both outdoor. It is worth pointing out that by computing the 2-d joint histograms of background versus frame intensities, we observed that S_1, S_5 are mostly characterized by linear intensity changes, while S_2–S_4 exhibit also non-linear changes. Background images together with sample frames from the sequences are shown in [10, 11]. Moreover, we point out here that, based on an experimental evaluation carried out on S_1–S_5, methods Q_f and $Q_{0,M}$ have been shown to deliver state-of-the-art performance in [10, 11]. In particular, Q_f and $Q_{0,M}$ yield at least equivalent performance compared to the most accurate existing methods while being, at the same time, much more efficient. Hence, since results attained on S_1–S_5 by existing linear and order-preserving methods (i.e. [5, 7, 8, 9]) are reported in [10, 11], in this paper we focus on assessing the relative merits of the 6 developed quadratic polynomial

Experimental results in terms of accuracy as well as computational efficiency are provided. As for accuracy, quantitative results are obtained by comparing the change masks yielded by each approach against the ground-truths (manually labeled for S_1-S_4, available online for S_5). In particular, we computed the True Positive Rate (TPR) versus False Positive Rate (FPR) Receiver Operating Characteristic (ROC) curves. Due to lack of space, we can not show all the computed ROC curves. Hence, we summarize each curve with a well-known scalar measure of performance, the Area Under the Curve (AUC), which represents the probability for the approach to assign a randomly chosen changed pixel a higher change score than a randomly chosen unchanged pixel [14].

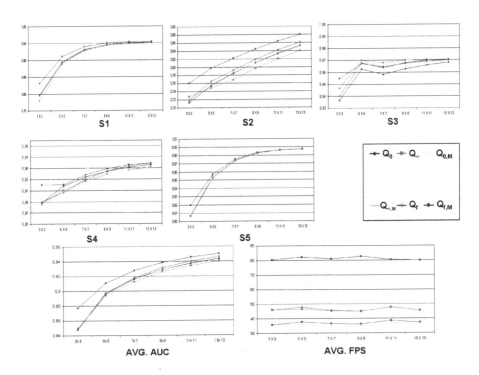

Fig. 2. Top: AUC values yielded by the evaluated algorithms with different neighborhood sizes on each of the 5 test sequences. Bottom: average AUC (left) and FPS (right) values over the 5 sequences.

Each graph shown in Figure 2 reports the performance of the 6 algorithms in terms of AUC with different neighborhood sizes (3×3, 5×5, $\cdots$, 13×13). In particular, the first 5 graphs are relative to each of the 5 testing sequences, while the two graphs on the bottom show, respectively, the mean AUC values and the mean Frame-Per-Second (FPS) values over the 5 sequences. By analyzing AUC values reported in the Figure, it can be observed that two methods yield overall a better performance among those tested, that is, $Q_{0,M}$ and $Q_{f,M}$, as also summarized by the mean AUC graph. In particular, $Q_{f,M}$ is the most accurate on S_2 and S_5 (where $Q_{0,M}$ is the second best), while $Q_{0,M}$ is the most accurate on S_1 and S_4 (where $G_{f,M}$ is the second best). The different results on S_3, where Q_∞ is the best performing method, appear to be mainly due to the presence of disturbance factors (e.g. specularities, saturation, ...) not well modeled by a quadratic transformation: thus, the best performing algorithm in such specific circumstance turns out to be the less constrained one (i.e. Q_∞).

As for efficiency, the mean FPS graph in Figure 2 proves that all methods are $O(1)$ (i.e. their complexity is independent of the neighborhood size). As expected, the more constraints are imposed on the adopted model, the higher the computational cost is, resulting in a reduced efficiency. In particular, an additional computational burden is brought in if a full quadratic form is assumed (i.e. not homogeneous), similarly if the transformation is assumed to be monotonic. Given this consideration, the most efficient method turns out to be Q_0, the least efficient ones $Q_{\infty,M}$, $Q_{f,M}$, with Q_∞, $Q_{0,M}$, Q_f staying in the middle. Also, the results prove that the use of a non-homogeneous form adds a higher computational burden compared to the monotonic assumption. Overall, the experiments indicate that the method providing the best accuracy-efficiency tradeoff is $Q_{0,M}$.

4   Conclusions

We have shown how background subtraction based on Bayesian second-order polynomial regression can be declined in different ways depending on the nature of the constraints included in the formulation of the problem. Accordingly, we have derived closed-form solutions for each of the problem formulations. Experimental evaluation show that the most accurate algorithms are those based on the monotonicity constraint and, respectively a null $Q_{0,M}$ or finite $Q_{f,M}$ variance for the prior of the constant term. Since the more articulated the constraints within the problem the higher computational complexity, the most efficient algorithm results from a non-monotonic and homogeneous formulation (i.e. Q_0). This also explains why $Q_{0,M}$ is notably faster than $Q_{f,M}$, so as to turn out the method providing the more favorable tradeoff between accuracy and speed.

References

1. Elhabian, S.Y., El-Sayed, K.M., Ahmed, S.H.: Moving object detection in spatial domain using background removal techniques - state-of-art. Recent Patents on Computer Sciences 1, 32–54 (2008)
2. Stauffer, C., Grimson, W.E.L.: Adaptive background mixture models for real-time tracking. In: Proc. CVPR 1999, vol. 2, pp. 246–252 (1999)
3. Elgammal, A., Harwood, D., Davis, L.: Non-parametric model for background subtraction. In: Proc. ICCV 1999 (1999)
4. Durucan, E., Ebrahimi, T.: Change detection and background extraction by linear algebra. Proc. IEEE 89, 1368–1381 (2001)
5. Ohta, N.: A statistical approach to background subtraction for surveillance systems. In: Proc. ICCV 2001, vol. 2, pp. 481–486 (2001)
6. Xie, B., Ramesh, V., Boult, T.: Sudden illumination change detection using order consistency. Image and Vision Computing 22, 117–125 (2004)
7. Mittal, A., Ramesh, V.: An intensity-augmented ordinal measure for visual correspondence. In: Proc. CVPR 2006, vol. 1, pp. 849–856 (2006)
8. Heikkila, M., Pietikainen, M.: A texture-based method for modeling the background and detecting moving objects. IEEE Trans. PAMI (2006)
9. Lanza, A., Di Stefano, L.: Detecting changes in grey level sequences by ML isotonic regression. In: Proc. AVSS 2006, pp. 1–4 (2006)
10. Lanza, A., Tombari, F., Di Stefano, L.: Robust and efficient background subtraction by quadratic polynomial fitting. In: Proc. Int. Conf. on Image Processing ICIP 2010 (2010)

11. Lanza, A., Tombari, F., Di Stefano, L.: Accurate and efficient background subtraction by monotonic second-degree polynomial fitting. In: Proc. AVSS 2010, (2010)
12. Crow, F.: Summed-area tables for texture mapping. Computer Graphics 18, 207–212 (1984)
13. MUSCLE Network of Excellence (Motion detection video sequences)
14. Bradley, A.P.: The use of the area under the ROC curve in the evaluation of machine learning algorithms. Pattern Recognition 30, 1145–1159 (1997)

Adaptive Background Modeling for Paused Object Regions

Atsushi Shimad, Satoshi Yoshinaga, and Rin-ichiro Taniguchi

Kyushu University, Fukuoka, Japan

Abstract. Background modeling has been widely researched to detect moving objects from image sequences. Most approaches have a false-negative problem caused by a stopped object. When a moving object stops in an observing scene, it will be gradually trained as background since the observed pixel value is directly used for updating the background model. In this paper, we propose 1) a method to inhibit background training, and 2) a method to update an original background region occluded by stopped object. We have used probabilistic approach and predictive approach of background model to solve these problems. The great contribution of this paper is that we can keep paused objects from being trained.

1 Introduction

A technique of background modeling has been widely applied to foreground object detection from video sequences. It is one of the most important issues to construct a background model which is robust for various illumination changes. Many approaches have been proposed to construct an effective background model; pixel-level approaches[1,2,3,4], region-level approaches[5,6], combinational approaches[7,8] or so on. Almost of these approaches have a common process of updating of background model. Actually, this process is very beneficial to adapt for various illumination changes. On the other hand, we can say that the traditional background model has an ability to detect "Moving Objects" only. In other words, it causes FN (false negative) problem when a foreground object stops in the scene. This is because the paused foreground object is gradually learned as background by blind updating process.

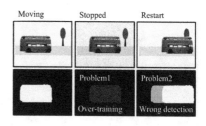

Fig. 1. Problem of blind updating of background model

Therefore, we have to handle following problems (also see Fig. 1) in order to keep detecting the paused object.

1. Over-training of foreground objects
2. Wrong detection of original background regions

R. Koch et al. (Eds.): ACCV 2010 Workshops, Part I, LNCS 6468, pp. 12–22, 2011.
© Springer-Verlag Berlin Heidelberg 2011

The first problem is caused by blind updating process of background model. Some researches tried to solve this problem by control learning rate of the background model. For example, decreasing the learning rate of some regions in which foreground objects probably stop[9] or utilizing two background model which have different learning rates[10] has been proposed. However, these approaches have not resolve the essential problem of over-training since they just extend the time for being learned as background.

The second problem is caused by a paused foreground object when it starts to move again. In such a case, an original background region hidden by the object might be detected wrongly since the paused foreground object has been included in the background model. Another possibility is that the FP problem will be caused when some illumination change occur while the foreground object stops. The hidden region will be detected wrongly since the background model does not know the illumination change occurred in the hidden region. A study which considers the illumination changes until a foreground object is regarded as paused object has been proposed[11], but it does not handle the illumination change (background change) in the region hidden by the paused foreground object.

In this paper, we propose a novel approach which use two different kinds of models; one is a pixel-level background model and the other is a predictive model. Two problems mentioned above can be resolved by utilizing these two models efficiently. The characteristics of our study are summarized as follows.

1. Our approach can control over-training of paused foreground objects without adjusting the learning rate.
2. Our approach can update the original background region hidden by paused objects.

In addition, our background model is robust against illumination changes by using two kinds of models in combination.

2   Framework

The processing flow of our proposed background model is shown in Fig. 2. At the first stage, background likelihoods of an observed image are calculated based on the probabilistic model(see section 3.1) and the predictive model(see 3.2). At the second stage, the foreground region is determined by integrating two background likelihoods evaluated by the pixel-level background model and the predictive model(see section 4). Finally, at the third stage, the parameters of both models are updated. Generally, the observed pixel

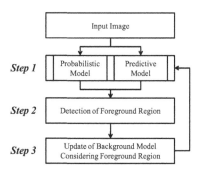

Fig. 2. Processing flow

value is directly used for updating the parameters. In our approach, meanwhile, when a pixel is judged as "foreground" at the second stage, we use alternative pixel value around the pixel which has similar background model. This process avoid the foreground object being trained as "background". We will give a detailed explanation in section 5.

3   Probabilistic Model and Predictive Model

3.1   Probabilistic Model Base on GMM

We have modified the GMM-based background model[2]. The modified background model consists of 2 steps; evaluation of background likelihood and update of model parameters .

Evaluation of Background Likelihood. Let x_i^t be a pixel value on a pixel i at frame t. For simple expression, we omit the notation i when we explain each pixel process. The background likelihood is represented as

$$P(x^t) = \sum_{k=1}^{K} \frac{w_k^t}{(2\pi)^{\frac{n}{2}} |\Sigma|^{\frac{1}{2}}} \exp\left(-\frac{1}{2}(x^t - \mu^t)^T \Sigma^{-1}(x^t - \mu^t)\right) \tag{1}$$

The original approach[2] judges whether or not an observed pixel value belongs to "background". Our approach does not output such a judgment result explicitly. Instead, we calculate the background likelihood at this processing stage.

Update of Model Parameters. The model parameters are updated in the same way as the original method[2].

The weights of the K distributions at frame t, w_k^t, are adjusted as follows

$$w_k^t = (1 - \alpha)w_k^{t-1} + \alpha M_k^t \tag{2}$$

where α is the learning rate and M_k^t is 1 for the model which matched and 0 for the remaining models. After this approximation, the weights are renormalized.

Every new pixel value x^t is examined against the existing K Gaussian distributions, until a match is found. A match is defined as a pixel value within 2.5 standard deviations of distribution. The parameters of unmatched distributions remain the same. When a match is found for the new pixel value, the parameters of the distribution are updated as follows.

$$\mu^t = (1 - \rho)\mu^{t-1} + \rho y^t, \quad \sigma^t = (1 - \rho)\sigma^{t-1} + \rho(y^t - \mu^t)^T(y^t - \mu^t) \tag{3}$$

where the ρ is the second learning rate, y^t is a pixel value which is used for update of model parameters. We purposely distinguish the notation y^t from x^t since the pixel value y^t depends on the judgment result explained in following section 5.

If none of the K distribution matches the current pixel value, a new Gaussian distribution is made as follows.

$$w_{k+1}^t = W, \quad \mu_{k+1}^t = y^t, \quad \sigma_{k+1}^t = \sigma_k^t \tag{4}$$

where W is the initial weight value for the new Gaussian. If W is higher, the distribution is chosen as the background model for a long time. After this process, the weights are renormalized. Finally, when the weight of the least probable distribution is smaller than a threshold, the distribution is deleted, and the remaining weights are renormalized.

3.2   Predictive Model Based on Exponential Smoothing

Exponential Smoothing. We use an exponential smoothing method[12] to acquire a predictive pixel value z^t. Exponential smoothing is a technique that can be applied to time series data, either to produce smoothed data for presentation, or to make forecasts. The simplest form of exponential smoothing is given by the following formula.

$$m^t = \beta x^t + (1 - \beta)m^{t-1} \tag{5}$$

where m^t is the estimate of the value, x^t is the observed value at frame t. β is the smoothing constant in the range $\beta(0 \leq \beta \leq 1)$. The forecast function, which gives an estimate of the series can be written as follows:

$$z^t = m^t + \frac{1 - \beta}{\beta}r^{t-1}, \qquad r^t = \beta(z^t - z^{t-1}) + (1 - \beta)r^{t-1} \tag{6}$$

where r^t is the current slope and z^t is the estimate of the value with a trend.

Evaluation of Background Likelihood. The predictive model mentioned above is used for two purposes. One is for searching a pixel which has a similar tendency with the pixel hidden by a foreground object, which will be explained in section 5. The other is for region-level background model explained in this section. Some literatures have reported that spatial locality information is effective for illumination changes[6,13]. This idea derives from a hypothesis that similar changes will be observed around the pixels when illumination change occurs. In the proposed method, we use not only the predictive value of target pixel but also the values of neighbor pixels simultaneously in order to evaluate background likelihood.

Let R be a set of neighbor pixels around pixel i, the background likelihood $Q(x^t)$ is calculated by following formula.

$$Q(x_p^t) = \frac{\sum_{i \in R} \phi(x_i^t, z_i^t)}{|R|}, \qquad \phi(x^t, z^t) = \begin{cases} 1 & \text{if } |x^t - z^t| < th \\ 0 & \text{otherwise} \end{cases} \tag{7}$$

The $\phi(x^t, z^t)$ is a range which allows predictive error.

Update of Model Parameters. The parameters of predictive model are updated by an observed pixel value. In the same way with the probabilistic background model, we decide whether or not to use the observed value directly. The detailed explanation will be given in section 5.

4   Foreground Detection Based on MRF

The background model and foreground model output the evaluation result of background and foreground likelihood. The final decision whether or not each pixel is foreground is determined by integrating each evaluation result. We define an energy function based on Markov Random Field (MRF) and give each pixel proper label (foreground or background) by minimizing the energy function. Our energy function is defined as

$$E(L) = \lambda \sum_{i \in \mathcal{V}} G(l_i) + \sum_{(i,j) \in \mathcal{E}} H(l_i, l_j) \tag{8}$$

where $L = (l_1, \ldots, l_N)$ is the array of labels, and N is the number of pixels. The $\mathcal{V}$ and $\mathcal{E}$ represent a set of all pixels and a set of all nearest neighboring pixel pairs respectively. The $G(l_i)$ and $H(l_i, l_j)$ represent the penalty term and smoothing term respectively and they are calculated as follows.

$$G(l_i) = \frac{P(x_i) + Q(x_i)}{2}, \quad H(l_i, l_j) = \frac{1}{\ln(\|x_i - x_j\| + 1 + \epsilon)} \tag{9}$$

We assign proper labels to pixels which minimize the total energy $E(L)$, and it is solved by a graph cut algorithm[14]. We make a graph which has two terminal nodes (Source (s) and Sink (t)) and some nodes corresponding to pixels. Edges are made between nodes. We give each edge a cost $u(i, j)$ defined as follows.

$$u(i,j) = H(l_i, l_j), \quad u(s, i) = \lambda(1 - G(l_i)), \quad u(i, t) = \lambda G(l_i) \tag{10}$$

5   Update of Model Parameters

If we directly use observed pixel values for model update process, not only background regions but also foreground regions are gradually trained by the model. It will cause FN (false negative) problem when an moving object stops in the scene (e.g. bus stop, intersection and so on). One of the solutions is to exclude foreground pixels from update process. However, such ad-hoc process will generate another problem that background model on the foreground pixel cannot adapt itself for illumination changes while the foreground object stops. As the result, when the paused object starts to move again, the occluded region will be detected wrongly (FP (false positive) problem). To solve this problem, our approach updates model parameters on the foreground pixels with the help of neighbor background pixels.

The specific update process of our proposed approach is as follows. Let F and B be a set of foreground pixels and background pixels judged in section 4 respectively, the pixel value y_i^t for model update is calculated as

$$y_i^t = \begin{cases} x_i^t & \text{if } i \in B \\ x_c^t & \text{if } i \in F \end{cases}, \quad c = \operatorname*{argmin}_{j \in B} f(\Theta_i, \Theta_j). \tag{11}$$

The Θ is a set of parameters of probabilistic model and predictive model on each pixel's. In our experiments, we set the Θ to be $\Theta^t = \{\mu_1^t, m^t, r^t\}$, which denotes the average background pixel value of the distribution which has the largest weight μ_1^t, exponential smoothing m^t and the slope of the observed value r^t. The most important contribution in this paper is to use x_c^t for model update. When a pixel is judged as foreground, our approach searches the model which has the most similar model parameters with the pixel. The similarity between model parameters is evaluated by the distance function $f(\Theta_i, \Theta_j)$, where we use the L1 norm in our experiments.

In this way, our approach does not use foreground pixel values to update model parameters. Alternatively, we use the pixel value on the background pixel whose model parameters are the most similar with the one on the foreground pixels. This procedure avoid the foreground object from being trained as background. Therefore, even if a foreground object stops in the scene, our approach keeps detecting the foreground object. In addition, the implicit update process of the background models hidden by the foreground object reduces FP problem when the paused object start to move again.

6   Experimental Results

We have used several public datasets to investigate the effectiveness of our proposed method. The computational speed of the proposed method was $7fps$ for QVGA image size by using a PC with a Core i7 3.07GHz CPU.

According to our preliminary experimental results, we have decided some parameters as follows; $\alpha = 0.5$, $\beta = 0.5$, $th = 15$. These parameters were common to following experiments.

6.1   Evaluation of Implicit Model Update

The dataset used in this section is released at PETS2001[1] including illumination changes in the outdoor scene. We have clipped two subscenes from the original image sequence; one is a scene in which illumination condition changes from dark to bright, and the other is a scene from bright to dark. The both scenes consist of about 600 frame images. Moreover, we have selected two 10×30 pixel areas; an area with simple background and an area with complex background. We have conducted a simulation experiment under the condition that the foreground object stopped on the 10×30 pixel region and evaluated how effective the proposed implicit update process mentioned in section 5 was.

Table 1 shows the error value and the number of FP pixels around illumination changes. The error value means the difference value between the estimate value of background model and the observed pixel value. Meanwhile, we counted up the number of pixels whose error value exceeded a threshold as FP pixels. This situation was under the assumption that paused object started to move again.

---

[1] Benchmark data of International Workshop on Performance Evaluation of Tracking and Surveillance. ftp://pets.rdg.ac.uk/PETS2001/

Table 1. Comparison of Model Update Methods: "B to D" denotes Bright to Dark, "D to B" denotes Dark to Bright

		B to D Simple BG	B to D Complex BG	D to B Simple BG	D to B Complex BG
Without	Error	102.8	60.9	105.5	63.2
Update	FP	250	99	250	106
Traditional	Error	9.2	8.4	12.0	10.0
Update	FP	0	0	0	0
Proposed	Error	14.0	23.8	14.8	29.6
Method	FP	7	6	0	13

We have compared out proposed method with two methods; without model update (Table 1:without update) and with model update by traditional method (Table 1:traditional update). Note that traditional method used the observed pixel value directly for model update.

When we didn't update model parameters, the error value was large and a lot of FP pixels were detected wrongly. The traditional update method could adapt for the illumination changes. As the result, the error value and the number of FP pixels were very small. On the other hand, our proposed method could also adapt the illumination changes even though the investigated area was occluded by the pseudo foreground object. The error value in the complex background became larger than those in the simple background. However, this didn't lead to a sensible increase of the number of FP pixels. These discussions applied to both scenes; scene from dark to bright and scene from bright to dark. Therefore, we could conclude that the implicit update process of the background model was effective to update the region occluded by paused foreground object.

6.2 Accuracy of Paused Object Detection

We have user three outdoor scenes[2] to investigate the detection accuracy of paused foreground object regions. The Scene 1, Scene 2 and Scene 3 in Fig. 3 shows the snapshot of about 100th frame, 60th frame and 150th frame after the moving object stopped. The illumination condition in Scene 1 is relatively stable compared with the other scenes. We have compared our proposed method with two representative methods; GMM based method[2] and fusion model of spatial-temporal features[7]. The parameters in these competitive methods were set to be the same as original papers. We have evaluated the accuracy by the precision ratio, recall ratio and F-measure given by following formulas.

$$\text{Precision} = \frac{\text{TP}}{\text{TP} + \text{FP}}, \text{Recall} = \frac{\text{TP}}{\text{TP} + \text{FN}}, \text{F} = 2/\left(\frac{1}{\text{Precision}} + \frac{1}{\text{Recall}}\right)(12)$$

The F-measure indicates the balance precision and recall. The larger value means better result. The TP, FP and FN denote the number of pixels detected correctly, detected wrongly, undetected wrongly respectively.

---

[2] We got ground truth dataset from http://limu.ait.kyushu-u.ac.jp/dataset/

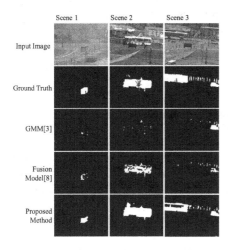

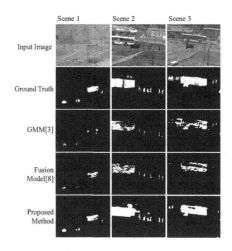

Fig. 3. Result of object detection after the moving object stopped. Scene 1: 100th frame after stopped, Scene 2: 60th frame after stopped, Scene 3: 150th frame after stopped

Fig. 4. Result of object detection after the object restarted to move

Fig. 3 shows the evaluated images, and Table 2 shows the evaluation results. The GMM based method[2] could detect just a few foreground pixels since it had learned the paused foreground object as "background". The fusion model[7] also gradually learned the foreground objects as "background". This is why the recall ratios of these methods were very low in all scenes. On the other hand, our proposed method gave much better recall ratio than competitive methods. The F-measure was also superior to the others.

Secondly, we have evaluated the precision ratio, recall ratio and F-measure with another scene in which the paused object had started to move again. The proposed method gave us better result than the other methods (See Table 3). The GMM based method[2] detected many FP pixels in the region where the foreground object had been paused(See Fig. 4). This is because illumination

Table 2. Accuracy evaluation of object detection after the moving object stopped

		Scene 1	Scene 2	Scene 3
GMM[2]	Precision	0.87	0.95	0.86
	Recall	0.13	0.05	0.16
	F-measure	0.23	0.10	0.27
Fusion Model[7]	Precision	0.98	0.95	0.94
	Recall	0.37	0.69	0.13
	F-measure	0.53	0.80	0.24
Proposed Method	Precision	0.90	0.85	0.87
	Recall	0.76	0.99	0.74
	F-measure	0.82	0.92	0.81

Table 3. Accuracy evaluation of object detection after the object restarted to move

		Scene 1	Scene 2	Scene 3
GMM[2]	Precision	0.95	0.93	0.80
	Recall	0.22	0.52	0.46
	F-measure	0.35	0.66	0.58
Fusion Model[7]	Precision	0.98	0.93	0.94
	Recall	0.48	0.61	0.46
	F-measure	0.65	0.73	0.61
Proposed Method	Precision	0.92	0.78	0.91
	Recall	0.78	0.98	0.82
	F-measure	0.85	0.87	0.86

change occurred during the period. Meanwhile, the fusion model[7] and the proposed method didn't detect the occluded region wrongly. However, the fusion model could not detect inside of the moving object because of over-training of foreground object. This is why the recall ratio of the fusion model was lower than the proposed method.

6.3 Evaluation of Robustness against Illumination Changes

We have used a outdoor image sequence in which illumination condition had sometimes changed rapidly, which was also used in the section 6.1. We have selected three images from 5,000 frames for evaluation. The parameters of background models including competitive methods were set to be the same as previous experiments.

The recall ratio, precision ratio and F-measure are shown in Table 4. In the case of FP or FN to be zero, we showed the F-measure "–" in Table 4 since it cannot be calculated. The illumination condition of scene # 831 was changed around the time. The GMM based method[2] detected many FP pixels since it was hard for GMM to adapt for rapid illumination changes. Meanwhile, our proposed method didn't detect any

Table 4. Accuracy evaluation with PETS2001 dataset

		# 831	# 1461	# 4251
GMM[2]	FN	0	211	234
	FP	1111	133	665
	F-measure	–	0.76	0.22
Fusion Model[7]	FN	0	450	311
	FP	0	41	1
	F-measure	–	0.57	0.24
Proposed Method	FN	0	82	120
	FP	0	478	422
	F-measure	–	0.71	0.47

FP pixels as good as the fusion model[7], which was reported that it is very robust against various illumination changes. The scene # 1461 included foreground objects under the stable illumination condition. The fusion model[7] detected the foreground object in the smaller size than the ground truth. This is because the fusion process was achieved by calculating logical AND operation between two kinds of background models. Therefore, the FN became large and the FP became smaller compared with the GMM based method. On the other hand, the proposed method detected the foreground objects including their shadow region. Note that shadow regions were not target to detect in the ground truth. This is why the FP became larger in the proposed method. To solve this problem, we are going to introduce a shadow detection method such as [15] in the future work. Finally, the scene # 4251 included foreground objects with illumination changes. This scene is one of the most difficult scenes for object detection. The proposed method gave better result than other two competitive methods. Note that the illumination change was not a factor of FP pixels. It was caused by shadow regions. Through above discussion, we are sure that our proposed method is very robust for illumination changes.

7   Conclusion

We have proposed a novel background modeling method. The proposed method could update a background region even when the region was occluded by a foreground object. This process was very effective for not only implicit background update but also keeping foreground object to being detected when the foreground object stopped in the scene. Through several experiments, we have confirmed the effectiveness of our approach from the viewpoints of robustness against illumination changes, handling of foreground objects and update of background model parameters. In our future works, we will study about efficiency strategy of initializing background model, complement of undetected pixels such as inside of the objects.

References

1. Stauffer, C., Grimson, W.: Adaptive background mixture models for real-time tracking. Computer Vision and Pattern Recognition 2, 246–252 (1999)
2. Shimada, A., Arita, D., Taniguchi, R.: Dynamic Control of Adaptive Mixture-of-Gaussians Background Model. In: CD-ROM Proceedings of IEEE International Conference on Advanced Video and Signal Based Surveillance 2006 (2006)
3. Elgammal, A., Duraiswami, R., Harwood, D., Davis, L.: Background and Foreground Modeling Using Non-parametric Kernel Density Estimation for Visual Surveillance. Proceedings of the IEEE 90, 1151–1163 (2002)
4. Tanaka, T., Shimada, A., Arita, D., Taniguchi, R.: A Fast Algorithm for Adaptive Background Model Construction Using Parzen Density Estimation. In: CD-ROM Proc. of IEEE International Conference on Advanced Video and Signal based Surveillance (2007)
5. Shimada, A., Taniguchi, R.: Hybrid Background Model using Spatial-Temporal LBP. In: IEEE International Conference on Advanced Video and Signal based Surveillance 2009 (2009)
6. Satoh, Y., Shun'ichi Kaneko, N.Y., Yamamoto, K.: Robust object detection using a Radial Reach Filter(RRF). Systems and Computers in Japan 35, 63–73 (2004)
7. Tanaka, T., Shimada, A., Taniguchi, R., Yamashita, T., Arita, D.: Towards robust object detection: integrated background modeling based on spatio-temporal features. In: Asian Conference on Computer Vision 2009 (2009)
8. Toyama, K., Krumm, J., Brumitt, B., Meyers, B.: Wallflower: Principle and Practice of Background Maintenance. In: International Conference on Computer Vision, pp. 255–261 (1999)
9. Basharat, A., Gritai, A., Shah, M.: Learning object motion patterns for anomaly detection and improved object detection. Computer Vision and Pattern Recognition, 1–8 (2008)
10. Porikli, F., Ivanov, Y., Haga., T.: Robust abandoned object detection using dual foreground. EURASIP Journal on Advances in Signal Processing (2008)
11. li Tian, Y., Feris, R., Hampapur, A.: Real-time detection of abandoned and removed objects in complex environments. In: International Workshop on Visual Surveillance - VS 2008 (2008)
12. Holt Charles, C.: Forecasting seasonals and trends by exponentially weighted moving averages. International Journal of Forecasting 20, 5–10 (2004)

13. Heikkilä, M., Pietikäinen, M., Heikkilä, J.: A texture based method for detecting moving objects. In: British Machine Vision Conf., vol. 1, pp. 187–196 (2004)
14. Boykov, Y., Kolmogorov, V.: An experimental comparison of min-cut/max-flow algorithms for energy minimization in computer vision. IEEE Transactions on Pattern Analysis and Machine Intelligence 26, 1124–1137 (2004)
15. Martel-Brisson, N., Zaccarin, A.: Moving cast shadow detection from a gaussian mixture shadow model. Computer Vision and Pattern Recognition, 643–648 (2005)

Determining Spatial Motion Directly from Normal Flow Field: A Comprehensive Treatment

Tak-Wai Hui and Ronald Chung

Department of Mech. and Automation Engineering,
The Chinese University of Hong Kong, Hong Kong
twhui1@mae.cuhk.edu.hk, rchung@cuhk.edu.hk

Abstract. Determining motion from a video of the imaged scene relative to the camera is important for various robotics tasks including visual control and autonomous navigation. The difficulty of the problem lies mainly in that the flow pattern directly observable in the video is generally not the full flow field induced by the motion, but only partial information of it, which is known as the normal flow field. A few methods collectively referred to as the direct methods have been proposed to determine the spatial motion from merely the normal flow field without ever interpolating the full flows. However, such methods generally have difficulty addressing the case of general motion. This work proposes a new direct method that uses two constraints: one related to the direction component of the normal flow field, and the other to the magnitude component, to determine motion. The first constraint presents itself as a system of linear inequalities to bind the motion parameters; the second one uses the rotation magnitude's globality to all image positions to constrain the motion parameters further. A two-stage iterative process in a coarse-to-fine framework is used to exploit the two constraints. Experimental results on benchmark data show that the new treatment can tackle even the case of general motion.

1 Introduction

A moving object or camera generally induces a certain apparent flow pattern in the acquired video. How the relative motion in space between the object and camera can be determined from the apparent flow is a classical problem whose solution has tremendous applications to autonomous navigation, visual control, robotics, human action understanding, and intelligent user interface.

Due to the well-known ambiguity between motion speed and object size-and-depth, from monocular video alone the translation magnitude of motion is generally not determinable and left as an overall arbitrary scale related to object depth. In other words, if we describe the spatial motion as consisting of a translation component $\mathbf{t}$ (as a 3D displacement vector) and a rotation component $\mathbf{w}$ (in rotation's angle-axis form), we are to determine the direction of $\mathbf{t}$ and the full $\mathbf{w}$. However, due to the familiar *aperture problem*, the full flow induced by the spatial motion at any image position is observable generally only partially, as a *normal flow*. This partial observability makes motion determination a challenge.

R. Koch et al. (Eds.): ACCV 2010 Workshops, Part I, LNCS 6468, pp. 23–32, 2011.

Classical solutions to motion determination are largely about establishing explicit motion correspondences across the image frames. One is the feature-based track, which tracks the distinct features in the image stream [1, 5]. Another track, originated from Horn and Schunck [10] as well as Lucas and Kanade [14], tracks practically all image positions. In [2], the multi-scale approach was extended by including the gradient constancy assumption to overcome the drawback of using the grey value constancy assumption. Higher order constancy assumptions had also been included into the variational model [15]. After estimating the full flow field, motion parameters are estimated in a subsequent stage. A method of using constraint lines to estimate the focus of expansion (FoE) was developed by [12]. However, the rotational component cannot be recovered. Linearity and divergence properties of the orthogonal cross-section of the projected flow fields were used in [7]. An iterative approach is used to de-rotate the image by feeding back the estimated rotational component in the next iterative cycle. A recent research work presents a linear formulation of the bilinear constraint [16]. It also pointed out that highly accurate estimate for full flow does not necessarily provide an accurate estimation of ego-motion because it also depends on the error characteristic within the estimated full flow field.

A few methods have been proposed to determine camera motion from the normal flow field directly. Such methods are often referred to as the direct methods. A classical direct method is [11], in which only camera motion having pure translation, pure rotation, or a motion with known rotation can be recovered. Another direct method that is based upon selection of image points that form global patterns in the image was developed by [8]. They transformed the parameter estimation problem to a series of boundary extraction problems. The method needs to determine boundaries between two sparsely labeled regions in the image domain. Another direct method utilizes a ψ-line searching algorithm to determine the direction of $\mathbf{t}$ [17]. However, only a limited number of normal flows could participate in the recovery. In a later work, the searching problem is formulated on the L-space [18]. More normal flows could contribute to the solution. Yet each set of affine parameters of r on each ψ-line of the L-space are still estimated by a limited number of normal flows. Both two methods require the application of minimization over unbounded decision variables.

The direct methods in the literature either cannot tackle the case of general motion in which both translation and rotation are present and unknown, or have to deal with the problem of identifying boundaries between image regions that are only sparsely labeled. In this paper, we describe a direct method that is free of such limitations. We propose the use of two constraints: one related to the direction component of the normal flows, and the other to the globality of the motion magnitude. While the former manifests as a system of linear inequalities that bind the motion parameters, the latter serves to constrain the motion parameters further by insisting that every image position must have a component of normal flow magnitude that is consistent with a global rotation magnitude. A two-stage iterative voting process, in a coarse-to-fine framework, is implemented to determine the motion parameters. Experimental results on benchmark data show that the method is capable of recovering general motion.

2 The Apparent Flow Direction (AFD) Constraint

An earlier report[4] has laid down the foundation on how normal flow direction imposes a constraint on the spatial motion parameters. Here we provide a more geometric intuition of the constraint as well as a more complete formulation of it. In general, optical flow $\dot{\mathbf{p}}$ at any image position $\mathbf{p}$ is not directly observable from the image because of the well-known aperture problem. Only the projected component of the flow to the spatial intensity gradient at the position, by the name of normal flow $\dot{\mathbf{p}}_n$, is directly observable, and it is generally computed from the spatial and temporal derivatives of the intensity profile at $\mathbf{p}$.

The constraint is powerful but simple to state. If normal flow must be a projected component of full flow, the full flow must have an orientation no different from that of the observed normal flow by more than 90°. This restriction on the direction of the full flow in turn constrains the motion parameters $(\mathbf{t}, \mathbf{w})$. The constraint at $\mathbf{p}$ with non-zero full flow and normal flow can be formulated as:

$$\dot{\mathbf{p}} \cdot \dot{\mathbf{p}}_n = (\dot{\mathbf{p}} \cdot \mathbf{n})^2 > 0 \tag{1}$$

where $\mathbf{n}$ is a unit vector in the direction of the local intensity gradient at $\mathbf{p}$.

2.1 Preliminaries

Consider a camera that has the image plane placed at focal length f from its optical center. Define a camera-centered coordinate system C, which has the Z-axis pointing along the optical axis toward the imaged scene. Consider any object point $\mathbf{P} = (X, Y, Z)^T$ of the scene with respect to C. Suppose that its projection onto the image plane is $\mathbf{p} = (x, y)^T$. Under perspective projection,

$$\mathbf{P} \cong \widetilde{\mathbf{p}} \tag{2}$$

where $\cong$ denotes equality up to arbitrary nonzero scale, and $\widetilde{\mathbf{p}} = (x/f, y/f, 1)^T$ represents projective coordinates of $\mathbf{p}$ with f-normalized x and y components.

Suppose that the camera undergoes a general motion with a translation $\mathbf{t}$ and a rotation $\mathbf{w}$. The motion of the point $\mathbf{P}$ relative to frame C is:

$$\dot{\mathbf{P}} = -\mathbf{t} - \mathbf{w} \times \mathbf{P} \tag{3}$$

Suppose also that the normal flow $\dot{\mathbf{p}}_n$ of orientation γ in the image space is observed at image position $\mathbf{p}$. The motion parameters $\mathbf{t}$ and $\mathbf{w}$ must be such that they induce a full flow (at $\mathbf{p}$) that has a direction no different from γ by more than 90°. A few further algebraic manipulations will turn Equation (1) to:

$$AFD(\mathbf{p}, \gamma) : \mathbf{t} \cdot \mathbf{a}_t / Z - \mathbf{w} \cdot \mathbf{a}_w < 0 \tag{4}$$

where $\mathbf{a}_t = \widetilde{\mathbf{p}} \times (\sin\gamma, -\cos\gamma, 0)^T$, $\mathbf{a}_w = \mathbf{a}_t \times \widetilde{\mathbf{p}}$, and Z is the scene depth at $\mathbf{p}$. This is the constraint imposed by normal flow direction γ. It is expressed in a form more precise than that in [4]. We refer to it as the Apparent Flow Direction (AFD) constraint.

2.2   The Special Case: Pure Translation

Suppose the camera undergoes a pure translation $\mathbf{t}$ that has a component toward the imaged scene. All the optical flows should be pointing away from the focus of expansion (FoE). The full flow $\dot{\mathbf{p}}$ induced by the FoE could be any vector of orientation between $(\gamma\text{-}90°)$ and $(\gamma\text{+}90°)$ in the image. The case that the camera undergoes a pure translation that has a component away from the imaged scene is similar, except that the FoE is replaced by the focus of contraction (FoC).

By setting the $\mathbf{w}$-component to zero in the $AFD(\mathbf{p}, \gamma)$ constraint, the locus of $\mathbf{t}$'s direction in space (regardless of whether it represents an FoE or FoC) is

$$AFD_t(\mathbf{p}, \gamma) : \widehat{\mathbf{t}} \cdot \widehat{\mathbf{a}}_t < 0 \tag{5}$$

where $\widehat{\mathbf{t}}$ and $\widehat{\mathbf{a}}_t$ are unit vectors of $\mathbf{t}$ and $\mathbf{a}_t$ respectively. In essence, $AFD_t(\mathbf{p}, \gamma)$ is a linear inequality on the direction of $\mathbf{t}$, representing exactly half of the of the of the parameter space of $\mathbf{t}$'s direction.

2.3   The Special Case: Pure Rotation

The case of pure rotation is analogous to that of pure translation. Suppose the camera rotates about an axis $\mathbf{w}$ (in the right-hand manner) with an angular velocity given by the magnitude of $\mathbf{w}$. By setting the $\mathbf{t}$-component to zero in the $AFD(\mathbf{p}, \gamma)$ constraint, the locus of $\mathbf{w}$ imposed by $\mathbf{p}$ and γ can be expressed as:

$$AFD_w(\mathbf{p}, \gamma) : \widehat{\mathbf{w}} \cdot \widehat{\mathbf{a}}_w > 0 \tag{6}$$

where $\widehat{\mathbf{w}}$ and $\widehat{\mathbf{a}}_w$ are unit vectors of $\mathbf{w}$ and $\mathbf{a}_w$ respectively. The locus is again in the form of a linear inequality that binds the direction of $\mathbf{w}$.

2.4   Solving the System of Linear Inequalities for the Two Special Cases

Suppose there are m data points (image positions have observable normal flows) in the image. They will each give rise to an inequality described in either Equation (5) or Equation (6), about the directions of $\mathbf{t}$ and $\mathbf{w}$ respectively. Define $\mathbf{A}_t = [\widehat{\mathbf{a}}_t]_{m \times 3}$, $\mathbf{A}_w = [-\widehat{\mathbf{a}}_w]_{m \times 3}$, the entire set of inequalities can be expressed as either

$$\mathbf{A}_t \widehat{\mathbf{t}} < 0 \text{ or } \mathbf{A}_w \widehat{\mathbf{w}} < 0 \tag{7}$$

The parameter space of $\mathbf{t}$ and $\mathbf{w}$ can be parameterized by the spherical coordinates $(\rho = 1, \phi, \theta)$. With a number of data points available, each supplying a different locus for $\mathbf{t}$ (or $\mathbf{w}$). Here we provide a geometric interpretation of the task, and supply an alternative solution mechanism that is computationally more efficient but also more accurate in its solution. The task in hand (as expressed by Equation (7)) is about seeking a 3-vector $\mathbf{n}$ (which is about either $\mathbf{t}$ or $\mathbf{w}$) that makes no acute angle with any row vector of a matrix $\mathbf{A}$ (which could be either $\mathbf{A}_t$ or $\mathbf{A}_w$, depending upon whether it is a case of pure translation or pure rotation). Notice that only the direction of $\mathbf{n}$, not its magnitude, is desired.

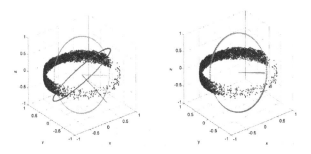

Fig. 1. Distribution of noisy $\hat{\mathbf{a}}_t$'s and the estimation of translation **t**.(a) Original solution. (b) Solution using a resampling-based method. Red and green lines are the estimated and true directions of **t** respectively.

If the row vectors of **A** are viewed as radial vectors of a unit sphere in 3-space, the solution is about designating a pole **n** of the sphere, so that if the pole is regarded as the north pole, a maximal number of the above radial vectors lie in the sphere's south hemisphere. In this light, an alternative way of acquiring the solution to the inequality system could be formulated as an optimization problem: given **A**, seek a unit 3-vector **n** such that

$$\mathbf{n} = \arg \min_{\text{subject to } \mathbf{n} \in \mathbf{R}^3, ||\mathbf{n}||=1} \sum (\hat{\mathbf{a}}_i \cdot \mathbf{n}) \qquad (8)$$

where $\mathbf{a}_i (i = 1, 2, ..., m)$ are normalized row-vectors of **A**.

To avoid solution of local minimum, at the end of each minimization process we randomly re-sample the space of decision variables around the stabilized values, and use such sampled values to initialize the decision variables for re-running the minimization process. We iterate the process until random distribution of the stabilized decision values does not lower the objective function further. Figure 1 shows example results of recovering **t** on a synthetic dataset. The normal flows were corrupted by an additive white Gaussian noise of 5dB signal-to-noise ratio. The above resampling-based minimization mechanism was able to supply a direction of **t** (Figure 1(b)) much closer to the ground truth than the one without using resampling-based minimization (Figure 1(a)). The angular error in **t** was reduced from 52.9° to 1.6°. The case of pure rotation is similar. It is omitted here.

2.5   The Case of General Motion

When there are both translation and rotation in the motion, normal flows are affected by both **t** and **w** in unknown proportions. By observing Equation (4), it can be deduced that it cannot be $\mathbf{t} \cdot \mathbf{a}_t$ being positive and $\mathbf{w} \cdot \mathbf{a}_w$ being negative at the same time, or else the inequality expressed by Equation (4) will be violated. From the geometric perspective, and in particular the principle of vector addition, the same conclusion can be reached, as illustrated by Figure 2.

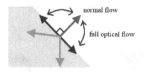

Fig. 2. Relationship between full flow and normal flow

Consider any image position **p** where the normal flow is observable. The direction γ of the normal flow introduces respective bounds, in the form of $AFD_t(\mathbf{p}, \gamma)$ and $AFD_w(\mathbf{p}, \gamma)$, to the directions of the vectors $\dot{\mathbf{p}}_t$ and $\dot{\mathbf{p}}_w$, which are optical flow components induced by the translation **t** and rotation **w** of the motion respectively. As the full flow must have an overall orientation between (γ-90°) and (γ+90°), it is impossible that both the two optical flow components $\dot{\mathbf{p}}_t$ and $\dot{\mathbf{p}}_w$ point to the shaded region shown in Figure 2. This imposes a constraint to (**t,w**), in the form of a logical OR operation over $AFD_t(\mathbf{p}, \gamma)$ and $AFD_w(\mathbf{p}, \gamma)$. More precisely, each of { **t,w** } ought to span exactly half of their own direction space, and with (**t,w**) (their directions only not magnitudes) considered together, they span three-quarter of a 4-D parameter space. In other words, the AFD constraint allows each single normal flow to trim away one-quarter of (**t,w**)-direction-space from consideration.

3   The Apparent Flow Magnitude (AFM) Constraint

The AFD constraint is applicable regardless of whether the motion is a specific one or a general one. However, whether it alone could lead to a unique result of the motion parameters depend upon how general is the orientation distribution of the normal flow field, which is a function of the intensity-gradient distribution of the image. Our experimental results show that for some specific scenes, the AFD constraint could only restrict the motion parameters to a possible set of solutions, not a unique one. In this work we explore also how the magnitude information of the field can be brought in to narrow down the motion values further.

Consider the case of pure rotation. At any image position **p** the rotation **w** induces an optical flow $\dot{\mathbf{p}}_w$. This optical flow $\dot{\mathbf{p}}_w$ then projects to the local intensity gradient to form the observable normal flow $\dot{\mathbf{p}}_n$. If the angle between $\hat{\dot{\mathbf{p}}}_w$ ($\|\mathbf{w}\|$-normalized $\dot{\mathbf{p}}_w$) and $\dot{\mathbf{p}}_n$ is α, we have

$$\|\mathbf{w}\| = \|\dot{\mathbf{p}}_n\|/(\|\hat{\dot{\mathbf{p}}}_w\| \cos \alpha) \tag{9}$$

In other words, any hypothesis of the rotation axis $\hat{\mathbf{w}}$ will allow the rotation magnitude $\|\mathbf{w}\|$ to be determined from any image position **p** where normal flow is observable. Since this rotation magnitude is a global quantity for the entire normal flow field, it should have the same value from any particular normal flow data point. As a consequence, the consistency over $\|\mathbf{w}\|$ of the normal flow field could serve to confirm if the hypothesized rotation axis is correct or not. We refer to this constraint on the motion parameters as the Apparent Flow Magnitude (AFM) constraint.

In our implementation, for any particular hypothesis of the rotation axis $\widehat{\mathbf{w}}$, the standard deviation (SD) of $||\mathbf{w}||$ computed from all the normal flow data points in the image space serves to indicate how likely the hypothesis is correct; an SD that is too large will make it justifiable to have the particular hypothesis of $\widehat{\mathbf{w}}$ ruled out. Notice that the AFM constraint also supplies the true value of $||\mathbf{w}||$ once the final answer on the rotation axis $\widehat{\mathbf{w}}$ is attained.

It is worth pointing out that, in Equation (9), an α close to 90° will amplify the uncertainty in estimating $||\mathbf{w}||$. We refer to such normal flow data points as the degenerate data points. It is necessary that in the above process such normal flow data points are excluded from computing the SD measure. Figure 3 shows the distribution of $||\mathbf{w}||$-estimation in a synthetic data experiment for a $\widehat{\mathbf{w}}$ that is only 0.1° away from the ground truth. Yet the SD of the distribution is still large, due mainly to the fact that the degenerate data points are not excluded from the estimation process. In the figure, those $||\mathbf{w}||$-estimates that are from the degenerate data points are marked with red circles. If such data points are excluded, the SD can be reduced from 0.183° to 0.0253°.

To extend the use of the AFM constraint to the case of general motion, we adopt a simple trick. For each particular set of $\left(\widehat{\mathbf{t}}, \widehat{\mathbf{w}}\right)$ (here we use $\widehat{\mathbf{t}}$ and $\widehat{\mathbf{w}}$, the normalized $\mathbf{t}$ and $\mathbf{w}$, to denote their directions), we find out which normal flow data point $\mathbf{p}$ has the intensity gradient direction $\widehat{\mathbf{u}} = \nabla I / ||\nabla I||$ (where $I(x,y)$ refers to the intensity distribution in the associated image) orthogonal to the optical flow component $\widehat{\dot{\mathbf{p}}}_t$ induced by the particular $\widehat{\mathbf{t}}$. At such $\mathbf{p}$'s, the optical flow $\dot{\mathbf{p}}$ and in turn the normal flow $\dot{\mathbf{p}}_n$ is solely governed by the optical flow component $\widehat{\dot{\mathbf{p}}}_w$ that is induced by $\widehat{\mathbf{w}}$. At such data points the above analysis could still apply. To summarize, by defining a set $\Pi_\perp \left(\widehat{\mathbf{t}}\right) = \left\{\mathbf{p} : \widehat{\dot{\mathbf{p}}}_t \perp \widehat{\mathbf{u}}(\mathbf{p})\right\}$, the AFM constraint for the case of general motion is expressible as:

$$AFM : \left(\widehat{\mathbf{t}}^T, \widehat{\mathbf{w}}^T\right)^T = \arg \min_{\widehat{\mathbf{t}}, \widehat{\mathbf{w}} \in \mathbf{R}^3, \ ||\widehat{\mathbf{t}}||, ||\widehat{\mathbf{w}}|| = 1 \ and \ \mathbf{p} \in \Pi_\perp(\widehat{\mathbf{t}})} SD||\mathbf{w}(\mathbf{p})|| \qquad (10)$$

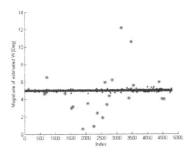

Fig. 3. The distribution of $||\mathbf{w}||$-estimates in a synthetic data experiment

4   Putting the Two Constraints Together

We adopt an iterative two-stage process to apply the AFD and AFM constraints alternately for determining motion. The first stage involves the use of AFD to constrain the directional components $\left(\widehat{\mathbf{t}}, \widehat{\mathbf{w}}\right)$ of the motion. Such a process could give rise to unique solution of the directional components, but not always. In case that it does not, the constrained $\left(\widehat{\mathbf{t}}, \widehat{\mathbf{w}}\right)$ will go through the second stage, which is about the application of the AFM constraint.

   To speed up the processing and to avoid the effect of local minima, a coarse-to-fine strategy over the bounded $\left(\widehat{\mathbf{t}}, \widehat{\mathbf{w}}\right)$-parameter space is used. A set of motion vectors $\left(\widehat{\mathbf{t}}, \widehat{\mathbf{w}}\right)$ parameterized by spherical coordinates ϕ and θ are first generated using a coarse sampling resolution. In the first stage, the AFD constraint is applied to reject impossible combinations of $\left(\widehat{\mathbf{t}}, \widehat{\mathbf{w}}\right)$. For each of such motion vectors, the total number of normal flow data points that fulfill the AFD constraint is also recorded. Those motion vectors that have too few of such data points are rejected. The AFM constraint is applied to refine the solution further in the second stage. The motion vectors that do not have small enough SD of $\|\mathbf{w}\|$ are rejected. The magnitude of $\mathbf{w}$ is also estimated at the same time. The result is then carried forward to serve as seeds of new sampling points in the next iteration that adopts a slightly finer resolution. The iterations continue until motion values of enough precision are attained.

5   Experimental Results on Benchmark Data

Two sets of data are both benchmark data used in the literature, namely the Fountain sequence (FS) [16] and the very widely used Yosemite sequence (YS). Both datasets are about general motions. One latest result is that of the method presented in [16], which makes use of high-accuracy optical flow estimation as input flow (meaning that the method has to use smoothness constraint of some sort for interpolating the full flows from the apparent flows), namely the methods of Brox *et al.* [2] and Farnebäck [6]. To make our experiments results directly comparable to those results, the same pair of image frames was used in each experiment. The ground truth of the FS is $\mathbf{t} = (\text{-0.6446 } 0.2179 \ 2.4056)^T$ pixels and $\mathbf{w} = (\text{-0.125 } 0.2 \text{ -0.125})^T$ deg/frame [16]. The ground truth of YS is $\mathbf{t} = (0 \ 0.17 \ 0.98)^T \times 34.8$ pixels and $\mathbf{w} = (1.33 \ 9.31 \ 1.62)^T \times 10^{-2}$ deg/frame [9].

   To determine the normal flow at any particular image position, we simply use the equality $\dot{\mathbf{p}}_n = -I_t \nabla I / \|\nabla I\|^2$. The spatial ($\nabla I$) and temporal ($I_t$) derivatives are calculated by the 8-point method described by Horn and Schunck [10]. To ease the differentiation process, the image data were smoothed by 2D Gaussian filter beforehand. FS and YS had 16.46% and 26.59% detectable normal flows respectively. The iterative two-stage process, with a coarse-to-fine strategy, was then applied to each of the datasets to determine the direction $\widehat{\mathbf{t}}$, the rotation axis $\widehat{\mathbf{w}}$, and the rotation magnitude $\|\mathbf{w}\|$. Tables 1 and 2 show the results in

Table 1. Errors of motion estimation on the Fountain sequence

Technique	Ang. error, t [deg]	Absolute error, w [deg/frame]		
		$|\Delta w_x|$	$|\Delta w_y|$	$|\Delta w_z|$
Raudies *et al.*[16] (Brox *et al.*(2D) [2])	4.395	0.001645	0.0286	0.02101
Raudies *et al.* [16]	6.841	0.01521	0.05089	0.025
(Farnebäck [6], 100% density)				
Raudies *et al.* [16]	1.542	0.0008952	0.01349	0.003637
(Farnebäck [6], 25% density)				
Our method	0.740115	0.0071692	0.010532	0.018084

Table 2. Errors of motion estimation on the Yosemite sequence without clouds

Technique	Ang. error, t [deg]	Absolute error, w [deg/frame]		
		$|\Delta w_x|$	$|\Delta w_y|$	$|\Delta w_z|$
Raudies *et al.*[16] (Brox *et al.*(2D) [2])	4.893	0.02012	0.1187	0.1153
Raudies *et al.* [16]	4.834	0.03922	0.00393	0.07636
(Farnebäck [6], 100% density)				
Lourakis [13]	3.7	0.038732	0.028419	0.011516
Heeger *et al.* [9]	3.5	0.0568	0.0344	0.0807
Lourakis [12]	3.1			
Raudies *et al.* [16]	1.208	0.007888	0.01178	0.02633
(Farnebäck [6], 25% density)				
Raudies *et al.* [16]	1.134	0.01261	0.008485	0.02849
(Farnebäck[6], 25% density, RANSAC)				
Our method	0.988969	0.0037171	0.012006	0.025528

comparison with those reported in the literature. Our method achieved better result in recovering the translational direction as compared with the existing methods. On the recovery of rotation, overall speaking, the proposed method also achieved results comparable to or better than those of the existing methods. Our method performed slightly less than [16] on estimating **w**, yet it is worth noticing that the method reported in [16] is a full flow-based one and requires to introduce the smoothness assumption somewhere in attaining the full flows, which could cause a problem for scenes that are not smooth enough.

6 Conclusion and Future Work

We have described two important constraints that allow the normal flow field to be used directly for motion determination. One constraint is related to the directional information of the normal flow field, and the other to its magnitude information. We have also outlined how the two constraints can be used alternately in an iterative fashion to determine the motion parameters. Notice also that the method, being a direct method itself, does not require the presence of distinctly trackable features in the imaged scene, nor does it require to assume smoothness about the imaged scene for interpolating the full flows. Experimental results over benchmark datasets show that not only can the method recover general motion, it also has a performance comparable even to those of methods that have to bring in the smoothness assumption as well. Future work will address how multiple motions in the imaged scene can be detected, and how depth can be recovered from visual motion.

Acknowledgement. The work described was partially supported by the Chinese University of Hong Kong 2009-2010 Direct Grant (Project No. 2050468).

References

1. Armangue, X., Araujo, H., Salvi, J.: A review on egomotion by means of differential epipolar geometry applied to the movement of a mobile robot. Patten Recognition 36(12), 2927–2944 (2003)
2. Brox, T., Bruhn, A., Papenberg, N., Weickert, J.: High accuracy optical flow estimation based on a theory for warping. In: Pajdla, T., Matas, J(G.) (eds.) ECCV 2004. LNCS, vol. 3024, pp. 25–36. Springer, Heidelberg (2004)
3. Bruss, A.R., Horn, B.K.P.: Passive navigation. In: Int'l Conf. on Computer Vision Graphic and Imagining, vol. 21, pp. 3–20 (1983)
4. Chung, R., Yuan, D.: Direct determination of camera motion from normal flows. In: Int'l Conf. on Visualization, Imaging, and Image Processing, pp. 153–157 (2009)
5. Cipolla, R., Okamoto, Y., Kuno, Y.: Robust structure from motion using motion parallax. In: Int'l Conf. Computer Vision, pp. 374–382 (1993)
6. Farnebäck, G.: Polynomial expansion for orientation and motion estimation. PhD thesis, Dept. of Electrical Engineering, Linköping University (2002)
7. Fejes, S., Davis, L.S.: What can projections of flow fields tell us about the visual motion. In: Int'l Conf. on Computer Vision, pp. 979–986 (1998)
8. Fermüller, C., Aloimonos, Y.: Qualitative egomotion. IJCV 15, 7–29 (1995)
9. Heeger, D.J., Jepson, A.D.: Subspace methods for recovering rigid motion I: algorithm and implementation. IJCV 7(2), 95–117 (1992)
10. Horn, B.K.P., Shunck, B.G.: Determining optical Flow. AI 17, 185–203 (1981)
11. Horn, B.K.P., Weldon, E.J.: Direct methods for recovering motion. IJCV 2, 51–76 (1988)
12. Lourakis, M.I.A.: Using constraint lines for estimating egomotion. In: Asian Conf. on Computer Vision, pp. 971–976 (2000)
13. Lourakis, M.I.A.: Egomotion estimation using quadruples of collinear image points. In: European Conf. on Computer Vision, pp. 834–848 (2000)
14. Lucas, B.D., Kanade, T.: An iterative image registration technique with an application to stereo vision. In: Int'l Conf. on Computer Vision, pp. 933–938 (1981)
15. Papenberg, N., Bruhn, A., Brox, T., Didas, S., Weickert, J.: Highly accurate optic flow computation with theoretically justified warping. IJCV 67(2), 141–158 (2000)
16. Raudies, F., Neumann, H.: An efficient linear method for the estimation of egomotion from optical flow. In: Denzler, J., Notni, G., Süße, H. (eds.) DAGM. LNCS, vol. 5748, pp. 11–20. Springer, Heidelberg (2009)
17. Silva, C., Santos-Victor, J.: Robust egomotion estimation from the normal flow using search subspaces. IEEE Trans. PAMI 19(9), 1026–1034 (1997)
18. Silva, C., Santos-Victor, J.: Egomotion estimation on a topological space. In: Int'l Conf. on Pattern Recognition, pp. 64–66 (1998)

Background Subtraction for PTZ Cameras Performing a Guard Tour and Application to Cameras with Very Low Frame Rate

C. Guillot[1], M. Taron[1], Patrick Sayd[1], Q.C. Pham[1],
C. Tilmant[2], and J.M. Lavest[2]

[1] CEA, LIST, Vision and Content Engineering Laboratory
BP 94, Gif-sur-Yvette, F-91191 France
[2] LASMEA UMR 6602, PRES Clermont Université/CNRS
63177 Aubière cedex, France

Abstract. Pan Tilt Zoom cameras have the ability to cover wide areas with an adapted resolution. Since the logical downside of high resolution is a limited field of view, a guard tour can be used to monitor a large scene of interest. However, this greatly increases the duration between frames associated to a specific location. This constraint makes most background algorithms ineffective. In this article we propose a background subtraction algorithm suitable to cameras with very low frame rate. Its main interest consists in the resulting robustness to sudden illumination changes. The background model which describes a wide scene of interest consisting of a collection of images can thus be successfully maintained. This algorithm is compared with the state of the art and a discussion regarding its properties follows.

1 Introduction

While the number of cameras used in public areas constantly increases, a strong effort is made to develop robust algorithm able to automate scene monitoring. Background subtraction is a popular pre-processing task often required to introduce scene understanding in video sequences.

Wide angle cameras can be used to monitor a wide scene, their interest is however limited by their low resolution when it comes to analysing the scene. Pan Tilt Zoom (PTZ) cameras have two rotation axis and a zoom function which enable focusing on a part of the scene at any suitable resolution. The obvious drawback of the PTZ sensor lies in its limited field of view.

When dealing with static camera, one of the usual approaches to issues such as tracking or object recognition is to build a background model. This model, which will have to be initialised and updated continuously, allows the detection of objects of interest by estimating a distance to the current image. As for PTZ camera, to maintain a whole background model is challenging since the necessary information is rarely available.

In this article, a PTZ camera performing a guard tour over a wide area is used to detect objects of interest. The camera follows a predefined set of positions

R. Koch et al. (Eds.): ACCV 2010 Workshops, Part I, LNCS 6468, pp. 33–42, 2011.

(pan, tilt, zoom) covering the area at an adapted resolution. For each of these positions it can be considered that we are in the case of a static camera suffering a very low frame rate (approximately 1 image every 10 to 20 seconds). Such a duration between frames constitutes a major difficulty since the background model will not be continuously updated and show important disparities in terms of illumination between the model and the current image.

This article presents a thorough study of a very low update rate background subtraction algorithm. It briefly reviews the related work (section 2), then presents a previous contribution of the authors (section 3) which has motivated this study. A comparison between local texture descriptors and the introduction of a more robust feature descriptor is then presented (section 4.2). A discussion regarding the background model update strategy follows (section 4.3) and additional experimental results are provided in section 5.

2   Related Work

There exist many background subtraction techniques in the literature, most of which are designed for static cameras with a frame rate above 12fps. Starting with basic frame differencing [1], it was soon necessary to build more evolved frameworks to describe the background. Stauffer and Grimson [2] first introduced a popular statistical approach based on a mixture of Gaussian distributions to model the luminance of each pixel. The model is updated at each frame to account for the variations of the background. An overview of background subtraction methods based on mixtures of Gaussians is given by Bouwmans *et al.* [3]. Elgammal *et al.* [4] have even achieved greater accuracy by substituting the MoG model with kernel density estimator.

Single pixel luminance does not carry sufficient information to address the complexity of outdoor scenes. It was therefore necessary to introduce spatial and temporal coherence in background subtraction algorithms. In [4], classification as background was enforced by considering the distribution model of neighbouring pixels. Background description models were also improved to carry dynamic information based on optical flow estimates [5]. Even when dealing with static backgrounds, accounting for sets of pixels provide better results. This motivated the work of Chen *et al.*'s [6], where texture descriptor is considered based on the tiling of the image with 8×8 blocks. This descriptor encodes a local colour contrast histogram with 48 parameters and increases robustness to illumination variations. This methods was proved to very efficient in [7] and is used in the remainder of this article to compare the performances of our background subtraction algorithm.

Zhu *et al.*[8] proposed a background subtraction algorithm based on the extraction of Harris keypoints and SIFT descriptors but which can only detect moving objects.

In the specific case of PTZ cameras, most approaches are based on the creation of a mosaic of the scene background. New images from the camera are registered on the mosaic as a prerequisite to background subtraction model update

[9,10,11,12]. The drawback of these methods is that there is no global update of the background model. There is no warranty that the model of an area that has not been visited for a while is usable. Therefore it turns out that these methods are more suitable to the tracking or moving object than the complete modelling of a large scene of interest.

3 Background Subtraction by Keypoint Density Estimation

In [13] we presented a background subtraction method based on the estimation of the density of non matching keypoints. The motivation for this method came from the fact that edge descriptor reveal themselves more robust to illumination changes than texture descriptors. This algorithm has been proved to be very effective in the experimental context of a PTZ camera.

We assume that keypoints which cannot be matched from the current image to the background image belong to objects of interest. Harris keypoints are extracted from both images and SURF [14] descriptor are computed on both images. Due to the mechanical error of the PTZ camera images are first registered using keypoints with the highest Harris score (strong edges).

Because Harris keypoints are not stable, corresponding points may not be present on both images for matching. We have used the union of keypoint location on both images prior to the matching and classification of points based on the Euclidean distance in the space of SURF descriptors.

Once we have a set of non matching keypoints, we use kernel smoothing techniques to estimate a continuous density $\hat{d}_h$:

$$\hat{d}_h(x) = \frac{1}{Nh} \sum_{i=1}^{N} K\left(\frac{\|x - p_i\|_{img}}{h}\right), \tag{1}$$

with $(p_1, \ldots, p_N)$ the set of N non matching keypoints, K a Gaussian kernel function and h a smoothing parameter which specifies the influence of each observation on its neighbourhood. Pixels are classified as foreground if $N\hat{d}_h > s$ and as background otherwise.

Background model is updated according to the following equation:

$$bg_n(x) = \begin{cases} bg_{t-1}(x) & \text{if } N\hat{d}_h > s \\ bg_{t-1}(x)\frac{N\hat{d}_h(x)}{s} + img_t(x)\left(1 - \frac{N\hat{d}_h(x)}{s}\right) & \text{otherwise} \end{cases} \tag{2}$$

At this point, it is important to note that this approach presents some limitations in terms of implementation:

- SURF Gradient is based on image gradient and normalised to achieve better robustness to changes in illumination. In poorly textured areas the normalisation step amplifies the influence of camera noise and the artefacts due to image compression. This prevents some keypoints from matching in homogeneous areas, which leads to false positives.

- Harris threshold is chosen especially to prevent keypoints candidates to be located in homogeneous areas. Setting this value is still empirical and sequence dependant.
- Considering a variable set of points for background subtraction has lead to the use of a plain image as a background model. This is not plainly satisfactory because the update (eq. 2) actually blurs the model, and might lead to the creation of ghosts and false detection.

The remainder of the article presents an update to the algorithm in order to address these limitations.

4   Texture Descriptors for Background Subtraction

We propose a background subtraction algorithm similar to the keypoint which relies on a regular grid of modified SURF descriptors as a background model instead of a variable set of keypoints. These descriptors can be computed on weakly textured areas (sec. 4.1). This algorithm is as effective as the keypoint density algorithm but Harris feature point extraction and its associated manual threshold is no longer necessary.

Background subtraction is performed by computing the distance of a descriptor from a point of the grid to the corresponding one in the background model. Once we have a set of matching and not matching points we use equations 1 as a post processing to smooth out the classification results. The threshold is set to avoid false alarms in case of isolated detection. Meanwhile isolated mis-detection are automatically filled in by the neighbouring detections.

4.1   Weighting the SURF Descriptor

The SURF descriptor has been thought to be discriminative when computed on textured zones, but the normalisation process renders this inefficient, when in low textured areas, noise overcomes gradient information.

The intuition is that if there is no gradient information we can't decide at a local level whether a pixel is foreground or background. Thus, we arbitrarily decide that two low textured areas (those where the SURF norm is low) should match.

To do so, we consider the distribution of the norm of the SURF descriptors on a set of sequences displaying texture and homogeneous areas (Fig. 1). The resulting distribution presents two modes. The lowest corresponding to homogeneous, it is removed with an appropriate weighting function applied to the SURF descriptor values.

$$D'_{SURF} = D_{SURF} * f(\|D_{SURF}\|) \quad \text{with} \quad f(x) = \begin{cases} 0 & \text{if } x < 120 \\ \dfrac{x - 120}{480 - 120} & \text{if } 120 \leq x \leq 480 \\ 1 & \text{if } 480 < x \end{cases}$$

$$(3)$$

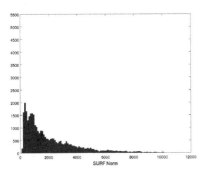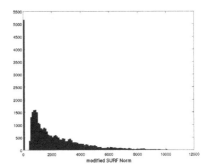

Fig. 1. Left: Histogram of the norm of SURF descriptor before normalisation. Right: Histogram of the norm of SURF descriptor before normalisation and after applying equation 3.

As a consequence, if the SURF descriptor is computed on a textured zone its norm remains 1. If it is computed on a zone which is not textured, it is set to 0, with a continuous transition between these two cases.

4.2   Evaluating the Quality of the Texture Descriptor

Chen *et al.*[6] have designed their own texture descriptor to perform background subtraction. It encodes a histogram of contrast between the different colour components. However we can question the choice of such a descriptor since there exists well known other descriptors used in other fields of computer vision.

We have compared the Chen descriptor to the SURF descriptor [14] on a sequence presenting challenging changes in illumination (Fig. 3a). To assess the quality of the descriptors only, we performed background subtraction on this sequence with no post processing of any kind. Descriptors are computed on the same regular grid and the classification as foreground or background is done only according to the distance toward the corresponding descriptor on a reference image (no statistical modelling in the space of descriptors). We Have computed ROC curves on this sequence with a variation of the classification threshold (Fig. 2).

Results are very poor if one consider a single frame as a reference (very strong disparities between images). However the obtained precision with SURF is always twice as better than the Chen Descriptor. If one considers consecutive images, there is a global increase in robustness with the use of SURF descriptors and the modified SURF descriptors.

4.3   Background Update

As descriptors are computed on a regular grid rather than a set of keypoints. It is now possible to handle the background model update in the space of descriptors rather than the image space. If $D_{\mathrm{Bkg},t}$ is a background descriptor and $D_{\mathrm{Img},t}$

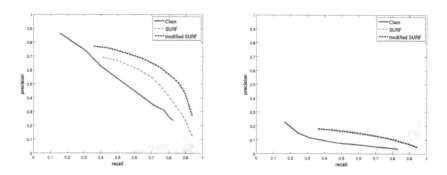

Fig. 2. Precision and recall curves of the Chen, SURF and modified SURF (sec. 4.1) descriptors computed on the sequence from Fig. 3a. Left: comparing two consecutive images. Right: comparing one specific image to all images of the sequence.

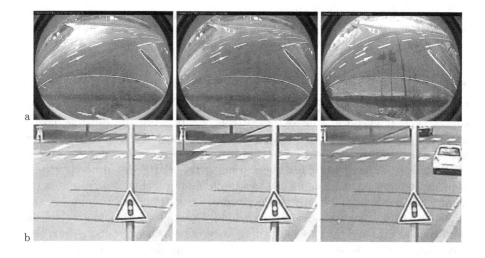

Fig. 3. Test Sequences. These sequences present important illumination changes, shadows and reflections on a rain-soaked road.

is a descriptor computed from the current image at time t and classified as background, then we use the following updating rule:

$$D_{\mathrm{Bkg},t+1} = \alpha D_{\mathrm{Bkg},t} + (1 - \alpha)D_{\mathrm{Img},t} \tag{4}$$

For our application the learning rate α is chosen rather high ($\alpha > 0.25$). As we consider sequences with very low frame rate, it is necessary to update the model quickly to follow the global illumination changes.

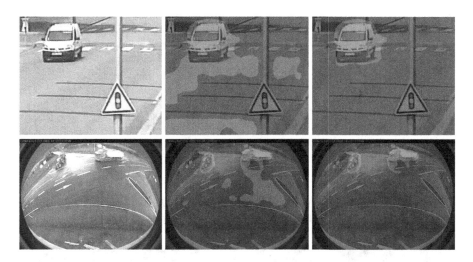

Fig. 4. Detection results on a sequence with very low textured zones and compression artefacts. First column: original image. Second column: segmentation result using keypoints density with a low threshold on the Harris score (Harris points can be located in homogeneous areas). Third column: grid of modified SURF descriptors.

Table 1. Comparison of detection results on various sequences

Method	Statistic	PTZ	Train	Outdoor1	PETS	Outdoor2
Modified SURF grid	Recall	0.74	0.77	0.69	0.75	0.63
	Precision	0.74	0.79	0.67	0.75	0.70
Keypoint density	Recall	0.61	0.83	0.53	0.8	0.64
	Precision	0.61	0.73	0.58	0.65	0.67
Chen *et al.*[6]	Recall	0.47	0.63	0.51	0.73	0.55
	Precision	0.24	0.61	0.69	0.84	0.60
Stauffer and Grimson	Recall	0.56	0.5	0.22	0.63	0.46
	Precision	0.12	0.44	0.46	0.6	0.55

5   Experimental Results

The first part of the experiments is devoted to the comparison of the SURF and modified SURF descriptor. Figure 4 shows the kind of issue which may arise on poorly textured areas and how the modified SURF descriptor deals with it. On these areas, the compression artefact create unstructured gradients which make the original SURF descriptor ineffective. Figure 4 shows that the modified SURF descriptor can be computed on uniform areas while not generating false mismatches. Figure 5 displays ROC curves which confirms quantitatively what can be observed on figure 4.

Figure 6 presents qualitative results for the case of a PTZ camera performing a guard tour. The time elapsed between consecutive frames is 24 seconds. Notice

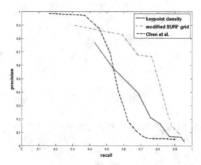

Fig. 5. Precision and recall curves computed on the *light change* sequence. The threshold for the keypoint density algorithm is the same as the one used in figure 4.

Fig. 6. PTZ Sequence

the green borders on the image due to the registration between images acquired during the tour.

The second part of the experiments compares the modified SURF grid background subtraction algorithm to Chen *et al.*'s [6] algorithm and our previous algorithm based on keypoints density estimation [13]. The application to PTZ cameras performing a Guard tour is equivalent to a fixed cameras with a very low frame rate. Therefore we have applied the algorithms to fixed cameras presenting challenging sequences and artificially lowered the frame rate to one image every 20 seconds.

Figure 7 shows qualitative results. Whereas the PETS sequence is not challenging in terms of illumination variation they show that our algorithm behaves well on weakly textured scenes. The *train* sequence is another example sequence for which our algorithm is stable even when sudden changes in illumination occur. On these sequences modified SURF grid behaves as well as the keypoint density algorithm.

Figure 8 shows quantitative results on two sequences where sudden changes in illumination occur. Table 1 sums up the results from various sequences and shows that our algorithm is more stable than others. The presented statistics may seem low at first sight, but these were computed in the most challenging experimental conditions. Moreover, as can be seen on figure 7, the loss of precision of our

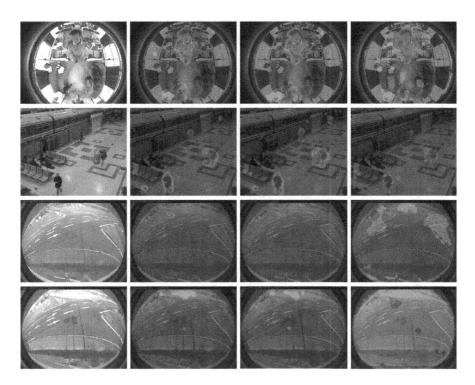

Fig. 7. Qualitative results obtained in various situations. First row is captured on board of a train. Second row is a sequence extracted from the PETS 2006 challenge (http://www.cvg.rdg.ac.uk/PETS2006/). Rows 3 and 4 are consecutive images extracted from the sequence in Fig. 3a. First column: original image. Second column: our algorithm. Third column: keypoint density algorithm [13]. Fourth column: Chen *et al.*'s algorithm.

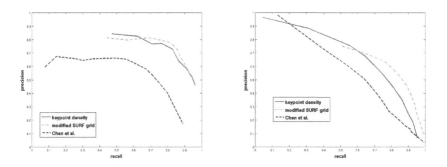

Fig. 8. Precision and recall curves. Left: *train* sequence. Right: *light change 2* sequence.

algorithm is inherent to the method and mainly due to the fact that it always over segment foreground blobs. In no case does it generate actual false alarms.

6 Conclusion

We have propose a simple yet efficient background subtraction algorithm. We use a modified version of the SURF descriptor which can be computed on weakly textured areas. We successfully apply our algorithm in the challenging context of PTZ cameras performing a guard tour and for which illumination issues are critical. Our algorithm successfully detects blobs with a sufficient accuracy used as a first step toward object detection application.

References

1. Jain, R., Nagel, H.: On the analysis of accumulative difference pictures from image sequences of real world scenes, vol. 1, pp. 206–213 (1979)
2. Stauffer, C., Grimson, W.E.L.: Adaptive background mixture models for real-time tracking. In: CVPR (1999)
3. Bouwmans, T., El Baf, F., Vachon, B.: Background Modeling using Mixture of Gaussians for Foreground Detection - A Survey. Recent Patents on Computer Science (2008)
4. Elgammal, A.M., Harwood, D., Davis, L.S.: Non-parametric model for background subtraction. In: Vernon, D. (ed.) ECCV 2000. LNCS, vol. 1843, pp. 751–767. Springer, Heidelberg (2000)
5. Mittal, A., Paragios, N.: Motion-based background subtraction using adaptive kernel density estimation. In: CVPR, vol. 2, pp. 302–309 (2004)
6. Chen, Y.T., Chen, C.S., Huang, C.R., Hung, Y.P.: Efficient hierarchical method for background subtraction. Pattern Recognition (2007)
7. Dhome, Y., Tronson, N., Vacavant, A., Chateau, T., Gabard, C., Goyat, Y., Gruyer, D.: A benchmark for background subtraction algorithms in monocular vision: a comparative study. In: IPTA (2010)
8. Zhu, Q., Avidan, S., Cheng, K.T.: Learning a sparse, corner-based representation for time-varying background modelling. In: ICCV, vol. 1, pp. 678–685 (2005)
9. Bhat, K., Saptharishi, M., Khosla, P.K.: Motion detection and segmentation using image mosaics. In: ICME (2000)
10. Cucchiara, R., Prati, A., Vezzani, R.: Advanced video surveillance with pan tilt zoom cameras. In: Proc. of Workshop on Visual Surveillance (VS) at ECCV (2006)
11. Azzari, P., Di Stefano, L., Bevilacqua, A.: An effective real-time mosaicing algorithm apt to detect motion through background subtraction using a ptz camera. In: AVSS (2005)
12. Robinault, L., Bres, S., Miguet, S.: Real time foreground object detection using ptz camera. In: VISAPP (2009)
13. Guillot, C., Taron, M., Sayd, P., Pham, Q.C., Tilmant, C., Lavest, J.M.: Background subtraction adapted to ptz cameras by keypoint density estimation. In: BMVC (2010) (to appear in bmvc 2010: supplied as supplementary material)
14. Bay, H., Ess, A., Tuytelaars, T., Gool, L.V.: Surf: Speeded up robust features. In: CVIU (2008)

Bayesian Loop for Synergistic Change Detection and Tracking

Samuele Salti, Alessandro Lanza, and Luigi Di Stefano

Computer Vision Lab
ARCES-DEIS, University of Bologna
Bologna, Italy
{samuele.salti,alessandro.lanza,luigi.distefano}@unibo.it

Abstract. In this paper we investigate Bayesian visual tracking based on change detection. Although in many proposals change detection is key for tracking, little attention has been paid to sound modeling of the interaction between the change detector and the tracker. In this work, we develop a principled framework whereby both processes can virtuously influence each other according to a Bayesian loop: change detection provides a completely specified observation likelihood to the tracker and the tracker provides an informative prior to the change detector.

1 Introduction and Related Work

Recursive Bayesian Estimation (RBE) [1] casts visual tracking as a Bayesian inference problem in state space given noisy observation of the hidden state. Bayesian reasoning has been used also to solve the problem of Change Detection (CD) in image sequences [2], and CD is at the root of many proposals in visual tracking. Nonetheless, interaction between the change detection and tracking modules is usually modeled heuristically. This negatively affects the quality of the information flowing between the two computational levels, as well as the soundness of proposals. Furthermore, the interaction can be highly influenced by heuristically hand-tuned parameters, such as CD thresholds. Hence, a first original contribution of this paper is a theoretically grounded and almost parameters-free approach to provide an observation likelihood to the RBE tracker from the posterior obtained by a Bayesian Change Detection (BCD).

Recently, Cognitive Feedback has emerged as an interesting and effective proposal in Computer Vision [3]. The idea is to let not only low-level vision modules feed high-level ones, but also the latter influence the former. This creates a closure loop, reminiscent of effects found in psychophysics. This concept has not been deployed for the problem of visual tracking yet. Nevertheless, it fits surprisingly well in the case of BCD, where priors can well model the *stimuli* coming from RBE. Hence, the second original contribution of this paper deals with investigating on using Cognitive Feedback to create priors from the state of an RBE tracker in the case of visual tracking based on change detection.

The third novel contribution deals with exploiting the synergy between the previously presented approaches, so as to obtain a fully Bayesian tracking system.

R. Koch et al. (Eds.): ACCV 2010 Workshops, Part I, LNCS 6468, pp. 43–53, 2011.

As a preliminary investigation into this novel approach, in this paper we have conducted the theoretical analysis and the experimental validation only for the simpler case of single-target tracking.

As for related work, a classical work on blob tracking based on background subtraction is W4 [4]. In this system the output of the change detector is thresholded and a connected component analysis is carried out to identify moving regions (blobs). However, the interaction between tracking and change detection is limited, tracking is not formalized in the context of RBE, CD depends on hard thresholds, no probabilistic reasoning is carried out to derive a new measure from the CD output or to update the object position. [5] and [6] are examples of blob trackers based on change detection where the RBE framework is used in the form of the Kalman filter. Yet, the use of this powerful framework is impoverished by the absence of a truly probabilistic treatment of the CD output. In practice, covariance matrices defining measurement and process uncertainties are constant, and the filter evolves toward its steady-state regardless of the quality of the measures obtained from change detection. [7] is one of the most famous attempt to integrate RBE in the form of a particle filter with a statistical treatment of background (and foreground) models. The main limitations are the use of a calibrated camera with reference to the ground plane and the use of a foreground model learned off-line. While the former can be reasonable,the use of foreground models is always troublesome in practice, given the high intra-class variability of human appearances. Moreover, no cognitive feedback is provided from the Particle Filter to influence the change detection.

2   Models and Assumptions

Recursive Bayesian Estimation [1] allows for hidden state estimation from noisy measures in discrete-time systems. From a statistical point of view, the problem of estimating the state translates into the problem of estimating a degree of belief in its possible values, *i.e.* its PDF, given all the available information, *i.e.* the initial state and all the measurements up to a given moment. The solution is seeked recursively: given the PDF of the state at time $k-1$ conditioned on all previous measurements, $p(\mathbf{x}_{k-1}|\mathbf{z}_{1:k-1})$, and the availability of a new measurement, $\mathbf{z}_k$, a new estimate for the PDF at time k is computed.

We assume a rectangular model for the tracked object, as done in many proposals such as *i.e.* [8]. Hence, the state of the RBE tracker, $\mathbf{x}_k$, comprises at least four variables

$$\mathbf{x}_k = \left\{ i_k^b, j_k^b, w_k, h_k, \ldots \right\} \tag{1}$$

where (i_k^b, j_k^b) are the coordinates of the barycenter of the rectangle and w_k and h_k its dimensions. These variables define the position and size at frame k of the tracked object. Of course, the state internally used by the tracker can beneficially include other cinematic variables (velocity,acceleration,...). Yet, change detection can only provide a measure and benefit from a prior of the position and size of the object. Hence, other variables are not used in the reminder of the paper, though they can be used internally by the RBE filter, and are used in our implementation (Sec. 5).

In Bayesian change detection each pixel of the image is modeled as a categorical Bernoulli-distributed random variable, c_{ij}, with the two possible realizations $c_{ij} = \mathcal{C}$ and $c_{ij} = \mathcal{U}$ indicating the event of pixel (i, j) being changed or unchanged, respectively.

In the following we refer to the matrix $\mathbf{c} = [c_{ij}]$ of all these random variables as the *change mask* and to the matrix $\mathbf{p} = [p(c_{ij} = \mathcal{C})]$ of probabilities defining the Bernoulli distribution of these variables as *change map*. The change mask and the change map assume values, respectively, in the $(w \times h)$-dimensional spaces $\Theta = \{\mathcal{C}, \mathcal{U}\}^{w \times h}$ and $\Omega = [0, 1]^{w \times h}$, with w and h denoting image width and height, respectively. The output of a

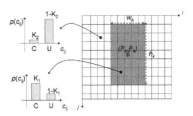

Fig. 1. Model for the change map given a bounding box

Bayesian change detector is the posterior change map given the current frame f_k and background model b_k, *i.e.* the value of the Bernoulli distribution parameter for every pixel in the image given the frame and the background:

$$p(c_{ij} = \mathcal{C}|f_k, b_k) = \frac{p(f_k, b_k|c_{ij} = \mathcal{C})p(c_{ij} = \mathcal{C})}{p(f_k, b_k)} \qquad (2)$$

Clearly, either a non-informative prior is used, such as a uniform prior, or this information has to flow in from an external module. We assume that the categorical random variables c_{ij} comprising the posterior change mask are independent, *i.e.* they are conditionally independent given f_k, b_k.

All the information that can flow from the RBE filter to the BCD and vice versa is in principle represented in every frame by the joint probability density function $p(\mathbf{x}_k, \mathbf{c})$ of the state vector and the change mask. Both information flows can be formalized and realized as its marginalization:

$$p(c_{ij}) = \int_{\mathbb{R}^4} \sum_{\mathbf{c}^{ij} \in \Theta^{ij}} p(\mathbf{x}_k, c_{ij}, \mathbf{c}^{ij}) \, d\mathbf{x}_k \quad (3) \qquad p(\mathbf{x}_k) = \sum_{\mathbf{c} \in \Theta} p(\mathbf{x}_k, \mathbf{c}) \quad (4)$$

where $\mathbf{c}^{ij}$ denotes the change mask without the (i, j)-*th* element, taking values inside the space $\Theta^{ij} = \{\mathcal{C}, \mathcal{U}\}^{w \times h - 1}$. The PDF computed with (3) defines an informative prior for the BCD algorithm, and the estimation of the state obtained with (4) can then be used as the PDF of a new measure by the RBE tracker, *i.e.* as $p(\mathbf{z}_k|\mathbf{x}_k)$. We detail in Sec. 3 and Sec. 4 the solutions for (3) and (4).

As we shall see in next sections, to use the above equations we need a statistical model that links the two random vectors $\mathbf{x}_k$ and $\mathbf{c}$. In agreement with our rectangular model of the tracked object, as shown in Fig. 1 we assume

$$p(c_{ij} = \mathcal{C}|\mathbf{x}_k) = \begin{cases} K_1 & \text{if } (i, j) \in R(\mathbf{x}_k) \\ K_2 & \text{otherwise} \end{cases} \qquad (5)$$

where $R(\mathbf{x}_k)$ is the rectangular region delimited by the bounding box defined by the state $\mathbf{x}_k$ and $0 \leq K_2 \leq K_1 \leq 1$ are two constant parameters specifying the probability that a pixel is changed inside and outside the bounding box, respectively. Moreover, we assume that the random variables c_{ij} are conditionally independent given a bounding box, *i.e.*

$$p(\mathbf{c}|\mathbf{x}_k) = \prod_{ij} p(c_{ij}|\mathbf{x}_k) \tag{6}$$

3   Cognitive Feedback

Given the assumptions of the previous section, we can obtain an exact solution for (3), *i.e.* , given the PDF of the state vector $p(\mathbf{x}_k)$, we can compute a prior $p(c_{ij})$ for each pixel of the frame that can then be beneficially used by the BCD algorithm. Starting from (3), we can rewrite it as

$$p(c_{ij}) = \int_{\mathbb{R}^4} \sum_{\mathbf{c}^{ij} \in \Theta^{ij}} p(\mathbf{x}_k, c_{ij}, \mathbf{c}^{ij}) \, d\mathbf{x}_k = \int_{\mathbb{R}^4} p(\mathbf{x}_k, c_{ij}) \, d\mathbf{x}_k = \int_{\mathbb{R}^4} p(c_{ij}|\mathbf{x}_k)p(\mathbf{x}_k) \, d\mathbf{x}_k \tag{7}$$

In the final marginalization we can recognize our model of the change map given a bounding box defined in (5) and the PDF of the state. Therefore, this equation provides a way to let the current estimation of the state computed by the RBE module influence the prior for the BCD algorithm, thereby realizing the Cognitive Feedback. In particular, as discussed above, we will use the prediction computed for the current frame using the motion model, *i.e.* $p(\mathbf{x}_k|\mathbf{z}_{1:k-1})$.

To solve (7) we have to span the space $\mathbb{R}^4$ of all possible bounding boxes $\mathbf{x}_k$. We partition $\mathbb{R}^4$ into the two complementary sub-spaces B_{ij} and $\bar{B}_{ij} = \mathbb{R}^4 \setminus B_{ij}$ of bounding boxes that contain or not the considered pixel (i, j), respectively. Given the assumed model (5), we obtain

$$p(c_{ij} = \mathcal{C}) = \int_{\mathbb{R}^4} p(c_{ij}|\mathbf{x}_k)p(\mathbf{x}_k) \, d\mathbf{x}_k = K_1 \int_{B_{ij}} p(\mathbf{x}_k) \, d\mathbf{x}_k + K_2 \int_{\bar{B}_{ij}} p(\mathbf{x}_k) \, d\mathbf{x}_k$$

$$= K_2 + (K_1 - K_2) \int_{B_{ij}} p(\mathbf{x}_k) \, d\mathbf{x}_k \ . \tag{8}$$

Since, obviously, $I_{ij} = \int_{B_{ij}} p(\mathbf{x}_k) \, d\mathbf{x}_k$ varies in $[0, 1]$, it follows that $p(c_{ij} = \mathcal{C})$ varies in $[K_2, K_1]$: if no bounding box with non-zero probability contains the pixel, we expect a probability that the pixel is changed equal to K_2, if all the bounding boxes contain the pixel the probability is K_1, it is a weighted average otherwise.

By defining new variables i_L, j_T, i_R, j_B to represent the current bounding box, more suitable for the next computations, as

$$\mathbf{A} = \begin{bmatrix} 1 & -\frac{1}{2} \\ 1 & \frac{1}{2} \end{bmatrix}, \quad \begin{bmatrix} i_L \\ i_R \end{bmatrix} = \mathbf{A} \begin{bmatrix} w_k \\ i_k^b \end{bmatrix}, \quad \begin{bmatrix} j_T \\ j_B \end{bmatrix} = \mathbf{A} \begin{bmatrix} h_k \\ j_k^b \end{bmatrix} \tag{9}$$

and assuming the newly defined random variables to be independent, the integral of the previous equation becomes

$$I_{ij} = \iiiint_{\substack{i_L \leq i \leq i_R \\ j_T \leq j \leq j_B}\}B_{ij}} p(i_L)p(i_R)p(j_T)p(j_B)\, di_L di_R dj_T dj_B$$

$$= \int_{-\infty}^{i} p(i_L)di_L \int_{i}^{+\infty} p(i_R)di_R \int_{-\infty}^{j} p(j_T)dj_T \int_{j}^{+\infty} p(j_B)dj_B$$

$$= F_{i_L}(i)\left(1 - F_{i_R}(i)\right) F_{j_T}(j)\left(1 - F_{j_B}(j)\right) \qquad (10)$$

where F_x stands for the CDF of the random variable x.

This reasoning holds for any distribution $p(\mathbf{x}_k)$ we might have on the state vector. If, for instance, we use a particle filter as RBE tracker, we can compute an approximation of the CDF from the approximation of the PDF provided by the weighted particles, after having propagated them according to the motion model and having marginalized them accordingly. In the case of the Kalman Filter all the PDFs are Gaussians, hence we can define all the factors of the product in (10) in terms of the standard Gaussian CDF, $\Phi(\cdot)$

$$I_{ij} = \Phi\left(\frac{i - \mu_L}{\sigma_L}\right)\Phi\left(\frac{\mu_R - i}{\sigma_R}\right)\Phi\left(\frac{j - \mu_T}{\sigma_T}\right)\Phi\left(\frac{\mu_B - j}{\sigma_B}\right) \qquad (11)$$

where μ_x and σ_x stand for the mean and the standard deviation of the random variable x. The factors of the product in (11) can be computed efficiently with only 4 searches in a pre-computed Look-Up Table of the standard $\Phi(\cdot)$ values.

4 Reasoning Probabilistically on Change Maps

Given the change map $\mathbf{p} = [p(c_{ij} = \mathcal{C})]$ obtained by the BCD algorithm, we aim at computing the probability density function $p(\mathbf{x}_k)$ of the current state of the RBE filter, to use it as the observation likelihood $p(\mathbf{z}_k|\mathbf{x}_k)$. To this purpose, from the marginalization in (4) we obtain:

$$p(\mathbf{x}_k) = \sum_{\mathbf{c} \in \Theta} p(\mathbf{x}_k, \mathbf{c}) = \sum_{\mathbf{c} \in \Theta} p(\mathbf{x}_k|\mathbf{c})p(\mathbf{c}) = \sum_{\mathbf{c} \in \Theta} p(\mathbf{x}_k|\mathbf{c}) \prod_{ij} p(c_{ij}) \qquad (12)$$

where the last equality follows from the assumption of independence among the categorical random variables c_{ij} comprising the posterior change map computed by BCD. To use (12), we need an expression for the conditional probability $p(\mathbf{x}_k|\mathbf{c})$ of the state given a change mask, based on the assumed model (5), (6) for the conditional probability $p(\mathbf{c}|\mathbf{x}_k)$ of the change mask given a state. Informally speaking, we need to find the inverse of the model (5), (6). By Bayes rule, eq. (6) and independence of the variables c_{ij}:

$$p(\mathbf{x}_k|\mathbf{c}) = p^*(\mathbf{x}_k)\frac{p(\mathbf{c}|\mathbf{x}_k)}{p^*(\mathbf{c})} = p^*(\mathbf{x}_k)\prod_{i,j}\frac{p(c_{ij}|\mathbf{x}_k)}{p^*(c_{ij})} \; . \qquad (13)$$

It is worth pointing out that we have used the notation $p^*(\mathbf{x}_k)$ and $p^*(c_{ij})$ in (13) since here these probabilities must be interpreted differently than in (12): in (12) $p(\mathbf{x}_k)$ and $p(c_{ij})$ represent, respectively, the measurement and the change map of the current frame, whilst in (13) both must be interpreted as priors that form part of our model for $p(\mathbf{x}_k|\mathbf{c})$, which is independent of the current frame. Furthermore, using as prior on the state $p^*(\mathbf{x}_k)$ the prediction of the RBE filter, as done in the Cognitive Feedback section, would have created a strong coupling between the output of the sensor and the previous state of the filter, that does not fit the RBE framework, where measures depend only on the current state, and could easily lead the loop to diverge. Hence, we assume a uniform non-informative prior $p^*(\mathbf{x}_k) = \frac{1}{\alpha}$ for the state. Instead, the analysis conducted for the Cognitive Feedback is useful to expand each $p^*(c_{ij})$ in (13). Since we are assuming a uniform prior on an infinite domain for the state variables, $i.e.$ a symmetric PDF with respect to $x = 0$, it turns out that its CDF is constant and equals to $\frac{1}{2}$:

$$CDF(x) = \frac{1}{\alpha}x + \frac{1}{2} \xrightarrow{\alpha \to +\infty} \frac{1}{2} \qquad (14)$$

Hence, every $p^*(c_{ij})$ in (13) can be expressed using (8) and (10) as:

$$p^*(c_{ij} = \mathcal{C}) = K_2 + (K_1 - K_2)\left(\frac{1}{2}\right)^4 = K_C. \qquad (15)$$

By plugging (13) in (12) and defining $K_U = p^*(c_{ij} = \mathcal{U}) = 1 - K_C$:

$$\alpha p(\mathbf{x}_k) = \prod_{i,j} \left(\frac{p(\mathcal{C}|\mathbf{x}_k)p(\mathcal{C})}{K_C} + \frac{p(\mathcal{U}|\mathbf{x}_k)p(\mathcal{U})}{K_U} \right) \qquad (16)$$

where, for simplicity of notation, we use $\mathcal{C}$ and $\mathcal{U}$ for $c_{ij} = \mathcal{C}$ and $c_{ij} = \mathcal{U}$, respectively. Since we know that $p(\mathcal{U}) = 1 - p(\mathcal{C})$ and $p(\mathcal{U}|\mathbf{x}_k) = 1 - p(\mathcal{C}|\mathbf{x}_k)$, we obtain:

$$\frac{p(\mathbf{x}_k)}{\beta} = \prod_{i,j} \left(p(\mathcal{C})\big(p(\mathcal{C}|\mathbf{x}_k) - K_C\big) + K_C\big(1 - p(\mathcal{C}|\mathbf{x}_k)\big) \right) \qquad (17)$$

with $\beta = 1/\alpha(K_C(1 - K_C))^{w \times h}$. By substituting the model (5) for $p(\mathcal{C}|\mathbf{x}_k)$ and taking the logarithm of both sides to limit round-off errors, after some manipulations we get:

$$\gamma + \ln p(\mathbf{x}_k) = h(\mathbf{x}_k, \mathbf{p}) = \sum_{(i,j) \in R(\mathbf{x}_k)} \ln \frac{p(\mathcal{C})K_3 + K_4}{p(\mathcal{C})K_5 + K_6} \qquad (18)$$

where $\gamma = -\ln\beta - \sum \ln\big(p(\mathcal{C})K_5 + K_6\big)$ and $h(\cdot)$ is a computable function of the state vector value $\mathbf{x}_k$ for which we want to calculate the probability density, of the change map $\mathbf{p}$ provided by the BCD algorithm, and of the constants

$$K_3 = K_1 - K_C \quad K_4 = K_C(1 - K_1) \quad K_5 = K_2 - K_C \quad K_6 = K_C(1 - K_2) \qquad (19)$$

Hence, by letting $\mathbf{x}_k$ vary over the space of all possible bounding boxes, (18) allows us to compute, up to the additive constant γ, a non-parametric estimation

$h(\cdot)$ of the log-PDF of the current state vector of the RBE tracker. This holds independently of the PDF of the state.

In the case of the Kalman Filter, the PDF of the state vector (i^b, j^b, w, h) is Gaussian. In such a case, the variables (i_L, j_T, i_R, j_B) are a linear combination of Gaussian Random Variables. Moreover, we are assuming that variables (i_L, j_T, i_R, j_B) are independent. Therefore, the variables (i_L, j_T, i_R, j_B) are jointly Gaussian and the mean $\boldsymbol{\mu}$ and the covariance matrix $\boldsymbol{\Sigma}$ of the state variables are fully defined by the four means μ_L, μ_R, μ_T, μ_B and the four variances σ_L^2, σ_R^2, σ_T^2, σ_B^2 of (i_L, j_T, i_R, j_B). To estimate these eight parameters, let us substitute the expression of the Gaussian PDF for $p(\mathbf{x}_k)$ in the left-hand side of (18), thus obtaining:

$$\delta - \ln(\sigma_L\sigma_R\sigma_T\sigma_B) - \frac{(i_L-\mu_L)^2}{2\sigma_L^2} - \frac{(i_R-\mu_R)^2}{2\sigma_R^2} - \frac{(j_T-\mu_T)^2}{2\sigma_T^2} - \frac{(j_B-\mu_B)^2}{2\sigma_B^2} = h(\mathbf{x}_k, \mathbf{p})$$

(20)

where $\delta = \gamma - 2\ln(2\pi)$. The eight parameters of the PDF and the additive constant δ might be estimated by imposing (20) for a number $N > 9$ of different bounding boxes and then solving numerically the obtained over-determined system of N non-linear equations in 9 unknowns. To avoid such a challenging problem, we propose an approximate procedure. First of all, an estimate $\widehat{\boldsymbol{\mu}}$ of the mean of the state vector $\boldsymbol{\mu} = (\mu_L, \mu_R, \mu_T, \mu_B)$ can be obtained by observing that, due to increasing monotonicity of logarithm, the mode of the computed log-PDF coincides with the mode of the PDF, and that, due to the Gaussianity assumption, the mode of the PDF coincides with its mean. Hence, we obtain an estimate $\widehat{\boldsymbol{\mu}}$ of $\boldsymbol{\mu}$ by searching for the bounding box maximizing $h(\cdot)$. Then, we impose that (20) is satisfied at the estimated mean point $\widehat{\boldsymbol{\mu}}$ and that all the variances are equal, i.e. $\sigma_L^2 = \sigma_R^2 = \sigma_T^2 = \sigma_B^2 = \sigma^2$, thus obtaining a functional relationship between the two remaining parameters δ and σ^2:

$$\delta = 2\ln\sigma^2 + h(\widehat{\boldsymbol{\mu}}, \mathbf{p})$$

(21)

By substituting in (20) the above expression for δ and the estimated $\widehat{\boldsymbol{\mu}}$ for $\boldsymbol{\mu}$, we can compute an estimate $\widehat{\sigma}^2(\mathbf{x})$ of the variance σ^2 by imposing (20) for whatever bounding box $\mathbf{x} \neq \widehat{\boldsymbol{\mu}}$. In particular, we obtain:

$$\widehat{\sigma}^2(\mathbf{x}) = \frac{1}{2} \frac{\|\widehat{\boldsymbol{\mu}} - \mathbf{x}\|_2^2}{h(\widehat{\boldsymbol{\mu}}, \mathbf{p}) - h(\mathbf{x}, \mathbf{p})}$$

(22)

To achieve a more robust estimate, we average $\widehat{\sigma}^2(\mathbf{x})$ over a neighborhood of the estimated mean bounding box $\widehat{\boldsymbol{\mu}}$. Finally, to obtain the means and covariance of the measurements for the Kalman Filter, we exploit the property of linear combinations of Gaussian variables:

$$\boldsymbol{\mu} = \begin{bmatrix} \mathbf{A}^{-1} & \mathbf{0} \\ \mathbf{0} & \mathbf{A}^{-1} \end{bmatrix} \widehat{\boldsymbol{\mu}} \ , \ \boldsymbol{\Sigma} = \widehat{\sigma}^2 \begin{bmatrix} \mathbf{A}^{-1} & \mathbf{0} \\ \mathbf{0} & \mathbf{A}^{-1} \end{bmatrix} \begin{bmatrix} \mathbf{A}^{-1} & \mathbf{0} \\ \mathbf{0} & \mathbf{A}^{-1} \end{bmatrix}^T$$

(23)

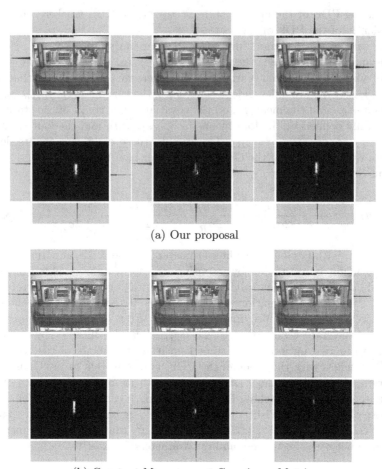

(a) Our proposal

(b) Constant Measurement Covariance Matrix

Fig. 2. The top row shows the frames and the next row the change maps. Along the sides of every picture we plot the marginal Gaussian probabilities of the four state variables $[i_L, i_R, j_T, j_B]$. Around the frames we report (in blue) the marginals of the Kalman prediction and around the change maps (in red) the marginals of the observation likelihood (Sec. 4). The means of the PDFs are drawn on the change maps.

5   Experimental Results

We have tested the proposed Bayesian loop on publicly available datasets with ground truth data: some videos from the CAVIAR[1] and ISSIA Soccer datasets [9]. We have used a Kalman Filter with constant velocity motion model as RBE tracker and the algorithm in [10] as BCD. The detection to initialize the tracker was done manually from the ground truth.

---

[1] http://homepages.inf.ed.ac.uk/rbf/CAVIAR/

To illustrate the benefits of the probabilistic analysis of the change map, we discuss some frames of a difficult part of a CAVIAR sequence. In this video, the tracked subject wanders in and out of the shop, passing in front of a pillar similar in color to his clothes. Fig. 2(a) shows 3 frames of the sequence, respectively before, during and after the camouflage. During the camouflage the background subtraction correctly computes high probabilities that a pixel is changed only for the pixels lying outside the pillar. Our rectangular model cannot fit such output and selects only the portion of the person on the left of the column as the mean of the current PDF (red bounding box). A sensor based on change detection will always likely fail to handle camouflage. Yet, the RBE tracker is conceived to work with a noisy sensor, provided that it is possible to evaluate the uncertainty of its output. Thanks to the procedure of Sec. 4, our method exploits this trait of the framework: as can be seen in the middle pictures, the uncertainty of the measure during the camouflage increases and gets similar to the Kalman prediction uncertainty, for the rectangular model leaves out portions of the change map with high probabilities of being foreground. With this configuration, the correction step of the filter decreases the contribution to the final state estimation of the shrunk measure coming from BCD with respect to the predicted state, thus correctly tracking the target. Had a constant uncertainty model been used (see Fig. 2(b)), by the time of the camouflage the filter would have reached the steady state and would follow the measures with a constant amount of confidence in them. This leads to an incorrect prior for the change detection on the next frame and thus to divergence, as shown in the third frame of Fig. 2(b) where the system cannot recover from wrong measurements.

To demonstrate its capabilities and robustness, the complete system has been used to track people wondering in a shopping mall using three sequences from the CAVIAR dataset and soccer players during a match in the sixth sequence of the ISSIA dataset. Tracking results on these video are shown in the supplementary material . Our system does not require to set a threshold to classify the output of the change detection, only the model for

Table 1. Performance scores. $^{(*)}$ indicates loss of target.

Seq.	Full Loop	Partial Loop	Kalm+MS
CAV1	0.553	0.298	0.208$^{(*)}$
CAV2	0.474	0.382	0.010$^{(*)}$
CAV3	0.500	0.055$^{(*)}$	0.012$^{(*)}$
LGK	0.457	0.011$^{(*)}$	0.581
LPE	0.474	0.012$^{(*)}$	0.492

$p(c_{ij} = C|\mathbf{x}_k)$ must be set. We used $K_1 = 0.5$, allowing for unchanged pixels into our bounding box (approximation of the rectangular model) and $K_2 = 0.1$ to allow for a small amount of errors of the BCD out of the bounding box. To quantitatively evaluate the performance we use the mean ratio over a sequence between the intersection and the union of the ground truth bounding box with the estimated bounding box.

As for the CAVIAR dataset, the main difficulties are changes in appearance of the target due to lightening changes inside and outside the shop, shadows, camouflage, small size of the target and, for sequence 2, dramatic changes in target size onto the image plane (he walks inside the shop until barely disappears). Despite all these nuisances our system successfully tracks all the targets.

The ISSIA Soccer dataset is less challenging as far as color, lightening and size variations are concerned, and the players cast practically no shadow. Yet, it provides longer sequences and more dynamic targets. We used our system to track the goalkeeper and a player: the goalkeeper allows to test our system on a sequence 2500 frames long; the player shows rapid motion changes and unpredictable poses (he even falls to the ground kicking the ball). Our tracker was able to successfully track both targets throughout the whole sequence. Quantitative evaluation is reported in Table 1. To highlight the importance of the Bayesian loop, we have performed the same experiments without considering the full PDF estimated during the change map analysis, but just the mean and a constant covariance matrix (i.e. the same settings as in Fig. 2(b)): results achieved by our proposal are consistently better throughout all the sequences. We also compare our performance against a Mean Shift tracker used in conjunction with a Kalman Filter [8]. The CAVIAR sequences are too difficult for a tracker based on color histograms, because of the reasons discussed above: the tracker looses the target in all the sequences. On the ISSIA sequences, instead, it obtains slightly better performances than our proposal. We impute this to the use of gray levels in our tracker: for example, yellow parts of the tracked players get really similar to the green background. We are developing a color BCD to solve the problem.

6   Conclusions

A principled framework to model the interaction between Bayesian change detection and tracking have been presented. By modeling the interaction as marginalization of the joint probability of the tracker state and the change mask, it is possible to obtain analytical expressions for the PDFs of the tracker observation likelihood and the change detector prior. Benefits of such interaction have been discussed with experiments on publicly available datasets.

References

1. Maskell, S., Gordon, N.: A tutorial on particle filters for on-line nonlinear/non-gaussian bayesian tracking. IEEE Trans. on Signal Processing 50, 174–188 (2001)
2. Lanza, A., Di Stefano, L., Soffritti, L.: Bayesian order-consistency testing with class priors derivation for robust change detection. In: AVSS, pp. 460–465 (2009)
3. Thomas, A., Ferrari, V., Leibe, B., Tuytelaars, T., Gool, L.V.: Depth-from-recognition: Inferring metadata by cognitive feedback. In: 3DRR (2007)
4. Haritaoglu, I., Harwood, D., Davis, L.S.: W4: Real-time surveillance of people and their activities. IEEE T-PAMI 22, 809–830 (2000)
5. Stauffer, C., Grimson, W.: Adaptive background mixture models for real-time tracking. In: Proc. of the CVPR 1999, pp. 246–252 (1999)
6. Harville, M., Li, D.: Fast, integrated person tracking and activity recognition with plan-view templates from a single stereo camera. In: CVPR, vol. 2, pp. 398–405 (2004)
7. Isard, M., MacCormick, J.: Bramble: A bayesian multiple-blob tracker. In: Proc. of the ICCV 2001 (2001)

8. Comaniciu, D., Ramesh, V., Meer, P.: Kernel-based object tracking. PAMI 25, 564–575 (2003)
9. D'Orazio, T., Leo, M., Mosca, N., Spagnolo, P., Mazzeo, P.: A semi-automatic system for ground truth generation of soccer video sequences. In: AVSS (2009)
10. Lanza, A., Di Stefano, L.: On-line learning of background appearance changes for robust background subtraction. In: IET Int. Conf. on Imaging for Crime Detection and Prevention (2009)

Real Time Motion Changes for New Event Detection and Recognition

Konstantinos Avgerinakis, Alexia Briassouli, and Ioannis Kompatsiaris

Informatics and Telematics Institute
Centre for Research and Technology, Hellas
6th km Charilaou-Thermis
Thermi, 57001, Thessaloniki, Greece

Abstract. An original approach for real time detection of changes in motion is presented, for detecting and recognizing events. Current video change detection focuses on shot changes, based on appearance, not motion. Changes in motion are detected in pixels that are found to be active, and this motion is input to sequential change detection, which detects changes in real time. Statistical modeling of the motion data shows that the Laplace provides the most accurate fit. This leads to reliable detection of changes in motion for videos where shot change detection is shown to fail. Once a change is detected, the event is recognized based on motion statistics, size, density of active pixels. Experiments show that the proposed method finds meaningful changes, and reliable recognition.

1 Introduction

Event and activity recognition have become particularly important in the recent years, as they provide valuable information for surveillance, traffic monitoring etc. The video segments processed are usually extracted by shot change detection, or have been segmented before the processing, possibly manually. Shot detection separates the video into subsequences filmed from the same camera/viewpoint and can achieve very high accuracy, but is based on appearance. Activity recognition takes place over video segments that are found by shot change detection in [1]. In practice, this may not always work, as shot detection is based on appearance, although different activities may take place in subsequences with the same appearance. This motivates us to propose a method for separating a video sequence based on motion, which would provide a more meaningful segmentation. Motion has been used for this in [2], where frames with low activity are separated from the others using MPEG-7 motion descriptors, but this is not generally applicable to the case of videos with different activities that need to be separated from each other.

In this work, binary masks of active pixels (Activity Areas) are initially extracted using a kurtosis-based method. The illumination variations over active pixels are processed in the sequel in order to detect changes in them. Statistical modeling of the data shows that the best probability distribution for the sequential likelihood testing is the Laplace. Sequential change detection is then

R. Koch et al. (Eds.): ACCV 2010 Workshops, Part I, LNCS 6468, pp. 54–63, 2011.

applied to the data to detect changes in it. Since only the currently available video frames are used, the change detection takes place in real time. Once the sequence is divided into subsequences containing different motion, recognition of the action can take place.

This paper is organized as follows. In Sec. 2, the method for extracting the Activity Areas is presented and the CUSUM change detection algorithm is presented in Sec. 3. The statistical modeling required for the CUSUM is included in 3.1. The methods employed for activity/event classification are described in Sec. 4. Experiments with a wide range of indoors and outdoors videos are analyzed in Sec. 5. Finally, conclusions and future work are discussed in Sec. 6.

2 Activity Area

A binary mask of the active pixels in the video, the Activity Area, is helpful in reducing the computational cost of the method and also reducing the possibility of having false alarms, by limiting the data to the truly active pixels. The Activity Area can be extracted at each frame by processing the data available until that moment, i.e. the inter-frame illumination variations until frame k, thus retaining the real time nature of the system. The data at frame k and pixel $\bar{r}$ is a $1 \times k$ vector that can be written as $\mathbf{v}_k(\bar{r}) = [v_1(\bar{r}), ... v_k(\bar{r})]$, where $v_n(\bar{r})$ is the illumination variation at frame $n, 1 \leq n \leq k$, caused either by actual motion or by measurement noise. Each pixel's illumination variation at frame n can be modeled by the following hypotheses:

$$H_0 : v_n(\bar{r}) = z_n(\bar{r})$$
$$H_1 : v_n(\bar{r}) = u_n(\bar{r}) + z_n(\bar{r}), \tag{1}$$

where $z_n(\bar{r})$ originates from measurement noise and $u_n(\bar{r})$ from actual motion. Additive measurement noise is often modeled as a Gaussian random variable [3], [4], so the active pixels can be discerned from the static ones as they are non-Gaussian. A classical non-Guassianity measure is the kurtosis, which can be employed to separate the active from static pixels, as its value is equal to zero for Gaussian data. For a random variable y, the kurtosis is given by $kurtosis[y] = E[y^4] - 3(E[y^2])^2$. The kurtosis of $\mathbf{v}_k(\bar{r})$ is estimated, to form a "kurtosis mask", which obtains high values at active pixels, and low values at the static ones. The kurtosis has been found to be very sensitive to outliers, and can detect them reliably even for non-Gaussian data [5], [6]. Thus, if the measurement noise deviates from the Gaussian model, the kurtosis will still lead to an accurate estimate of the active pixels. The robustness of the kurtosis for extracting Activity Areas has been analyzed in [7] as well, where it is shown to provide accurate activity areas even for videos with slightly varying backgrounds (e.g. backgrounds with moving trees). The activity areas for some videos used in the experiments in this work are shown in Fig. 1, where it can be seen that the regions of motion are accurately localized. Other foreground extraction methods could also be employed to extract activity areas from the video, such as the

Fig. 1. Activity Areas superposed on frames of videos examined

Gaussian Mixture models of [8], [9]. The method used should be computationally efficient, like the one proposed here, in order to allow operation in real time.

In practice, there may be errors in an activity area, e.g. a sudden illumination change may cause the entire video frame to be "active". This does not negatively affect the results, since in that case static pixels will also be included in the test, whose flow estimates do not significantly affect the change detection performance. It is also possible that there may be a local occlusion over a few frames that introduces errors in the flow estimates. In most cases, the errors introduced by the occlusion can be overcome because data is collected over a window of frames in which correct (unoccluded) flow values will also be included (see Sec. 3). If, nonetheless, a false alarm is caused by this occlusion, it can be eliminated at a post-processing stage that examines the motion data before and after each change: in the case of false alarms, the motion before and after the false alarm remains the same, so that detected change is ignored.

3   Change Detection

Sequential change detection methods are perfectly suited for designing a real time system for detecting changes, as they are specifically designed for this purpose. Additionally, methods like the CUSUM have been shown to provide the quickest detection of changes in the distribution of a data stream [10], [11]. The data used in this context are the illumination variations of the active pixels in each video frame, which have been extracted using only the currently available video frames. The method used here is the CUSUM (Cumulative Sum) approach developed by Page [12], based on the log-likelihood ratio test statistic at each frame k:

$$T_k = \ln \frac{f_1(\mathbf{V}_k)}{f_0(\mathbf{V}_k)}. \tag{2}$$

Here, $\mathbf{V}_k = [v_1(\bar{r}_1), ..., v_1(\bar{r}_{N_1}), ..., v_k(\bar{r}_1), ..., v_k(\bar{r}_{N_k})]$ represents the illumination of all active pixels over frames 1 to k, assuming that the activity area of each frame n contains N_n pixels. The data distribution before a change is given by $f_0(\mathbf{V}_k)$ and after a change it is $f_1(\mathbf{V}_k)$, so the test statistic of Eq. (2) becomes:

$$T_k = \sum_{i=1}^{k} \sum_{j=1}^{N_i} \ln \frac{f_1(v_i(\bar{r}_j))}{f_0(v_i(\bar{r}_j))}. \tag{3}$$

The log-likelihood ratio uses $\sum_{i=1}^{k} \times \sum_{j=1}^{N_i}$ samples. This is a large number of samples, which provides a good approximation of the data distributions and is expected to lead to reliable detection performance.

In this problem, neither the data distributions before and after a change, nor the time of change are known. In order to find the moment of change using Eq. (2), the distributions f_0 and f_1 have to be approximated. The initial distribution f_0 can be approximated from the first w_0 data samples [13], under the assumption that no changes occur in the first w_0 frames. This is a realistic assumption and does not significantly affect the real time nature of the approach, as errors of 10 frames around a change are almost always difficult to discern visually. The distribution f_1 is approximated at each time instant k using the most recent data available, namely the w_1 most recent frames, in order to avoid a bias towards the baseline pdf f_0. The size of the windows w_0, w_1 is determined by using training data, and it is found that $w_0 = 10$, $w_1 = 1$ led to good distribution approximations and accurate change detection for most videos. These windows are sufficient in size, because they contain all the pixels inside the activity area, which lead to a sufficiently large sample size.

The data is assumed to be independent and identically distributed (i.i.d.) in Eq. 3, an assumption that is common in such problems [14], as joint data distributions can be quite cumbersome to determine in practice. The CUSUM algorithm has been shown to be asymptotically optimal even for data that is not independent [15], so deviations from the i.i.d. assumptions are not expected to introduce noticeable errors. Indeed, in the experiments changes are detected with accuracy under the i.i.d. assumption, under which the test can become computationally efficient, as Eq. 3 obtains the following recursive form:

$$T_k = \max\left(0, T_{k-1} + \ln \frac{f_1(\mathbf{V}_k)}{f_0(\mathbf{V}_k)}\right) = max\left(0, T_{k-1} + \sum_{i=1}^{k} \sum_{j=1}^{N_i} \ln \frac{f_1(v_i(\bar{r}_j))}{f_0(v_i(\bar{r}_j))}\right). \quad (4)$$

The test statistic T_k is compared at each frame with a threshold to find if a change has occurred at that frame. The related literature recommends using training data to find a reliable threshold for good detection performance [11]. We have found that at each time instant k, the threshold can be estimated from $\eta_k = mean([T_{k-1}] + c \times std[T_{k-1}]$, where $mean[T_{k-1}]$ is the mean of the test statistic's values until frame $k - 1$ and $std[T_{k-1}]$ is the standard deviation of those values. Very reliable detection results are found for $c = 5$ for the videos used in these experiments.

3.1 Statistical Data Distribution Modeling

The test of Eq. (3) requires knowledge of the family of data probability distributions before and after a change. In the literature, the data has been assumed to follow a Gaussian distribution [3] due to lack of knowledge about its nature. We propose finding a more accurate model for the pdf, in order to achieve optimal detection results. The data under consideration are the illumination variations of each active pixel over time. These variations are expected to contain outliers, as a pixel is likely to be inactive over several frames, and suddenly become active. Data that contains outliers is better modeled by a heavy-tailed distribution, such as the Laplace, the generalized Gaussian or the Cauchy, rather than the Gaussian. We compare the statistical fit achieved by the Laplace and Gaussian distributions, as their parameters can be estimated quickly, without affecting the real time character of the proposed approach. The Laplace pdf is given by:

$$f(x) = \frac{1}{2b} \exp\left(-\frac{|x - \mu|}{b}\right), \tag{5}$$

where μ is the data mean and $b = \sigma/\sqrt{2}$ is its scale, for variance σ^2, which can be directly estimated from the data.

The histogram of the data (illumination variations) is estimated to approximate the empirical distribution. The data mean and variance are also estimated and used to estimate the parameters for the Gaussian and Laplace pdfs. The resulting pdfs are compared both visually and via their mean squared distance from the empirical distribution for the videos used in the experiments. As Fig. 2 shows for several videos, the empirical data distribution is best approximated by the Laplace model. This is expected, since the Gaussian pdf does not account for the heavy tails in the empirical distribution, introduced by the data outliers. The average mean squared error for the approximation of the data by Gaussian and Laplace pdfs is 0.04 for the Laplace distribution, while it is 0.09 for the Gaussian model, verifying that the Laplace is better suited for our data.

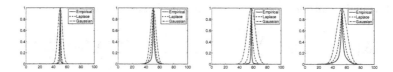

Fig. 2. Statistical modeling using Gaussian, Laplace distributions for traffic videos

The CUSUM test based on the Laplace distribution then becomes:

$$T_k = \sum_{i=1}^{k} \sum_{j=1}^{N_i} \left(\ln \frac{b_0}{b_1} - \frac{v_i(\bar{r}_j) - \mu_1}{b_1} + \frac{v_i(\bar{r}_j) - \mu_0}{b_0} \right), \tag{6}$$

so the CUSUM test now is:

$$T_k = \max\left(0, T_{k-1} + \sum_{i=1}^{k} \sum_{j=1}^{N_i} \left(\ln \frac{b_0}{b_1} - \frac{v_i(\bar{r}_j) - \mu_1}{b_1} + \frac{v_i(\bar{r}_j) - \mu_0}{b_0} \right) \right) \tag{7}$$

and can be applied to each current data sample after the estimation of the distribution parameters as described in Sec. 3.

4   Recognition

For surveillance videos in various setups, the event of interest focuses on the arrival or departure of people or other entities from the scene. When someone enters a scene, the activity area becomes larger, and when they exit, the activity area size decreases. The experiments show that this leads to correct annotation of such events in a variety of indoors and outdoors scenarios. Additional information can be extracted by examining the velocity before and after a detected change: if it decreases, the activity taking place is slowing down, and may even come to a stop if the velocity after a change becomes zero. Similarly, an increase of speed can easily be detected after a change.

For traffic videos, the events to be recognized are transitions between heavy, medium and light traffic. When there is heavy traffic, the activity area consists of many small connected components, originating from the small vehicle motions. Here, connected component refers to active pixels that are continuous in space, forming coherent groups of pixels. During light traffic, the cars move fast, so the activity areas comprise of fewer connected components. Medium traffic leads to more connected components than light traffic, but fewer than heavy traffic. Training videos of traffic are examined, and it is determined that heavy traffic occurs when there are more than 60 connected components in the activity area, there are $30 - 60$ for medium traffic, and less than 30 for light traffic (Fig. 3). This indeed leads to recognition of the varying traffic conditions, and can be achieved in real time.

(a)   (b)   (c)   (d)   (e)   (f)

Fig. 3. Light, medium and heavy traffic. The connected components of the active regions increase as the traffic gets heavier.

5   Experiments

Experiments take place with various videos to examine the accuracy of the change detection results, for surveillance and traffic applications. The method is also compared to shot change detection.

Surveillance

A variety of surveillance videos, indoors and outdoors, from banks, entrances, train stations and others, are examined for detection of changes. In all cases, the change points are detected correctly. Figs. 4(a),(b) show the frames before and after a new robber enters to rob an ATM (video duration 1 min 39 sec, at 10 fps). Figs. 4(c),(d) show a security guard before and after he jumps over a gate

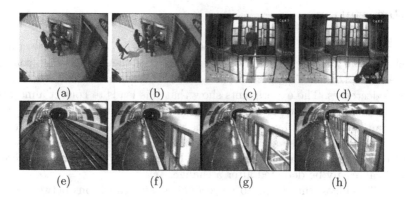

Fig. 4. Frames before/after a change. ATM robbery: new robber enters. Guard: guard enters, jumps over gate. Train station: train appears, slows down, stops.

(video duration 9 sec, at 10 fps). In Figs. 4(e), (f) a train station is shown before and after the train enters, and Figs. 4(g), (h) show the train before and after it slows down (video duration 10 sec, at 10 fps). The examined videos can be seen in the supplementary material, with the moments of change highlighted in red, showing that the changes are correctly detected.

In these videos, recognition consists of finding whether a moving entity is entering or exiting the scene, and if its speed changes. An entrance is detected when the size of the activity area increases in the frames after the detected change, while an exit occurs if the activity area decreases in size. This makes intuitive sense, since more pixels become active as someone enters a scene, and vice versa, and leads to the correct annotation of these events, as can be seen in the supplementary material. Additional information about the activity taking place can be extracted by examining the motion magnitude before and after a change. For the video with the train entering the station, after the fist change, it is found to be slowing down, and after the second change, it stops completely. This method leads to correct annotations that can be seen in the corresponding result videos in the supplementary material. For high-level annotations, additional information about the scene needs to be known, for example context information that the video is of a bank and the location of the ATM can help identify a robbery. Such information can be provided a priori, or extracted from the scene with additional visual processing. In practice, it is likely that contextual information will be available, as a system is designed for a particular application.

Traffic

Traffic videos (of duration 10 sec at 10 fps) are also examined, to detect changes between heavy, medium and light traffic. As can be seen in Fig. 5 the test statistics provide a clear indication of the moment of change. Videos of the highway traffic with these changes highlighted in red are provided in the supplementary material. The proposed method detects the changes correctly: as seen in Fig. 5, the frames before and after the change point clearly contain a different amount

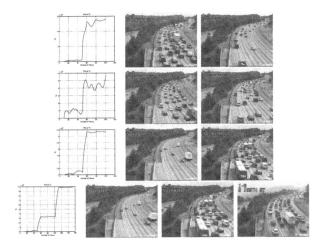

Fig. 5. First column: CUSUM test statistic. Columns 2-4: frames before/after changes.

of traffic. It finds two changes in the last video, although the last two subsequences in it both contain heavy traffic. This error is introduced because they are filmed in very different weather conditions: the second video is filmed on a rainy day, which changes the motion estimates significantly. However, the recognition stage that follows corrects this false alarm by correctly characterizing both segments as having heavy traffic. The recognition of the type of traffic in each video subsequence takes place based on the number of connected components in the corresponding activity area, as described in Sec. 4, and leads to correct results in all cases.

Comparison with shot change detection

The usefulness of the proposed approach can be better determined when comparing it to traditional shot change detection methods, such as that of [1]. Shot change detection can find changes between shots introduced by variations in appearance, rather than in motion. We apply this method to the videos on which sequential change detection is performed. The ground truth for the changes in motion is found by visually observing the videos. Table 1 shows that the shot change detection is unable to detect most changes in the motion, even when these have caused a slight change in the video appearance. For example, in traffic videos the change from light to heavy traffic may be accompanied by the appearance of more cars in the scene. Nonetheless, shot change detection cannot discern this change, whereas the proposed approach finds it. Similarly, when robbers enter or exit the scene, the proposed method finds these changes, but shot change detection does not. In the last two traffic videos, there are some significant appearance changes that coincide with the motion changes (see supplementary material). These changes are detected by the shot change detection as well, as expected.

Table 1. Comparison with shot change detection

Video	True Changes	Our method	Shot ch. det.
ATM robbery	38, 55, 100, 450, 520, 651	42, 58, 102, 458, 530, 654,	-
ATM robbery	685, 729, 790, 814, 891, 908	690, 733, 794 , 818, 896, 913	-
Police Station	20, 35, 100, 140, 167, 210	21, 37, 110, 145, 170, 216	-
Train station	8, 100, 232	10, 104, 237	-
Heavy-Light	50	51	-
Heavy-Medium	50	51	-
Light-Heavy	51	52	51
Light-Heavy-Medium	50, 100	52, 104	103

Finally, the proposed approach has a lower computational burden than traditional shot change detection. It runs completely in real time, whereas the shot change detection requires several minutes to run on the same videos, in C++. The results for both methods and the corresponding ground truth presented in Table 1 demonstrate that the sequential change detection approach correctly finds frames at which changes occur in the video. This can be used to signal alarms in a security setup, or divide the video into subsequences which can be used at a latter stage as input to an event recognition system.

6 Conclusions

In this work, a novel, real time approach for separating videos into meaningful subsequences with different events is proposed. The active regions of the video are localized using higher order statistics, and a binary mask, the activity area, is produced. Only the motion in pixels inside the activity area is processed, in order to minimize computational cost and probability of false alarms. Sequential change detection, specifically the CUSUM method, is applied to the motion vectors of the video, to detect changes in them in real time. The Laplace model is used to describe the motion vectors, as it accurately describes the outliers in them. Once the video is separated into subsequences containing different activities, recognition is applied to the subsequences to characterize the events taking place in them. The recognition uses information from the activity areas, as well as the motion taking place in them. Comparisons take place with shot change detection, where it is shown that they are unable to detect changes in motion, and therefore different events, which the proposed method can find. Additionally, shot change detection requires significant computational time, whereas the system presented here operates in full time. Experiments with surveillance and traffic videos demonstrate that it provides reliable detection of changes and recognition of the events taking place, making it a reliable tool for numerous applications. Future work includes working with more complex sequences, containing more than one activities which undergo changes.

Acknowledgements. This research has received funding from the European Community's 7th Framework Programme FP7/2007-2013 under grant agreement FP7-214306 - JUMAS.

References

1. Chavez, G.C., Cord, M., Philip-Foliguet, S., Precioso, F., de Araujo, A.: Robust scene cut detection by supervised learning. In: EUPISCO (2006)
2. Ajay, D., Radhakrishan, R., Peker, K.: Video summarization using descriptors of motion activity: a motion activity based approach to key-frame extraction from video shots. J. Electronic Imaging 10, 909–916 (2001)
3. Aach, T., Kaup, A., Mester, R.: Statistical model-based change detection in moving video. Signal Processing 31, 165–180 (1993)
4. Wren, C., Azarbayejani, A., Darrell, T., Pentland, A.: Pfinder: real-time tracking of the human body. IEEE Transactions on Pattern Analysis and Machine Intelligence 19, 780–785 (1997)
5. Hassouni, M., Cherifi, H., Aboutajdine, D.: Hos-based image sequence noise removal. IEEE Transactions on Image Processing 15, 572–581 (2006)
6. Giannakis, G., Tsatsanis, M.K.: Time-domain tests for Gaussianity and time-reversibility. IEEE Transactions on Signal Processing 42, 3460–3472 (1994)
7. Briassouli, A., Kompatsiaris, I.: Robust temporal activity templates using higher order statistics. IEEE Transactions on Image Processing (to appear)
8. Stauffer, C., Grimson, W.: Adaptive background mixture models for real-time tracking. In: Proceedings of IEEE Computer Society Conference on Computer Vision and Pattern Recognition, CVPR 1999 (1999)
9. Zivkovic, Z., van der Heijden, F.: Efficient adaptive density estimation per image pixel for the task of background subtraction. Pattern Recogn. Lett. 27, 773–780 (2006)
10. Dragalin, V.P.: Optimality of a generalized cusum procedure in quickest detection problem. In: Statistics and Control of Random Processes: Proceedings of the Steklov Institute of Mathematics, pp. 107–120. Providence, Rhode Island (1994)
11. Moustakides, G.V.: Optimal stopping times for detecting changes in distributions. Ann. Statist. 14, 1379–1387 (1986)
12. Page, E.S.: Continuous inspection scheme. Biometrika 41, 100–115 (1954)
13. Muthukrishnan, S., van den Berg, E., Wu, Y.: Sequential change detection on data streams. In: ICDM Workshop on Data Stream Mining and Management, Omaha NE (2007)
14. Lelescu, D., Schonfeld, D.: Statistical sequential analysis for real-time video scene change detection on compressed multimedia bitstream. IEEE Transactions on Image Processing 5, 106–117 (2003)
15. Bansal, R.K., Papantoni-Kazakos, P.: An algorithm for detecting a change in a stochastic process. IEEE Transactions on Information Theory 32, 227–235 (1986)

Improving Detector of Viola and Jones through SVM

Zhenchao Xu, Li Song, Jia Wang, and Yi Xu

Institute of Image Communication and Information Processing
Shanghai Jiao Tong University, Shanghai, PRC 200240
{xuzhenchao,song_li,jiawang,xuyi}@sjtu.edu.cn

Abstract. Boosted cascade proposed by Viola and Jones is applied to many object detection problems. In their cascade, the confidence value of each stage can only be used in the current stage so that interstage information is not utilized to enhance classification performance. In this paper, we present a new cascading structure added SVM stages which employ the confidence values of multiple preceding Adaboost stages as input. Specifically, a rejection hyperplane and a promotion hyperplane are learned for each added SVM stage. During detection process, negative detection windows are discarded earier by the rejection SVM hyperplane, and positive windows with high confidence value are boosted by promotion hyperplane to bypass the next stage of cascade. In order to construct the two distinct hyperplanes, different cost coefficients for training samples are chosen in SVM learning. Experiment results in UIUC data set demonstrate that the proposed method achieve high detection accuracy and better efficiency.

1 Introduction

Object detection is popular and significant issue in computer vision and pattern recognition. Examples include vehicle, face, and pedestrian detection. Many approaches have been proposed to solve detection problem in different circumstance. The majority of them use machine learning to construct a detector from a large number of training examples. Then the detector is scanned over the entire input image in order to find a pattern of intensities which is consistent with the target object. In smart video surveillance systems, object detection are usually integrated with object tracking and the methods for the two tasks can be facilitated by each other. In order to provide a real-time assistance for tracking process, a both accurate and rapid detection method is essential in integration object detector into a tracking algorithm.

A great number of algorithms have been proposed to address the problem of object detection. At the beginning, some researchers present models based on background subtraction to solve detection task, but it is difficult for them to identify a special class object from a crowd foreground. In [1], Viola and Jones describe a boosted cascade based on haar features for rapid face detection. Rotated haar-like features is introduced by Lienhart and Maydt [2] for

R. Koch et al. (Eds.): ACCV 2010 Workshops, Part I, LNCS 6468, pp. 64–73, 2011.

better detection. Viola and Jones also improve their proposal by integrating image intensity information with motion information. Wang and Jia [3] propose a cascading structure using boosted HOG features. There work focus on boosting classification performance by improving feature pool, but optimization in cascade structure is ignored. Wu and Brubaker [4] present a asymmetric learning for cascade to reduce the training time of detector. Chen [5] speed up detection process by combining cascade Adaboost with linear SVM, their framework improve the efficiency of negative detection windows but take no action for positive windows.

In this paper, we describe a new boost cascaded classifier added SVM stages which can reduce the detection time. Each efficient SVM stage is composed of a rejection hyperplane and a promotion hyperplane, which are both learned by SVM, but with different cost coefficients for positive examples and negative examples. Some negative detection windows are discarded earlier by the rejection SVM hyperplane to save the time in rejecting negatives. With the help of promotion hyperplane, those positive windows with high confidence don't enter the next stage as normal, but jump to the following one of the next stage so that they can be detected faster through bypassing some stages. The experiment results show that the proposed method can get better efficiency and achieve approximate accuracy in detection.

The paper is organized as follows. In section 2, we review the Viola and Jones's cascading structure and SVM learning. In Section 3, we introduce our improving cascading structure and discuss the training process of rejection hyperplane and promotion hyperplane. The experiment results are reported in section 4. Finally, we summarize and conclude the paper in Section 5.

2   Preliminaries

2.1   Basic Cascading Classifier

Viola and Jones's detector based on cascade structure (see Fig. 1) is extensively used in many researcher's work. The cascade consists of several Adaboost classifiers which are arranged in order of complexity. In this cascade structure, the output of previous stage classifiers is used as the input of the subsequent stages of cascade, and each successive classifier is trained only those samples which pass through the preceding stages. Detection windows are thought to be positive only when they can pass all the stages of cascade. While those windows do not contain object are rejected in the early stage of cascade.

The cascade can achieve real-time in detection, which is because that if at any point in the cascade a classifier rejects the sub-window under inspection, no further processing is performed and the search moves on to the next sub-window. The cascade therefore has the form of a degenerate decision tree. The performance of the entire cascade is closely related with each individual stage classifier, because the activation of each stage depends completely on the behavior of its predecessor. The overall detection rate D and false positive rate F for an entire cascade can be estimated as follows:

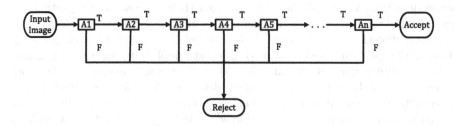

Fig. 1. Schematic depiction of basic cascading classifier, where "A" denotes the Adaboost stages. The output of previous stage classifiers is used as the input of the subsequent stages of cascade.

$$D = \prod_i^n d_i \qquad (1)$$

$$D = \prod_i^n f_i \qquad (2)$$

where n denotes the number of stages in this cascade, d_i and f_i denote the detection rate and false positive rate of the ith stage, respectively.

2.2 Adaboost Learning

Generally, the stage classifier of cascade is constructed by strong learning algorithms, which are used to select a small set of features and enhance the performance of classifier. Adaboost learning algorithm is proposed by Freund and Schapire [6] and is proved that the learning error of the strong Adaboost classifier approaches zero exponentially in the number of training rounds.

Adaboost learning is an adaptive machine learning algorithm in the sense that subsequent classifiers built are tweaked in favor of those examples misclassified by previous classifiers. A few weak classifiers are selected by Adaboost learner in a series rounds. Given example images $\{x_i, y_i\}$, i=1,...,n, where n is the number of examples, yi= -1,1 for negative and positive examples respectively. Weights of each example $w_{1,i}$ is initialized to be $\frac{1}{n}$. For week classifier j, the threshold classification function is $h_j(x_i)$, and the error is evaluated as follows:

$$e_j = \sum_i \frac{1}{2} w_i \mid h_j(x_i) - y_i \mid \qquad (3)$$

On every round, the classifier with lowest error e_k is extracted from weak classifier set. Then update and normalize the weights of each examples as follows:

$$w_{k+1,i} = w_{k,i} \gamma_k^{1-a_i} \qquad (4)$$

$$w_{k+1,i} = \frac{w_{k+1,i}}{\sum_{i=1}^n w_{k+1,i}} \qquad (5)$$

where k denotes the kth round, $a_i = 0$ if example x_i is classified correctly, $a_i = 1$ otherwise, and $\gamma_k = \frac{e_k}{1-e_k}$. According to the adaptive process of weights adjustment, the week classifier selected in the next round focuses more on incorrect samples. Hence, each round of the boosting process, which selects a new weak classifier, can be viewed as a feature selection process. Adaboost provides an effective learning algorithm and strong bounds on generalization performance [7][8].

3   Improving Cascade through SVM

In Viola and Jones's cascade structure, the classification result relies on the confidence value of the stage classifier for data. During detection process, each stage classifier compares the confidence value for a detection window with its threshold, and then decides to accept or reject the detection window. Normally, the decision of the current stage is only related with its corresponding confidence value, which can not be used by other stages. In that case, interstage information is not utilized by cascade to make decision of classification.

Actually, it is feasible to exploit both stage-wise information and cross-stage information to boost the performance of detector, which is implemented by creating some new stages for original cascade, and the input vector of each new stage is composed of the confidence values of multiple preceding Adaboost stages. In order to own ability to make a further decision to those detection windows which have passed the preceding stages, the new stage added after several Adaboost stages in original cascade will be trained to be high precision based on SVM. In the following of this paper, the new high precision stage is called "H" stage.

3.1   Improving Cascading Structure

An efficient cascade added one "H" stage after every two Adboost stages is illustrated in Fig. 2. The cascade structure is defined as "AAHAHAH...AH". In this structure, we employ the interstage cross-reference information of neighboring stages to boost the detection performance. The confidence values of the preceding two Adaboost classifiers are used as the input of the "H" classifier.

In our algorithm, "H" stage makes a decision with three choices by learning two SVM hyperplanes, which are formulated as $H_- : \mathbf{w}_- \cdot \mathbf{x} + b_- = 0$, and $H_+ : \mathbf{w}_+ \cdot \mathbf{x} + b_+ = 0$, respectively. We add a new "jump" choice besides simply accepting and rejecting windows. The decision of "H" stage is based on the SVM confidence value of "H" stage for detection window x. We will give a detailed discussion about cost coefficients selection strategy in next section.

3.2   Optimization of "H" Stage

For each stage of our cascade, we need train two different SVM hyperplanes to further reject negative samples and accelerate positives to pass cascade classifier. Unbalanced cost coefficients for positive and negative training examples are used

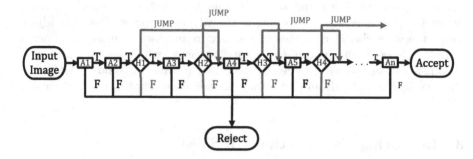

Fig. 2. Illustration of a efficient cascade structure. In the cascade, "A" denotes the Adaboost stages and "H" denotes the high efficiency SVM stages. The confidence value of the preceding two Adaboost stages are as the input of "H"stage. Red arrows show that detection window is promoted to bypass the next stage, and purple lines demonstrate that detection window is rejected in advance.

for finding most efficient negative rejection hyperplanes and positive promotion hyperplane. The training of this two hyperplanes can be formulated as:

$$\min_{w,b,x_i} \quad \frac{1}{2}\| \mathbf{w} \|^2 + C_+ \sum_{k=1}^{n_+} \xi_i + C_- \sum_{k=1}^{n_-} \xi_i \tag{6}$$

$$s.t.: \quad y_i[K(\mathbf{w},\mathbf{x_i}) + b] - 1 \geq -\xi_i \tag{7}$$

$$\xi_i \geq 0, \; for \; i = 1 \ldots n \tag{8}$$

In our implementation, we expect that the rejection hyperplane to allow all the positive training samples to be classified correctly and get highest rejection rate for negatives. In order to achieve the expectation, training positives examples are given a large cost coefficient C_+, while negative examples are given a quite small cost coefficient C_-, which means $C_+ \gg C_-$. Therefor, any training positives located in incorrect side of rejection hyperplane will bring a bigger penalty for objective function Eq. 6, while negatives located in wrong side bring minor penalty. As show in Fig. 3 (a), all positive data are above the rejection hyperplane trained in this condition of unbalanced cost coefficients, while negatives are distributed in both sides of the rejection hyperplane.

Accelerating positives examples to go through cascade classifier can save time in detecting positive windows. Because of only positives with quite high confidence value have potential to bypass the next stage. In training of promotion hyperplane, incorrect decision for negative will be penalized greatly but misclassification of positives makes a little sense. The unequally importance of samples motivate us to set the small cost coefficient C_+ for positives but give negatives a quite big cost coefficient C_- , which means $C_- \gg C_+$. As show Fig. 3 (b), all negatives examples are below the promotion hyperplane, while positives are distributed in both sides of the hyperplane.

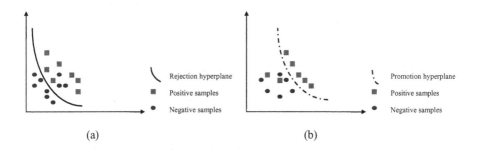

Fig. 3. SVM hyperplane learning in "H" stage. (a) Rejection hyperplane, red box denote positive samples, and blue box denote negatives. (b) Promotion hyperplane, red box denote positive samples, and blue box denote negatives.

3.3   Training Process of Improving Cascading Structure

In our proposal, there are two steps in training the improving cascading structure. The first step is training of original Viola and Jones's boosted cascade classifier. The second step is construction of novel "H" stage based on SVM. Table. 1 is the input parameters for the whole training process. The pseudo-code for learning the improving cascading structure is given in algorithm 1 and algorithm 2.

Table 1. Input parameters for training original cascade and "H" stage

notation	definition or explanation
d_{min}	Minimal desired hit rate of Adaboost stage
f_{max}	Maximal desired false alarm rate of Adaboost stage
$F_{overall}$	Overall false positive rate of cascade
f_i	Current alarm rate of Adaboost stage
φ_i	Current threshold of Adaboost stage
N	The number of cascading stages
$V_{i,j}$	Confidence value of Adaboost stage
$x_{i,j}$	Input data of "H" stage
d_{min}^{H}	Minimum detection rate of "H" stage
f_{max}^{H}	Maximum false alarm rate of "H" stage
d_i^{H}	Detection rate of "H" stage
f_i^{H}	False alarm rate of "H" stage
α_i	Ratio of cost coefficient in rejection hyperplane
β_i	Ratio of cost coefficient in promotion hyperplane

Algorithm 1. Training of Original Cascade

Initilization: $N = log_{f_{max}}^{F_{overall}}, f_i = 1$.

for i=1:N

    ◇ **while**$(f_i > f_{max})$

        •Add a week learner to the Adaboost classifier, and make sure it have lowest error as Eq. 3.

        •Update the threshold φ_i to guarantee detection rate d_{min} is satisfied.

        •Calculate the false alarm rate f_i.

        •Modify the weights of each training samples and normalize them according to Eq. 4 and Eq. 5 respectively.

end for

Algorithm 2. Training of "H" Stage
---

(1) Train Reject SVM separating hyperplane

Initialization: $C_i^+ = \alpha_1 C_i^-$, $\alpha_1 = 5$.

for i=1:N-1

    ◇Use the confidence value of preceding two stages as the input of current SVM stage, $x_{i,j} = (V_{i,j}, V_{i+1,j})$.

    ◇ **while**$(d_i^H > d_{min}^H)$

        •Training SVM separating hyperplane H_i^+ using optimization function.

        •Recalculate detection rate d_i^H for H_i^+.

        •Modify α_i: $\alpha_i \leftarrow 1.5\alpha_i$.

(2) Train Reject SVM separating hyperplane

Initialization: $C_i^- = \beta_1 C_i^+$, $\beta_1 = 5$.

for i=1:N-1

    ◇Use the confidence value of preceding two stages as the input of current SVM stage, $x_{i,j} = (V_{i,j}, V_{i+1,j})$.

    ◇ **while**$(f_i^H > f_{max}^H)$

        •Training SVM separating hyperplane H_i^- using optimization function.

        •Recalculate false positive rate f_i^H for H_i^-.

        •Modify β_i: $\beta_i \leftarrow 1.5\beta_i$.

end for

---

4   Experiment Results

In order to evaluate our efficient cascade classifier, we applied it in a challenging data set, the UIUC Image Database for car detection in our experiment. In total, we use 550 positive car samples and 550 non-car samples in training process. The size of all training images is 50×20. In addition, a test dataset which contain car images or non-car images is used to analyze performance of our classifier.Both the training dataset and test dataset are appropriate for our experiment, they contain cars in distinct backgrounds and different categories negative samples.

Haar rectangle features, including horizontal-edge, vertical-edge and titled rectangle features, are emplyed for our detector. This is because Haar features can acquire the crucial information of object and be calculated quickly through the integral-image. The feature set for our experiment consists of 344630 features for each 50×20 detection window.

Intel OpenCV library and Lin Chih-Jen's LIBSVM are employed to construct our experiment system. In the following, we will demonstrate the benefits of the presented approach by comparing our improving cascade with basic cascade. At the beginning, we set the minimum detection rate d_{min} of each Adaboost stage classifier to be 99.95% and the maximum false positive rate f_{max} of each Adaboost classifier to be 50%, and original Adaboost cascade is created using Viola and Jones's method. Ratio α for rejection hyperplane and ratio β for promotion hyperplane are initialized to be 5. Then, we adjust α and β to construct "H" stages with the optimal rejection hyperplane and promotion hyperplane.

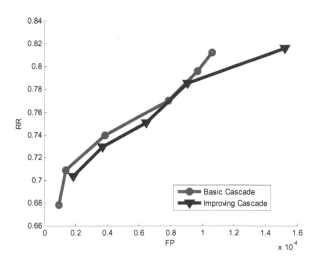

Fig. 4. Accuracy Performance contrast of basic cascade structure and improving cascade, where the horizontal axis denotes false positive rate, the vertical axis denotes recall rate. Blue curve and red curve denote the performance of basic cascade and improving cascade, respectively.

We show the superiority of our method by comparing the accuracy and efficiency performance of two kinds of cascade structure. We use the recall rate (RR) versus false positive rate (FP) curve to reflect the accuracy of the detector. The recall rate describes the ratio of the number of positives samples that were correctly classified to the total number of objects, whereas the false positive rate describes the ratio of the negatives that were incorrectly classified to the total number of testing negative windows. The accuracy performance of various detectors is demonstrated in Fig. 4.

Table 2. Efficiency performance contrast of basic cascade and improving cascade. BN and IN are the number of desired stages to reject all negative windows for basic cascade and improving cascade, respectively. BP and IP are the number of desired stages to detect all positive windows for basic cascade and improving cascade, respectively. $\Delta 1$ and $\Delta 2$ are the reduced stages for using our cascade.

Image	BN	IN	$\Delta 1$	BP	IP	$\Delta 2$
Image 1	640	588	52	140	92	48
Image 2	425	388	37	70	46	24
Image 3	592	535	57	70	48	22
Image 4	467	422	45	98	64	34
Image 5	388	363	25	98	67	31
Image 6	736	689	47	126	84	42

Fig. 5. Some detection results in UIUC data set

There are several positive windows and negative windows in each test image in UIUC. We compare the efficiency performance of basic cascade and improving cascade according to the number of desired stage classifiers for all positive and negative windows in one image. In our experiment, we use BN and IN to denote the number of desired stages to reject all negative windows in one image for basic cascade and improving cascade, respectively, and use BP and IP to denote the number of desired stages to detect all positive windows in one image for basic cascade and improving cascade, respectively. $\Delta 1$ and $\Delta 2$ are the reduced stages for using our cascade to reject negatives and detect positives, respectively. The efficiency contrast of our detector with basic cascading structure for several images in UIUC is demonstrated in Table. 2.

In average for totally 170 test images, 6.9% stages are reduced for using rejection hyperplane, and 20.9% stages are reduced for using promotion hyperplane. The experiment results show that our cascade structure with "H" stage achieve approximately the same detection accuracy as basic cascade classifier and better efficiency performance. Some detection results are illustrated in Fig. 5.

5 Conclusion

In this paper, we present a new cascaded structure by added high efficient stages, each of them provide a rejection hyperplane and a promotion hyperplane for vehicle detection. The structure can help keep the detection accuracy and have better detection efficiency. The efficient structure can also be applied to other object detection problems. In future, we will try to extend this structure to multiclass classification and integrate our detection method with some tracking algorithms.

Acknowledgements

This work was supported in part by NSFC(60702044,60625103,60632040),MIIT of China(2010ZX03004-003) and 973(2010CB731401,2010CB731406).

References

1. Viola, P., Jones, M.: Rapid object detection using a boosted cascade of simple features. In: IEEE Proc. Int. Conf. Computer Vision and Pattern Recognition, pp. 511–518 (2001)
2. Lienhart, R., Maydt, J.: An extended set of haar-like features for rapid object detection. In: IEEE ICIP, pp. 900–903 (2002)
3. Wang, Z.-R.: Pedestrian detection using boosted HOG features. Intelligent Transportation Systems, 1155–1160 (2008)
4. Wu, J., Brubaker, S.C., Mullin, M.D., Rehg, J.M.: Fast asymmetric learning for cascade face detection. Pattern Analysis and Machine Intelligence, 369–382 (2008)
5. Chen, Y.-T., Chen, C.-S.: Fast Human detection using a novel boosted cascading structure with meta stages. IEEE Transactions on Image Processing, 1452–1464 (2008)
6. Freund, Y., Schapire, R.E.: A decision-theoretic generalization of on-line learning and an application to boosting, pp. 23–37 (1995)
7. Papageorgiou, C., Oren, M., Poggio, T.: A general framework for object detection. In: International Conference on Computer Wsion, pp. 555–562 (1998)
8. Schapire, R.E., Freund, Y., Bartlett, P., Lee, W.S.: Boosting the margin: A new explanation for the effectiveness of voting methods. In: Proceedings of the Fourteenth International Conference on Machine Learning, pp. 1651–1686 (1997)
9. Burges, J.C.: A Tutorial on Support Vector Machines for Pattern Recognition. Data Mining and Knowledge Discovery, 121–167 (1998)
10. Tu, P.H., Rittscher, J., Krahnstoever, N.: Simultaneous estimation of segmentation and shape. Computer Vision and Pattern Recognition, 486–493 (2005)

Multi-camera People Localization and Height Estimation Using Multiple Birth-and-Death Dynamics

Ákos Utasi and Csaba Benedek

Hungarian Academy of Sciences, Computer and Automation Research Institute
Distributed Events Analysis Research Group
Kende u. 13-17. H-1111 Budapest, Hungary

Abstract. This paper presents a novel tool for localizing people in multi-camera environment using calibrated cameras. Additionally, we will estimate the height of each person in the scene. Currently, the presented method uses the human body silhouettes as input, but it can be easily modified to process other widely used object (*e.g.* head, leg, body) detection results. In the first step we project all the pixels of the silhouettes to the ground plane and to other parallel planes with different height. Then we extract our features, which are based on the physical properties of the 2-D image formation. The final configuration results (location and height) are obtained by an iterative stochastic optimization process, namely the multiple birth-and-death dynamics framework.

1 Introduction

Detecting and localizing people are key problems in many surveillance applications and are still challenging tasks in cluttered, crowded scenes due to the high occlusion rate caused by other people and static scene objects. Therefore, one object silhouette mask cannot be assumed to belong to only one person, and body masks can also break apart. Under such conditions single view localization or tracking might be impossible. The presented method is capable of accurately localizing individuals on the 3-D ground plane using multiple cameras. Hence, it can be used for many other high level machine vision tasks, such as scene understanding, multiple object tracking, or group/crowd behavior analysis. In addition, our method will also estimate the height of each individual. The proposed method assumes that the scene is monitored by multiple calibrated cameras, and the extracted human body silhouettes are available. These silhouettes are projected on the ground and multiple parallel planes. The presented method does not use any color or shape models for distinguishing multiple people in the scene. Instead, we will exploit the advantage of multiple cameras, and from the result of the multi-camera projection two similar geometric features are extracted in each 2-D position: one on the ground plane, and one on the other planes. Finally, the extracted features are used in a stochastic optimization process with geometric constraints to find the optimal configuration of multiple people.

The rest of the paper is organized as follows. In Sec. 2 we briefly present the related work in multi-camera people detection. The proposed method is discussed in Sec. 3. In Sec. 4 we evaluate our method using a public dataset. Finally, Sec. 5 concludes the paper.

R. Koch et al. (Eds.): ACCV 2010 Workshops, Part I, LNCS 6468, pp. 74–83, 2011.

2   Related Work

In the last decades single-camera person detection and tracking has undergone a great evolution. See [1] for an extensive review of state-of-the-art methods. However, all of these methods have limited ability to handle crowded and cluttered scenes, where the occlusion rate is high. In such situations multi-view approaches provide a better solution, that can accurately estimate the position of multiple people. Mikic *et al.* [2] proposed a blob based approach (one object is represented by one blob on each view), where they estimated the 3-D centroid of an object by deriving a least squares solution of an over-determined linear system, where the measurements were the image coordinates of multiple views. [3] models the appearance (color) and locations of the people, to segment people on camera views. This helps the separation of foreground regions belonging to different objects. [4] extracts moving foreground blobs, and calculates the centroid of the blob's lowest pixels, which is projected on the ground plane. This information, in addition to the 2-D bounding box corners, is then used in a motion model. The method in [5] assumes that the objects are observed by multiple cameras at the head level. The ground plane is discretized into a grid, and from each grid position a rectangle (having the size of an average pedestrian) is projected to the camera views to model human occupancy. The method in [6] fuses evidence from multiple views to find image locations of scene points that are occupied by people. The homographic occupancy constraint is proposed, which fuses foreground likelihood information from multiple views to localize people on multiple parallel planes. This is performed by selecting one reference camera view and warping the likelihoods from the other views. Multi-plane projection is used to cope with special cases, when occupancy on the scene reference plane is intermittent (*e.g.* people running or jumping). In our method we also use multi-plane projection, but with a different purpose. We use the foreground masks from each camera, which are projected to the ground plane and to other parallel planes, and are used for feature extraction. Our hypothesis on the person's location and height is always a combination of evidences from two planes, the ground and the hypothetical head plane to form a discriminative feature. This is done by utilizing the 2-D image formation of the projected 3-D object. The method in [7] applies long–term statistical learning to make the spatial height distribution, which is used to estimate the height of a moving object. In our method such a long–term learning process is not needed, since the height of each person will be estimated during the optimization along with the position.

Another important issue is related to object modeling. *Direct* techniques construct the objects from primitives, like silhouette blobs [8] or segmented object parts. Although these methods can be fast, they may fail if the primitives cannot be reliably detected. On the other hand, *inverse methods* [9] assign a fitness value to each possible object configuration and an optimization process attempts to find the configuration with the highest confidence. In this way, flexible object appearance models can be adopted, and it is also straightforward to incorporate prior shape information and object interactions. However, search in the high dimensional population space has a high computational cost and the local maxima of the fitness function can mislead the optimization.

In the proposed model we attempt to merge the advantages of both low level and object level approaches. The applied Multiple Birth-and-Death (MBD) technique [9] evolves the population of objects by alternating object proposition (*birth*) and removal (*death*) steps in a simulated annealing framework and the object verification follows the robust *inverse* modeling approach.

3   Proposed Method

The input of the proposed method consists of human body silhouette masks extracted from multiple calibrated camera views (using Tsai's camera model[10]), monitoring the same scene. In our current implementation the foreground masks are obtained by first estimating a mixture of Gaussians (MoG) in each pixel [11], then the resulting models are used in the method of [12] without updating the model parameters. The main idea of our method is to project the extracted silhouettes both on the ground plane, and on the parallel plane shifted to the height of the person (see Fig. 1). This projection will create a distinct visual feature, and is visible from a virtual top viewpoint in the ground plane direction. However, no prior information of the persons height is known, and the height of different people in the scene may also be different. Therefore, we project the silhouette masks on multiple parallel planes with heights in the range of typical human height. In crowded scenes the overlapping rate is usually high, which would corrupt our hypothesis. We will solve this problem by fusing the projected results of multiple camera views on the same planes. The proposed method can be separated into the following three main steps and will be discussed in the subsequent sections in detail:

1. *Multi-plane projection:* The silhouettes are projected to the ground and to several parallel planes at different height.
2. *Feature extraction:* At each location of each plane we extract features that provide positive output for the real height and real location by using the physical properties of the 2-D image formation.
3. *Stochastic optimization:* We search for the optimal configuration in an iterative process using the extracted features and geometrical constraints.

3.1   Multi-plane Projection

Let us denote by P_0 the ground plane, and by P_z the parallel plane above P_0 at distance z. In the first step of the proposed method we project the detected silhouettes to P_0 and

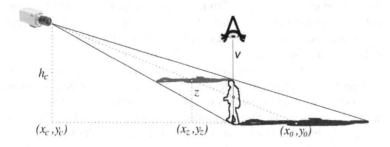

Fig. 1. Silhouettes are projected on the ground plane (blue) and on parallel planes (red)

(a) Projection for z equals to the person's real height

(b) Projection for z lower than the person's real height

(c) Projection for z higher than the person's real height

Fig. 2. Our features are based on the 2-D image formation properties and on the multi-plane projection representation. The ground plane projection of one silhouette is marked with blue, and the P_z plane projection for three different z values (z is the distance from the ground) with red.

to different P_z planes (with different $z > 0$ offsets) by using the model of the calibrated cameras. As shown Fig. 1, this can be efficiently performed by projecting on P_0 only, then using the following relationship. Let (x_c, y_c) denote the position of an arbitrary camera and h_c its height, and let (x_0, y_0) denote the position of a selected point of the silhouette projected to the ground plane (*i.e.* $h_0 = 0$). Then the (x_z, y_z) position of the same point projected on a parallel plane at z height can be expressed as

$$x_z = x_0 - (x_0 - x_c)\, z/h_c \tag{1}$$

$$y_z = y_0 - (y_0 - y_c)\, z/h_c \tag{2}$$

In Fig. 1 and later in the text the projection of the silhouette to the P_0 ground plane is marked with blue, and to one P_z plane with red color.

3.2 Feature Extraction

Our hypothesis on the location and height of a person is based on the physical properties of the 2-D image formation of a 3-D object. Consider the person with height h presented in Fig. 1, where we projected the silhouette on the P_0 ground plane (marked with blue) and the P_z plane with the height of the person (*i.e.* $z = h$, marked with red). Also consider the v vertical axis of the person that is perpendicular to the P_0 plane. We can observe that from this axis, the silhouette points projected to the $P_z|_{z=h}$ plane lie in the direction of the camera, while the silhouette print on P_0 is on the opposite side of v. For more precise investigations, in Fig. 2 the scene is visualized from a viewpoint above P_z, looking down on the ground plane in a perpendicular direction. Here, the silhouette prints from P_z and P_0 are projected to a common $x - y$ plane and jointly shown by red and blue colors, respectively (overlapping areas are purple). We can observe in Fig. 2(a), that if the height estimation is correct (*i.e.* $z = h$), the two prints just touch each other in the $p = (x, y)$ point which corresponds to the ground position of the person. However, if the z distance is underestimated (*i.e.* $z < h$), the two silhouette prints will overlap as shown in Fig. 2(b), and when the distance is overestimated (*i.e.* $z > h$), the silhouettes will move away, see Fig. 2(c).

Next, we derive a fitness function which evaluates the hypothesis of a proposed scene object with ground position $p = (x, y)$ and height h, using the information from multiple cameras. Let (x_c^i, y_c^i) denote the projected position of the ith camera on the P_0

ground plane. We describe with angle $\varphi^i(p)$ the horizontal direction of the ith camera from p in the ground plane, calculated as:

$$\varphi^i(p) = \arctan\left(\frac{y - y_c^i}{x - x_c^i}\right) . \tag{3}$$

We will also use the definition of 'opposite' direction $\bar{\varphi}^i(p) = \varphi^i(p) + \pi$. The two directions are illustrated in Fig. 3(a).

Based on the above observations, an object hypothesis (x, y, h) is relevant according to the ith camera data if the following two conditions hold. *Firstly*, we should find projected silhouette points on the P_0 plane (*i.e.* blue prints) around the $p = (x, y)$ point in the $\bar{\varphi}^i(p)$ direction, but penalize such silhouettes points in the $\varphi^i(p)$ direction of the same neighborhood. Considering these constraints, we define the $f_0^i(p)$ ground plane feature as:

$$f_0^i(p) = \frac{\mathbf{Area}\left(A_0^i \cap S(\bar{\varphi}^i(p), \Delta, p, r)\right) - \alpha \cdot \mathbf{Area}\left(A_0^i \cap S(\varphi^i(p), \Delta, p, r)\right)}{\mathbf{Area}\left(S(\bar{\varphi}^i(p), \Delta, p, r)\right)} , \tag{4}$$

where A_0^i denotes the set of silhouettes projected to plane P_0 using the ith camera model; $S(\bar{\varphi}, \Delta, p, r)$ and $S(\varphi, \Delta, p, r)$ denote the circular sectors with center p in the $[\bar{\varphi} - \Delta; \bar{\varphi} + \Delta]$ resp. $[\varphi - \Delta; \varphi + \Delta]$ angle range (marked with green on Fig. 3(a)), and r is a constant radius parameter being set a priori.

With notations similar to the previous case, we introduce the $f_z^i(p)$ feature on the P_z plane around the $p = (x, y)$ point in the $\varphi^i(p)$ direction as:

$$f_z^i(p) = \frac{\mathbf{Area}\left(A_z^i \cap S(\varphi^i(p), \Delta, p, r)\right) - \alpha \cdot \mathbf{Area}\left(A_z^i \cap S(\bar{\varphi}^i(p), \Delta, p, r)\right\}}{\mathbf{Area}\left(S(\varphi^i(p), \Delta, p, r)\right)} . \tag{5}$$

Both $f_0^i(p)$ and $f_z^i(p)$ are then truncated to take values in the $[0, \bar{f}]$ range, and are normalized by $\bar{f}$. Here, $\bar{f}$ controls the area ratio required to produce the maximal output.

If the object defined by the (x, y, h) parameter set is fully visible for the ith camera, both the $f_0^i(p)$ and $f_z^i(p)$ features should have *high* values in point $p = (x, y)$ and height $z = h$. Unfortunately in the available views, some of the legs or heads may be partially or completely occluded by other pedestrians or static scene objects, which will strongly corrupt the feature values. Although $f_0^i(p)$ and $f_z^i(p)$ features are weak in the individual cameras, we can construct a strong classifier if we consider all the camera data simultaneously and calculate the product of the average of the calculated feature values over the different views, *i.e.*

$$f(p, z) = \sqrt{\frac{1}{N}\sum_{i=1}^{N} f_0^i(p) \times \frac{1}{N}\sum_{i=1}^{N} f_z^i(p)} . \tag{6}$$

After the above feature definition, finding all the pedestrians in the scene is done by a global optimization process. Since the number of people is also unknown, and each person should be characterized by its x, y and h parameters, the configuration space has a high dimension, therefore an efficient optimization technique should be applied.

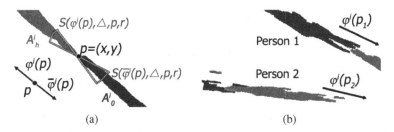

Fig. 3. (a) Notations and areas used for the calculation of the $f_0^i(p)$ and $f_z^i(p)$ features. (b) Silhouette prints to P_0 and P_z at a given z distance from a scenario with two people. Person 1's height has been accurately found ($h_1 = z$), however Person 2's one is underestimated ($z < h_2$).

3.3 Marked Point Process Model

Our goal is to detect and separate the people in the scene, and provide their position and height parameters. For this reason, we will use a simplified object model: we describe the people by their bounding cylinders in the 3-D space. Let us assume that the ground plane is flat and the people are standing on the ground. Thus, a given object-cylinder u is defined by its $x(u)$ and $y(u)$ coordinates in the ground plane and the $h(u)$ height of the cylinder, as shown in Fig. 4(a).

Let $\mathcal{H}$ be the space of u objects. The Ω configuration space is defined as [9]:

$$\Omega = \bigcup_{n=0}^{\infty} \Omega_n, \quad \Omega_n = \left\{ \{u_1, \dots, u_n\} \in \mathcal{H}^n \right\}. \tag{7}$$

Let ω denote an arbitrary object configuration $\{u_1, \dots, u_n\}$ in Ω. We define a $\sim$ neighborhood relation in $\mathcal{H}$: $u \sim v$ if the cylinders intersect. We refer to the global input data with $\mathcal{D}$ in the model which consists in the foreground silhouettes in all camera views and the camera matrices.

We introduce a non-homogeneous input-dependent energy function on the configuration space: $\Phi_{\mathcal{D}}(\omega)$, which assigns a *negative likelihood* value to each possible object population. The energy is divided into data dependent ($J_{\mathcal{D}}$) and prior (I) parts:

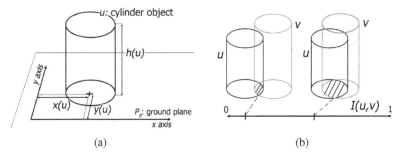

Fig. 4. (a) Cylinder objects are used to model persons in the 3-D space. Their ground plane position and height will be estimated. (b) Intersection of cylinders in the 3-D space is used as geometrical constraint in the object model.

$$\Phi_D(\omega) = \sum_{u \in \omega} J_D(u) + \gamma \cdot \sum_{\substack{u,v \in \omega \\ u \sim v}} I(u,v) , \qquad (8)$$

where $J_D(u) \in [-1,1]$, $I(u,v) \in [0,1]$ and γ is a weighting factor between the two terms. We derive the optimal object configuration as the maximum likelihood configuration estimate, which can be obtained as $\omega_{ML} = \arg\min_{\omega \in \Omega} [\Phi_D(\omega)]$.

The next key task is to define the I prior and J_D data-based potential functions appropriately so that the ω_{ML} configuration efficiently estimates the true group of people in the scene. First of all, we have to avoid configurations which contain many objects in the same or strongly overlapping positions. Therefore, the $I(u,v)$ *interaction* potentials realize a prior geometrical constraint: they penalize intersection between different object cylinders in the 3-D model space (see Fig. 4(b)) :

$$I(u,v) = \mathbf{Area}(u \cap v)/\mathbf{Area}(u \cup v) . \qquad (9)$$

On the other hand, the $J_D(u)$ *unary* potential characterizes a proposed object candidate segment $u = (x,y,h)$ depending on the local image data, but independent of other objects of the population. Cylinders with negative unary potentials are called *attractive objects*. Considering (8) we can observe that the optimal population should consist of attractive objects exclusively: if $J_D(u) > 0$, removing u from the configuration results in a lower $\Phi_D(\omega)$ global energy.

At this point we utilize the $f_u = f(p(u), h(u))|_{p(u)=(x(u),y(u))}$ feature in the MPP model, which was introduced in Sec. 3.2. Let us remember, that the f_u fitness function evaluates a person-hypothesis for u in the multi-view scene, so that 'high' f_u values correspond to efficient object candidates. For this reason, we project the feature domain to $[-1,1]$ with a monotonously decreasing function (see also Fig. 5):

$$J_D(u) = Q(f_u, d_0, D) = \begin{cases} \left(1 - \frac{f_u}{d_0}\right) & \text{if } f_u < d_0 \\ \exp\left(-\frac{f_u - d_0}{D}\right) - 1 & \text{if } f_u \geq d_0 \end{cases} \qquad (10)$$

where d_0 and D are parameters. Consequently, object u is attractive according to the $J_D(u)$ term iff $f_u > d_0$, while D performs data-normalization.

3.4   Optimization by Multiple Birth-and-Death Dynamics

We estimate the optimal object configuration by the Multiple Birth and Death Algorithm [9] as follows:

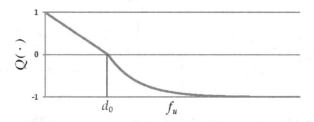

Fig. 5. Plot of the $Q(f_u, d_0, D)$ function

Initialization: start with an empty population $\omega = \emptyset$, and fit a 2-D pixel lattice to the P_0 ground plane. Let s denote a single pixel of this lattice.

Main program: set the birth rate b_0, initialize the inverse temperature parameter $\beta = \beta_0$ and the discretization step $\delta = \delta_0$, and alternate birth and death steps.

1. *Birth step*: Visit all pixels on the ground plane lattice one after another. At each pixel s,if there is no object with ground center s in the current configuration ω, choose birth with probability δb_0.

   If birth is chosen at s: generate a new object u with ground center $[x(u), y(u)] :=$ s, and set the height parameter $h(u)$ randomly between prescribed maximal and minimal height values. Finally, add u to the current configuration ω.
2. *Death step*: Consider the configuration of objects $\omega = \{u_1, \ldots, u_n\}$ and sort it by decreasing values of $J_{\mathcal{D}}(u)$. For each object u taken in this order, compute $\Delta\Phi_\omega(u) = \Phi_{\mathcal{D}}(\omega/\{u\}) - \Phi_{\mathcal{D}}(\omega)$, derive the *death rate* as follows:

$$d_\omega(u) = \frac{\delta a_\omega(u)}{1 + \delta a_\omega(u)}, \quad \text{with} \quad a_\omega(u) = e^{-\beta \cdot \Delta\Phi_\omega(u)}$$

and remove u from ω with probability $d_\omega(u)$.

Convergence test: if the process has not converged yet, increase the inverse temperature β and decrease the discretization step δ with a geometric scheme, and go back to the birth step. The convergence is obtained when all the objects added during the birth step, and only these ones, have been killed during the death step.

4   Experiments

To test our method we used the *City center* images of the PETS 2009 dataset [13] containing 400 video frames, and selected cameras with large fields of view (View_001, View_002, and View_003). In our experiments the projections were limited to a manually selected rectangular area on the ground plane, visible from all cameras. The MoG background model was defined in the CIE $L^\star U^\star V^\star$ color space, and after the parameter estimation process the covariances were manually increased to have a minimum value of 25.0 (chroma channels) or 49.0 (luma channel) to reduce the effects of cast shadow. Finally, to separate the foreground from the background the technique of [12] was used with the following settings: modality parameter $T = 0.6$, matching criterion $I = 3.0$.

In the feature extraction step (Sec. 3.2) we assumed that $r = 25cm$, Δ was set to constant $30°$, the penalty parameter to $\alpha = 1.0$, and the area ratio threshold to $\bar{f} = 0.75$. To set the parameters of the optimization process we assumed that at least one view should correctly contain the feet and another one the head of a person, which implies a $d_0 = 1/3$ threshold for object candidate acceptance. However, due to the noisy foreground masks, in our experiments we used a less restrictive value of $d_0 = 0.28$. D was set to constant 8, and we assumed a minimum distance constraint of $50cm$ between two people (*i.e.* the radius of the cylinder in Fig. 4(a)). As for the parameters of the Multiple Birth and Death optimization process, we followed the guidelines provided in [9], and used $\delta_0 = 20000$, $\beta_0 = 50$, and geometric cooling factors $1/0.96$. For each video frame

82   Á. Utasi and C. Benedek

Fig. 6. Top: result of the foreground-background separation. Bottom: estimated ground position and height of each person represented by a line. The monitored area is represented by a red rectangle.

we limited the optimization process to a maximum of 20 iterations, and did not use the result for the subsequent time step. For visualizing the results, we backprojected the estimated positions on the first camera view and draw a line between the ground plane and the estimated height (see Fig. 6 bottom), the monitored area boundary is represented by a red rectangle. Figure 6 top contains the results of the foreground-background separation. Finally, we visually evaluated the inaccuracy rate of the results (*i.e.* of positive detections with under- or overestimated height, being 6.27%), and we also calculated the rate of missed detections (being 1.75%). Further experimental results may be found at http://web.eee.sztaki.hu/~ucu/vs10-location-results.avi.

5   Conclusion

In this paper we presented a novel method to localize people in multiple calibrated cameras. For this tasks we extracted a feature, which is based on the physical properties of the 2-D image formation, and produces high response (evidence) for the real position and height of a person. To get a robust tool for cluttered scenes with high occlusion rate, our approach fuses evidences from multi-plane projections from each camera. Finally, the positions and heights are estimated by a constrained optimization process, namely the Multiple Birth-and-Death Dynamics. In the current implementation we use foreground-background separation [12] to extract foreground pixels. For evaluation we used the images of a public outdoor dataset, containing three camera views. According to our tests, the proposed method produces accurate estimation, even in cluttered environment, where full or partial occlusion is present. In the future we will investigate the effects of the different parameter settings of the feature extraction and the optimization steps. Moreover, we will examine the advance of using the optimization result for the estimation process of the subsequent time step. Another possible improvement might be the use of a robust body part detector (*e.g.* [14]) for creating evidence. This can be easily integrated in the proposed algorithm with minimal modification.

Acknowledgments

This work was partially supported by the THIS (Transport Hub Intelligent video System) project of the EU.

References

1. Yilmaz, A., Javed, O., Shah, M.: Object tracking: A survey. ACM Computing Surveys 38, 13 (2006)
2. Mikic, I., Santini, S., Jain, R.: Video processing and integration from multiple cameras. In: Proc. of the Image Understanding Workshop, pp. 183–187 (1998)
3. Mittal, A., Davis, L.S.: M2tracker: A multi-view approach to segmenting and tracking people in a cluttered scene using region-based stereo. Int. J. of Computer Vision 51, 189–203 (2002)
4. Kang, J., Cohen, I., Medioni, G.: Tracking people in crowded scenes across multiple cameras. In: Proc. of the Asian Conf. on Computer Vision (2004)
5. Fleuret, F., Berclaz, J., Lengagne, R., Fua, P.: Multicamera people tracking with a probabilistic occupancy map. IEEE Trans. on Pattern Analysis and Machine Intelligence 30, 267–282 (2008)
6. Khan, S.M., Shah, M.: Tracking multiple occluding people by localizing on multiple scene planes. IEEE Trans. on Pattern Analysis and Machine Intelligence 31, 505–519 (2009)
7. Havasi, L., Szlávik, Z.: Using location and motion statistics for the localization of moving objects in multiple camera surveillance videos. In: Proc. of the IEEE Int. Workshop on Visual Surveillance (2009)
8. Benedek, C., Szirányi, T.: Bayesian foreground and shadow detection in uncertain frame rate surveillance videos. IEEE Trans. on Image Processing 17, 608–621 (2008)
9. Descombes, X., Minlos, R., Zhizhina, E.: Object extraction using a stochastic birth-and-death dynamics in continuum. J. of Math. Imaging and Vision 33, 347–359 (2009)
10. Tsai, R.Y.: A versatile camera calibration technique for high-accuracy 3D machine vision metrology using off-the-shelf tv cameras and lenses. IEEE J. of Robotics and Automation 3, 323–344 (1987)
11. Dempster, A.P., Laird, N.M., Rubin, D.B.: Maximum likelihood from incomplete data via the EM algorithm. J. of Royal Statistical Society, Series B 39, 1–38 (1977)
12. Stauffer, C., Grimson, W.E.L.: Learning patterns of activity using real-time tracking. IEEE Trans. on Pattern Analysis and Machine Intelligence 22, 747–757 (2000)
13. PETS: Dataset - Performance Evaluation of Tracking and Surveillance (2009), http://www.cvg.rdg.ac.uk/PETS2009/a.html
14. Wu, B., Nevatia, R.: Detection and segmentation of multiple, partially occluded objects by grouping, merging, assigning part detection responses. Int. J. of Computer Vision 82, 185–204 (2009)

Unsupervised Video Surveillance

Nicoletta Noceti and Francesca Odone

DISI - Università degli Studi di Genova, Italy
{noceti,odone}@disi.unige.it

Abstract. This paper addresses the problem of automatically learning common behaviors from long time observations of a scene of interest, with the purpose of classifying actions and, possibly, detecting anomalies. Unsupervised learning is used as an effective way to extract information from the scene with a very limited intervention of the user. The method we propose is rather general, but fits very naturally to a video-surveillance scenario, where the same environment is observed for a long time, usually from a distance. The experimental analysis is based on thousands of dynamic events acquired by three-weeks observations of a single-camera video-surveillance system installed in our department.

1 Introduction

A primary goal of research in the video-surveillance field is to devise methods able to cope *automatically* with variable scene complexities and with environment changes. Within this ambitious framework, this paper presents a modular approach to learning common behaviors from an observed scene, starting from an unlabeled set of dynamic events gathered during a training phase. The proposed pipeline is entirely driven by data and starts from an intermediate representation of temporal data based on the use of strings, whose main aim is to make the learning process independent from the specific initial description. Then, a recursive implementation of spectral clustering allows us to learn behavior models and controlling their complexity. At run time we associate new observations to the estimated behaviors, if possible. An updating procedure allows us to evolve the behavior models when the percentage of associated events degrades.

We test the method on a single-camera video-surveillance system installed in our department; the monitored environment is rather complex (a public hall, illuminated by both natural and artificial lights) and hosts a variety of dynamic events. The data ambiguity caused by the loss of information of a single-camera acquisition system is attenuated by the adoption of a multi-cue initial data representation. The experimental assessment is carried out on a set of thousands of data gathered over three weeks observations: such data are used to first build an initial model and then to keep it updated, allowing for anomaly detection with respect to the current model during the day and an appropriate batch analysis and update during night hours. The reported experiments show the capability of the proposed method to (i) model frequent motion patterns, (ii) highlight anomalies, and (iii) deal with changes caused by scene variations.

R. Koch et al. (Eds.): ACCV 2010 Workshops, Part I, LNCS 6468, pp. 84–93, 2011.

Event analysis and recognition to the purpose of obtaining "intelligent" video-surveillance solutions have been addressed by many authors [1]. Learning from examples is a rather conventional way to deal with data complexity. If the available examples are labeled, i.e., to each one of them we may associate a label of a known behavior, then supervised algorithms lead to effective behavior categorization methods [2,3]. Since labeled data are not always available the general (unsupervised) goal is to model normal behaviors from (possibly big) sets of unlabeled observations. Among the first contributions relevant to behavior analysis, [4] proposes an approach based on learning from data a codebook derived from data quantization. More recent works specifically focusing on video-surveillance field are [5,6,7] where the reference applications is traffic monitoring, therefore the amount of variability on the potential behaviors is limited. Among the different tasks related to behavior analysis, anomaly detection has been constantly focus of attention in the last years [5,6,7,8].

From the algorithmic standpoint a reference to temporal series clustering may be found in [9,10,11], while [12] offers a complete survey to the topic of finding a good similarity measure for time-series. A rather complete account of the open issues related to events classification in an unsupervised setting is reported in [13]. Also, it is worth mentioning that recently clustering algorithms able to adapt to evolutions of the available data have been presented [14,15].

For what concerns clustering performance evaluation issues, there is no standard way to assess the obtained clusters both with respect to the initial set of data and with respect to possible future observations. Most of the work is based on the use of a ground-truth [16,17] .

This work contributes to the behavior analysis literature in many ways: it discusses the use of clustering methods for multi-cue data endowed with a rather characteristic internal structure induced by temporal coherence. It suggests a practical solution for model selection in the unsupervised case. Also, it proposes an effective strategy for updating the dataset used to model behaviors, which feeds updated information to a stable set of previous observations. Finally, from the application standpoint it suggests a very modular video mining pipeline that can be effectively applied on top of rather general low level video analysis modules with the final aim of classifying common events and anomalies.

2   The Proposed Method for Behavior Analysis

This section summarizes the proposed method, highlighting the behaviors modeling phase, the data pruning phase, and finally the behaviors update phase.

2.1   Batch Training Phase

This phase aims at obtaining an initialization of models of common behaviors in the observed scene, from unlabeled data. An early version of this phase was presented in [18,19]. The method, entirely data-driven, is organized in two abstraction levels: *(i)* we automatically compute a (set of) alphabet(s) that allows us to obtain a representation of the temporal series which does not depend from the original measurements; *(ii)* then we represent the time series with respect to the alphabet(s) and estimate groups of coherent temporal sequences, leading to a model of frequent behaviors.

We start from a set of video sequences (e.g. a few days of observations) to initialize our behavior models. Dynamic events (trajectories of moving objects) are extracted by a motion analysis module and represented as *temporal series* x_i of instantaneous observations x_i^t. Each observation is a feature vector in the input space $\mathbb{R}^d$ describing the object at time t: in our case $x_i^t = [\mathbf{P}_i^t, S_i^t, M_i^t, D_i^t]$ where $\mathbf{P}_i^t$ represents the 2-dim object position on the image plane, S_i^t the object size at time t, M_i^t and D_i^t velocity magnitude and direction at time t.

Alphabet construction. We start from a set of N temporal series $\mathbf{X} = \{x_i\}_{i=1}^N$, where each x_i is a sequence of k_i vectors in some Euclidean space $\mathbb{R}^d$, i.e. $x_i = (x_i^1, x_i^2, \ldots, x_i^{k_i})^T$ and $x_i^t \in \mathbb{R}^d$, $t = 1, \ldots, k_i$. An appropriate alphabet can be *built automatically* by partitioning the input space, where each state of the partition will represent a character of the alphabet. We partition the space by clustering training data with spectral clustering, using a recursive version of [20]. The procedure recursively bipartites the similarity graph until a condition on the normalized cut is reached. The granularity of the solution is controlled by a threshold, τ_{pts}, in the range $[0,1]$.

To handle the fact that different measurements of the feature vector may take values on different value ranges, instead than data normalization, we use a convex combination of kernels on sub-sets of coherent cues (multi-cue kernel [19]): given two observations in $\mathbb{R}^d$, x and y, their similarity is computed as $K(x,y) = \sum_{i=1}^{N_f} W_i K_i(x_i, y_i, \theta_i)$ where N_f is the number of sub-sets $N_f \leq d$, and W_i sum up to 1, θ_i are the parameters of kernel K_i. In this work we use Gaussian kernels and the multi-cue kernel is defined as $G(x_1, x_2) = w_P G_P(\mathbf{P}_1, \mathbf{P}_2) + w_S G_S(S_1, S_2) + w_M G_M(M_1, M_2) + w_D G_D(D_1, D_2)$. By associating a label to each state in the partition an alphabet is finally obtained. Changing the values of the weights vector $\mathbf{W} = [w_P, w_S, w_M, w_D]$ relates to considering different alphabets, i.e. different views on the same data set. Optimal values for the weights can be chosen either with prior information or with a model selection procedure (in Sec. 3.1 we discuss the latter).

Behaviors clustering. Once an alphabet is built, a temporal series x_i may be translated into a string s with an association of each element $x_i^t \in \mathbb{R}^d$ to the state of the partition it belongs to. To obtain compressed descriptions that capture the peculiarities of each behavior we consider only transitions between states (e.g. "aaabcc" becomes "abc"). After all training trajectories have been represented as strings, spectral clustering is applied to them to identify the most meaningful clusters corresponding to frequent and well defined behaviors, according to the initial representation, the alphabet chosen, and a cut threshold τ_{str} which, again, controls the granularity of the solution. The similarity matrix is built by using a *string kernel*, the *P-spectrum* kernel K_P [21] that counts how many substrings of fixed length P the two strings have in common. Since we focus on transitions between atomic symbols of the alphabet, we set $P = 2$.

The clusters detected via this clustering step represent frequent motion patterns in the input data, and can thus be interpreted as our *behavioral models*.

Association of new data. Given the behavioral models we need a procedure to associate new (test) data to one of the clusters in real time. Instead of interpolation methods (such as Nyström) that are usually computationally expensive, we adopt a very efficient method based on computing a *cluster candidate*, selected as follows: let $C_k = \{s_i\}_{i=1}^{n_k}$ be a cluster with n_k strings. For each string s_i we define a voting function v_i such that $v_i(s_j) = K_P(s_i, s_j), \quad s_j \in C_k, j \neq i$. The string s_i^* voted by s_i can now be selected as $s_i^* = v_i^*(\mathbf{S}) = \text{argmax}_{s_j \in C_k, i \neq j} v_i(s_j)$. By putting together the contributes of all strings in the cluster, the candidate is the one receiving the highest number of votes. Thus the N_B behavioral models $\{B_i\}_{i=1}^{N_B}$ are described by a set of candidate strings $\{s_i^*\}_{i=1}^{N_B}$, one for each cluster.

This technique allows for a rapid visual inspection of the obtained results, since archetypical sequences can be easily visualized and compared with data waiting to be associated with clusters. Given a new string s_t for each behavior model we compute $Simil(s_t) = \{K_P(s_t, s_j^*)\}_{i=1}^{N}$ and then we associate to s_t the label of the candidate with the highest similarity if it is greater than a given threshold τ_t. Otherwise, it is classified as an anomaly.

Data pruning. This step is motivated by the purpose of controlling the size of the training set as time goes by and new observations need to be inserted. Also, it is well known that an appropriate data pruning can actually improve generalization performances [22]; in the unsupervised case this relates to the ability of a given data partitioning to reflect the probability distribution that generated the data. The procedure we follow for pruning data in our unsupervised setting aims at improving the clusters compactness according to the following procedure:

- We train a multi-class classifier (we used Regularized Least Squares - RLS) on the data labeled according to the clustering results (i.e. by associating a label to each cluster which is inherited by the trajectories associated to it).
- We apply Leave One Out (LOO) cross validation to classify each datum, associating to it the most probable estimated label.
- Then we discard data whose label, estimated by the classifier, do not correspond to the ones estimated by clustering. We order the remaining data with respect to the classifier output in descending order.
- Finally, we discard the last p percent of the sorted data.

We first observe that using the labels induced by the clustering within a supervised approach we are implicitly relying on the quality of the clusters. The first data discarded by the pruning procedure are the ones that do not fit entirely the estimated clusters. Then, as p grows, we should eliminate redundant elements, and later we should be deleting important information. p is selected considering average LOO error and the compactness of the pruned clusters. This procedure is applied after both initialization and updates of behavior models. At the end of pruning clusters candidates will be recomputed on the survived data.

2.2 Behavior Models Update

The estimated behavior models are used at run time to build statistics on the observed behaviors and to highlight unusual events. As time goes by such behaviors could become obsolete and would need to be updated. In an unsupervised setting there is no direct way to evaluate performance degradation. We estimate it *indirectly* by monitoring the amount of data associations to known behaviors — if the scenario is stable this percentage should also be stable. A severe decrease might testify that static changes occurred in the environment and thus the models need to be updated. Also, we experimentally observed a slow degradation occurring over time. To address these events we update the learned behaviors during night hours (or when the density of dynamic events is small) by updating the behavioral models applying spectral clustering to a set of old and new data. We first refer to the previously selected alphabet: if, after update, the percentage of associated events does not improve, we also update the alphabet.

3 Application to a Video-Surveillance Scenario

In this section we experimentally evaluate the proposed method on a video-surveillance system monitoring a busy hall of our department (Fig. 1, above) [1] . Our experiments consider events occurring at peak times over three consecutive weeks. The first week (Fig. 1, above, left) has been used to initialize the behavior models. The other two weeks are used to collect test data on which the models are evaluated and updated, if needed. On the second week the scene conditions do not change, so that the dynamic events that are observed remain rather pertaining to the models. Instead, on the first days of the third week a special event occurred (the department opened to high school students and the hall was used as a meeting point) and the scene layout changed (panels have been added and desks moved) as well as the expected behaviors (Fig. 1, above, middle). At the end of the week, the scene goes back to the usual regime with some stable changes with respect to the original layout (observe, for instance, the desk in the bottom right zone - Fig. 1, above, right).

3.1 Training and Model Selection via Loose Annotation

We first consider the batch training phase performed of data from one week acquisitions (see Fig. 1, below, right). In this phase the set of parameters to be tuned comes from both alphabet construction and the actual behavior modeling phase. In unsupervised learning model selection is a hard problem [23] unless a ground truth is available (which is not the case in real applications). Lacking a proper ground truth (and after we verified the inappropriateness of most quality indices available in the literature for our rather complex model selection procedure) we propose to start from a *manual annotation of the environment,*

---

[1] The Imanalysis suite, we obtained within a technology transfer program with the company Imavis srl, http://www.imavis.com/

Fig. 1. Above, sample frames describing the environment changes over the three weeks considered. Below: left, regions selected with the manual environment annotation and used as source/sink regions to loosely annotate the training trajectories, shown on the right, according to the 8 patterns denoted as red arrows.

performed by a user and guided by prior knowledge on the presence of natural source and sink points such as doors and rest areas (see Fig. 1, below, left). The environment annotation induces a *loose annotation* on the data, grouped accordingly to first and last points. The term loose refers to two main aspects:

- Since the coherence criteria depends only on first and last points of trajectories, very different patterns can belong to the same group.
- The manual annotation reflects the spatial properties of the trajectories, thus from the point of view of the other features (target size, velocity expressed in terms of magnitude and direction of motion) the obtained groups are heterogeneous.

The derived annotation is a coarse ground truth that we *only* use for model selection, thus on a verification stage. It is not included in the modeling pipeline which is entirely unsupervised. In our training set, made of 1205 events, 8 main behavioral patterns have been annotated (see Fig. 1, below, left). The parameters participating to model selection are the kernel weights $\mathbf{W}$ and spectral clustering cut thresholds τ_{pts} and τ_{str}. After a qualitative analysis of the similarity matrix, the values in $\mathbf{W}$ are chosen in $[0, 1]$ with sampling step $\Delta = 0.1$ so that they sum up to 1. τ_{pts} and τ_{str} are chosen in $[0.6, 1.]$. We experimentally observed that values lower than 0.6 result in small alphabets and, thus, poor behavioral patterns.

Fig. 2. Top row: the 7 estimated common behaviors from the initialization phase. Below: clusters induced by the association of test data.

We evaluate a clustering instance C_i produced by a selection of parameters by solving the assignment problem among estimated and (loosely) annotated clusters via the Hungarian algorithm and computing the Correct Clustering Rate (CCR) [13]: $CCR(C_i) = \sum_{j=1}^{n_i} c_j$ where n_i is the number of clusters of instance C_i and c_j is the number of trajectories of cluster j correctly associated. When computing the rate we admit that a loosely annotated behavior might correspond to more than one clusters, and vice-versa. This is because properties not modeled by the loose annotation (which just derived from spatial coherence) could be captured by the clustering. Model selection is finally performed as $C^* = \text{argmax}_{i=1...N_c} CCR(C_i)$. In our experiments, this process selects a clustering instance with 7 patterns of activities with the parameters $\mathbf{W} = [0.3, 0, 0.3, 0.4]$, $\tau_{pts} = 0.9$ and $\tau_{str} = 1$ that produced the best CCR (equal to 76.2%). Fig 2 (top row) shows the data associated to the estimated clusters. Notice that this visualization is purely based on spatial information (for instance behaviors 1 and 4 look similar but their average velocity are 2.9 and 3.8 respectively, with small standard dev).

3.2 Experiments on Data Pruning

To test the appropriateness of data pruning, first, we recompute the candidates with an increasing percentage p of pruned data, and evaluate the percentage of associated events on the training set (an event of the training set is associated if the highest similarities of its string representation is with the candidate of the cluster it was associated to). The trend, shown in Fig. 3, left above, proves that the performances remain rather stable and slightly higher around $p = 40\%$. This suggests an appropriate value for p and gives a coarse estimate of the noisy information included in the training set that can be profitably discarded without loosing in discriminative power of the candidates.

A second analysis relies on evaluating the trend of intra-clusters cohesiveness, as follows. Given a cluster C, let m_C be the number of all possible pair-wise similarities among elements of C. The cohesiveness of C is estimated by the ratio of the number of similarities above a given threshold ($t = 0.8$ in our experiment) to m_C. The plot in Fig 3, left below, nicely shows how mean and standard deviation of the quality index improves as p increases.

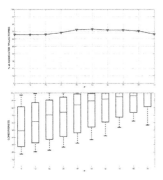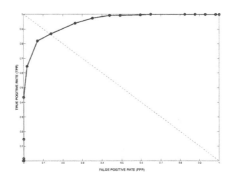

Fig. 3. Left, evaluation of data pruning: above, percentages of associated trajectories as p increases; below, mean and standard deviation of intra-cluster cohesiveness. Right, the roc curve describing the performance of the binary classifier of normal/abnormal events on the validation set.

3.3   Model Evolution

We finally evaluate the proposed pipeline at run-time following the procedure described in Sec. 2.1. To choose the threshold, τ_t, we gather a validation set including 570 events, equally distributed between known patterns (coherent with the training set) and anomalies. The ROC curve (Fig. 3) shows the performance of discriminating between common behaviors and anomalies we obtain on the validation set by varying τ_t. We select the threshold corresponding with the equal error rate (e.e.r.), $\tau_t = 0.65$.

Fig. 4 summarizes the association percentages over the three weeks. The first week (training period) is drawn in black and give us an intuition on the amount of noise in the observed scene, and an estimate of the percentage of association to known behaviors that we may expect at run time (the green lines describe mean and standard deviation of the associations). Blue plot refers to the test analysis performed with a fixed model. During the first test week the percentages are affected by a physiological decreasing due to the time evolution. However, between days 6 and 14, the percentages are rather stable. At days 15 and 16 an abrupt variation can be observed (caused by the start of the special event). With the fixed behavior model, when the event ends (day 17) the association percentage goes back to normal. The decreasing number of associations of day 15 suggests that an evolution of the model might be advisable. A new training set is thus collected which is composed of *(i)* trajectories belonging to the current training set that overcome the pruning process, and *(ii)* trajectories observed during the last day. If the association process is replicated employing this new model (red plot) the percentage of associated events reaches the usual trend, testifying the capability of the evolved model of appropriately describing the current scenario. Fig. 2 (bottom) shows the trajectories correctly associated by the evolving method over the test period (only a random sampling of the data is shown for readability).

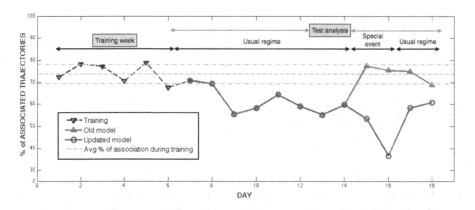

Fig. 4. Trend of the association percentages over the 3 weeks considered in our experiments. Blue plot refers to the adoption of a fixed model, that shows inappropriateness when the video content significantly changes. If the model is evolved accordingly to our pipeline, it shows the capability of being adaptive to temporal changes.

4   Discussion

This paper presented a method for modeling common behaviors from long-time observations and keeping this model up-to-date with respect to an evolving reference scenario. We refer specifically to an unsupervised setting, therefore the available data do not need to be labeled, but will be automatically analyzed by a spectral clustering algorithm, after a mapping on an appropriate feature space (the space of strings). Thanks to this intermediate description, provided that an appropriate low-level video analysis module is available, the devised pipeline can be applied to rather general data and scenarios.

Once a behavioral model is estimated, it can be applied to new data and used at run time to check the presence of unusual events. To counter the fact that such model would become obsolete due to scene changes or simple environment conditions (due to seasonal changes, for instance, if the environment is naturally illuminated), an update process to be run batch is provided. An experimental validation of the proposed method is shown on rather complex data coming from a commercial video-surveillance system.

References

1. Special issue on event analysis in videos. IEEE Trans on Circuits and Systems for Video Technology 18 (2008)
2. Pittore, M., Campani, M., Verri, A.: Learning to recognize visual dynamic events from examples. IJCV (2000)
3. Bashir, F., Khokhar, A., Schonfeld, D.: Object trajectory-based activity classification and recognition using hidden markov model. IEEE Trans. on IP 16 (2007)
4. Stauffer, C., Grimson, E.: Learning patterns of activity using real-time tracking. IEEE Transactions on PAMI 22 (2000)

5. Hu, W., Xiao, X., Fu, Z., Xie, D., Tan, T., Maybank, S.: A system for learning statistical motion patterns. IEEE Trans on PAMI 28 (2006)
6. Piciarelli, C., Micheloni, C., Foresti, G.L.: Trajectory-based anomalous event detection. IEEE Trans on Circuits and Systems for Video Technology 18 (2008)
7. Anjum, N., Cavallaro, A.: Multifeature object trajectory clustering for video analysis. IEEE Trans. on Circuits and Systems for Video Technology 18 (2008)
8. Hamid, R., Johnson, A., Batta, S., Bobick, A., Isbell, C., Colenam, G.: Detection and explanation of anomalous activities: representing activities as bags of event n-grams. In: Proc. CVPR (2005)
9. Jebara, T., Song, Y., Thadani, K.: Spectral clustering and embedding with hidden markov models. In: Kok, J.N., Koronacki, J., Lopez de Mantaras, R., Matwin, S., Mladenič, D., Skowron, A. (eds.) ECML 2007. LNCS (LNAI), vol. 4701, pp. 164–175. Springer, Heidelberg (2007)
10. Liao, T.W.: Clustering of time series data: A survey. Patt. Recogn. 38 (2005)
11. Niebles, J.C., Wang, H., Fei-Fei, L.: Unsupervised learning of human action categories using spatial-temporal words. In: Proc. of BMVC (2006)
12. Rieck, K., Laskov, P.: Linear-time computation of similarity measures for sequential data. JMLR 9, 23–48 (2008)
13. Morris, B., Trivedi, M.M.: Learning trajectory patterns by clustering: Experimental studies and comparative evaluation. In: Proc. CVPR (2009)
14. Ning, H., Xu, W., Chi, Y., Gong, Y., Huang, T.S.: Incremental spectral clustering by efficiently updating the eigen-system. Pattern Recogn. 43, 113–127 (2010)
15. Chi, Y., Song, X., Zhou, D., Hino, K., Tseng, B.L.: On evolutionary spectral clustering. ACM Trans. Knowl. Discov. Data 3, 1–30 (2009)
16. Halkidi, M., Batistakis, Y., Vazirgiannis, M.: On clustering validation techniques. Journal of Intelligent Information Systems 17, 107–145 (2001)
17. Günter, S., Bunke, H.: Validation indices for graph clustering. Pattern Recogn. Lett. 24, 1107–1113 (2003)
18. Noceti, N., Santoro, M., Odone, F.: String-based spectral clustering for understanding human behaviours. In: THEMIS-BMVC (2008)
19. Noceti, N., Odone, F.: Towards and unsupervised framework for behavior analysis. In: Proc. of AI*IA (PRAI*HBA) (2010)
20. Shi, J., Malik, J.: Normalized cuts and image segmentation. IEEE Trans. on PAMI 22, 888–905 (2000)
21. Taylor, J.S., Cristianini, N.: Kernel Methods for Pattern Analysis. Cambridge University Press, Cambridge (2004)
22. Angelova, A., Abu-Mostafa, Y., Perona, P.: Pruning training sets for learning of object categories. In: Proc CVPR (2005)
23. Grira, N., Crucianu, M., Boujemaa, N.: Unsupervised and semi-supervised clustering: a brief survey. In Muscle VIFP EU NoE (2005)

Multicamera Video Summarization from Optimal Reconstruction

Carter De Leo and B.S. Manjunath

University of California, Santa Barbara

Abstract. We propose a principled approach to video summarization using optimal reconstruction as a metric to guide the creation of the summary output. The spatio-temporal video patches included in the summary are viewed as observations about the local motion of the original input video and are chosen to minimize the reconstruction error of the missing observations under a set of learned predictive models. The method is demonstrated using fixed-viewpoint video sequences and shown to generalize to multiple camera systems with disjoint views, which can share activity already summarized in one view to inform the summary of another. The results show that this approach can significantly reduce or even eliminate the inclusion of patches in the summary that contain activities from the video that are already expected based on other summary patches, leading to a more concise output.

1 Introduction

Many domains, from surveillance to biology, can benefit from collecting large quantities of video data. However, long recordings over many deployed cameras can easily overwhelm a human operator's ability to review, preventing the data from being as useful as possible. In many applications with stationary cameras, much of the recorded video is uninteresting, so time spent having a human review it is wasted. Video summarization aims to highlight the most important segments of an input video, helping to focus reviewing time where it is most beneficial.

The concept of extracting the important portions of a video is not usually well defined, since importance is a subjective notion. While looking at motion or color contrast can serve as an approximation to importance, these methods take an indirect approach to the summarization problem. Instead, we propose a method that formulates the problem in a more principled way that easily generalizes to multiple cameras.

Videos from within a single camera or from close-by cameras in a network also exhibit redundancy in what they display, since activities in one region are often closely related to activities in another. For example, refer to Figure 1(a), which shows a network of two cameras positioned along a bike path. When a person leaves the view of camera 1 traveling to the right, it is expected that they will appear in camera 2 after a delay. If the delay does not greatly deviate from the average trip time observed over many people, showing the person in both views 1 and 2 is redundant; if a human observer sees the person in one view, they already

R. Koch et al. (Eds.): ACCV 2010 Workshops, Part I, LNCS 6468, pp. 94–103, 2011.

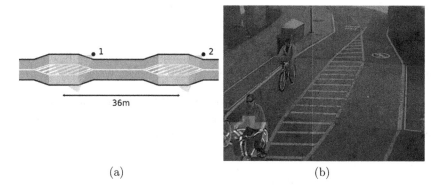

(a)   (b)

Fig. 1. Layout of the two-camera network used for the experiments and a sample spatio-temporal patch drawn from camera 1 to be highlighted in the summary

have a good understanding of what happened in the other. In this case, a good summary should devote less time to the appearance of the person in one view after establishing the person's presence in the other. However, if the travel time does significantly differ from what is expected, the summary should spend extra time presenting this anomaly. This shows that a good summary would respond not just to motion, but also to whether that motion is already expected.

We view the output summary video as a set of observations on the original input video. Since the summary video is a reduced form of the input video, many possible observations are missing. The best set of observations to chose for the summary can be understood as those that, taken alone, would allow us to best reconstruct the missing data. These observations take the form of a spatio-temporal patch highlighted in the summary output, such as the example patch in Figure 1(b). Reconstruction requires a predictive model to describe how an observation at one spatio-temporal location influences the state at others, which the system can learn over local regions of the video itself since the camera viewpoint is static. This captures the intuition that if a reviewer is familiar with what normally occurs within a scene, they have effectively learned a predictive model themselves. As such, a summary consisting of the observations that give the best reconstruction of the missing data, or the rest of input video, would also give a reviewer the best mental reconstruction of what occurred.

2   Related Work

Video summarization, as well as the related problem of video anomaly detection, has been well studied in the literature, so we discuss only a subset of the past work here. Approaches tend to be divided between methods using tracked object paths and those that use features that do not rely on tracking. Systems that use tracking [1,2,3] attempt to extract the trajectories of objects of interest within a scene, then cluster those trajectories to identify outliers. Objects following unusual trajectories are then assumed to be interesting. In visually challenging scenes, however, extracting suitable object trajectories may be difficult, degrading performance.

Many other systems rely on determining the similarity between frames using other features. Examples include gradient orientations [4,5], local motion [6], and color and texture [7,8]. Another approach is to apply seam carving techniques to videos [9] to remove regions with smooth colors. While these systems can yield satisfying results, they are not directly attempting to make the most interesting or representative portions of the input video appear in the summary, instead relying on related indicators.

The system of Simakov et al.[10] does approach the summarization problem more directly by trying to choose patches of the input video to include in the summary to simultaneously maximize measurements of completeness, or how much data from the input is present in the output, and coherence, or that everything in the output was also in the input. This takes the viewpoint that a good summary is one that includes as much of the input data as possible within a constrained space without introducing artifacts, whereas our proposed approach considers a good summary as one that best allows for data missing from the summary to be inferred, thus representing the entire input.

3   Approach

Our goal is to analyze a set of input videos and determine the subset of spatio-temporal patches from them that would best summarize their contents. These can be packed into a shorter output video for a human operator to review. The first step is a scene decomposition to group camera views into regions, followed by feature clustering and region linking to cluster activities occurring in each region and determine region topology. The system learns occurrence models for the activities and then uses a genetic algorithm to seek the summary that best represents the activity sequence occurring in a region. Here, a summary refers to any selected subset of key patches. The algorithm grades the fitness of a candidate summary by finding the error of the resulting reconstruction, defined as the estimate of the complete sequence of activity labels given the subset in the candidate summary.

3.1   Scene Decomposition

Our system starts with a scene decomposition to spatially divide the input videos into regions that tend to move similarly, based on the work by Loy[11]. We follow their approach except for a change in the activity feature used. Unlike the low framerate videos presented in those experiments, the videos used here have an average frame rate around 15-20 fps. This allows the use of optical flow as the activity feature instead of the features used by Loy to accommodate low temporal resolution. We calculate the affinity matrix $\mathbf{A}$ between the 10x10 pixel, non-overlapping subblocks with sufficient activity in the video. Spectral clustering

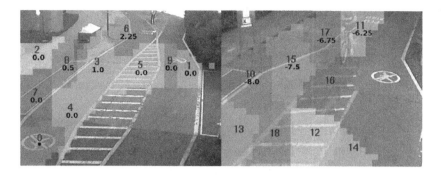

Fig. 2. Discovered links and time shifts to region 0. (Best viewed in color).

on **A** by the method presented by Zelnik-Manor[12] gives the segmented regions. An example segmentation for a video in our data set appears in Figure 2. Notice that the segmentation has separated the regions covering the bike path from the pedestrian areas on either side, giving the regions semantic meaning.

3.2 Feature Clustering and Region Linking

With the scene decomposition done, the average optical flow vector over a region can be calculated at each frame and then clustered by fitting a GMM. The number of clusters K_i for the ith region is determined automatically using the Akaike information criterion[13]. Now the activity in each region can be succinctly represented by a single sequence $\mathbf{y}_i$ with $y_{i,t} \in [0, K_i)$ consisting of cluster indices over time.

This representation also allows discovery of the linkages between regions and the typical time lag between activity in one region leading to activity in another. For a proposed linkage between regions i and j for time lag τ, we can calculate the Time Delayed Mutual Information[14]:

$$\mathbf{I}_{i,j}(\tau) = \sum_{y_i} \sum_{y_j} p(y_{i,t}, y_{j,t+\tau}) \ln \left[\frac{p(y_{i,t}, y_{j,t+\tau})}{p(y_{i,t})p(y_{j,t+\tau})} \right] \quad (1)$$

The probability distributions are estimated by counting activity occurrences over the length of the videos. For each local maxima of $\mathbf{I}_{i,j}(\tau)$ for $\tau \in [-\tau_{max}, \tau_{max}]$ that exceeds a threshold I_{min}, define a link between regions i and j with a time shift of τ. We do not consider regions within a camera view differently from regions appearing in different camera views, so this linkage discovery naturally extends to a multicamera network. As an example, we use video segments collected from the two cameras shown in Figure 1(a). Figure 2 shows the resulting linkages from region 0, in the lower left corner of the first camera, to all other regions. The labels show the relative time shift τ in seconds for the link between that region and region 0; regions without labels are not connected to region 0.

3.3   Learning Occurrence Models

The system uses the set of component index sequences $\{\mathbf{y}_i \forall i\}$ to learn a set of occurrence models. For each region, estimate $p(y_{i,t})$, $p(y_{i,t+1}|y_{i,t})$, and $p(y_{j,t+\tau}|y_{i,t})$ over $(j,\tau) \in \mathcal{L}(i)$, the set of regions and corresponding time shifts that form links to region i. From these, compute the negative-log costs for assigning indices to patches:

$$c_i^p(q) = -\ln(p(y_{i,t} = q))$$
$$c_i^f(r|q) = -\ln(p(y_{i,t+1} = r|y_{i,t} = q)) \tag{2}$$
$$c_{ij,\tau}^l(r|q) = -\ln(p(y_{j,t+\tau} = r|y_{i,t} = q)) \quad (j,\tau) \in \mathcal{L}(i)$$

c^p is the prior cost, or the cost of assigning component index q to a patch without knowledge of surrounding patches. c^f is the forward cost, or the cost of assigning index r to a patch when its temporal predecessor has index q. Finally, c^l is the lateral cost, or the cost of assigning index r to a patch in region j when it is linked with time shift τ to a patch in region i that has index q.

3.4   Single Region Activity Reconstruction

Our goal is to reconstruct an index sequence by selecting a subset of the patches from the corresponding region to include in the summary. Selected key patches in the sequence act as observed states, while the remaining patches act as missing observations. As such, the system uses a modified Viterbi algorithm with the prior and forward cost models to choose the most likely sequence of indices that explain the chosen observations. This Viterbi lattice is illustrated in Figure 3(a), where the columns correspond to steps in time and the rows correspond to the possible activity indicies for the region in the range $[0, K_i)$. The costs defined in the previous section determine the costs used to label the lattice edges. Specifically, for region i, the edge from the starting node to the node for activity q in the t=0 layer is labeled using $c_i^p(q)$. For an edge from the node for activity q to activity r in the next layer, use $c_i^f(r|q)$. In this example, the optimal path through the lattice is shown in bold.

Choosing a patch as a key patch amounts to forcing a step in the lattice to take the state seen in the input video, as in Figure 3(b). This choice updates the optimal path between states. Call $\hat{\mathbf{y}}_{i|\mathcal{P}_i}$ the reconstructed sequence after choosing to force the patches of region i in set $\mathcal{P}_i$ to their correct values. The error for this choice of key patches is:

$$E_i(\mathcal{P}_i) = \sum_t ec(y_{i,t} = q, \hat{y}_{i,t|\mathcal{P}_i} = r) \tag{3}$$
$$ec(q,r) = \sqrt{\boldsymbol{\mu}_{i,q} \mathbf{C}_{i,q}^{-1} \boldsymbol{\mu}_{i,r}}$$

where the error cost ec of reconstructing a patch as having index r when it was actually q is the Mahalanobis distance from the correct GMM cluster, with mean $\boldsymbol{\mu}_{i,q}$ and covariance $\mathbf{C}_{i,q}$, to $\boldsymbol{\mu}_{i,r}$, the mean of the classified cluster.

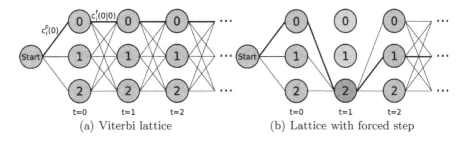

(a) Viterbi lattice   (b) Lattice with forced step

Fig. 3. Viterbi lattice for region activity reconstruction before and after forcing. Columns represent time and rows represent cluster indices. The optimal path is shown in bold.

3.5 Key Patch Selection

To form the summary, we would ideally like to find $\mathcal{P}_i$ such that:

$$\mathcal{P}_i = \operatorname*{argmin}_{\mathcal{P}'_i} \mathrm{E}_i(\mathcal{P}'_i) \tag{4}$$

However, there are many possible choices for $\mathcal{P}_i$; even a one minute sequence from our data set has about 1200 frames, so choosing 10% of them to include in a summary would give around 10^{168} choices. Since it is not imperative that we find the globally best $\mathcal{P}_i$ instead of a merely good one, a genetic algorithm is an appropriate way to examine such a large search space. A proposed $\mathcal{P}_i$ can be naturally represented as a binary string with length equal to the number of frames and ones in the positions corresponding to patches included in the summary, so this problem maps directly to a genetic approach. We use a modified version of the CHC algorithm[15], which stands for *cross-generational elitist selection, heterogeneous recombination, and cataclysmic mutation*. The CHC algorithm employs an aggressive search that ensures non-decreasing fitness of the best solution between generations, offset by periodic reinitialization of the population of solutions to discourage convergence on local maxima. For a proposed $\mathcal{P}_i$, we evaluate its fitness as:

$$\mathrm{F}(\mathcal{P}_i; \alpha, \beta) = \frac{\mathrm{E}_i(\emptyset) - \mathrm{E}_i(\mathcal{P}_i)}{|\mathcal{P}_i|} \cdot \exp\left(-\frac{(\alpha - |\mathcal{P}_i|)^2}{2\beta^2} \cdot 1(\mu - |\mathcal{P}_i|)\right) \tag{5}$$

This consists of two terms. The first is an efficiency term, which rates solutions higher that have achieved a large reduction in the reconstruction error per patch that it has forced. The second is a falloff term that penalizes solutions that are more concise than the target level of summarization α, but has no effect on longer solutions. Empirically, shorter solutions tend to be more efficient, so this term prevents selective pressure from creating a summary that is much more concise that the user wishes. Instead, we favor solutions that spend extra forced patches reducing the reconstruction error even modestly instead of forgoing them all together. The factor β controls how steep this penalty should be and is set such that $\sqrt{2}\beta = \alpha/10$ for all of our experiments.

3.6   Extension to Multiple Regions

If two regions i and j were found to be linked in the preceding steps, then knowing the activity present in i should also tell a human viewer something about the activity occurring in j, with some possible time shift. In our example videos, if the generated summary for region 6 establishes that a bicyclist is traveling to the left along the path, the viewer already assumes that the same bicyclist will shortly appear in region 0; if this occurs, the summary does not need to choose as many patches in region 0 to make this clear. This intuition naturally extends across cameras as well; after seeing the bicyclist leave region 0, the viewer can expect to see the same person again in region 11. If the reappearance happens close to the time shift discovered for that region link, showing that activity is largely redundant. If the actual delay differs significantly from τ, then something unusual may have happened, and the summary should spend additional summary patches illustrating this.

Formally, we can incorporate information coming from a linked region within the lattice framework by altering the transition costs for a time step using the lateral cost models learned earlier:

$$c_{i,t}(r) = c_i^f(r|y_{i,t-1}) + \sum_{(j,\tau)\in\mathcal{L}(i)} c_{ji,-\tau}^l(r|y_{j,\hat{t}-\tau}) \tag{6}$$

This represents the cost for selecting cluster label r for frame t in region i. The first term in the sum is the existing cost based on the intra-region forward model. The second term has been added to account for influence from other regions on the current region's lattice solution, based on the inter-region lateral model.

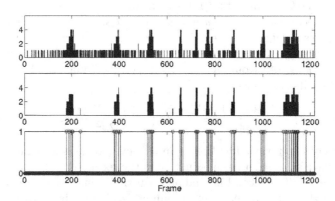

Fig. 4. Single region summarization for region 0 for a 5% target length. Top: Actual sequence. Middle: Reconstructed sequence. Bottom: Chosen summary patches.

4   Experiments

4.1   Single Region

Figure 4 shows the resulting summaries generated by the system for a one minute sample of the video corresponding to region 0 for a 5% summary length target. The top row shows the actual activity sequence for the region and the second row shows the reconstructed sequence. The third row shows spikes corresponding to the patches chosen for the summary. These are the observations in time used to generate the reconstruction. Notice that the density of the spikes is greatest where the activity indicies change, which corresponds to bicyclists moving through the region in the original video.

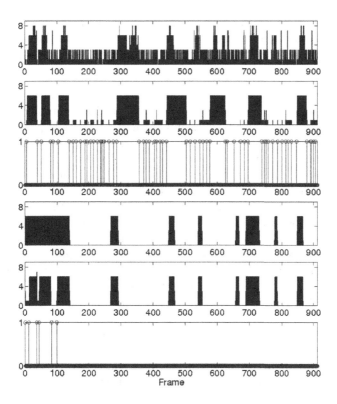

Fig. 5. Summarization of region 15 with and without information from region 0. Row 1: Actual sequence. Row 2: 5% reconstruction of region 15 in isolation. Row 3: Patches chosen for the reconstruction in the row above. Row 4: Reconstruction of region 15 incorporating information from region 0. No patches from region 15 have have been chosen. Row 5: Reconstruction of region 15 with region 0 information and choosing patches to give error equal to 5% reconstruction in isolation. Row 6: Patches chosen for the reconstruction in the row above.

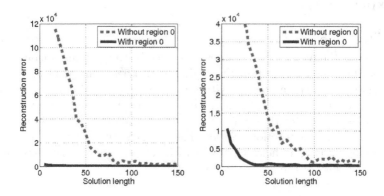

Fig. 6. Difference in reconstruction error versus summary length when excluding and including information from region 0. Left plot is for region 4, right plot is for region 15.

4.2   Multiple Regions

Figure 5 shows the effect on a neighboring region's information on the reconstruction of region 15. The first row shows the actual activity sequence for region 4. The second and third rows show the single region reconstruction of region 15 with a 5% length target and the chosen patches, as in the previous section. The fourth row shows the unforced reconstruction of region 15, or the reconstruction before any key patches have been chosen from 15 when its link to region 0 is included in the costs calculated from Equation 6. To produce this, we first generate the 5% length reconstruction for region 0 as shown in the previous section and then use the resulting $\hat{y}_0$ in Equation 6 to determine the optimal path through the lattice for region 15. This shows that even if the summary did not include any patches from region 15, seeing region 0, which is in a different camera view, has already provided an idea of its activity. The fifth and sixth rows show the reconstruction and chosen patches of region 15 incorporating information from region 0 and choosing enough patches to make the total error equal to that from the reconstruction in isolation seen in the second row. Here, the system can provide observations on region 15 to correct deviations in its activity from what would be predicted by region 0. In this example, the system reaches the same total error as it did with 60 patches in isolation with only 6 patches when inter-region information is incorporated.

The benefit gained from considering information from linked regions reaches a saturation point as the algorithm includes additional patches in the summary. Figure 6 shows the total reconstruction error for regions 4 and 15 versus the summary target length, both when no inter-region information is included and when region 0 is included. Notice that using inter-region information helps provide a lower error reconstruction for a given summary length. However, since the system only needs to correct deviations from expected activity when using information from region 0's summary, it experiences less benefit by allowing a longer summary. The horizontal distance between the two curves shows the decrease in the number of frames that need to be displayed to the user after including inter-region information. Notice that for region 4, the unforced reconstruction using region 0's summary already has lower error than a reconstruction in isolation for many target lengths, so it could be excluded from the summary completely.

5   Conclusion

We proposed a technique for video summarization that takes a principled approach to creating an output summary video. By viewing the spatio-temporal patches that are retained for the output summary as observations of the local motion of the input video, our system attempts to optimally construct the summary to best allow inference of the missing input data. This allows it to choose key patches not just based on motion, but on a viewer's expectation of what motion will occur. Our results show the validity of this approach and its ability to generalize to camera networks with disjoint views by allowing motion shown in one region to inform what is shown in another, creating a more concise summary.

References

1. Wang, X., Tieu, K., Grimson, W.: Correspondence-free activity analysis and scene modeling in multiple camera views. PAMI (2009)
2. Wang, X., Ma, K., Ng, G., Grimson, W.: Trajectory analysis and semantic region modeling using a nonparametric bayesian model. In: CVPR (2008)
3. Piciarelli, C., Micheloni, C., Foresti, G.L.: Trajectory-based anomalous event detection. IEEE Trans. Circuits Systems Vid. Tech. 18, 1544–1554 (2008)
4. Breitenstein, M., Grabner, H., Gool, L.V.: Hunting nessie – real-time abnormality detection from webcams. In: ICCV WS on VS (2009)
5. Pritch, Y., Ratovitch, S., Hendel, A., Peleg, S.: Clustered synopsis of surveillance video. In: AVSS (2009)
6. Adam, A., Rivlin, E., Shimshoni, I., Reinitz, D.: Robust real-time unusual event detection using multiple fixed-location monitors. PAMI (2008)
7. Zhong, H., Shi, J., Visontai, M.: Detecting unusual activity in video. In: CVPR (2004)
8. Zhu, X., Wu, X., Fan, J., Elmagarmid, A., Aref, W.: Exploring video content structure for hierarchical summarization. Multimedia Systems 10, 98–115 (2004)
9. Chen, B., Sen, P.: Video carving. Eurographics (2008)
10. Simakov, D., Caspi, Y., Irani, M., Shechtman, E.: Summarizing visual data using bidirectional similarity. In: CVPR (2008)
11. Loy, C., Xiang, T., Gong, S.: Multi-camera activity correlation analysis. In: CVPR, pp. 1988–1995 (2009)
12. Zelnik-Manor, L., Perona, P.: Self-tuning spectral clustering. In: NIPS (2004)
13. Akaike, H.: A new look at the statistical model identification. IEEE Trans. Automatic Control 19, 716–723 (1974)
14. Loy, C., Xiang, T., Gong, S.: Modelling activity global temporal dependencies using time delayed probabilistic graphical model. In: ICCV (2009)
15. Eshelman, L.: The chc adaptive search algorithm. Foundations of Genetic Algorithms, 256–283 (1991)

Noisy Motion Vector Elimination by Bi-directional Vector-Based Zero Comparison

Takanori Yokoyama and Toshinori Watanabe

Graduate School of Information Systems,
University of Electro-Communications,
1-5-1 Chofugaoka, Chofu-shi, Tokyo 182-8585, Japan
{yokotaka,watanabe}@sd.is.uec.ac.jp

Abstract. Network cameras are becoming increasingly popular as surveillance devices. They compress the captured live video data into Motion JPEG and/or MPEG standard formats, and they transmit them through the IP network. MPEG-coded videos contain motion vectors that are useful information for video analysis. However, the motion vectors occurring in homogeneous, low-textured, and line regions tend to be unstable and noisy. To address this problem, the noisy motion vector elimination using vector-based zero comparison and global motion estimation was proposed. In this paper, we extend the existing elimination method by introducing a novel bi-directional vector-based zero comparison to enhance the accuracy of noisy motion vector elimination, and we propose an efficient algorithm for zero comparison. We demonstrate the effectiveness of the proposed method through several experiments using actual video data acquired by an MPEG video camera.

1 Introduction

Surveillance cameras are being widely employed both commercially and privately to safeguard property and to monitor suspicious activities. Previously, analog cameras, which output analog video signals, were used for video surveillance. In recent years, network cameras that connect to the IP network are going widespread use owing to their high flexibility and scalability. Network cameras capture live video data, and they generally compress the captured video data into Motion JPEG and/or MPEG standard formats; the compressed video data is then transmitted through the IP network. As network cameras prevail, opportunities to directly handle the compressed data increase.

The Moving Pictures Experts Group (MPEG) was established in 1988, and the MPEG standards have been extensively adopted worldwide for the compression of multimedia data. MPEG-coded videos contain motion vectors to compress video data by using motion compensation. These motion vectors contain useful information for video analysis, and a number of video analysis methods using motion vectors have been proposed [1,2,3,4,5]. However, the motion vectors occurring at low-textured, homogeneous, and line regions are noisy for motion analysis. Although these noisy motion vectors adversely affect motion

R. Koch et al. (Eds.): ACCV 2010 Workshops, Part I, LNCS 6468, pp. 104–112, 2011.

analysis, effective reduction is not always adopted in conventional motion analysis methods. Colace *et al.* proposed noisy motion vector reduction for camera parameter estimation by using the discrete cosine transform (DCT) coefficients of MPEG video data [6]. They assumed that the energy of homogeneous and low-textured regions is concentrated at low frequencies of the DCT coefficients, and they eliminated them by thresholding. Hessel and Eickeler also employed the DCT coefficients to detect low-textured areas [7]. Eng and Ma proposed a trajectory extraction method [8] whereby a standard vector median filter and fuzzy set approach are adopted for the reduction of irregular motion vectors. The methods described above can reduce the noisy motion vectors occurring in homogeneous and low-textured regions, but they cannot reduce the ones occurring in line regions. To address this shortcoming, Yokoyama *et al.* proposed a method [9] for elimination of noisy motion vectors occurring in all the homogeneous, low-textured, and line regions by introducing global motion estimation [10] and zero comparison [11]. By using this method, it is possible to obtain stable local motion vectors and to estimate global motion, which can be used for the motion analysis of video data acquired by a moving camera.

In this paper, we propose a novel bi-directional vector-based comparison to enhance the accuracy of noisy motion vector elimination. In the proposed method, we introduce an additional region to enable bi-directional examination of the target region movement; this enhances the accuracy of noisy motion vector elimination. We also propose an efficient algorithm for bi-directional zero comparison.

2   Noisy Vector Elimination

In this section, we briefly introduce a noisy motion vector elimination method using global motion estimation and zero comparison [9]. Figure 1 shows the two main processes and the flow of motion vector sets in this elimination method. The global motion estimation can utilize the motion vectors that correctly reflect global motion even occurring at homogeneous and low textured regions. The n-th frame input motion vector set MV_n is separated into the motion vector set $\widehat{\mathrm{GMV}}_n$ for estimating the global motion parameter set m_n and other motion vectors. As a result of global motion estimation, the global motion vector set GMV_n is generated by the estimated parameter set m_n, and the tentative local motion vector set $\widehat{\mathrm{LMV}}_n$ is obtained by subtracting GMV_n from MV_n. $\widehat{\mathrm{LMV}}_n$ includes both useful local motion vectors and noisy motion vectors for motion analysis. Therefore, the vector-based zero comparison process eliminates the noisy motion vectors NMV_n from $\widehat{\mathrm{LMV}}_n$ on the basis of their movement, and it outputs the stable local motion vectors LMV_n.

2.1   Zero Comparison

The original zero comparison method [11] is used to determine whether a target region is moving or stationary without any prior knowledge or assumption regarding the moving object. It searches the best matched region for the target

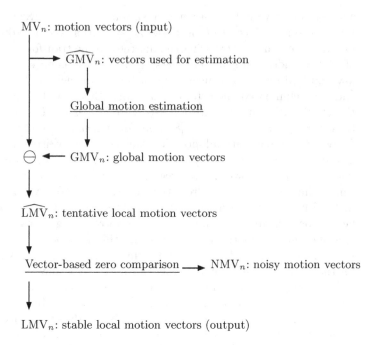

MV_n: motion vectors (input)

$\widehat{GMV}_n$: vectors used for estimation

Global motion estimation

GMV_n: global motion vectors

$\widehat{LMV}_n$: tentative local motion vectors

Vector-based zero comparison ⟶ NMV_n: noisy motion vectors

LMV_n: stable local motion vectors (output)

Fig. 1. Main processes and flow of motion vectors in noisy motion vector elimination

region out of the next frame and examine the difference distributions (i.e., a correlation map) are examined. When a target region that is neither homogeneous nor low-textured is moving, the correlation map tends to have a peak at a different position from zero (i.e., the origin of the map). The zero comparison method checks this fact by calculation of two distances D_0 and D_1; D_0 is the sum of absolute distances (SAD) between the target region R_o and its corresponding region R_n in the next frame, and D_1 is the SAD between R_o and its best match region R_b in the next frame (see Fig. 2). If the difference between D_0 and D_1 exceeds a predefined threshold value t_z, the correlation map has a peak at a different position from zero due to its movement (Fig. 2(b)). Otherwise, we can conclude that the target region is stationary, homogeneous, or low-textured. If the target region R_t is stationary, both the position of D_0 and D_1 become equal as shown in Fig. 2(a). If the target region is homogeneous or low textured whether moving or not, the difference between D_0 and D_1 becomes low due to their region property.

2.2   Vector-Based Zero Comparison

The vector-based zero comparison method proposed in [9] is developed by extending the original zero comparison method in order to adapt motion vectors including global motion. Figure 3 shows an example: a stationary line is falsely

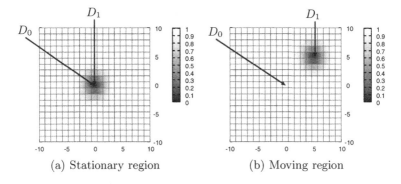

(a) Stationary region   (b) Moving region

Fig. 2. Correlation map examples

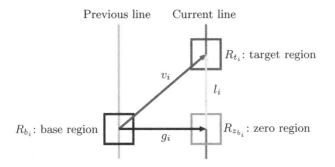

Fig. 3. Abstract of vector-based zero comparison under pseudo motion of a stationary line caused by global motion due to camera movement

moving to right by a global motion due to camera movement. In this case, an MPEG encoder outputs v_i that connects a current frame region R_{t_i} and its reference frame region R_{b_i}. In the vector-based zero comparison method, the target region to be examined its movement is R_{t_i}, and R_{b_i} corresponding to the region R_o in the original zero comparison method becomes a base region for calculating distances D_0 and D_1. The motion vector v_i is decomposed into a local motion vector l_i and its global motion vector g_i by the global motion estimation process. If the norm of l_i exceeds zero or a predefined threshold value, it can be assumed that R_{t_i} is moving despite the whole line is not moving. The global motion shifts R_{b_i} from its current frame position to $R_{z_{b_i}}$ position that corresponds to R_n in the original method. Therefore, we calculate D_0 and D_1 as follows:

$$D_0 = d(R_{b_i}, R_{z_{b_i}}), \tag{1}$$
$$D_1 = d(R_{b_i}, R_{t_i}). \tag{2}$$

If $D_0 - D_1 \leq t_z$, we can conclude that the target region R_{t_i} is stationary, homogeneous or low-textured. When $D_0 - D_1 > t_z$, we can conclude that R_{t_i} has local motion. By applying this vector-based zero comparison method to

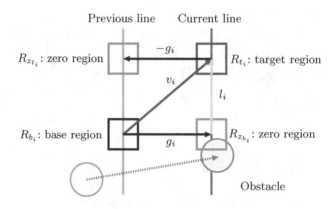

Fig. 4. Abstract of bi-directional zero comparison under pseudo motion of a stationary line caused by global motion due to camera movement; the zero region corresponding to the base region is affected by obstacle

$\widehat{\text{LMV}}_n$, we can identify and eliminate the motion vectors occurring in stationary, homogeneous, and low textured regions.

3   Bi-directional Vector-Based Zero Comparison

In the original vector-based zero comparison, three regions R_{t_i}, R_{b_i}, and $R_{z_{b_i}}$ are used to determine whether the target region is stationary or moving. The bi-directional method is a simple expansion of the original zero comparison method by addition of one examined region $R_{z_{t_i}}$ in the previous frame and calculation of the distance

$$D_0' = d(R_{t_i}, R_{z_{t_i}}). \tag{3}$$

Here, $R_{z_{t_i}}$ is a backward shifted region from the R_{t_i} position by the global motion g_i. The effectiveness of this expansion can be explained using the example shown in Fig. 4. In this figure, we assume that an obstacle overlaps with the region $R_{z_{b_i}}$. The distance $D_0 = d(R_{b_i}, R_{z_{b_i}})$ increases due to the presence of the obstacle, and we incorrectly conclude that the target region has local motion because $D_0 - D_1 > t_z$. However, the distance $D_0' = d(R_{t_i}, R_{z_{t_i}})$, which is the zero region corresponding to R_{t_i} in backward direction, decreases; hence, we can correctly conclude that the target region is stationary because $D_0' - D_1 \leq t_z$.

We also propose an early termination strategy for zero comparison. R_{b_i} and R_{t_i}, which are directly connected by v_i, constitute the best match region pair. The distance between R_{b_i} and R_{t_i} becomes smaller than that between other regions; this implies that $D_1 \leq D_0$ and $D_1 \leq D_0'$. Therefore, D_0 and D_0' should be calculated before D_1 because we can conclude that the target region doesn't have local motion when $D_0 \leq t_z$ or $D_0' \leq t_z$.

Bi-directional vector-based zero comparison algorithm:

Give the tentative local motion vectors $\{v_i\}_{i=1\ldots m}$ and their previously estimated global motion vector $\{g_i\}_{i=1\ldots m}$ as inputs. Apply the following steps to each vector v_i that connects the target region R_{t_i} in the current frame and the base region R_{b_i} in the previous frame:

1. Calculate the distance $D_0 = d(R_{b_i}, R_{z_{b_i}})$.
2. If $D_0 \leq t$, the motion vector v_i is stationary; terminate the process.
3. Calculate the distance $D_1 = d(R_{t_i}, R_{b_i})$.
4. If $D_0 - D_1 \leq t$, v_i is stationary; terminate the process.
5. Calculate the distance $D'_0 = d(R_{t_i}, R_{z_{t_i}})$.
6. If $D'_0 \leq t$, v_i is stationary; terminate the process.
7. If $D'_0 - D_1 \leq t$, v_i is stationary; else v_i has local motion.

Steps 1, 2, 3, and 4 analyze motion in forward direction, whereas Steps 5, 6 and 7 analyze motion in the backward direction. Steps 1, 3, and 5 are calculate the distance between regions. Steps 2, 4, and 6 denote termination of the process for each vector.

4   Experiments

In this section, we show the experimental results obtained by the proposed method; we used an actual video recording of three walkers acquired by an MPEG-4 camera with pan, tilt, zoom-in, and zoom-out movements. The video data contains 716 frames; each frame encoded by the MPEG-4 simple profile has a resolution of 640×480 pixels and frame rate of 30 fps. We used C/C++ environments for system implementation: gcc 4.4.1 on Linux 2.6.31-22-generic with FFmpeg SVN-r13022 and OpenCV 1.0 libraries. FFmpeg is an audio/video codec library, and OpenCV is a computer vision library that we used to display video sequences and matrix operations. All experiments were performed on a laptop computer with a 1.66 GHz Intel Core Duo CPU L2400 and 1.0 GB RAM.

Figure 5 shows the processed motion vectors drawn on the original frame. In this figure, all the lines show the tentative local motion vectors $\widehat{LMV}_n$ obtained after global motion estimation; the red lines denote the noisy motion vectors NMV_n eliminated by the proposed elimination process, and the yellow lines denote the local motion vectors LMV_n. Figure 5(e) shows incorrect elimination from the failure of global motion estimation due to suddenly movements of the camera, which occurred between Frames #332 and #333. In following frame #334, global motion estimation was captured by global motion estimation, and the elimination results were improved as shown in Fig. 5(f). From the results presented above, we can visually confirm that the proposed method effectively eliminated the noisy motion vectors occurring in the background.

Table 1 lists the processing details for each tentative local motion vectors. In this table, "Forward" denotes the same processing as the original method; whereas "Backward" denotes the additional processing of the proposed method.

(a) Frame #102   (b) Frame #265   (c) Frame #322

(d) Frame #332   (e) Frame #333   (f) Frame #334

(g) Frame #400   (h) Frame #472   (i) Frame #600

Fig. 5. Elimination result examples

Table 1. Processing details

	Forward		Backward		
	Step 2	Step 4	Step 6	Step 7	Moving
Calls	51,745	45,400	39,100	36,456	–
Outputs	6,345	6,300	2,644	4,108	32,348
Output ratio to the function calls	0.123	0.122	0.051	0.079	0.625
	0.245		0.130		
Output ratio to the stationary outputs	0.327	0.325	0.136	0.212	–
	0.652		0.348		

For 13.0% of the total function calls, the backward processing determined that target region is stationary. This implies that backward processing reduced the noisy motion vectors that could not be examined by the original method. For 17.4% of the total function calls, early termination occurred. Thus, the early termination strategy was effective. The duration of the video data was 24 s.

Table 2. Performance of moving region detection

	Original	Bi-Directional
Recall	0.833 (0.132)	0.814 (0.144)
Precision	0.661 (0.133)	0.715 (0.121)
F-measure	0.728 (0.120)	0.753 (0.115)

The processing time for elimination was measured as 2.5 s by the gettimeofday function.

We evaluated the performance of the proposed elimination method through motion region detection. The detection results were obtained by the method [12] that was adapted to evaluate the performance of the original zero comparison based elimination method. We manually generated the ground truth and evaluated the detection results on the basis of three performance indices: recall, precision, and F-measure. Table 2 shows the performance of the moving region detection results using stable local motion vectors obtained by the original zero-comparison method and the proposed bi-directional zero comparison method. As seen in the table, the recall performance was slightly worse because of the excess elimination of motion vectors occurring in the homogeneous regions inside moving objects (see the eliminated motion vectors on the back of the walking person in Fig. 5(c)). However, the proposed method outperformed the original method in terms of precision and F-measure.

5   Conclusion

In this paper, we have proposed a method to eliminate noisy motion vectors by using bi-directional zero-comparison. The original vector-based zero comparison is useful to determine whether the target region is stationary or moving without any prior knowledge or assumption. We improved upon the original method by introducing a bi-directional mechanism and an effective algorithm to implement it. We demonstrated the elimination process in details and the effectiveness of the proposed method through a moving region detection application. We also confirmed that the proposed method is a potential alternative to the original elimination method. Furthermore, we believe that the proposed method for noisy motion vector elimination can be one of the fundamental pre-processing frameworks for MPEG video analysis using motion vectors. In the future, we plan to enhance the performance of the proposed method by introducing concepts of mathematical morphology in order to recover the incorrectly eliminated motion vectors of moving object regions.

Acknowledgement

This work was supported by KAKENHI (21700100).

References

1. Lie, W.N., Chen, R.L.: Tracking moving objects in mpeg-compressed videos. In: IEEE International Conference on Multimedia and Expo, ICME 2001, August 22-25, pp. 965–968 (2001)
2. Khan, J.I., Guo, Z., Oh, W.: Motion based object tracking in mpeg-2 stream for perceptual region discriminating rate transcoding. In: MULTIMEDIA 2001: Proceedings of the ninth ACM international conference on Multimedia, pp. 572–576. ACM, New York (2001)
3. Eng, H.L., Ma, K.K.: Motion trajectory extraction based on macroblock motion vectors for video indexing. In: Proceedings of International Conference on Image Processing, ICIP 1999, vol. 3, pp. 284–288 (1999)
4. Achanta, R., Kankanhalli, M., Mulhem, P.: Compressed domain object tracking for automatic indexing of objects in mpeg home video. In: Proceedings of IEEE International Conference on Multimedia and Expo, ICME 2002, vol. 2, pp. 61–64 (2002)
5. Favalli, L., Mecocci, A., Moschetti, F.: Object tracking for retrieval applications in mpeg-2. IEEE Transactions on Circuits and Systems for Video Technology 10, 427–432 (2000)
6. Colace, F., De Santo, M., Molinara, M., Percannella, G.: Noisy motion vectors removal for reliable camera parameters estimation in mpeg coded videos. In: Proceedings of International Conference on Information Technology: Research and Education, ITRE 2003, pp. 568–572 (2003)
7. Hesseler, W., Eickeler, S.: Mpeg-2 compressed-domain algorithms . EURASIP J. Appl. Signal Process 2006, 186–186 (2006)
8. Eng, H.L., Ma, K.K.: Motion trajectory extraction based on macroblock motion vectors for video indexing. In: Proceedings of International Conference on Image Processing, ICIP 1999, vol. 3, pp. 284–288 (1999)
9. Yokoyama, T., Ota, S., Watanabe, T.: Noisy mpeg motion vector reduction for motion analysis. In: AVSS 2009: Proceedings of the 2009 Sixth IEEE International Conference on Advanced Video and Signal Based Surveillance, pp. 274–279 (2009)
10. Su, Y., Sun, M.T., Hsu, V.: Global motion estimation from coarsely sampled motion vector field and the applications. IEEE Trans. Circuits Syst. Video Techn. 15, 232–242 (2005)
11. Morita, T.: Motion detection and tracking based on local correlation matching. Transactions of the Institute of Electronics, Information and Communication Engineers, D-II J84-D-II 299–309 (2001) (in Japanese)
12. Iwasaki, T., Yokoyama, T., Watanabe, T., Koga, H.: Motion object detection and tracking using mpeg motion vectors in the compressed domain. IEICE Transactions on Information and Systems 91, 1592–1603 (2008) (in Japanese)

Spatio-Temporal Optimization for Foreground/Background Segmentation*

Tobias Feldmann

Karlsruhe Institute of Technology (KIT), Germany
feldmann@kit.edu

Abstract. We introduce a procedure for calibrated multi camera setups in which observed persons within a realistic and, thus, difficult surrounding are determined as foreground in image sequences via a fully automatic purely data driven segmentation.

In order to gain an optimal separation of fore- and background for each frame in terms of Expectation Maximization (EM), an algorithm is proposed which utilizes a combination of geometrical constraints of the scene and, additionally, temporal constraints for a optimization over the entire sequence to estimate the background. This background information is then used to determine accurate silhouettes of the foreground.

We demonstrate the effectiveness of our approach based on a qualitative data analysis and compare it to other state of the art approaches.

1 Introduction

Image based analysis of movements of gymnasts during their exercises at World Championships is a challenging goal for many applications, e.g. automatic judgement of the performances, objective comparisons between rivals, evaluation of training improvement, and visualization possiblilities during television broadcast. Markerless pure image based pose and movement analysis is highly desireable in the above mentioned scenarios. Unfortunately, in realistic scenarios, the image data of the gymnasts is affected by a lot of surrounding noise, e.g. movement of spectators and judges and changing lighting conditions which do not appear in laboratory like environments. Additionally, the recorded sequences are usually short and the distracting background in the recordings is often not exactly known a priori because of a possible permanently changing environment with potential occlusions and other disturbances. Thus, algorithms have to be developed which can cope with the above mentioned influences by identifying and removing the perturbances to allow for precise results in later steps.

Markerless video based human motion capture is an extensively studied area of interest. Many approaches try to reconstruct the 3d pose of the observed humans with monocular [1], stereo [2], multi view [3] or multi view stereo setups [4]. Different approaches are used to segment humans in the image data. One

---

* This work was partially supported by a grant from the *Ministry of Science, Research and the Arts of Baden-Württemberg.*

idea is to separate fore- and background by differencing over time [3,5]. Another idea is to use the color coherence of the surrounding with [6] or without an explicit human model [7].

In case of image differencing, the fore- and background is usually modeled by Codebooks [8] or statistically with e.g. Gaussian Mixture Modells (GMM) [9] and EM update schemes, where the discrimination between fore- and background depends on temporal thresholds [10]. This can be extended by the integration of topological knowledge via graphcuts to estimate nonstationary backgrounds as presented in [11,12]. In [13] a method has been proposed which exploits geometrical constraints in terms of disparity maps of stereo setups to cope e.g. with lighting influences in surveillance tasks. To overcome problems with non-moving foreground objects and moving backgrounds [14] proposed a method which replaces temporal thresholds by utilizing the geometric constraints of multi camera setups for foreground/background adaption and probabilistic segmentation. When using multi camera setups, the information of all cameras is usually utilized to create a probabilistic 3d reconstruction [15,16,17] for further analysis of the surface and pose of observed objects or humans.

2   Contribution

The main focus of this paper is on segmentation of gymnasts in short image sequences in realistic, often cluttered and uncontrolled environments with partially moving backgrounds using a multi camera setup. Although usual foreground/background segmentation algorithms often produce artifacts due to disturbances of the surrounding, Feldmann et al. showed in [14] that by using geometric constraints this approach leads to proper segmentation results.

The approach in [14] needs empty background images of the scene during intialization. Perfect empty background images are not available in many cases in environments where gymnasts usually perform their exercises. Thus, our approach is motivated by the idea of spatio-temporal optimization to examine if it is possible to determine per frame optimal fore- and background models by examining all frames of given video sequences back and forth in time. We show that this is possible given a) a roughly matching background image, e.g. created by the approach in [11], b) an image sequence including the foreground and c) geometric constraints regarding to the reconstruction area (cf. Fig. 1).

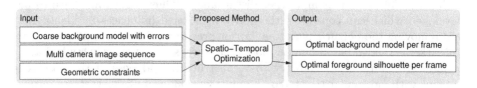

Fig. 1. Input data, proposed optimization method and output data

We demonstrate that our approach leads to improved silhouettes in which on the one hand wrongly classified foreground will be removed, and on the other hand erroneous silhouette holes have a chance to get closed by learning better differentiation models.

3   Spatio-temporal Optimization

Spatio-temporal optimization for foreground segmentation is based on an idea of two nested loops. The idea of the first loop is: First, foreground in image data is used to generate a 3d reconstruction; Second, the 3d reconstruction leads the approach to identify foreground in other images. While operating the first loop, knowledge about the fore- and background can be accumulated. If the input data at the beginning does not fit well, this first loop alone gives only mediocre results. Now the second closed loop comes into play: Roll the first loop back and forth in time to aquire knowledge of the past and the future and try to optimize the solution for each frame and all camera views. In contrary to the common use of temporal constraints, where time is used to detect changes and label it as foreground, in our approach time is additionally used to describe the context of the background over time.

Several building blocks are needed for the inner loop. First of all, models for the representation of fore- and background (cf. section 3.1) are necessary. Furthermore, a strategy is needed to update these models. Hence, a link has to be found, to geometrically distinguish foreground and background (cf. section 3.2). Finally, temporal coherence has to be used for the outer loop, to integrate all other building blocks in a reasonable manner (cf. section 3.3).

3.1   Fore- and Background Models

We introduce the random variable $\mathcal{F} \in \{0, 1\}$ to decide, whether a pixel at a given time t is fore- ($\mathcal{F} = 1$) or background ($\mathcal{F} = 0$). Given a color value c, the probability distributions $p(c|\mathcal{F} = 1)$ and $p(c|\mathcal{F} = 0)$ are used to model fore- and background color distributions which are used to infer the conditional probability $P(\mathcal{F} = 1|c)$ that an observation of the color c belongs to the foreground.

The foreground segment of the image sequence is modeled by a per frame Gaussian Mixture Model (GMM) with the gaussian density function $\eta(c, \mu, \Sigma)$ where c is a color value and μ^k and Σ^k are mean and variance of the kth component of the mixture (cf. eq. 1).

$$p(c|\mathcal{F} = 1) = (1 - P_{\mathrm{NF}}) \underbrace{\sum_{k=1}^{K} \omega^k \eta(c, \mu^k, \Sigma^k)}_{\text{learned fg GMM}} + P_{\mathrm{NF}} \underbrace{\mathcal{U}(c)}_{\text{unknown fg}} . \tag{1}$$

Additionally, a weight ω^k for each component is introduced. Assuming, that the pixels of the foreground segment are known, the k-*means* algorithm is used to partition the foreground colors and to derive mean, variance and weight of the components. Unless the foreground may change very fast due to new foreground objects, and the color distribution is, thus, unknown, two new variables will be introduced. 1. The probability of new foreground: P_{NF} which couples already learned and new unknown foreground. 2. The distribution of new colors, which are distributed uniformly: $\mathcal{U}(c)$. This foreground model is updated continuously by clustering the foreground colors during consecutive frames of the sequence.

To model the background we again use a GMMs with a gaussian density function $\eta(c, \mu, \Sigma)$, where $\mu_t^{k,d}$ and $\Sigma_t^{k,d}$ are mean and variance of the dth color channel with $d \in \{1, 2, 3\}$, c_t^d is the current color and k is the gaussian component of the mixture to model the color distribution of the image background. Each component has an additional weight w_t^k (cf. eq. 2). Since we have a static camera setup, we model the color distribution (in contrast to the foreground) of each pixel position as a separate GMM.

The model is extended by a directly integrated shadow and highlight model to be independent from illumination changes, where $\mathcal{S}_t \in \{0, 1\}$ models the change of the illumination (cf. eq. 2).

The background color model and the highlight/shadow model are used to compose the complete background model. Both parts of the model are coupled by the probability of shadow P_{S} which we set to $P_{\text{S}} = \frac{1}{2}$. The complete background model is, hence, defined by

$$p(c_t|\mathcal{F}_t = 0) = (1 - P_{\text{S}}) \underbrace{\sum_{k=1}^{K} w_t^k \eta(c_t, \mu_t^k, \Sigma_t^k)}_{\text{learned bg GMM}} + P_{\text{S}} \underbrace{\sum_{k=1}^{K} w_t^k p(c_t|\mathcal{S}_t^k = 1)}_{\text{shadow/highlight model}} . \quad (2)$$

The highlight/shadow model is built analogously to the background color model. The weightings of the color model are reused for the components of the shadow model. The definition of highlights and shadows is performed in the YUV colorspace. The luminance ratio λ is calculated in the Y channel by $\lambda = \frac{Y_t}{Y_B} = \frac{c_t^1}{\mu_t^{k,1}}$ and then using simple thresholds of the luminance ratio with $\tau_S < 1$ for shadows and $\tau_H > 1$ for highlights. The density of a certain component k of the highlight/shadow model is then defined by

$$p(c_t|\mathcal{S}_t^k = 1) = \begin{cases} \frac{1}{(\tau_H - \tau_S)\mu_t^{k,1}} \prod_{d=2,3} \eta(c_t^d, \mu'^{k,d}_t, \Sigma_t^{k,d}) & \text{if } \tau_S \leq \lambda_t^k \leq \tau_H \\ 0 & \text{else} \end{cases} , \quad (3)$$

where $\frac{1}{(\tau_H - \tau_S)\mu_t^{k,1}}$ is a scale factor to achieve the density's integration to be 1.

The background model needs to be updated in case of changes. Assuming that we are able to identify regions which are definitely not foreground and differ from the background model (cf. section 3.2), the GMMs of the associated locations have to be updated. The update process is done continuously over time by utilizing the online Expectation Maximization (EM) approach from [10].

3.2   3d Reconstruction by Probabilistic Fusion

As stated in section 3.1, a method is needed to identify the foreground in camera images. We think methods exploiting the strong prior of geometric coherence should be used in case of multiple views. Thus, we use the already estimated fore- and background models and infer a probabilistic voxel reconstruction analogue to [16]. This means that the space between all cameras is discretized into a 3d voxel space. For each voxel $\mathcal{V} \in \{0,1\}$, the probability of occupation is infered by using a shape from silhouette approach based on the probabilities of fore- and background in each camera image. The probabilistic approach helps to overcome problems caused by misclassified pixels in certain camera images. The causal chain of voxel reconstruction is given by $\mathcal{V} \to \mathcal{F}_n \to \mathcal{C}_n$ where $\mathcal{F}_n$ is the random variable of foreground in nth view and $\mathcal{C}_n$ the random variable of a pixel's color in nth view. The causal chain implies that the voxels have to be projected into the images for voxel reconstruction in order to determine whether they are occupied by foreground or not. For simplification of the following steps, we assume that a voxel projects to exactly one pixel.

To model errors, which might apear during reconstruction, additional probabilities are introduced for the probabilistic reconstruction. The probability, whether a voxel is occupied or not is initially set to $P(\mathcal{V}) = \frac{1}{2}$. Furthermore, *three* erroneous cases are considered.

First case: The voxel should be occupied, but is erroneously not. This probability of a *detection failure* P_{DF} is caused by an erronously classified background in camera n, i.e. $\mathcal{V}$: $P(\mathcal{F}_n = 0 | \mathcal{V} = 1)$. *Second case:* The voxel should not be occupied, but is erroneously occupied. This probability of a *false alarm* P_{FA} is caused by an erronously classified foreground in camera n, i.e. $\mathcal{V}$: $P(\mathcal{F}_n = 1 | \mathcal{V} = 0)$. *Third case:* The voxel should not be occupied, but is erroneously occupied because another voxel on the same line of sight is occupied. This is defined as the probability P_O of an *obstruction*.

The conditional probability of foreground of an unoccupied voxel is, hence, $\mathcal{V}$: $P(\mathcal{F}_n = 1 | \mathcal{V} = 0) = P_O(1 - P_{DF}) + (1 - P_O)P_{FA}$ and of background of an unoccupied voxel it is $\mathcal{V}$: $P(\mathcal{F}_n = 0 | \mathcal{V} = 0) = 1 - [P_O(1 - P_{DF}) + (1 - P_O)P_{FA}]$.

The voxel occupation can now be marginalized (c.f. [14] for more details) over the unknown variables $\mathcal{F}_n$ by observing the colors $c_1, \ldots, c_N$ at the corresponding pixels in the images of the cameras $1, \ldots, N$ by calculating

$$P(\mathcal{V} = 1 | c_1, \ldots, c_N) =$$

$$\frac{\prod_{n=1}^{N} \sum_{f \in \{0,1\}} P(\mathcal{F}_n = f | \mathcal{V} = 1) p(c_n | \mathcal{F}_n = f)}{\sum_{v \in \{0,1\}} \prod_{n=1}^{N} \sum_{f \in \{0,1\}} P(\mathcal{F}_n = f | \mathcal{V} = v) p(c_n | \mathcal{F}_n = f)}. \tag{4}$$

The resulting voxel reconstruction is then used to identify fore- and background segments in the camera images in the next model update step (c.f. 3.1).

3.3 Integrating Iterations

The proposed method needs a rough initialization of the background model. Hence, an image with a somewhat similar content is used for initialization. The foreground model is initially distributed equally.

After the initialization phase, the iteration phase starts. The idea is derived from a global bundle adjustment. Other than in a bundle adjustment, the iterations in this approach are bound to the constraints predefined by the continuity of time. Hence the optimization can not be done by skipping in between different time steps, but only by running along the image sequence (forward and backward). Each eradication of one direction is defined as to be one iteration step. The following five consecutive steps are executed in each iteration as long as additional images in this direction are available:

1. Use the current foreground and background model and the current images to calculate a 3d reconstruction by probabilistic fusion (cf. subsection 3.2).
2. Project the probabilistic reconstruction into each camera image to generate foreground masks.
3. Use the thresholded, inverted foreground mask to update the background (cf. subsections 3.1-3.1) at positions far from the foreground.
4. Use the current foreground and background model and the current images to calculate a probabilistic segmentation.
5. Use a logical AND operation on the segmentation result and the probabilistic projection to generate a mask to learn the current foreground in the foreground model (cf. subsection 3.1).

The iterations are repeated until the number of pixels in each camera image reach a stable state, i.e. the absolute difference of the number of foreground pixels of two consecutive iterations is smaller than a given epsilon. We define a state to be stable if 99.99% of the pixels do not change any more between two consecutive steps.

4 Evaluation

In a first step, we evaluated the algorithm qualitatively on a series of 200 sequences recorded in a gym. A gymnast is performing exercises, e.g. back handsprings along the floor mat, while others inside the same room take care of the recordings. We used a static calibrated camera system with seven VGA cameras which were set up in a circle like configuration. We present the results on the images of one randomly chosen recording with 204 images per camera. We show the first frame of the camera 4 with the most disturbances to demonstrate the effectiveness of our algorithm.

In Fig. 2 the gymnast can be seen in the foreground of the scene. Additionally, the image for background initialization is shown which was used to generate the initial background model. Please note that the background in the first two images differs in many details because the background image was taken after the

Fig. 2. Images of camera 4, from left to right: 1. First image of scene with gymnast (foreground), 2. Input background image recorded earlier, 3. Automatically generated most probable background

Fig. 3. First image of test sequence, from left to right: 1. Difference between input fore- and background images (binary threshold), 2. Proposed algorithm after one iteration without spatio-temporal optimization (gray values correspond to infered probabilities), 3. Result of proposed algorithm with spatio-temporal optimization after two iterations

recordings. However, these errors do not compromise our algorithm. To show this, we present the automatically generated most probable background generated automatically by our algorithm after two iterations as the third image of Fig. 2. Please note the shadows on the floor, which are mostly probable in the first frame due to the temporal constraints and the shadow model of our approach.

The results of the foreground segmentation are presented in Fig. 3. The left-most image shows a simple binarized difference between the first and second image of Fig. 2. All changes between the input images now become obvious. In the second image, we present the results of our algorithm, after one itera-tion. It is visible that many artifacts still appear even though the result is much better than the simple differencing. The third image finally shows the results of the proposed algorithm. It is obvious that the gymnast has been identified very precisely as foreground and all previous artifacts have been removed. The second and third images contain gray values which correspond directly to the probability of the pixel to be foreground scaled between $0 \ldots 255$.

Holes in the silhouette of the foreground (Fig. 3, left) could be closed in many cases with our approach. The data driven hole closing is possible, because by identifing the foreground utilizing the geometrical and temporal constraints, the color distributions of fore- and background could be learned more precisely over time. Unfortunately, there are still locations where the background looks exactly like the foreground. In these cases, our purely data driven approach has no knowledge about higher level information, like connected components etc. and thus, has no chance to close such holes (even though this could be done in an additional post processing step). We decided to present the results of the second

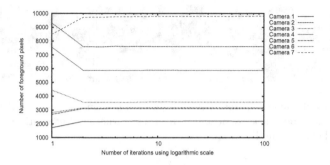

Fig. 4. On the gymnast sequence, the proposed algorithm obtains a stable state after two iterations (sharp bend at iteration 2)

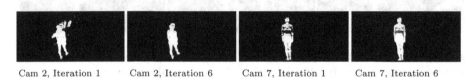

Cam 2, Iteration 1   Cam 2, Iteration 6   Cam 7, Iteration 1   Cam 7, Iteration 6

Fig. 5. Example silhouettes from camera 2 and 7 at first and sixth iteration to reason decreasing and increasing foreground pixels during iterations (cf. Fig. 4). Camera 2: Erroneous foreground gets removed. Camera 7: Silhouette holes are getting closed.

iteration firsthand, because we found the algorithm to reach a stable state very fast after two iterations in all seven cameras as it can be seen in Fig. 4.

Pixels with a probability higher than 75% have been counted as foreground. Two cases of optimizations can be seen in the plot. The first case is a drastical reduction of false positives pixels e.g. in camera 2 (cf. Fig. 5, left two images) due to background changes which our method removes almost completely within 2 iterations. The contrary case is the removal of false negatives, e.g. in camera 7 (cf. Fig. 5, two images on the right), where due to ambiguities a lot of foreground pixels have been erroneously labeled as background initially. After only a few iterations, the proposed method is able to learn better models and is hence able to identify the foreground more correctly.

In a second evaluation we compared our algorithm with the approaches of [8] and [10]. The results of random frames of a randomly selected camera 3 can be seen in Fig 6. It is obvious that the proposed algorithm outperforms the other approaches regarding the foreground silhouette extraction in difficult scenarios. The algorithm of [8] is not able to adopt over time in a meaningful manner. The algorithm of [10] is able to cope with the intially wrong backgrounds but also learns the correct foreground into the background model. Thus, in contrast to the proposed approach, only fast motions can be segmented as expected and motions of persons in the background result in additional unintentional foreground segments.

Summing up we were able to show, that the presented approach works reliable in realistic, difficult scenarios and is able to greatly enhance the segmentation

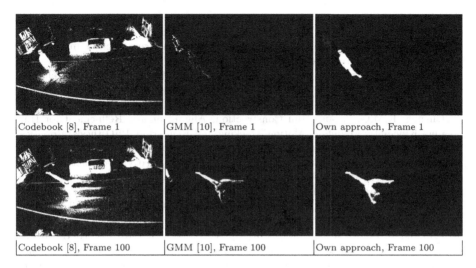

Fig. 6. Top row: Results of frame 1. Bottom row: Results of frame 100. From left to right: Results of the algorithm of [8], [10] and the results of our own approach.

results over time. Compared to state of the art approaches, we found that only our algorithm gains satisfying results in such realistic but difficult scenarios. For scientific purposes and better verifiability, we provide two demo sequences and an implementation of the algorithm at `http://human.informatik.kit.edu`

5   Conclusion

We presented a fully automatic approach to segment fore- and background in recorded video sequences by using a calibrated environments instead of artificial laboratory environments. In contrast to other approaches, our method is completely data driven and does neither make assumptions about homogenous backgrounds which are often not the case, nor does it need an explicit model of humans, no matter if these are shape models or kinematic and/or volumetric models. Hence, our approach could also be used for silhouette extraction in similar scenarios with objects inside a camera volume where no explicit model of the observed object is available.

We combined spatial constraints defined by the camera setup and temporal constraints in our approach. We were able to show that better background models could be learned with our approach and that we can find a stable state very fast after only a few iterations as well as it leads to locally optimal results in the sense of EM.

We tested the algorithm on over 200 sequences and found it to produce very good results in foreground segmentation as shown in an exemplary qualitative data analysis. We found the proposed approach to significantly improve the segmentation results. Additionally, the presented approach gains better results as state of the art algorithms in realistic scenarios. We think, that our approach should be, thus, prefered in multi camera setups recording realistic video footage.

References

1. Grauman, K., Shakhnarovich, G., Darrell, T.: A bayesian approach to image-based visual hull reconstruction. In: CVPR 2003, vol. 1, pp. 187–194 (2003)
2. Gordon, G., Darrell, T., Harville, M., Woodfill, J.: Background estimation and removal based on range and color. In: CVPR 1999, vol. 2, pp. 459–464. IEEE Computer Society, Los Alamitos (1999)
3. Cheung, K.M.G.: Visual Hull Construction, Alignment and Refinement for Human Kinematic Modeling, Motion Tracking and Rendering. PhD thesis, Robotics Institute, Carnegie Mellon University, Pittsburgh, Pennsylvania 15213 (2003)
4. Vogiatzis, G., Torr, P.H.S., Cipolla, R.: Multi-view stereo via volumetric graph-cuts. In: CVPR 2005, vol. 2, pp. 391–398 (2005)
5. Mikić, I.: Human Body Model Acquisition and Tracking using Multi-Camera Voxel Data. Electrical and computer engineering (image and signal processing) University of California, San Diego (2002)
6. Rosenhahn, B., Kersting, U.G., Smith, A.W., Gurney, J., Brox, T., Klette, R.: A system for marker-less human motion estimation. In: Kropatsch, W.G., Sablatnig, R., Hanbury, A. (eds.) DAGM 2005. LNCS, vol. 3663, pp. 230–237. Springer, Heidelberg (2005)
7. Lee, W., Woo, W., Boyer, E.: Identifying foreground from multiple images. In: Yagi, Y., Kang, S.B., Kweon, I.S., Zha, H. (eds.) ACCV 2007, Part II. LNCS, vol. 4844, pp. 580–589. Springer, Heidelberg (2007)
8. Kim, K., Chalidabhongse, T., Harwood, D., Davis, L.: Background modeling and subtraction by codebook construction. In: ICIP 2004, vol. 5, pp. 3061–3064 (2004)
9. Stauffer, C., Grimson, W.: Adaptive background mixture models for real-time tracking. In: CVPR 1999, vol. 2, pp. 2246–2252. IEEE Computer Society, Los Alamitos (1999)
10. Kaewtrakulpong, P., Bowden, R.: An improved adaptive background mixture model for realtime tracking with shadow detection. In: Proc. 2nd European Workshop on Advanced Video Based Surveillance Systems. Kluwer Academic Publishers, Dordrecht (2001)
11. Russell, D., Gong, S.: Minimum cuts of a time-varying background. In: BMVC 2006, vol. 2, pp. 809–818 (2006)
12. Russell, D.M., Gong, S.G.: Segmenting highly textured nonstationary background. In: BMVC 2007, (2007)
13. Lim, S.N., Mittal, A., Davis, L.S., Paragios, N.: Fast illumination-invariant background subtraction using two views: Error analysis, sensor placement and applications. In: CVPR 2005, pp. 1071–1078. IEEE Computer Society, Washington, DC, USA (2005)
14. Feldmann, T., Dießelberg, L., Wörner, A.: Adaptive foreground/background segmentation using multiview silhouette fusion. In: Denzler, J., Notni, G., Süße, H. (eds.) DAGM. LNCS, vol. 5748, pp. 522–531. Springer, Heidelberg (2009)
15. Broadhurst, A., Drummond, T.W., Cipolla, R.: A probabilistic framework for space carving. In: ICCV 2001, pp. 388–393 (2001)
16. Franco, J.S., Boyer, E.: Fusion of multi-view silhouette cues using a space occupancy grid. Technical Report 5551, INRIA (2005)
17. Kolev, K., Brox, T., Cremers, D.: Robust variational segmentation of 3d objects from multiple views. In: Franke, K., Müller, K.-R., Nickolay, B., Schäfer, R. (eds.) DAGM 2006. LNCS, vol. 4174, pp. 688–697. Springer, Heidelberg (2006)

Error Decreasing of Background Subtraction Process by Modeling the Foreground

Christophe Gabard[1], Laurent Lucat[1], Catherine Achard[2],
C. Guillot[1], and Patrick Sayd[1]

[1] CEA LIST, Vision and Content Engineering Laboratory, Point Courrier 94,
Gif-sur-Yvette, F-91191 France
[2] UPMC Univ Paris 06, Institute of Intelligent Systems and Robotics (ISIR),
4, place Jussieu; 75252 Paris Cedex 05, France

Abstract. Background subtraction is often one of the first tasks involved in video surveillance applications. Classical methods use a statistical background model and compute a distance between each part (pixel
or bloc) of the current frame and the model to detect moving targets. Segmentation is then obtained by thresholding this distance. This commonly
used approach suffers from two main drawbacks. First, the segmentation
is blinded done, without considering the foreground appearance. Secondly, threshold value is often empirically specified, according to visual
quality evaluation; it means both that the value is scene-dependant and
that its setting is not automated using objective criterion.

In order to address these drawbacks, we introduce in this article a foreground model to improve the segmentation process. Several segmentation
strategies are proposed, and theoretically as well as experimentally compared. Thanks to theoretical error estimation, an optimal segmentation
threshold can be deduced to control segmentation behaviour like hold an
especially targeted false alarm rate. This approach improves segmentation results in video surveillance applications, in some difficult situations
as non-stationary background.

1 Introduction

In many computer vision systems, moving object detection is often a required
task before higher level processes like event understanding. Background (here
after noted BG) removal techniques are then used to obtained a first moving
object extraction before adding temporal consistency, e.g. involving a tracking
step. According to [1], the commonly used BG removal processes are composed of
three main steps: an optional pre-processing task, a BG modeling and updating
step, and a FG (Foreground) detection.

A large number of BG representations are proposed in the literature, as recently summarized in [1,2]. Beside non-recursive techniques which involve a component selection from a temporal buffer (frame differencing, average, median,
...), some statistical algorithms are proposed, aiming to model the probabilistic
data distribution. Wren *et al.* [3] describe each BG pixel by a single Gaussian

R. Koch et al. (Eds.): ACCV 2010 Workshops, Part I, LNCS 6468, pp. 123–132, 2011.

distribution. This modelisation, well adapted to static BG, does not handle dynamic BG nor illumination changes. In [4], Stauffer and Grimson used a mixture of K Gaussian distributions, updated by a recursive version of the EM algorithm. When data can not be handled by Gaussian distribution, non-parametric BG modeling can be considered. Elgammal *et al.* [5] propose using Kernel density estimators. This method involves high computational cost and needs to specify parameters (mainly kernel distribution). When BG is correctly modeled, the next main step consists in segmenting images and detecting moving object. This is usually processed by thresholding a data probability, a specific distance to BG model, or an error to a predictive model. This threshold, often set empirically, is specific to each scene.

In this work, a FG model is considered in addition to common BG model. This combined information is jointly used in the segmentation formulation in order to improve segmentation results. This allows to specify an automatic threshold according to statistic rates (true positive rate, accuracy, precision, ...)

This paper is organized as follows: in the next Section, some recent related works are presented on both FG modeling associated to Bayesian criterion and automatic threshold computation. Section 3 details the proposed approach and, in particular, a mathematical computation of some statistical rates with application to Gaussian modelisation. In this Gaussian case, two segmentation strategies are presented and compared. The following Section presents an extension of the proposed approach to a more generic FG/BG data modeling, as well as another possible use of the error analysis. The proposed algorithms are evaluated in Sections 5 and 6.1, based on, respectively, synthetic and real sequences. Finally, we conclude on the proposed method and present promising extensions.

2 Related Works

Although the most largely used approaches for BG subtraction remains the sole use of a BG model, some recent papers consider both FG and BG models. For example Elgammal *et al* [5] use a KDE-based blob model for each people in the sequence. In this work, FG models are not used in the segmentation process, but only in a posterior tracking phase. A FG model has also been introduced with Bayes criterion in a probabilistic classification framework [6,7,8,9,10]. In [6], the observation probability of each pixel according to BG/FG models is compared in order to set parameters according to misclassification rate. But, like in [9], a basic uniform *pdf*[1] is used instead of a real FG model. Consequently this segmentation is equivalent to the common thresholding process. In [8], a FG model is built thanks to a recursive spatial analysis of segmentation. It allows authors to obtain a FG probability that improves segmentation results. However, the FG probability is not linked to the true FG appearance but to a recursive segmentation analysis, limiting the potential enhancement. In [10] a Gaussian FG model is used and updated under low motion assumption, leading, for each new observation, to accurate probabilities. But, again, a threshold equivalent to

---

[1] Probability density function.

an *a priori* probability is necessary to take the decision and is manually tuned. A FG model is also introduced in [7] and a full Bayes criterion is formulated to deduce the posterior probabilities.

The use of a FG model seems to improve segmentation results by adding local appearance adjustment. However, the probabilistic segmentation formulation requires the knowledge of an prior probability that is often not available. The choice of arbitrary values is not highly suitable because of the intrinsic scene dependency. As this specification is often critical in terms of performances, some works, like [11,12,13], focus on an automatic threshold tuning according to misclassification. The main idea is to extract some parameters that directly control detection result quality. In [11], Gao and Coetzee propose a full modeling of the BG subtraction process (BG modeling, segmentation process, unknown post-processing, . . .) and analyze the error. According to a specific scene, this allows to plot theoretical ROC curves and choose the best parameters. But this method requires intensive computation and is a pure off-line learning approach, aiming to get a global threshold. This one is optimized according to a particular learning scene, and is not updated during the sequence to take into account scene changes. For satellite imagery detection, Smith and Annoni evaluate in [12] some distances between two Gaussian probabilistic distributions. The error analysis allows selecting the optimal threshold that leads to equal misdetection and false alarm rates. This approach is also inline used in [13] to tune threshold and ensure a given false-alarm rate. However the classification is processed without considering FG model and then important information is ignored.

Our objective in this paper is to simultaneously exploit all these information. We are motivated by the fact that FG model improves the segmentation process, especially if it is associated to an adaptive threshold setting, which can be specified thanks to the theoretical error estimation.

To evaluate the improvement of the proposed approach, a comparison with the well known Stauffer and Grimson methods [4] is proposed. That is equivalent to threshold the commonly used Mahalanobis distance where only BG model is available. For a Gaussian BG distribution, $\mathcal{N}(\mu_B, \sigma_B^2)$, the observed data x is classified as BG according to the threshold Th if:

$$\mu_B - Th * \sigma_B < x < \mu_B + Th * \sigma_B \tag{1}$$

3   The Proposed Approach

In this section, under the assumption that data follows a Gaussian distribution, some suitable characteristics are deduced. For simplicity purposes, we mainly focus this work on one dimensional data as for instance brightness or special gradient, but we expected that this study could be quite easily extended to N-dimensional models.

This method is based on a Bayesian decision criterion including both BG and FG models like in [6,9]. Then, x is classified as BG if :

$$P_B(x) > \beta . P_F(x) \tag{2}$$

where $\beta > 0$, $P_B(x)$ and $P_F(x)$ are the BG and FG observed data probabilities, respectively. If (2) is not satisfied, x is classified as FG.

BG subtraction, or BG/FG segmentation, can be actually formulated as a binary classification of the observation in FG or BG labels. To evaluate classification performance, some statistical measures are usually computed like false alarm rate or sensitivity.

Let $p_B(x)$ and $p_F(x)$ denote the *pdf* of the BG and FG data distribution respectively and $\Omega(T, p_B, p_F)$ the subspace of the parameter T. It is possible to define, from this information, some detection rates, as for example the true positive rate (*Sensitivity*) defined by the FG distribution part correctly classified:

$$TPR = \int_{\Omega(T,p_B,p_F)} p_F(x).dx \tag{3}$$

With the same statement, True Negative Rate (*specificity*), False Positive Rate and False Negative Rate can be computed:

$$TNR = \int_{\Omega(T,p_B,p_F)} p_B(x).dx \quad FPR = \int_{\overline{\Omega(T,p_B,p_F)}} p_B(x).dx$$
$$FNR = \int_{\overline{\Omega(T,p_B,p_F)}} p_F(x).dx \tag{4}$$

If FG and BG data distributions are available, all the error rates only depend on the classification parameter T. Then, for each BG/FG couple, this parameter can be locally determined according to a given detection rate. In a general case, the computation is quite expensive but this complexity can be significantly decreased with some modeling assumptions.

First we need to specify the data subspace where the segmentation process generates a BG classification $\Omega(\beta, p_B, p_F)$. Solving $P_B(x) > \beta.P_F(x)$ is equivalent to resolve the second order equation :

$$a.x^2 + b.x + c > 0 \quad \text{with} \quad \begin{cases} a = \sigma_B^2 - \sigma_F^2 \\ b = 2.\left(\mu_B.\sigma_F^2 - \mu_F.\sigma_B^2\right) \\ c = \mu_F^2.\sigma_B^2 - \mu_B^2.\sigma_F^2 - 2.\sigma_B^2.\sigma_F^2.\ln\left(\beta.\frac{\sigma_B}{\sigma_F}\right) \end{cases} \tag{5}$$

Solving this equation allows us to find $\Omega(\beta, p_B, p_F)$ and determine the TPR:

$$TPR = \begin{cases} if \ \sigma_B > \sigma_F & \begin{cases} F_F(x_1) - F_F(x_2) & if \ \beta > \beta_l \\ 0 & if \ \beta \le \beta_l \end{cases} \\ if \ \sigma_B < \sigma_F & \begin{cases} 1 + F_F(x_1) - F_F(x_2) & if \ 0 < \beta < \beta_l \\ 1 & if \ \beta \ge \beta_l \end{cases} \end{cases} \tag{6}$$

where $x_{\{1,2\}}$ denote the roots of (5) and $F_F(x)$ the repartition function of the Foreground Gaussian distribution :

$$x_{\{1,2\}} = \mu_F.\frac{\sigma_B^2}{\sigma_B^2 - \sigma_F^2} - \mu_B.\frac{\sigma_F^2}{\sigma_B^2 - \sigma_F^2}\{+,-\}\frac{\sigma_F.\sigma_B}{\sigma_B^2 - \sigma_F^2}.\sqrt{(\mu_F - \mu_B)^2 + 2.(\sigma_B^2 - \sigma_F^2).\ln(\beta\frac{\sigma_B}{\sigma_F})}$$
$$x_0 = \frac{\mu_F.\sigma_B^2 - \mu_B.\sigma_F^2}{\sigma_B^2 - \sigma_F^2}$$
$$\beta_l = \frac{\sigma_F}{\sigma_B}\exp\left(\frac{-(\mu_B - \mu_F)^2}{2.(\sigma_B^2 - \sigma_F^2)}\right)$$

$$\tag{7}$$

$$F_F(x) = \tfrac{1}{2}\left(1 + \mathrm{erf}\left(\tfrac{x-\mu_F}{\sqrt{2}.\sigma_f}\right)\right) \ \text{ with } \ \mathrm{erf}(x) = \tfrac{2}{\sqrt{\pi}} \int_0^x e^{-t^2} dt \qquad (8)$$

From this error analysis, two segmentation strategies are proposed which involve the BG/FG model mentioned above.

3.1 First Segmentation Strategy: Targeted TPR or TNR

In the first segmentation strategy the detection threshold is estimated for each image part (pixel, bloc) in order to satisfy a given error rate according to BG and FG models, for example finding the threshold β (equation 2) that targets a fixed true positive rate:

$$TPR = \gamma \Leftrightarrow \beta = \begin{array}{l} \arg_\beta \mathrm{solve}\,(\gamma - (0 + F_F(x_1) - F_F(x_2)) = 0) \ \ if \ \sigma_B > \sigma_F \\ \arg_\beta \mathrm{solve}\,(\gamma - (1 + F_F(x_1) - F_F(x_2)) = 0) \ \ if \ \sigma_B < \sigma_F \end{array} \quad (9)$$

Since formulation of the optimized threshold is not trivial, an optimization routine has been proposed. This is however not a major issue since the expression can be differentiated and then the solution is found with fast solving routine.

This strategy, that allows controlling a statistical rate, requires specifying a parameter (for example the TPR goal). This formulation is however preferable than the traditional ones (aiming to specify β) because the new specified parameter is scene independent.

3.2 Second Segmentation Strategy: Best TNR/TPR Ratio

The previous formulation can be easily extended to obtain a fully-automatic detection. For example, β can be set to minimize the distance to a perfect segmentation ($TNR = 1, TPR = 1$) in the ROC space:

$$\beta = \arg_\beta \min\left((1 - TNR(\beta))^2 + (1 - TPR(\beta))^2\right) \qquad (10)$$

This allows us to reach a fully automated segmentation strategy which is locally tuned to achieve the best compromise between TPR and TNR. With this new strategy, the TPR is then different for each pixel. This formulation can also be tuned with specific weight on TNR and TPR

4 Extension to More General Data Distribution

The presented approach can be easily extended to any kind of distribution using histograms. The problem is then reduced to the distance estimation between histograms that can be efficiently performed. Let $H_B[i]$ and $H_F[i]$ are the N-bins normalized histograms describing $p_B(x)$ and $p_F(x)$ respectively, and $n(x)$ the application allowing to select the histogram bin. The proposed segmentation strategy compares the corresponding bin of BG and FG histograms, according to a parametric factor:

$$x \in B \Leftrightarrow H_B[n(x)] > \beta * H_F[n(x)]$$ (11)

All the previously expressed rates can be easily computed as for example :

$$TNR(\beta) = \sum_{i=1}^{N} \mathbf{1}_{+}(H_B[i] - \beta * H_F[i]) * H_B[i]$$ (12)

with $\mathbf{1}_{+}(x) = 1$ when $x \geq 1$, and 0 otherwise. Then the optimization process becomes quite simple and efficient using histogram modelisation.

5 Validation on Synthetic Sequences

In order to determine which method provides the best results, their performances are evaluated using equivalent input conditions, for instance, reach a given TNR. General formal demonstration seems to be quite difficult to achieve, so, in order to get an overview of the comparative performances, we conducted intensive and exhaustive simulations. Results emphasize that, for a given TNR, the FNR of the proposed method is always smaller or equal to the FNR of the classical one. In Figure 1, the FNR difference, between the classical Stauffer and Grimson approach and the proposed one is plotted for a same $TNR = 0.9$ and different parameters of BG/FG models (In the Stauffer and Grimson formulation, specifying a TNR is close to select a common threshold $Th = \sqrt{2}.\mathbf{erf}^{-1}(\mathbf{TNR})$). To achieve this, known Gaussian BG and FG models are used. Then, for each couple of BG/FG models, some statistical rates are computed, like the TPR. First, we verify that the new strategy has always at least the same result than the first one. Then computations highlight that when the BG model has a higher variance than the FG model, the new segmentation strategy provides a significant gain.

The concept is then evaluated on a synthetic sequence. A BG image with different mean and noise values is generated. On the abscissa axis of the BG image, the intensity value ranges from 0 to 255 to cover all grey levels. Gaussian noise whose standard deviation increases from 1 to 70 (ordinate axis of the BG image) is added. A FG mobile square, moving in all the space is then included. It follows a Gaussian distribution with $\mu_F = 127$ and $\sigma_F = 25$. A TNR is set to 0.9 for both methods and long time statistics are estimated. Models are assumed to be known by the segmentation process. The added object is really

Table 1. Results of the first proposed method with a TNR to 0.9

Method	TNR	TPR	Accuracy	Precision
Traditional	0.90	0.548	0.724	0.846
Proposed method	0.90	0.686	0.793	0.873

present 50% of the time at each pixel. Statistical results presented in Table 1 show that the TNR is properly equal to 0.9, validating the proposed segmentation strategy. Furthermore, for all statistical measures, the new segmentation strategy improves results since the TPR is greater. Moreover, the gains locally obtained correspond to those simulated in Figure 1.

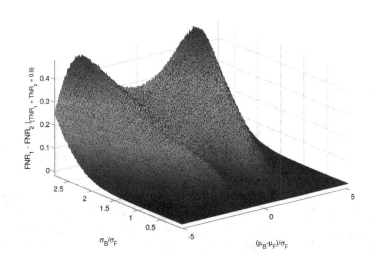

Fig. 1. $FNR_1 - FNR_2$ for $TNR_1 = TNR_2$. Left axis is the ratio of standard deviation: σ_B/σ_F. Right axis is relative mean difference: $(\mu_B - \mu_F)/\sigma_F$.

6   Results on Real Sequences

Algorithms have also been tested on real video. Two sequences are presented here. First the classical "bottle on water". This sequence is of great interest since BG is not stationary, thus increasing the segmentation complexity. Secondly a sequence, from [14], where people are moving inside a room. On these sequences, we evaluate the improvements of the proposed method compared to the Stauffer and Grimson one, thanks to the ground truth on the whole sequences.

6.1   Classification Improvement

Some results are presented in Figure 2 with a Gaussian modelisation of data distribution. For each sequence, a complete ROC curve is drawn and some segmented frames are presented. For the first segmentation strategy, ROC curves are drawn by setting different TPR. For the second method, weight associated to TNR and TPR are modified to get a complete ROC curve. The point with the same weight on TNR and TPR is drawn with a circle. TPR and FPR on ROC curve axis are obtained with complete ground truth on whole sequences. Experiment show that the two proposed segmentation strategies lead to better results than the traditional Stauffer and Grimson approach. For a same FPR,

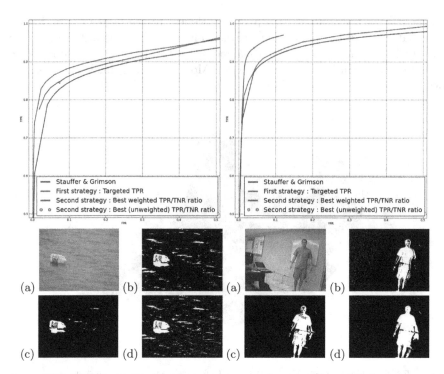

Fig. 2. Gaussian data distribution: Left: "Bottle on water", Right: sequence from [14]. Top : ROC curve, Bottom: Example of segmented frames. (a) Source frame, (b) Stauffer & Grimson, (c) Targeted TPR, (d) best unweighted ratio.

new segmentation strategy allows gains about 3-4%. It can be noticed that, in these experiments, a single Gaussian was used to model the target which is a rough approximation since it is composed of several colors. Nevertheless, new segmentation strategies improve the results, thus emphasizing the robustness of the proposed approach.

Same results are presented in Figure 3 for Histogram data distribution. They are close to the previous one with small improvement due to a better model estimation. For the same FPR, new segmentation strategy allows gains about 4-5%.

6.2   Targeting Rate

The proposed method is not only interesting to improve segmentation results. Another interesting setting is its ability to select threshold according to a target statistical rate (and not an arbitrary and scene-dependant threshold, as usually). For example the first proposed strategy allows to control the TPR of the detection. With Gaussian data modelisation, on all these experiments, for a targeted TPR Θ, resulting TPR are on the range $[\Theta - 1\%, \Theta + 15\%]$. With Histogram data modelisation, the target rates are on the range $[\Theta - 0.05\%, \Theta + 2\%]$. Gaussian

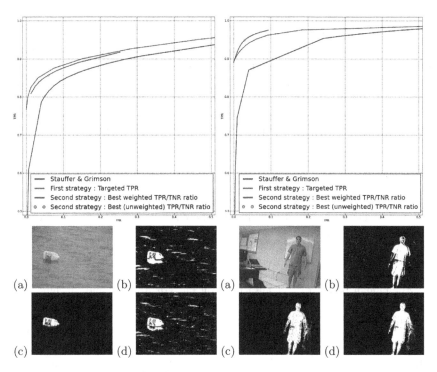

Fig. 3. Histogram data distribution: Left: "Bottle on water", Right: sequence from [14]. Top : ROC curve, Bottom: Example of segmented frames. (a) Source frame, (b) Stauffer & Grimson, (c) Targeted TPR, (d) best unweighted ratio.

models are less accurate than Histogram models because real data don't verify a Gaussian distribution.

7 Conclusion

In this article, we proposed a new BG subtraction algorithm that combines BG and FG models and performs an automatic threshold selection for each pixel of the image. The optimal threshold is estimated on each pixel in order to reach a targeted global detection error. In a first step, considering Gaussian distributions, we theoretically evaluated the improvement due to the combinated FG and BG approach, in comparison to the conventional pure BG approach. This demonstration has been validated on synthesis sequences simulating all possible configurations of Gaussian parameters, by varying gray level and noise in the image. Another benefit of the proposed method is that the threshold can be estimated according to several statistical rates (TPR, accuracy, precision,...), depending on the application, while traditional methods work only with the true negative rate or without any statistical links. For more complex sequences where the Gaussian assumption fails, the proposed method has been extended to more general distributions using histograms.

This new method has been validated on both synthetic and real sequences. Results emphasize that, when models are correctly estimated, gains are obtained using a FG model in the segmentation process. Moreover, these results have validated the automated threshold selection based on statistical error prediction.

Future works will be focused on extending the proposed approach to multidimensional image descriptors, which capture more robust information.

References

1. Elhabian, S., El-Sayed, K., Ahmed, S.: Moving object detection in spatial domain using background removal techniques-state-of-art. Recent Patents on Computer Science 1, 32–54 (2008)
2. Bouwmans, T., El Baf, F., Vachon, B.: Background modeling using mixture of gaussians for foreground detection-a survey (2008)
3. Wren, C., Azarbayejani, A., Darrell, T., Pentland, A.: Pfinder: Real-time tracking of the human body. IEEE Transactions on Pattern Analysis and Machine Intelligence 19, 780–785 (1997)
4. Stauffer, C., Grimson, W.: Adaptive background mixture models for real-time tracking. In: IEEE Computer Society Conference on Computer Vision and Pattern Recognition 1999, vol. 2 (1999)
5. Elgammal, A., Duraiswami, R., Harwood, D., Davis, L.S.: Background and foreground modeling using nonparametric kernel density estimation for visual surveillance. Proceedings of the IEEE 90, 1151–1163 (2002)
6. Withagen, P., Groen, F., Schutte, K.: Emswitch: A multi-hypothesis approach to em background modelling. Proceedings of the IEEE Advanced Concepts for Intelligent Vision Systems, ACIVS 1, 199–206 (2003)
7. Lindstrom, J., Lindgren, F., Astrom, K., Holst, J., Holst, U.: Background and foreground modeling using an online em algorithm. In: Proceedings of the Sixth IEEE International Workshop on Visual Surveillance (2006)
8. McHugh, J.M., Konrad, J., Saligrama, V., Jodoin, P.M.: Foreground-adaptive background subtraction. IEEE Signal Processing Letters 16, 390–393 (2009)
9. Landabaso, J., Pardas, M.: Cooperative background modelling using multiple cameras towards human detection in smart-rooms. In: In Proceedings of European Signal Processing Conference 2006 (2006)
10. Sheikh, Y., Shah, M.: Bayesian modeling of dynamic scenes for object detection. IEEE Transactions on Pattern Analysis and Machine Intelligence 27, 1778–1792 (2005)
11. Gao, X., Boult, T., Coetzee, F., Ramesh, V.: Error analysis of background adaption. Proc. IEEE Comput Soc Conf Comput Vision Pattern Recognit. 1, 503–510 (2000)
12. Smits, P., Annoni, A.: Toward specification-driven change detection. IEEE Transactions on Geoscience and Remote Sensing 38, 1484–1488 (2000)
13. Mittal, A., Paragios, N.: Motion-based background subtraction using adaptive kernel density estimation. In: Proceedings of the IEEE Computer Society Conference on Computer Vision and Pattern Recognition, CVPR 2004, vol. 2 (2004)
14. http://www.derektanderson.com/fallrecognition/datasets.html

Object Flow: Learning Object Displacement

Constantinos Lalos[1], Helmut Grabner[2], Luc Van Gool[2,3], and Theodora Varvarigou[1]

[1] School of Electrical & Computer Engineering, NTUA, Greece
lalosc@mail.ntua.gr, dora@telecom.ntua.gr
[2] Computer Vision Laboratory, ETH Zurich, Switzerland
{grabner,vangool}@vision.ee.ethz.ch
[3] ESAT-PSI/IBBT, K.U. Leuven, Belgium
luc.vangool@esat.kuleuven.be

Abstract. Modelling the dynamic behaviour of moving objects is one of the basic tasks in computer vision. In this paper, we introduce the *Object Flow*, for estimating both the displacement and the direction of an object-of-interest. Compared to the detection and tracking techniques, our approach obtains the object displacement directly similar to optical flow, while ignoring other irrelevant movements in the scene. Hence, *Object Flow* has the ability to continuously focus on a specific object and calculate its motion field. The resulting motion representation is useful for a variety of visual applications (e.g., scene description, object tracking, action recognition) and it cannot be directly obtained using the existing methods.

1 Introduction

Visual applications often rely on the information extracted by the moving objects inside a scene (e.g. cars, humans, machines etc.) These objects usually interact with other objects or the environment, thus modelling their dynamic behaviour is one of the basic tasks in computer vision.

The estimation of the motion field for the whole scene is typically performed using optical flow methods. Works on optical flow start in the early 80's [1,2] and target on establishing region correspondence between subsequent images[1]. Over the years a significant progress has been made, both in improving computational speed (e.g., [5]) and in dealing with large region displacements, (e.g., [6]). Recently, learning (e.g., [7]) and context [8] based approaches are taken into account in order to overcome the limitations of the classical optical flow formulation. In general, optical flow techniques have many possible applications, such as motion segmentation [9], object tracking [10], collection of statistics of the scene [11] or acting as human computer interface [12].

On the other hand, detection and tracking of individual objects (e.g., persons, cars) is important for several real-life applications including visual surveillance and automotive safety (e.g., [13]). In the last years, a lot of attention is paid to tracking by detection approaches (e.g., [14,15]). Hereby, a pre-trained object detector is applied on every frame

---

[1] Analogous to optical flow, where images are aligned based on a temporal adjacency, SIFT flow [3] can be exploited to match similar structures across different scenes. Recently it has been shown that parametric models such as affine motion, vignetting, and radial distortion can be modelled using the concept of Filter Flow [4].

R. Koch et al. (Eds.): ACCV 2010 Workshops, Part I, LNCS 6468, pp. 133–142, 2011.

(a) image pairs   (b) object det.   (c) optical flow   (d) *Object Flow*

Fig. 1. Video captured at Abbey road in London (a). An appearance based object detector (b) can localize the human, however gives no information about its movement. On the other hand, optical flow (c) approaches cannot distinguish between object movement and other irrelevant movements in the scene. Hence, we propose a motion representation (d), which has the ability to focus only on moving objects-of-interest in the scene.

and then the obtained detections are associated together across images. Furthermore, on-line learning methods (e.g., [16]) can be also used to dynamically update the object model and to cope with the variations of the object appearance. The data association problem is further simplified, since a discriminative model is trained in advance, for distinguishing the object appearance from its surrounding background. However, due to the self-learning strategy in place, such approaches might suffer from drifting (see [17] for a recent discussion).

Contribution. We introduce a method for obtaining the displacement of an object – the *Object Flow* – directly whereas other irrelevant movements inside the scene (e.g., other objects or moving background) are ignored (see Fig. 1). Since no on-line learning is performed during runtime, the results are stable (i.e. do not suffer from drifting). Hence, the resulting motion representation is useful for a variety of visual applications and cannot be directly obtained using the existing methods such as optical flow or object detection/tracking.

The remainder of the paper is organized as follows. Firstly, the idea of training a classifier on object displacement is described in detail at Section 2. Then the experimental results and the conclusions are elaborated at Sections 3 and 4 respectively.

2   *Object Flow*

In this section, we first formulate the learning problem for training a model (classifier), which is then used to deliver the *Object Flow*.

2.1   Problem Formulation and Learning

The goal of object detection is to find a required object in an image. Most state-of-the art methods (e.g., [18]), train a classifier with the appropriate samples in order to distinguish the object-of-interest from the background, i.e. formulate the task as a binary classification problem. In comparison with the typical object detection approaches, we consider the problem of detecting the displacement and the direction of a moving object locally, i.e. within a certain region Ω, (see Fig. 2). Within this region, pairs of patches

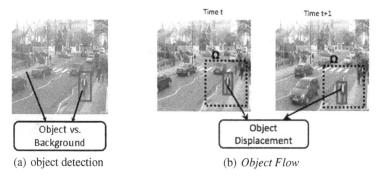

Time t   Time t+1

Object vs.
Background

Object
Displacement

(a) object detection   (b) *Object Flow*

Fig. 2. Object detection (a) is usually formulated as a binary classification problem distinguishing the object of interest from the background class. In contrast, *Object Flow* considers the problem of learning object displacement locally.

from different time intervals are classified. Nevertheless, the size of the search region Ω has its own role in the fulfilment of the object localization and direction estimation task. Especially in the case of abrupt motion or low frame rate video (see Sec. 3.1) an optimal estimation can be achieved by having a quite large search region. However, this size comes in contrast with the required computational complexity and might yield to ambiguities when more than one object are present in the scene.

Problem Formulation. We formulate the learning problem as a problem of learning a distance function, (see [19] for a recent overview). Our technique was inspired by the work of Hertz et al. [20], which learns a distance function for image retrieval by training a margin-based binary classifier (such as Support Vector Machines or Boosting methods) using pairs of samples. Positive pairs derived from the "same" class whereas negative pairs are samples drawn from two "different" classes. The learning problem is then formulated on the product space, i.e., $C : \mathcal{X} \times \mathcal{X} \rightarrow \mathcal{Y} = [-1, 1]$. Thus, the trained classifier $C(\mathbf{x}_1, \mathbf{x}_2)$ is supposed to give high confidence if the two samples $\mathbf{x}_1$ and $\mathbf{x}_2$ are similar, and low confidence otherwise.

Learning *Object Flow*. The overall learning approach is depicted at Fig. 3. For training a maximum margin classifier on object displacement in an off-line manner, a pool of appropriate samples has to be created. These samples should contain temporal information from pairs of images from the positive $\mathcal{X}^+$ and the negative $\mathcal{X}^-$ set respectively.

Positive set $\mathcal{X}^+$. A positive sample contains information about the way that object appearance transforms through time. Therefore, this sample is created by collecting two patches that derive from two different frames and contain the object under study i.e.

$$\mathcal{X}^+ = \{\langle \mathbf{x}_t^\star, \mathbf{x}_{t+1}^\star \rangle \mid \mathbf{x}_t^\star, \mathbf{x}_{t+1}^\star \in \Omega^{(i)} \text{ and correspond to an object}\} \tag{1}$$

The labelling of the object represented by the rectangles $\mathbf{x}^\star$ and $\mathbf{x}_{t+1}^\star$ can be accomplished using some reliable information, such as human labelling (ground truth), or the output from a high precision/recall detector or tracker.

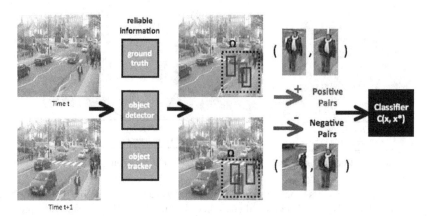

Fig. 3. Learning object's displacement is achieved by training a classifier with positive and negative labelled samples, which are locally extracted and contain temporal information

Negative set $\mathcal{X}^-$. The negative set is divided into two subsets, i.e. $\mathcal{X}^- = \mathcal{X}^-_{obj} \cup \mathcal{X}^-_{back}$. The first subset of negative samples contains the object in the current frame with a patch that contains a portion of it in a different frame i.e.

$$\mathcal{X}^-_{obj} = \{\langle \mathbf{x}^\star_t, \mathbf{x}^{(i)}_{t+1} \rangle \mid \mathbf{x}^\star_t, \mathbf{x}^{(i)}_{t+1} \in \Omega^{(i)} \text{ and } \mathbf{x}^\star_{t+1} \text{ correspond to an object}\} \quad (2)$$

These training samples assist the classifier to suppress local maxima around the real object region. On the other hand, the second subset of negative samples contains regions from the background. These samples are particularly useful, when dealing with difficult scenarios, since they can force the classifier to respond with low confidence values on empty regions i.e.

$$\mathcal{X}^-_{back} = \{\langle \mathbf{x}^{(i)}_t, \mathbf{x}^{(j)}_t \rangle \mid \mathbf{x}^{(i)}_t, \mathbf{x}^{(j)}_t \in \Omega^{(i)}\} \quad (3)$$

Examples of a positive and negative samples are depicted in Fig. 4.

Classifier. In this paper we use the approach of boosting for feature selection. A classifier can be trained in an off-line [21] or in an on-line [16] manner. In order to use pairs of images as an input, we follow the empirical tests of possible adaptations, proposed by Hertz et al. [20]. An approach for learning how the object appearance alters through

(a) pos. samples $\mathcal{X}^+$ (b) neg. samples $\mathcal{X}^-_{obj}$ (c) background samples $\mathcal{X}^-_{back}$

Fig. 4. Illustrative example of the typical training samples for training a classifier on *Object Flow*

time is the concatenation of the two patches. Another intuitive approach is by finding the absolute difference of the vectors representing the two patches. Our empirical tests indicate that this classifier works better with the first approach. As features we use the classical Haar-like features [21].

2.2 Flow Estimation

Object Flow is a vector field. In order to estimate it, for each point x, y in the image a local image patch $\mathbf{x}$ is extracted and the displacement magnitude $D_{obj}(\mathbf{x})$ and the angle $\phi_{obj}(\mathbf{x})$ can be calculated. More specifically, let $C(\mathbf{x}, \mathbf{x}')$ be the classifier response for a pair of patches, where $\mathbf{x}$ is a patch in the current image and $\mathbf{x}'$ is a patch belonging to the neighbourhood region of local patches Ω in the previous image. We define the displacement Δx and Δy of an object on the x and y directions respectively, as the weighted sum of distances within the local region Ω. More formally,

$$\begin{pmatrix} \Delta x_{obj}(\mathbf{x}) \\ \Delta y_{obj}(\mathbf{x}) \end{pmatrix} = \frac{1}{\sum_{\mathbf{x}' \in \Omega} C(\mathbf{x}, \mathbf{x}')} \sum_{\mathbf{x}' \in \Omega} C(\mathbf{x}, \mathbf{x}') \begin{pmatrix} dx \\ dy \end{pmatrix} \tag{4}$$

where, dx and dy are the x and y axis distances of the patch $\mathbf{x}$ from $\mathbf{x}'$. Based on this, magnitude and angle can be calculated as,

$$D_{obj}(\mathbf{x}) = \sqrt{\Delta x_{obj}(\mathbf{x})^2 + \Delta y_{obj}(\mathbf{x})^2}, \quad \phi_{obj}(\mathbf{x}) = \tan^{-1}\left(\frac{\Delta y_{obj}(\mathbf{x})}{\Delta x_{obj}(\mathbf{x})} \right). \tag{5}$$

In order to reduce outliers, local region displacements within the region Ω have to extend a significant positive classifier response i.e.,

$$\bar{C}_{obj}(\mathbf{x}) = \frac{1}{|\Omega|} \sum_{\mathbf{x}' \in \Omega} \hat{C}(\mathbf{x}, \mathbf{x}')^2, \quad \text{where} \quad \hat{C}(\mathbf{x}, \mathbf{x}') = \max(0, C(\mathbf{x}, \mathbf{x}')). \tag{6}$$

Summarizing, *Object Flow* is only reported, if the average classifier response is above some user defined threshold, which controls the sensitivity, i.e., $\bar{C}_{obj}(\mathbf{x}) > \theta$.

Illustrative Example. Fig. 5 depicts the *Object Flow* and the details for two specific regions. The trained classifier is evaluated on pairs of patches, using a reference patch at time t and patches from the corresponding local regions, $\Omega^{(1)}$ and $\Omega^{(2)}$, respectively at time $t+1$. We use a grid of overlapping patches of the same size, centred at the reference patch. As we can observe in the resulting 3-D plot for the region $\Omega^{(2)}$, high confidence values represent the regions, on which the object is likely to occur at time $t + 1$. On the other hand, the confidence values are very low for the region $\Omega^{(1)}$, since there are no objects inside. For visualizing the angle $\phi_{obj}(\mathbf{x})$ and the displacement $D_{obj}(\mathbf{x})$ (see Eq. (5)), we use the hue and saturation channel from HSV color space respectively.

3 Experimental Results

In this section we present qualitative and quantitative experimental results of the *Object Flow* on different objects and datasets, including walking pedestrians, faces and moving

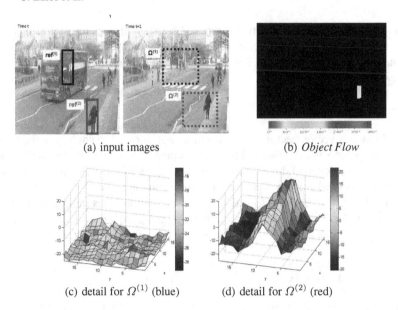

(a) input images   (b) *Object Flow*

(c) detail for $\Omega^{(1)}$ (blue)   (d) detail for $\Omega^{(2)}$ (red)

Fig. 5. Classifier responses for the regions $\Omega^{(1)}$ and $\Omega^{(2)}$ (a). Low classification responses are obtained if no object is present (c). In contrast, a clear peak, which shows the displacement of a particular object, is shown (d). The final *Object flow* field (b) is based on these local responses.

coffee mugs. The efficiency of our approach is demonstrated using difficult scenarios that involve low frame rate and motion blurring from a moving camera. Furthermore, we compare our results with common methods, such as an object detector, tracker and optical flow. The proposed motion representation can be used either in a static or in a moving camera configuration. Throughout the experiments we use a dense Grid that comprises of 81×81 overlapping and equally sized cells and we set a threshold $\theta = 0.35$ (see Sec. 2.2). All experiments are performed on a 2.67 GHz PC with 4 GB RAM.

3.1 *Object Flow* for Pedestrians

We captured a dataset from a public camera located on Abbey road in London[2], which consists of 49,000 frames. This dataset, obtained at a resolution of 384×284, is a challenging low frame rate scenario ($\sim$ 6 fps) that contains a complex background with various moving objects (e.g., cars). Therefore, we use a region Ω that comprises of 12×12 cells, since object motion is quite abrupt due to low frame rate. The first 40,000 frames of this dataset are used for collecting the appropriate training samples (see Sec. 2.1) and the remaining ones are used for evaluation. More specifically, the results described in this section can be produced using a single classification approach that is trained off-line using a pool of $|\mathcal{X}^+| \approx 2,000$ positive, $|\mathcal{X}_{obj}^-| \approx 15,000$ negative object samples and numerous negative samples $\mathcal{X}_{back}^-$ from the background.

---

[2] http://www.abbeyroad.com/webcam/, 2010/03/03.

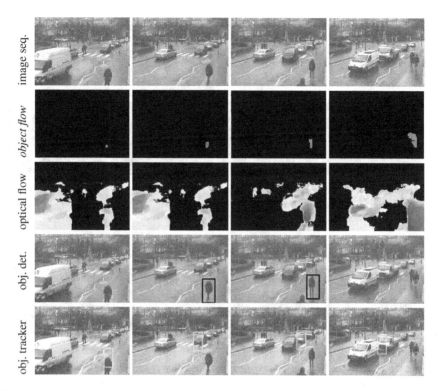

Fig. 6. In this experiment we present the benefits of *Object Flow*. Optical flow approaches disorient when similar objects are moving in the same/different direction with the object-of-interest (third row). In addition, human detection approaches do not have a constant detection rate (fourth row). Object tracking also suffers from drifting in complex environments (fifth row). *Object Flow* (second row) can simulate the motion field of a moving object correctly, by being able to focus only on the object under study.

We perform illustrative comparisons with optical flow, human detection and object tracking methods. We use the approach described at [5] to calculate optical flow, in order to evaluate its performance against the proposed *Object Flow* technique. For human detection and tracking we adopt the approaches described at [22] and [16] respectively. All the competing techniques are used without modifying any of the input parameters given in their original implementation.

As it can be observed at Fig. 6, our approach has a good performance in human localization. In addition, direction estimation for the moving objects-of-interest (second row) is the same with the one provided by the aforementioned optical flow approach, which focuses in all the moving objects in the scene (e.g. cars, third row). On the other hand, combining optical flow with an object detection approach (fourth row) may lead to possible pitfalls, since detection approaches do not have a constant detection rate, and thus have limited effectiveness in difficult environments. Similarly, tracking approaches (fifth row) disorientate on complex backgrounds, since objects of similar color or structure may appear inside the scene.

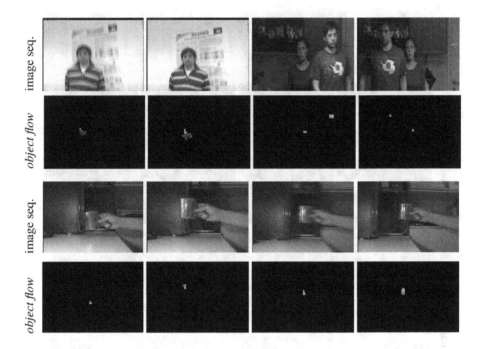

Fig. 7. In this experiment we train the classifier to deliver *Object Flow* for different object classes including faces and a coffee mug. The first and third row depict frames from different test sequences (camera movement, motion blur and multiple objects). The second and fourth row present the estimated *Object Flow*, respectively. (Video is available at the authors' web-page.)

3.2   *Object Flow* for Different Objects

The performance of *Object Flow* is also tested using two different object classes, i.e. faces and a specific mug. The algorithm is evaluated on scenarios that contain abrupt motion and on a moving camera configuration. In detail, we use three different video sequences that consist of $1,200$ frames, where $1,000$ frames are used for training the classifier and 200 frames for testing. Two sequences were captured from a moving indoor camera and contain a moving face and mug respectively. Another sequence was captured from a static indoor camera and contains two moving faces. These datasets were taken at 25 fps with a 704×576 resolution using an AXIS 213 PTZ camera.

We evaluate our approach for each patch $\mathbf{x}$ using a region Ω that comprises of 6×6 cells. For the mug and the face sequence the classifier was trained using a pool of $|\mathcal{X}^+| \approx 1,000$ positive, $|\mathcal{X}^-_{obj}| \approx 4,000$ negative object samples and numerous negative samples from the background. For creating face samples, an off-the-shelve face detection approach is adopted[3].

Illustrative results are depicted in Fig. 7. As it can been seen, the *Object Flow* has the ability to remain focused on the face and the mug even in cases of abrupt camera

---

[3] http://opencv.willowgarage.com/wiki/, 2010/04/28

motion (second and fourth row). Furthermore, the proposed method can deal with more than one objects-of-interest present at the same scene (see Fig. 7 first and second row, in third and fourth column).

3.3 Quantitative Comparison

We adopt the coffee mug dataset, which is a moving camera scenario and consists of 200 frames (see Sec. 3.2). On this sequence, ground truth is created by manually labelling the values for the angle and the displacement. For each frame we calculate the absolute error between the ground truth and the values provided by our approach. Since there is only one object-of-interest in the scene, we consider the angle $\phi_{obj}(\mathbf{x})$ and displacement magnitude $D_{obj}(\mathbf{x})$ of the patch $\mathbf{x}$, for which the classifier has the maximum response $\bar{C}_{obj}(\mathbf{x})$, according to Eq.(6).

For comparison we also implemented a simple baseline approach that combines object detection and optical flow. In that case, the displacement and angle are estimated by finding the average optical flow within the region of a detection (i.e. if a detection is present). Therefore, we first, train a classifier [21] using $1,000$ positive samples for the object and a negative set that contains numerous object-free samples from the background. The resulting detections are fused together by applying non-maximal suppression. Finally, Lucas-Kanade method [2] for optical flow estimation is adopted.

For all the frames in the sequence, we calculate the mean absolute displacement and angle error. More specifically, the average displacement error is decreased from 12 pixels for the baseline approach to 9 pixels using our approach. Similarly, the mean angle error is decreased from $75°$ to $62°$, respectively. The angle error seems to be quite large, which is quite reasonable, by taking into account that the object and camera change abruptly their direction in the chosen test sequence.

4   Conclusions

In this paper, we present the *Object Flow*, a method for estimating the displacement of an object-of-interest directly. Our approach is similar to optical flow, but it has the additional ability to ignore other irrelevant movements in the scene. This is achieved by training a classifier on the object displacement.

Experimental results demonstrate that the proposed approach achieves robust performance for different object classes, including pedestrians and faces. We are confident that *Object Flow* is useful for a variety of applications, such as object tracking or scene understanding. However, one current limitation is the computational complexity, which is going to be addressed in a future work.

Acknowledgments. This research was supported by the European Community Seventh Framework Programme under grant agreement no FP7-ICT-216465 SCOVIS.

References

1. Horn, B., Schunck, B.: Determining optical flow. Artificial Intelligence 17, 185–203 (1981)
2. Kanade, T., Lucas, B.: An iterative image registration technique with an application to stereo vision. In: Proc. Int. Joint Conf. on Artificial Intelligence, pp. 674–679 (1981)

3. Liu, C., Yuen, J., Torralba, A., Sivic, J., Freeman, W.: SIFT flow: Dense correspondence across different scenes. In: Forsyth, D., Torr, P., Zisserman, A. (eds.) ECCV 2008, Part III. LNCS, vol. 5304, pp. 28–42. Springer, Heidelberg (2008)
4. Seitz, S., Baker, S.: Filter flow. In: Proc. ICCV (2009)
5. Werlberger, M., Trobin, W., Pock, T., Wedel, A., Cremers, D., Bischof, H.: Anisotropic Huber-L1 optical flow. In: BMVC, London, UK (2009)
6. Brox, T., Bregler, C., Malik, J.: Large displacement optical flow. In: Proc. CVPR (2009)
7. Sun, D., Roth, S., Lewis, J., Black, M.: Learning optical flow. In: Proc. ECCV (2008)
8. Wu, Y., Fan, J.: Contextual flow. In: Proc. CVPR (2009)
9. Shi, J., Malik, J.: Motion segmentation and tracking using normalized cuts. In: Proc. ICCV (1998)
10. Odobez, J.-M., Daniel Gatica-Perez, S.B.: Embedding motion in model-based stochastic tracking. IEEE Transactions on Image Processing 15, 3515–3531 (2006)
11. Ali, S., Shah, M.: A lagrangian particle dynamics approach for crowd flow segmentation and stability analysis. In: Proc. CVPR (2007)
12. Santner, J., Werlberger, M., Mauthner, T., Paier, W., Bischof, H.: FlowGames. In: 1st Int. Workshop on CVCG in conjunction with CVPR (2010)
13. Stauffer, C., Grimson, W.: Adaptive background mixture models for real-time tracking. In: Proc. CVPR, vol. II, pp. 246–252 (1999)
14. Leibe, B., Schindler, K., Gool, L.V.: Coupled detection and trajectory estimation for multi-object tracking. In: Proc. ICCV (2007)
15. Breitenstein, M.D., Reichlin, F., Leibe, B., Koller-Meier, E., Gool, L.V.: Robust tracking-by-detection using a detector confidence particle filter. In: Proc. ICCV (2009)
16. Grabner, H., Bischof, H.: On-line boosting and vision. In: Proc. CVPR, vol. 1, pp. 260–267 (2006)
17. Stalder, S., Grabner, H., Gool, L.V.: Beyond semi-supervised tracking: Tracking should be as simple as detection, but not simpler than recognition. In: Proc. IEEE WS on On-line Learning for Computer Vision (2009)
18. Dalal, N., Triggs, B.: Histograms of oriented gradients for human detection. In: Proc. CVPR, vol. 1, pp. 886–893 (2005)
19. Yu, J., Amores, J., Sebe, N., Radeva, P., Tian, Q.: Distance learning for similarity estimation. IEEE Trans. on PAMI (2008)
20. Hertz, T., Bar-Hillel, A., Weinshall, D.: Learning distance functions for image retrieval. In: Proc. CVPR, vol. 2, pp. 570–577 (2004)
21. Viola, P., Jones, M.: Rapid object detection using a boosted cascade of simple features. In: Proc. CVPR, vol. I, pp. 511–518 (2001)
22. Prisacariu, V., Reid, I.: fasthog - a real-time gpu implementation of hog. Technical Report 2310/09 (Department of Engineering Science, Oxford University)

HOG-Based Descriptors on Rotation Invariant Human Detection

Panachit Kittipanya-ngam and Eng How Lung

Institute for Infocomme Research, 1 Fusionopolis Way,
21-01 Connexis (South Tower), Singapore 138632

Abstract. In the past decade, there have been many proposed techniques on human detection. Dalal and Triggs suggested Histogram of Oriented Gradient (HOG) features combined with a linear SVM to handle the task. Since then, there have been many variations of HOG-based detection introduced. They are, nevertheless, based on an assumption that the human must be in *upright* pose due to the limitation in geometrical variation. HOG-based human detections obviously fails in monitoring human activities in the daily life such as sleeping, lying down, falling, and squatting. This paper focuses on exploring various features based on HOG for rotation invariant human detection. The results show that square-shaped window can cover more poses but will cause a drop in performance. Moreover, some rotation-invariant techniques used in image retrieval outperform other techniques in human classification on *upright* pose and perform very well on various poses. This could help in neglecting the assumption of *upright* pose generally used.

1 Introduction

Because of the demand of smart surveillance system, the research on human detection has gained more attention. Not only is it a fundamental function required in most of surveillance system but also a challenging task in computer vision. In the past decade, there have been many proposed techniques on human detection. Enzweiler and Gavrila [1] review and decompose human detection into three stages: the generation of initial object hypotheses (ROI selection), verification (Classification) and temporal integration (Tracking). They also evaluate the state-of-the-art techniques in human detection: Haar wavelet-based AdaBoost cascade [2], Histogram of Oriented Gradient (HOG) features combined with a linear Suport Vector Machine (SVM) [3], neural network using local receptive fields [4] , and combined hierarchical shape matching and texture-based Neural Network using local receptive fields classification [5]. In this paper, we focus on studying characteristic of features passed to classifiers in the stage of classification.

In 2005, Dalal and Triggs [3] suggested to use HOG features combined with a linear SVM to handle the human body detection. Since then, there have been many variations of HOG-based detection introduced. They are, however, all based on a major assumption that the target human must be in *upright* pose.

R. Koch et al. (Eds.): ACCV 2010 Workshops, Part I, LNCS 6468, pp. 143–152, 2011.

It is mentioned in [3] that HOG descriptor is limited to a certain range of geometrical variation not bigger than the bin size. While the transition variation can be solved by scanning detection windows through whole image and scale variation can be managed by multi-scale methods, rotation variation is still in doubt. Therefore HOG-based body detection is limited to only such applications as detecting human in group photos, detecting and tracking human walking in the scene, and detecting human actions in *upright* pose [6]. This is why the HOG-based human detection fails in the task of monitoring human activities in the daily life such as sleeping, lying down, falling, and squatting.

This paper is focusing on exploring various HOG-based descriptors on rotation invariant human detection. Section 2 briefly explains HOG and discusses why its variations can not be invariant to rotation transformation. A review of rotation invariant features is described in Section 3. Finally, exploratory experiments on various HOG-based features for rotation invariant human detection are illustrated in Section 4 5 and 6 before they are discussed and concluded in section 7.

2   HOG: Histogram of Oriented Gradients

In 2005 [3], Navneet Dalal and Bill Triggs proposed a descriptor representing local object appearance and shape in an image, called Histogram of Oriented Gradient (HOG). The HOG descriptor is described by the distribution of edge directions in the histogram bins. The common implementation begins by dividing the detection window into small a square pixels area, called cells, and for each cell estimating a histogram of gradient directions for those pixels within the cell. The final descriptor is the combination of all histograms in the detection window. In [3], Dalal and Triggs suggest to use 64x128 pixel detection window and 8x8 pixel cell. For detecting human in an image, Support Vector Machine (SVM) is introduced to handle the task of human/non-human classification in each detection window by training SVM with human and non-human images. Since HOG was introduced in CVPR 2005, it has been widely used for detecting human, modified to obtain a better performance and extended to various applications [7,6,8,9]. Though there have been many variations of HOG human detectors proposed, those techniques assume the *upright* position of human because they can not handle the rotation variation. Here we suggest two possible explanations why HOG-based techniques are not rotation invariant: the features change a lot when image is rotated and the shape of detection window does not support the oriented version of object.

In [3], Dalal and Triggs also explain that they chose to define the detection window at 64 x 128 pixels, a rectangle shape because it includes more or less 16 pixels around the person from every side. They have shown that the size of border is important to the performance as it is expected to provide the right amount of context which can help in detection. They tried to reduce the size of pixels around the person from 16 to 8 and obtain the detection window of size 48 x 112 pixels. They have found that the performance of the detector with

48 x 112 detection window is 6% worse than detection window of size 64 x 128 pixels. They also tried to maintain the size of window at 64 x 128 pixels while increasing the person size in it and they have found that it also causes a similar drop of performance. Later, every work on HOG follows this idea of the shape and the size of detection window.

The *upright* rectangle shape of detection window is obviously a reason why HOG can not handle rotation transformation because this shape can not contain other poses of human inside especially rotated version of human such as sleeping, lying down and falling. In this paper, we suggest to use square-shaped detection window as the square window can contain more variations of human. However, we have to be awared that the area of context pixels of square shape will be more than that of *upright* rectangle. We studied how the bigger amount of context information would effect the performance in 5.

3 Review of Rotation Invariant Features

Though not many techniques tackling on orientation in human detection have been proposed, there have been many suggestions on rotation invariant features on other objects especially for the task of image retrieval. Mavandadi *et.al* [10] suggest to construct a rotation invariant features from magnitude and phase of Discrete Fourier Transform of polar-transformed image. Islam *et.al* [11] transform image using curvelet transform, an extended of 2-d ridgelet transform. Then the transformed features were aligned to the main dominant direction to generate the output which is invariant to the rotation of image. Jalil *et.al* [12] align the features by maximise the probability of the vectors obtained from Radon Transform(RT) and Discrete Differential Radon Transform(DDRT). Marimon and Ebrahimi [13] introduce Circular Normalised Euclidean Distance(CNED) to help in aligning the image orientation based on histogram of gradient orientation. Izadinia *et.al* [14] assume the boundary of object is clearly given and then using relative gradient angles and relative displacement vectors to construct a look up table (R-table) in Hough Transform, which is rotation-invariant. Arafat *et.al* [15] study the geometrical transformation invariance of several descriptors in the task of logo classification. Four features compared in this paper are Hu's Invariant Moment[16], Hu's moments of log-polar transformed image, Fourier transformation of log-polar transformed image and Gradient Location-Orientation Histogram (GLOH, a SIFT descriptor on log-polar coordinate). The similarity measure used in this paper is Euclidean Distance. Peng [17] applies Discrete Fourier Transformation (DFT) on HOG to reduce the circular shift in the image and measure the similarity by L1 metric distance equation.

Pinheiro introduces Edge Pixel Orientation Histogram (EPOH) in [18]. This technique divides each image into NxN subimages. On each subimage, HOG is applied to extract the distribution and then feature vector of each subimage is concatenated to construct the final feature. In this technique, the angle considered is within 0 and 180 degree. Therefore, the pixel of edge orientation outside this range will be counted in the bin of opposite angle and the normalised by the number of edge pixels in the subimage. EPOH is quite similar to HOG but they are different in the number of blocks and the way to construct histogram. Later in [18], they suggest to use Angular Orientation Partition Edge Descriptor (AOP) in image retrieval as it is invariant to rotation and translation. Given a centre point of image, AOP divides the gradient image into N angular sectors of the surrounding circle. In each angular division, the orientation of edge pixel is adjusted by using the angle of the radial axis as the reference to construct local angular orientation. Hence, the radial axis is the line drawn from centre point of the image to the centre point of the sector. Next, histogram is applied to each angular division to extract the local distribution of angular orientation before the feature vector of each angular sector is concatenated to construct 2-D vector $f(n_0, n_a)$ where n_0 is the bin of local angle and n_a is the angle of the radial axis for each sector. The final descriptor will be the absolute value of 1-D Fourier Transform of $f(n_0, n_a)$ relatively to the angular dimension n_a.

In this paper, we studied on five HOG-based techniques as follows:

HOGwoBlk: Histogram of Oriented Gradients without block division
HOGwoBlk-FFT: Amplitude of Fourier Coefficients of HOGwoBlk[17]
HOGwoBlk-FFTp: Phase of Fourier Coefficients of HOGwoBlk[17],
EPOH: Edge Pixel Orientation Histogram[18].
AOP: Angular Orientation Partition Edge Descriptor[18].

Hence, in our implementation, HOGwoBlk does not divide image in blocks and cell as in [3] because EPOH is considered a kind of HOG with divisions. This is for studying the effect of structural information on the performance. Additionally, on *upright* rectangle window, AOP was divided into 2 x 4 block divisions instead of angular divisions. However, the idea of angular sector and the idea of distribution of local orientation were still maintained by assigning the radial axis of each block to be the line drawn from centre point of the image to the centre point of the block. The rest of process remained the same as the original AOP in [18].

4   Various Features on Rotation Invariance

Here we assume that if the feature is rotation invariant, the extracted features of the target image and any of rotated versions should be very similar. In this section, the similarity between the target image and rotated versions of images is studied through various features. Here we selected six square images, 128 x 128 pixels, shown in the top line of figure 1. Five features mentioned in 3 were then extracted and used as reference features. Next, each image was rotated 15 degree counter clockwise and processed to obtain the features. In this section, each image was rotated for 24 times, 15 degree counter clockwise each time, to reach 360 degree. Finally, similarity values between reference images and rotated images on five different features were measured, recorded and plotted against the number of time they were rotated in figure 1. The similarity measurement used in this section is 1D correlation as this measure is similar to the way a linear SVM classifier constructs kernel matrix. In figure 1, the diagrams of original image and image subtracted foreground are very similar on both upright and lying down images while they are very much different from those of image subtracted background. Clearly, the background information dominate in the feature vectors. It is also noticed that HOGwoBlk-FFT and AOP can highly conserve the similarity over the change of orientation, EPOH is more sensitive than HOGwoBlk-FFT and AOP while HOGwoBlk and HOGwoBlk-FFTp can barely maintain the similarity when rotated. This explains and supports that only histogram of oriented gradients alone is not rotation invariant and why most of work assume *upright* pose but adding some strutural information like EPOH or AOP can improve the performance. Moreover, Figure 1 provides some hints that HOGwoBlk-FFT and AOP would make better features in human detection in the scenario of activities in daily life as they could handle the variation on rotation.

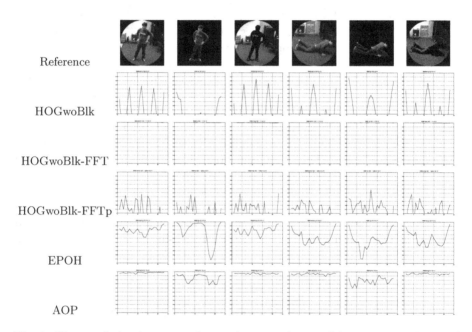

Fig. 1. The correlation between reference image and rotated images, over 24 step of 15 degree, on various HOG-based features

5 Effect of Shape of Detection Window on Presence/Absence of Human Classification

Mentioned previously, the square shape of detection window can cover more poses of human than *upright* rectangle. In this section, the comparison between *upright* rectangle and square shape of detection window is studied on a linear SVM classifier for the task of Presence/Absence of human classification. For each image, the image was cropped and resized into two difference shapes of image, 64x128 rectangle and 128x128 square. For positive images, the cropped area is based on centroid and boundary where annotations come with the database. For negative images, the windows were randomly cropped and resized. Next, five selected processors mentioned in previous section were applied to extract features from cropped images before the features were used to train and test on classification. Then, the performance of each processors on diffefent shapes of detection window was measured and shown in figure 2 via ROC curve and accuracy of classification in table 1. The database used in this section is called INRIA Person database [3]. The database includes two sets of *upright* photos, for training and for testing and each set is consist of positive images and negative images. Each positive image was flipped around vertical axis for increasing number of possible images. However, some images were discarded in this section because the square window, expanded from *upright* rectangle, covers the area outside the image where there is no pixel info. Therefore the number of images used for training is 2258 images in total, of which 1040 images are positive and 1218 images are negative while the number of images used for testing is 963 images, 510 positive and 453 negative. Some of cropped and resized images used in this section are shown in figure 2. In figure 2(a) and table 1, it is noticed that EPOH and AOP are nearly perfect in the classification when using *upright* rectangle detection window. When using square-shaped detection window, the performance in figure 2(b) and table 1 show that HOGwoBlk-FFT and HOGwoBlk-FFT phase are worse than others. The performance of HOGwoBlk, EPOH and AOP are overall so close to each other but AOP is significantly sensitive as it does not perform well on negative images. The results show that using square-shaped window can cause a drop in overall performance on HOG-based human classification. Probably, the increase of context information in square-shaped window adds more variations in the features and makes the task more difficult to classify for classifiers. Though AOP is slightly better than EPOH when using rectangle-shaped window, EPOH is overall better than AOP when using square-shaped window.

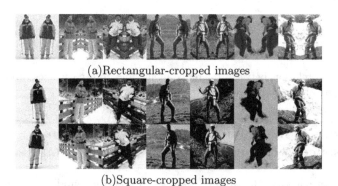

(a)Rectangular-cropped images

(b)Square-cropped images

Fig. 2. Examples of cropped and resized positive images from INRIA Person Database

Table 1. Percentage of correct classification(True Positive(TP) and True Negative(TN)

Type	Rectangle			Square		
	TP	TN	Total	TP	TN	Total
HOGwoBlk	87.84%	80.13%	84.22%	78.04%	76.16%	77.15%
HOGwoBlk-FFT	87.84%	71.74%	79.96%	73.53%	68.65%	71.24%
HOGwoBlk-FFTp	57.25%	79.47%	67.71%	37.45%	77.26%	56.18%
EPOH	98.24%	98.01%	98.13%	78.82%	**84.77%**	**81.62%**
AOP	**100%**	**100%**	**100%**	**82.35%**	68.21%	75.70%

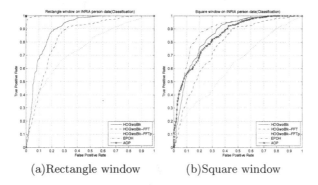

(a)Rectangle window   (b)Square window

Fig. 3. ROC of the classification on INRIA person data

6   Rotation Invariant Classification of the Presence of Human

In this section, five focused features will be tested on images of various poses of the activities in daily life, here called CVIU LAB. While INRIA person database was used to train the classifiers as in section 5, the set of various poses used for testing was recorded and annotated by the author, shown in figure 4. In thispart, each image was cropped and resized into square shape of image, 128x128 pixels. Positive images in the test set were sampled from a sequence of human doing a normal daily activities such as stretching arms, falling, lying down and squatting. The cropped area is based on centroid and boundary of foreground obtained from the background subtraction algorithm by Eng [19]. There are 118 images with horizontal flipped versions of them experimented, 236 images in total. For the 220 negative images, the windows were randomly cropped and resized from images without human. After cropped and resized, the feature vector of each selected technique was extracted and pass to pre-trained linear SVM classifiers to decide whether there is a human inside. Then, the performance of classification was measured, shown in figure 5 and discussed.

Figure 5 shows that AOP and EPOH can handle image of various poses when the HOGwoBlk is completely lost. The performance of features can be ranked

Fig. 4. Examples of cropped and resized images of various poses

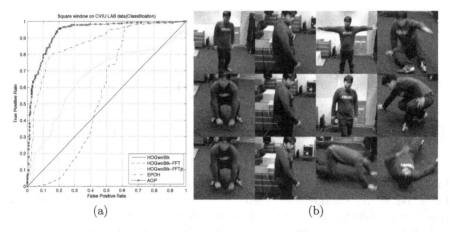

(a)   (b)

Fig. 5. ROC of the classification on CVIU LAB data

from the best as follows: AOP, EPOH, HOGwoBlk-FFT, HOGwoBlk-FFTp, and HOGwoBlk but HOGwoBlk-FFT is more sensitive in classifying human and non-human. This rank corresponds to the figure 1 showing that AOP, EPOH, and HOG-FFT can maintain similarity over rotation transformation and HOGwoBlk-FFTp and HOGwoBlk change much when image is rotated. Figure 5 displays the images which AOP missed in detecting human inside. It is still in doubt why the AOP can not detect human inside these image while it can handle the other similar images in the data set such as images in figure 4. Hence, all of images in figure 4 are the those AOP could handle.

7   Discussion and Conclusion

Hitogram of orientation of gradients alone can not be used on rotation invariant human detection because of the shape of windows and the significant change of feature over rotation variation. Here we suggest to use square shape of detection window and other kind of features to allow human detection to detect human in various poses. Though it causes a drop in performance.

Applying Fourier Transform relatively to the angular dimension to edge-orientation seems to be about to make edge-orientation features tolerant to rotation change when the similarity is measure by 1-D correlation.

Dividing image into subimages either in angular divisions or blocks can improve overall performance. Probably dividing allows features to include global structural information and the local distribution of edge orientation. This could be a reason why EPOH and AOP outperform other features in human classification.

AOP is the only feature in this study applyinh fouriere transform to reduce the effect of rotation and divided into subimages. This could be the reason why AOP outperform others in the human classification on both *upright* pose and various poses. Its performance convinces that features these characteristics could help neglecting the assumption of *upright* pose generally used in human detection. Though AOP looks nearly perfect to be used for human detection, AOP have higher rate in false positive than EPOH with square-shape window. Presumably, AOP is sensitive to context information in background.

References

1. Enzweiler, M., Gavrila, D.: Monocular pedestrian detection: Survey and experiments. IEEE Transactions on Pattern Analysis and Machine Intelligence 31, 2179–2195 (2009)
2. Viola, P., Jones, M.: Robust real-time face detection. In: Proceedings of the Eighth IEEE International Conference on Computer Vision ICCV 2001, vol. 2, p. 747 (2001)
3. Dalal, N., Triggs, B.: Histograms of oriented gradients for human detection. In: IEEE Computer Society Conference on Computer Vision and Pattern Recognition CVPR 2005, vol. 1, pp. 886–893 (2005)
4. Wohler, C., Anlauf, J.: An adaptable time-delay neural-network algorithm for image sequence analysis. IEEE Transactions on Neural Networks 10, 1531–1536 (1999)
5. Gavrila, D.M., Munder, S.: Multi-cue pedestrian detection and tracking from a moving vehicle. International Journal of Computer Vision 73, 41–59 (2007)
6. Ferrari, V., Marin-Jimenez, M., Zisserman, A.: Progressive search space reduction for human pose estimation. In: IEEE Conference on Computer Vision and Pattern Recognition CVPR 2008, pp. 1–8 (2008)
7. Lu, W.L., Little, J.: Simultaneous tracking and action recognition using the pca-hog descriptor. In: The 3rd Canadian Conference on Computer and Robot Vision 2006, p. 6 (2006)
8. Li, M., Zhang, Z., Huang, K., Tan, T.: Rapid and robust human detection and tracking based on omega-shape features. In: 16th IEEE International Conference on Image Processing (ICIP 2009), pp. 2545–2548 (2009)

9. Kaaniche, M., Bremond, F.: Tracking hog descriptors for gesture recognition. In: Sixth IEEE International Conference on Advanced Video and Signal Based Surveillance AVSS 2009, pp. 140–145 (2009)
10. Mavandadi, S., Aarabi, P., Plataniotis, K.: Fourier-based rotation invariant image features. In: 16th IEEE International Conference on Image Processing (ICIP 2009), pp. 2041–2044 (2009)
11. Islam, M., Zhang, D., Lu, G.: Rotation invariant curvelet features for texture image retrieval. In: IEEE International Conference on Multimedia and Expo, ICME 2009, pp. 562–565 (2009)
12. Jalil, A., Cheema, T., Manzar, A., Qureshi, I.: Rotation and gray-scale-invariant texture analysis using radon and differential radon transforms based hidden markov models. Image Processing, IET 4, 42–48 (2010)
13. Marimon, D., Ebrahimi, T.: Orientation histogram-based matching for region tracking. In: Eighth International Workshop on Image Analysis for Multimedia Interactive Services WIAMIS 2007, p. 8 (2007)
14. Izadinia, H., Sadeghi, F., Ebadzadeh, M.: Fuzzy generalized hough transform invariant to rotation and scale in noisy environment. In: IEEE International Conference on Fuzzy Systems FUZZ-IEEE 2009, pp. 153–158 (2009)
15. Arafat, S., Saleem, M., Hussain, S.: Comparative analysis of invariant schemes for logo classification. In: International Conference on Emerging Technologies ICET 2009, pp. 256–261 (2009)
16. Hu, M.K.: Visual pattern recognition by moment invariants. IRE Transactions on Information Theory 8, 179–187 (1962)
17. Peng, J., Yu, B., Wang, D.: Images similarity detection based on directional gradient angular histogram. In: Proceedings of 16th International Conference on Pattern Recognition, vol. 1, pp. 147–150 (2002)
18. Pinheiro, A.: Image descriptors based on the edge orientation. In: 4th International Workshop on Semantic Media Adaptation and Personalization SMAP 2009, pp. 73–78 (2009)
19. Eng, H.L., et al.: Novel region-based modeling for human detection within highly dynamic aquatic environment. In: Proceedings of the IEEE Computer Society Conference on Computer Vision and Pattern Recognition CVPR 2004, vol. 2, pp. II-390–II-397 (2004)

Fast and Accurate Pedestrian Detection Using a Cascade of Multiple Features

Alaa Leithy, Mohamed N. Moustafa, and Ayman Wahba

Computer Engineering, Faculty of Engineering, Ain Shams University, Cairo, Egypt
alaa.leithy@eng.asu.edu.eg, moustafa@ieee.org, ayman.wahba@eng.asu.edu.eg

Abstract. We propose a fast and accurate pedestrian detection frame-work based on cascaded classifiers with two complementary features. Our pipeline starts with a cascade of weak classifiers using Haar-like features followed by a linear SVM classifier relying on the Co-occurrence Histograms of Oriented Gradients (CoHOG). CoHOG descriptors have a strong classification capability but are extremely high dimensional. On the other hand, Haar features are computationally efficient but not highly discriminative for extremely varying texture and shape information such as pedestrians with different clothing and stances. Therefore, the combination of both classifiers enables fast and accurate pedestrian detection. Additionally, we propose reducing CoHOG descriptor dimensionality using Principle Component Analysis. The experimental results on the DaimlerChrysler benchmark dataset show that we can reach very close accuracy to the CoHOG-only classifier but in less than 1/1000 of its computational cost.

1 Introduction

Detecting pedestrians is a fundamental problem in image surveillance and analysis, it is essential in many applications such as automatic driver assistance. Extensive variety of postures, color and style of pedestrian clothing, illumination and weather conditions make this problem challenging. It can provide an initialization for human segmentation. More importantly, robust pedestrian identification and tracking are highly dependent on reliable detection and segmentation in each frame [8,6].

Recently, Watanabe et al. [13] reported one of the most accurate approaches for pedestrian detection using the "Co-occurrence histograms of oriented gradients (CoHOG)" feature descriptor. The CoHOG can precisely express local and global shape information at multiple scales since its building blocks have an extensive vocabulary. However, the CoHOG is an extremely high dimensional pattern descriptor, thus very computationally expensive and not suitable for real time applications.

On the other hand, Viola and Jones presented a fast and robust face detector using Haar-like features and AdaBoost [11] which has been applied to pedestrian detection [12] in the form of a static detector and also a dynamic one based on both motion and appearance information to detect a walking person.

R. Koch et al. (Eds.): ACCV 2010 Workshops, Part I, LNCS 6468, pp. 153–163, 2011.
© Springer-Verlag Berlin Heidelberg 2011

Although the Haar-like features are so simple and computationally efficient, it faces problems when representing data with high range of textural variations.

Therefore, our motivation was to get over the drawback of high dimensionality in CoHOG descriptor by reducing its computational cost and needed resources while preserving its accurate detection results.

In this paper, we propose a pipeline of classifiers for pedestrian detection with two contributions. The first contribution is combining the joint Haar-like features [9] and the CoHOG descriptors [13] to achieve a fast and accurate pedestrian detection system. This idea has been successfully reported in cat face detection [7]. So, our contribution here is introducing it to the pedestrian detection problem. Our second contribution is reducing the CoHOG descriptor dimensionality with Principle Component Analysis (PCA).

The rest of this paper is organized as follows: Section 2 surveys briefly the related work with focus on CoHOG descriptors and Joint Haar-like features with AdaBoost. Section 3 explains our proposed pedestrian detection approach. Section 4 shows experimental results on the DaimlerChrysler benchmark dataset, and we conclude in section 5.

2   Related Work

Many types of feature descriptors have been proposed for pedestrian detection. Viola and Jones proposed Haar-like features with a cascaded AdaBoost classifier [11] and extended them using appearance and motion features for pedestrian detection [12], which have improved detection accuracy. Dollar proposed a feature mining paradigm for image classification and used Haar wavelet with AdaBoost [3] to get better accuracy.

Recently, using gradient-orientation-based feature descriptors is a trend in pedestrian detection; Dalal et al. proposed "Histograms of Oriented Gradients (HOG)" in combination with SVM classifier [2] and also HOG is extended to motion feature descriptors with excellent detection ability. Watanabe et al. proposed a multiple-gradient-orientation-based feature descriptor "Co-occurrence histograms of oriented gradients (CoHOG)" [13]. Lin proposed a shape-based, hierarchical part-template matching approach to simultaneous human detection and segmentation [8] combining local part-based and global shape-template-based schemes. Yamauchi et al. presented a pedestrian detector based on co-occurrence of appearance and spatio-temporal features [10].

The rest of this section will briefly review studies that utilize the two main building blocks of our pedestrian detection framework, the CoHOG descriptors [13] and joint Haar-like features with AdaBoost [9].

2.1   CoHOG Descriptors

The CoHOG descriptor is based on a co-occurrence matrix which is constructed from a 2D histogram of pairs of gradient orientations [13]. The co-occurrence matrix expresses the distribution of the paired gradient orientations at a given offset over an image (see Fig. 1).

Offset(3,1)

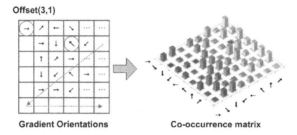

Gradient Orientations   Co-occurrence matrix

Fig. 1. Example of the co-occurrence matrix of gradient orientations with offset (3, 1)

The gradient orientations are extracted at each pixel by

$$\theta = \arctan\frac{v}{h} \tag{1}$$

where v and h are vertical and horizontal gradient respectively. The orientation labels are divided into eight orientation groups at 45 degree intervals. An input pattern is divided into small regions. The co-occurrence matrices are calculated for each region with various offsets (x, y) and the paired orientations (i, j) are voted into the corresponding component of the co-occurrence the matrix $C_{(x,y)}(i, j)$ [13].

The offsets are chosen within a 4 pixel radius and the number of the combinations is 30 after redundant offsets are removed. Therefore, 31 co-occurrence matrices are obtained after including a zero offset per each small region [13] and the dimensionality of the CoHOG descriptor is $(30 \times 64 + 8) \times N \times M$ for an input pattern divided into $N \times M$ small regions.

2.2   Joint Haar-Like Features with AdaBoost

Haar-like features have been successfully applied to many problems, e.g., face detection [11], and pedestrian detection [12]. Those features are usually arranged in a form of cascade of weak classifiers trained by AdaBoost.

A cascade of classifiers is like a decision tree where a classifier is trained at each stage to detect objects of interest (pedestrians) and reject a fraction of the non-pedestrian patterns [12] (see Fig. 3).

AdaBoost is a powerful machine learning algorithm that can learn a strong binary classifier $H(x)$ based on a linear combination of T weak classifiers, $h_t(x)$ by re-weighting the training samples:

$$H(x) = sign\left(\sum_{t=1}^{T} \alpha_t h_t(x)\right) \tag{2}$$

where α_t is the weight of the training data, and t is number of round.

The joint Haar-like feature is based on the co-occurrence of Haar-like features which are quantized to binary values [9]. The joint Haar-like feature is represented by a F-bit binary number which combines the binary variables computed

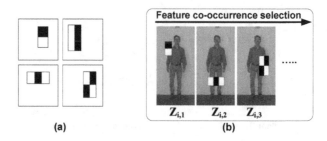

Fig. 2. Haar-like features [12]: (a) example of rectangle Haar-like features relative to the enclosing detection window. The sum of the pixels which lie within the white rectangles are subtracted from the sum of pixels in the black rectangles. (b) A joint Haar-like feature $h_i(x)$.

from F Haar-like features: $(Z_{t,1}, Z_{t,2}, \ldots, Z_{t,F})$ as shown in Fig. 2 (for more details see [11,12,9]).

3   The Proposed Pedestrian Detection System

Fig. 3 depicts our proposed pipeline of features and classifiers highlighting our two main contributions for pedestrian detection. The first contribution is to reduce the high dimensional CoHOG descriptors [13] with PCA. The second contribution is to combine the joint Haar-like features [9] and the CoHOG descriptors in a single cascade.

3.1   CoHOG Descriptors with PCA

In this paper, we divided each input pattern into 3×6 small regions to extract the CoHOG feature descriptor and chose the offsets within a 4 pixel radius. Therefore, the dimensionality of the CoHOG descriptor is $(30 \times 64 + 8) \times 3 \times 6 = 34,704$. Consequently, the CoHOG is an extremely high dimensional pattern descriptor precisely captures local shape information at multiple scales and has a strong classification capability. Therefore the CoHOG feature vectors which are strongly correlated could be reduced by conventional Principle Component Analysis (PCA) as shown in Fig. 4.

In the training process, the training samples are passed to the CoHOG descriptor block to calculate CoHOG features. Subsequently, we build the PCA

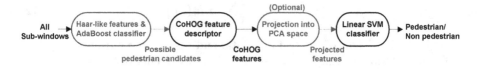

Fig. 3. Process flow of our pedestrian detection system. PCA block is optionally used.

Fig. 4. Process flow of CoHOG features descriptor and PCA

Fig. 5. Cascading distinct classifiers with two heterogeneous features. (a) Extract possible pedestrian candidates by simple joint Haar-like features. (b) Verify each candidate by high dimensional CoHOG descriptors.

space using b eigen vectors. After building the PCA space, the CoHOG features of training samples are projected into PCA space and then the projected Co-HOG features are used to train a linear classifier obtained by Support Vector Machines (SVM) [1], e.g., LIBLINEAR [4] or MATLAB SVM classifier [14].

In the testing process, all testing sub-windows are passed to CoHOG descriptor block and then CoHOG features are projected into PCA space. Finally, the simple linear SVM classifier is used to classify the projected CoHOG features whether they representing pedestrian or non-pedestrian.

3.2 Cascading Haar and CoHOG Descriptors

Our second contribution mainly consists of concatenating the CoHOG (and optionally the PCA compressed) classifier and the popular Haar features cascade. These two classifiers have complementary characteristics and give us the desirable outcomes. The first classifier consists of joint Haar-like feature with AdaBoost, which is used to extract possible pedestrian candidates from an input image as shown in Fig. 5. The second classifier uses the high dimensional (or optionally the PCA compressed) CoHOG descriptors with linear SVM classifier to verify and evaluate each detected candidate from the first classifier. This idea has been successfully reported in cat face detection [7] without assuming any cat-specific characteristics. So, our contribution here is about introducing it to the pedestrian detection problem in addition to applying the PCA dimensionality reduction.

The joint Haar-like feature can be computed very fast due to the integral image technique [11]. A strong classifier is learned by cascading multiple weak classifiers selecting the joint Haar-like features according to the AdaBoost algorithm. The first classifier in pipeline allows small number of false positives (non-pedestrians) without missing true candidates (pedestrians). In our implementation, we have set the first classifier performance goal as a minimum hit rate of 0.995 and a maximum false alarm rate of 0.35 at each stage and built a 10-stage detector. Therefore we built a detector in the first step with an estimated overall hit rate of $(0.995)^{1}0 \approx 0.951$ and an estimated overall false alarm rate of $(0.35)^{1}0 \approx 2.759 \times 10^{-5}$ (see Fig. 3).

The CoHOG descriptor is calculated from co-occurrence matrices for only samples that pass through the Haar cascade. The overall computational cost for CoHOG descriptors in our system are expected to be minor since they are extracted for only a small fraction of the total search space. Final classification decision is taken by the linear SVM classifier [4,14].

4   Experiments

In this section, we describe the experiments done to evaluate the performance of our proposed methods.

4.1   Experimental Setup

We evaluated the performance of our proposed methods against the benchmark DiamlerChrysler dataset [5]. The details of this dataset are shown in Tab. 1.

In CoHOG descriptor training, we used 22,000 images (18 × 36 pixels) of DiamlerChrysler dataset; images are divided into 3 × 6 small regions. Thus the dimension of our feature is 34,704. We used a 2.33 GHz Intel Xeon processor. While in AdaBoost classifier training, 30,560 images were used of the same

Table 1. DaimlerChrysler pedestrian detection benchmark dataset. (a) Benchmark training data used to train the CoHOG and Haar-AdaBoost classifiers. (b) Benchmark testing data sets used to evaluate the performance.

(a) Training dataset

	CoHOG Classifier	Haar-AdaBoost Classifier
Training data	4,000 x 3 Pedestrian	15560 Pedestrian
	5,000 x 2 Non-Pedestrian	15000 Non-Pedestrian
Image Size	18 x 36 pixels (cut-outs)	14 x 28 pixels (cut-outs)
Distribution site	http://www.science.uva.nl/research/isla/downloads/pedestrians/	

(b) Testing dataset

	Test set(1)	Test set(2)
Test data	4,800 x 2 Pedestrian	21790 (full images): 14132 fully
	5,000 x 2 Non-Pedestrian	visible pedestrian labels and
		37236 partial pedestrian labels.
Image Size	18 x 36 pixels (cut-outs)	640 x 480 pixels

dataset (resized into 14 × 28 pixels). We trained a 10 stage detector on a 2.66 GHz Intel Core 2 Duo processor.

To evaluate the performance of our method in terms of accuracy and computational cost, two testing sets were used from DiamlerChrysler dataset. Test set (1) contains 19600 cropped images (18 × 36 pixels) while test set (2) contains 21790 full images (640 × 480 pixels). In order to achieve scale and shift invariant detection in test set (2), we built a pyramid using different templates from 230 × 460 to 14 × 28 with scaling factor 1.1, shift in y-direction equals to max(1, 0.1 × template-height) and shift in x-direction equals to max(1, 0.1 × template-width). Therefore, each image in test set (2) is decomposed into 196622 samples. In the case of test set (2), we measured the detection accuracy based on the area of overlap between a detected region A_d and a ground truth region A_{gt}. The correct detection is defined as determining whether the overlap ratio $(A_d \cap A_{gt})/(A_d \cup A_{gt})$ exceeds 50% [15].

We compare our proposed methods to CoHOG of Watanabe [13] as it produces the most accurate results on the DaimlerChrysler benchmark dataset. We also include the static detector of simple Haar-like features and AdaBoost of Viola [12] as a baseline, and the more accurate version of Haar wavelet and AdaBoost of Dollar [3].

Our performance comparisons are shown in Tab. 2 in terms of computational cost and in Fig. 6 and Fig. 7 in terms of accuracy. We will discuss those results in the rest of this section.

Table 2. Average computational cost (measured in milliseconds) per testing image

		Test set 1 (cut-outs 18×36)	Test set 2 (full images 640×480)
	No. of Images	19600	21790
	Samples/Images	1	196622
CoHOG [13]	CoHOG features	4.80	943.79×10^3
	SVM Classifier	2.49	488.64×10^3
	Total time/Image	**7.29**	**1432×10^3**
Haar Cascade [12]	**Total time/Image**	**2.32×10^{-3}**	**456**
CoHOG + PCAb500 (Proposed)	CoHOG features	4.80	943.79×10^3
	PCA Projection	0.35	68.28×10^3
	SVM Classifier	31.02×10^{-3}	6100
	Total time/Image	**5.18**	**1018×10^3**
Harr + CoHOG (Proposed)	Harr Cascade	1.65×10^{-3}	324.65
	CoHOG features	2.23	547.20
	SVM Classifier	1.15	283.86
	Total time/Image	**3.38**	**1155.71**
Harr + CoHOG + PCAb500 (Proposed)	Harr Cascade	1.65×10^{-3}	324.65
	CoHOG features	2.23	547.20
	PCA Projection	0.16	39.60
	SVM Classifier	14.29×10^{-3}	3.55
	Total time/Image	**2.41**	**915**

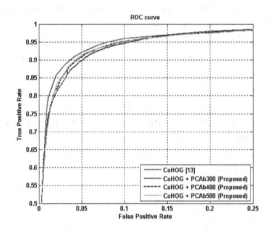

Fig. 6. CoHOG + PCA compared to CoHOG with different number of eigen basis

4.2   CoHOG with PCA

We mentioned that CoHOG descriptor is used with a dimension of 34,704, when a PCA dimensionality reduction is applied with a number of eigen basis b of 500, the feature dimension is reduced to 500 and the processing time of SVM classification in CoHOG will be replaced by the time for PCA projection added to SVM classification in our method. The computational cost, as shown in Tab. 2, will be reduced by 84.8% in SVM classification step and by 28.9% in total processing time. While the accuracy is still comparable to the standalone 'CoHOG [13]' (the true positive rate (TPR) is reduced from 92.1% to 91.2% at false positive rate (FPR) of 0.05%) as shown in Fig. 6. The ROC curve of our 'CoHOG + PCA' method is still better than 'Haar wavelet + AdaBoost [3]'.

We tested the 'CoHOG + PCA' method using different values for the number of eigen basis b (500, 400 and 300). As expected, we got the best ROC curve for this method at $b = 500$ as shown in Fig. 6. In the training process, we needed only 41.96 MB (compared to 2.84 GB of the standalone CoHOG) allocated for training samples and the time needed for training is reduced by 79.6%.

4.3   AdaBoost with CoHOG

The joint Haar-like features extract possible pedestrian candidates and the Co-HOG descriptors accurately eliminate false positives. We evaluated the performance of 'Haar + CoHOG' method on both testing sets.

The number of pedestrian candidates per image in test set (2) is 114 on the average (only 0.06% of the total testing samples per image). Only those samples succeed to pass from the Haar cascade and fed to the CoHOG classifier which takes about 831 msec (547 msec to extract CoHOG features and 284 msec for SVM classifier) to process them. Thus the total processing time for a full

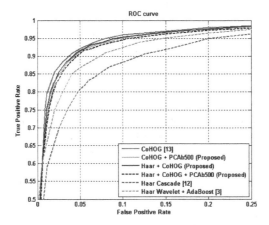

Fig. 7. Performance of proposed methods in comparison to the previous state-of-the-art methods

image (640 × 480 pixels) is reduced to about 1.1 sec which is about 0.1% of the 23.87 minutes needed for the standalone 'CoHOG [13]'.

The number of pedestrian candidates in test set (1) is about 9091 (46.4% of the total testing samples), the CoHOG classifier takes about 1.1 minutes to process them which is larger in comparison to test set (2) due to the larger number of positive samples (about 49% of the total testing samples) in test set (1). Thus the total processing time in test set (1) is reduced by 53.6% in comparison to 'CoHOG [13]' method and also by 34.8% in comparison to the proposed 'CoHOG + PCA' method with b=500. The accuracy of our 'Haar + CoHOG' method is still near and comparable to 'CoHOG [13]' method (TPR is reduced from 92.1% to 91.6% at FPR of 0.05%) as shown in the ROC in Fig. 7.

Adding the proposed PCA dimensionality reduction to the pipeline, we get the ROC curve of 'Haar + CoHOG + PCA' shown in Fig. 7, which is less accurate than our two proposed methods while it reduced the computational cost by 67% in test set (1) and by more than 99% in test set (2) in comparison to the standalone 'CoHOG [13]'. In summary, the results show that our two proposed contributions can improve the computational cost of CoHOG method with a little loss in accuracy. Meanwhile, we still perform better than other pyrevious the state-of-the-art methods.

5   Conclusions

We proposed two methods for an efficient pedestrian detection framework based on the high-dimensional CoHOG feature descriptor. The first proposed method, 'CoHOG + PCA', uses Principle Component Analysis (PCA) to reduce the high dimensionality of CoHOG. The second proposed method, 'Haar + CoHOG', cascades distinct classifiers with two heterogeneous features, the first one is the joint Haar-like feature with AdaBoost, which is fast to compute and enables to extract possible positions of pedestrian candidates from an input image at multiple scales, and then the second feature is the CoHOG descriptor which has a strong classification capability with a linear classifier and accurately eliminates false positives.

We compared the performance of our method and other previous methods in terms of computational cost and accuracy on the benchmark DaimlerChrsysler pedestrian dataset. The experimental results show that by applying our two proposed contributions, the computational cost can be reduced to 1/1000 of the standalone CoHOG while experiencing very close accuracy and still better than the other state-of-the-art methods.

It should be mentioned that our cascade pipeline can be easily applied to other detection problems because it is quite general and does not assume any specific characteristics for pedestrians.

Future work involves extending our experiments to another famous benchmark dataset like INRIA dataset[16], and replacing simple Haar features in our cascade with the more descriptive Haar feature mining of Dollar [3].

References

1. Cortes, C., Vapnik, V.: Support-vector networks. J. of Machine Learning 20(3), 273–297 (1995)
2. Dalal, N., Triggs, B.: Histograms of oriented gradients for human detection. In: CVPR, pp. 886–893 (2005)
3. Dollar, P., Tu, Z., Tao, H., Belongie, S.: Feature mining for image classification. In: CVPR, pp. 1–8 (2007)
4. Fan, R.-E., Chang, K.-W., Hsieh, C.-J., Wang, X.-R., Lin, C.-J.: LIBLINEAR: A library for large linear classification. J. of Machine Learning Research 9, 1871–1874 (2008)
5. Gavrila, D.M., Munder, S.: An experimental study on pedestrian classification. IEEE Tran. of PAMI 28(11), 1863–1868 (2006)
6. Gernimo, D., Lpez, A.M., Sappa, A.D., Graf, T.: Survey of Pedestrian Detection for Advanced Driver Assitance Systems. IEEE Tran. of PAMI 32(7), 1239–1258 (2010)
7. Kozakaya, T., Ito, S., Kubota, S., Yamaguchi, O.: Cat Face Detection with Two Heterogeneous Features. In: ICIP, pp. 1213–1216 (2009)
8. Lin, Z., Davis, L.S.: Shape-Based Human Detection and Segmentation via Hierarchical Part-Template Matching. IEEE Tran. of PAMI 32(4), 604–618 (2010)
9. Mita, T., Kaneko, T., Stenger, B., Hori, O.: Discriminative feature co-occurrence selection for object detection. IEEE Tran. of PAMI 30(7), 1257–1269 (2008)
10. Yamauchi, Y., Fujiyoshi, H., Iwahori, Y., Kanade, T.: People detection based on co-occurrence of appearance and spatio-temporal features. J. of NII Transactions on Progress in Informatics 7, 33–42 (2010)
11. Viola, P., Jones, M.: Robust real-time face detection. Int. J. of Computer Vision 57(2), 137–154 (2004)
12. Viola, P., Jones, M., Snow, D.: Detecting pedestrians using patterns of motion and appearance. In: ICCV, pp. 734–741 (2003)

13. Watanabe, T., Ito, S., Yokoi, K.: Co-occurrence histograms of oriented gradients for pedestrian detection. In: Wada, T., Huang, F., Lin, S. (eds.) PSIVT 2009. LNCS, vol. 5414, pp. 37–47. Springer, Heidelberg (2009)
14. MATLAB SVM classifier: , http://www.mathworks.fr/access/helpdesk/help/toolbox/bioinfo/ ref/svmtrain.html (last visited, June 2010)
15. The PASCAL Visual Object Classes Challenge (VOC), http://pascallin.ecs.soton.ac.uk/challenges/VOC/ (last visited, June 2010)
16. INRIA Person Dataset:, http://pascal.inrialpes.fr/data/human/ (last visited, June 2010)

Interactive Motion Analysis for Video Surveillance and Long Term Scene Monitoring

Andrew W. Senior[1], YingLi Tian[2], and Max Lu[3]

[1] Google Research,
76 Ninth Ave, New York, NY 10011
andrewsenior@google.com
[2] Department of Electrical Engineering
The City College, City University of New York,
160 Convent Ave., New York, NY 10031
ytian@ccny.cuny.edu
[3] IBM Global Technology Services,
17 Skyline Drive, Hawthorne, NY 10532
maxlu@us.ibm.com

Abstract. In video surveillance and long term scene monitoring applications, it is a challenging problem to handle slow-moving or stopped objects for motion analysis and tracking. We present a new framework by using two feedback mechanisms which allow interactions between tracking and background subtraction (BGS) to improve tracking accuracy, particularly in the cases of slow-moving and stopped objects. A publish-subscribe modular system that provides the framework for communication between components is described. The robustness and efficiency of the proposed method is tested on our real time video surveillance system. Quantitative performance evaluation is performed on a variety of sequences, including standard datasets. With the two feedback mechanisms enabled together, significant improvement in tracking performance are demonstrated particularly in handling slow moving and stopped objects.

Keywords: video surveillance, slow-moving and stopped object tracking, motion analysis, interaction, background subtraction.

1 Introduction

Automatic video surveillance is a rapidly expanding field, driven by increases in the affordability of technology and the perceived need for security. Demand and the constrained domain make it one of the most commercially viable application areas for computer vision technology. Many applications in the field require the tracking of moving objects (usually people and vehicles), so that events (such as entering a secure zone) can be detected or those objects can be found through a search interface.

In most automatic surveillance systems, objects of interest are first detected, usually by background subtraction (BGS) which will find moving objects [2, 3, 17]. Detected objects are then tracked by a tracking module [1, 4]. Most surveillance video analysis systems operate in a feed-forward manner to pass detections from

R. Koch et al. (Eds.): ACCV 2010 Workshops, Part I, LNCS 6468, pp. 164–174, 2011.

background subtraction to the tracker and then tracks are stored or processed further, for instance by behavior analysis modules. Such a system provides an efficient mechanism for detecting moving objects, but practical implementations suffer from a number of limitations when exposed to particular conditions (lighting variations, weather, heavy occlusion, crowding, non-rigid objects). A rich literature attempts to deal with each of these problems. In this paper, we concern the problems that arise in scenes with slow moving objects and where objects stop for significant periods of time. In particular these scenes challenge the fundamental assumption of a strict differentiation between foreground and background, and the pragmatic choice of using motion, or its proxies, to distinguish between the two. A given object may change from foreground to background and vice versa. For instance, a moving car may park and for all practical purposes needs to be treated as "background" — at least until it starts moving again.

Background subtraction algorithms are generally designed to be adaptive to be able to deal with scene changes (changing lighting; backgrounds whose appearance changes, such as trees and water; static objects). However, a slow moving, or stopped, object can lead to just such repeated observations, and result in the object being adapted piecemeal into the background. This leads to errors in tracking, as the object dissolves into multiple fragments, and false "ghost" fragments appear where the background contains the object after it moves away.

The tracking process usually treats groups of pixels collectively, as unitary objects, and this higher-level information derived by the tracker can be used to inform the process of background subtraction. The tracker explicitly models the objects, whose behaviors are subject to physical constraints (such as rigid motion) in ways different to the physical constraints that control the appearance of individual pixels. Many object tracking techniques focus on handling occlusions but neglect how to track slow moving or stopped objects. Boult et al. [4] describe a system that performs well at detecting slow moving objects.

There have been a few systems that have investigated the possibility of feedback from the tracker to the background subtraction module. In order to improve the robustness and efficiency of background subtraction methods, some papers [3, 5, 6] introduced feedback from the frame level and some papers employed the feedback from the tracker [7–11]. Abbott et al. [7] proposed a method to reduce computational cost in visual tracking systems by using track state estimates to direct and constrain image segmentation via background subtraction and connected components analysis. Harville [8] used application-specific high level feedback (frame level, person detector and tracker, and non person detector) frame to locally adjust sensitivity to background variation. Senior [12] suggests recalculating the background and foreground segmentation using the model of the tracked object after the background subtraction stage. Wang et al. [11] proposed a unified framework to address detection and tracking simultaneously to improve the detection results. They feed the tracking results back to the detection stage.

The interaction between the tracking and background subtraction can also be used to improve the tracking of the slow moving and stopped objects. Venetianer et al. [13] examine a way of pushing foreground objects into the background and vice versa. Yao and Odobez [14] use a similar layered background mechanism to remember stopped objects. Taycher et al. [15] proposed an approach that incorporates background modeling and object tracking to prevent stationary objects fading into the background. Our approach is most closely related to that of Pnevmatikakis and Polymenakos [9], who to overcome the problem of stationary targets fading into the background, propose a system combining a mixture of Gaussians background subtraction algorithm and a Kalman tracker in a feedback configuration. They control the learning parameters of the background adaptation on a pixel level in elliptical regions around the targets based on the tracking states from the Kalman tracker. A smaller learning parameter was used for a slow moving object. However, this mechanism will fail when the targets stay stationary for a long period. They will gradually fade into the background even with very small learning parameters.

In contrast, we create two feedback mechanisms that allow the tracker to suppress background updating for slow moving objects that are being tracked. Further, we introduce an active, tracker-driven, object-level healing process where whole objects are pushed to the background to solve the challenges in tracking caused by the stopped objects.

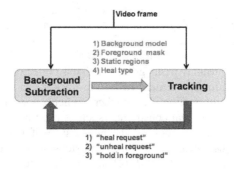

Fig. 1. Diagram of the interaction of background subtraction and tracking, showing the passing of metadata messages

2   Interaction between BGS and Tracking

2.1   Feedback Mechanisms for Interactions between BGS and Tracking

In order to improving tracking accuracy, we create two feedback mechanisms that allow interactions between BGS and tracking. The feedback required to handle slow and stopped objects are implemented by adding information to metadata of tracking observations which are accessible by BGS processing through following three requests: 1) "heal request"—tracking requests BGS to push the region back to background model; 2) "unheal request" "—tracking requests BGS to convert the background model of a healed region back to that before the heal happened; 3) "hold in foreground"—tracking requests BGS to hold a region without updating. Figure 1 shows the diagram of the interaction between BGS and tracking with the passing of metadata messages.

2.2 BGS Adaption Suppression for Tracking Slow Moving Objects

Slow Moving Objects Tracking Problem: Slow moving objects can present a significant problem to conventional background subtraction algorithms. In multiple Gaussian mixtures based BGS algorithms [2, 5], on which many current systems are based, each pixel is modeled by a mixture of Gaussians distribution. Observations of a pixel's color are assigned to the closest mode of the mixture, or to a newly created mode (replacing the least observed previous mode). The most frequently observed mode is considered the "background" mode, and observations matching that are considered to be background. Other values are flagged as foreground. When an object moves slowly or stops, any pixel may fall on the object for many frames and, if it is of a consistent color, that pixel will eventually be considered background. If multiple pixels are affected in the same way, parts of the object will be considered to be background and the object will progressively be "lost".

Previous systems have partially addressed this problem by detecting groups of foreground pixels that are being adapted into the background, and actively push the whole group in to the background [5]. Here, however the problem is that the detection may come only after some pixels have already been adapted into the background, and may only affect part of the object. Thus, while the switch to background is no-longer independent for each pixel, it may still occur in several fragments, and results in part of an object being background and part being foreground.

BGS Adaption Suppression: To deal with this situation, we institute a feedback mechanism that allows the tracker to suppress background updating for slow moving objects that are being tracked. When the tracker detects a slow-moving object (based on conditions of centroid movement ≤ 3 pixels in 0.5s; number of observations > 30, and no recent splitting behavior), it flags the object observations as "slow moving" and the background subtraction algorithm suppresses the adaptation in the region where the slow moving object was observed (as indicated by a mask passed in the metadata).

Typically adaptation will already have been carried out by the background subtraction (as the video frame was received), although some algorithms may wait until the video frame processing has completed. According to the algorithm used, adaptation is suppressed in the region of a slow moving object by copying pixels from a copy of the model saved before adaptation, or by carrying out the inverse operation on those pixels (for instance decreasing the observation counts).

Suppressing adaption in this way has the effect of maintaining the tracked object in the foreground, and uses object level information from the tracker — that the pixels belong to a known object that is moving slowly and has been reliably detected and tracked for some period — to which the background subtraction module by itself does not have access.

A drawback of this mechanism is that it inhibits the process by which false alarm foreground objects are removed. For instance a shadow or a reflection which appears but is tracked for a while, would ordinarily quickly be forgotten as the background model adapts, but, if the "hold in foreground" method engages then these objects can be preserved indefinitely. However, the following mechanism can prevent this from happening.

2.3   Tracking-Based BGS Healing for Stopped Objects

Stopped Objects Tracking Problem: Stopped objects lead to a different problem, and a dilemma for the design of a tracking system. Background modeling needs to adapt to changes in order to ignore "irrelevant" changes such as lighting. In a simple adaptive background subtraction system, when an object stops, as with slow moving objects above, then it will become part of the background and cease to be tracked. However the object is still present in the scene, and for some purposes (for instance the query "show me all cars present at 3p.m.") the system needs to explicitly represent that presence. A further problem is that since background subtraction algorithms typically operate independently on each pixel, then different pixels of the object will be declared background at different times, resulting in a progressive fragmentation as the object is incorporated into the background.

Fig. 2. Selected frames demonstrate ghosting. The car starts in the background and moves forward, leading to multiple foreground fragments and ultimately a large "ghost" or "hole" where it had been covering up the "true" background by using mixture of Gaussians BGS method.

When a static object starts moving, the background subtraction algorithm detects difference regions around the edges of an object, and as the original background is revealed, those pixels are detected as "foreground regions" and a "ghost" of revealed background is detected as foreground along with the true moving object, as shown in Figure 2. Toyama *et al.* describe this as the "waking person" problem, and conclude that it is not solvable in a self-contained background subtraction module. This presents several challenges to a tracking algorithm: (1) the object appears as many small foreground fragments; (2) the growing object is made up of a moving component and a static region; (3) the true object eventually separates from the static "ghost" region. Some background subtraction methods explicitly tackle this problem [3].

Tracking-based BGS Healing: With the adaptation-inhibition described in Section II.B, slow moving and stationary objects are not adapted into the background at all, so healing and fragmentation are no longer a problem. However static objects will now be held indefinitely in the foreground. As a parking lot fills slowly with cars, the number of "tracked" objects increases and their interactions and mutual occlusions become progressively more complex and unmanageable.

Consequently, we introduce an active, tracker-driven, object-level healing process where whole objects are pushed to the background. In this process, the tracker tracks whole objects and monitors their movement. When an object is stationary for a sufficient period (dependent on the scene context, for example dependent on the amount of activity in the scene and typical behaviors — whether objects stop for long or short periods) then the tracker determines that the object can be pushed to the background. The tracker sends a "heal request" message to the background subtraction algorithm, including a foreground mask indicating which pixels belong to the object.

On receiving the message, the BGS algorithm takes the selected pixels and adjusts the background model so that the currently observed pixels become categorized as background. The original contents of the region's background model are sent back to the tracker in a "heal" message. The heal message can also incorporate a categorization of the region, indicating whether it looks like a foreground object or a hole, based on integral of the edges in the object perimeter. On receiving the heal message, the tracker can optionally keep the track in a suspended state, ready to be reactivated if the object moves again. Alternatively (if the region was classified as a "hole") the entire track can be discarded as a false positive.

In this manner, stopped objects are quickly pushed to the background and cease to need active tracking. This reduces the complexity of the tracking problem since fewer tracked objects leads to fewer occlusions and difficult tracking situations, and also reduces the computational load by not "tracking" objects once they are stationary.

When the stopped object begins to move, the background subtraction will detect motion in the region and generate one or more foreground regions in or around the object. Any otherwise unexplained foreground region is compared to the stack of suspended tracks and if a matching track is found it is popped. The background subtraction module is sent an "unheal" request, with the old, stored background appearance, which is pushed into the background model, causing the entire object to again be detected as foreground in the following frame, and thus avoiding the ghosting of Figure 2.

Depending on the scene and typical behavior, the suspended "parked" tracks can be maintained indefinitely or forgotten when too old, invalid or too numerous. A grocery parking lot with rapid turnover may warrant keeping the suspended tracks until a car moves again, but an airport lot where cars are parked for days will not. Lighting changes can lead to significant changes in the background appearance while a stopped object is present, and make the stored background patches invalid. The age of a suspended track may also be of interest — for instance picking out parking violations or understanding parking behavior.

This layered approach will also fail in complex environments. Consider an oblique view, looking along a row of vehicles in a parking lot. As vehicles come and go, many different foreground layers will obscure a particular pixel, and the background exposed by an object's departure may be different from the background that was covered by its arrival. A more complex management of layers is imaginable for this scenario, but was not thought likely to be robust.

3 Interaction Mechanisms Implementation

In this section we describe a publish-subscribe architecture that supports the feedback mechanisms described above. The system processes video through a number of self-contained modules that are linked together through a publish-subscribe framework. Each component receives and transmits metadata packets exclusively through a "first-in, first-out" queue of messages managed by the framework. A metadata packet is taken from the front of the queue and is offered in turn to each of the components for processing before being discarded. While a component is processing a piece of metadata, it may add result metadata to the end of the queue. Most metadata packets are ignored by most components, and many packets will only be relevant to one other "downstream" component, but the architecture allows for considerable flexibility for broadcasting and feedback mechanisms in addition to a simple pipeline model. Components are able to request a priority, which allows the correct ordering of processing for metadata that is processed by multiple components.

The publish-subscribe system encapsulates the functionality of each component and allows for great flexibility in customizing processing on each channel of video, independently selecting one or more detection, tracking and classification algorithms and allowing optional components, such as camera stabilization or performance analysis modules to be added.

In practice the system processes multiple channels of video on a single machine, and each channel is handled by a single "engine" operating in a separate thread but with all engines managed by a single framework. The framework thus scales automatically to multiple processors, and can also handle load-sharing onto embedded coprocessors. The architecture also makes some processing amenable to pipelining of video frames (*e.g.* running BGS on one frame while tracking is executing on the previous frame in a separate thread), though the feedback mechanisms complicate this. Network relaying of selected metadata between processors permits multi-camera operations on a distributed system such as camera hand-off and multi-camera compound alerts.

The framework initiates processing of video by sending a "grab frame" message, which is handled by the video source, which responds by putting a video frame onto the queue. Where appropriate the first component may be stabilization which compensates for motion of the camera (from vibration, wind or active control) and can suspend other processing operations when motion is too great. The background subtraction algorithm operates on the video frame and outputs a foreground mask to the back of the queue. References to the video frame are held by all the components that will subsequently need it, but most other components require further metadata to begin their processing.

The tracker can begin processing when it receives the foreground mask, and it outputs a variety of result metadata, including "track start", "track end" and "track observation". Subsequent plug-ins such as object classifier, color classifier and alert detector all process the output of the background subtraction and tracker, and finally the index writer plug in sends information to be stored in a database.

Before issuing another "grab frame" message, the framework will issue an "end of sample" message to allow components to clean up before the next frame.

As shown in Figure 1, the feedback required to handle slow and stopped objects has been implemented by adding "heal request" and "unheal request" metadata to the original architecture. "hold in foreground" was implemented by adding a flag to the existing "track observation" metadata which were previously ignored by the background subtraction system, but are now acted upon when flagged in this way.

4   Experimental Results

The feedback mechanism of interactions between BGS and tracking is tested and evaluated on our surveillance system. The quantitative evaluation is performed on a set of six video sequences include four videos from the PETS2001 dataset [16] of cars and pedestrians crossing a university campus (about 2800 frames each) and two our own sequences: a top-down view of a four way intersection with cars stopping and waiting for a traffic light to change (Figure 3(a)) and an overhead view of a retail store taken through a fish-eye lens (Figure 3(b)).

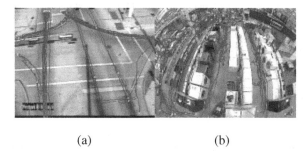

(a)   (b)

Fig. 3. Camera views of the test data with tracker output. (a) the traffic intersection (3400 frames, 64 tracks), (b) the store view (3100 frames, 23 tracks). The paths of object centroids are shown, fading from blue to red from start to finish.

The feedback mechanism was tested using simple tracking performance metrics comparing the tracker output to hand-labeled ground truth. The ground truth for each sequence consists of bounding boxes drawn around each object approximately every 30 frames, with labeling to associate a particular object's bounding boxes over time. Since the task requires tracking, evaluation is based on track-level rather than BGS level.

The performance analysis processing matches each ground truth track to the tracker's outputs by comparing the distance between the object centroids at each frame (linearly interpolating between the sparse ground truth points), with hysteresis. When at any time t, an object lies close to a ground truth track (within r, here 20, pixels) then the tracks are considered to match for the entire period around t where the tracks lie within $2r$ pixels. Trivial matches (where the match interval between an output track and ground truth track is a subset of the match for another output track, for instance when two tracks cross) are removed.

The track matching was verified to correctly match intervals of output tracks to ground truth tracks. The performance tool produces a variety of statistics, including the number of false positives (output tracks not corresponding to any ground truth track) and false negatives (ground truth tracks that have no corresponding output track); the "underrepresentation"—the proportion of ground truth track frames with no correspondence in an output track (e.g. because the object was not detected); and the fragmentation — the average number of output tracks matched to each ground truth track (because of gaps in detection, or identity confusion during occlusions).

Quantitative analysis results of performance on six sequences of video, from PETS 2001 and two proprietary datasets for particular scenarios, are shown in Table 1. The comparison between experimental results and ground truth averaged across all the six sequences shows that there is a 39% reduction in false negatives (ground truth tracks that are not matched in the tracker output) with a 2.7% increase in the number of false positives (tracker output tracks that do not match any ground truth).

Table 1. Tracking performance results on 4 sequences from the PETS2001 dataset and two other datasets. "Under" is the percentage of ground truth frames missing and "Frag" is the average number of tracks matched to a ground truth track.

Sequences	Without feedback		With feedback	
	Under%	Frag	Under%	Frag
PETS D1 C1	21.2	1.56	8.1	1.22
PETS D1 C2	27.8	1.36	12.3	1.36
PETS D2 C1	20.3	1.93	17.6	2.27
PETS D2 C2	8.1	1.50	4.6	1.50
Intersection	33.7	1.04	24.3	1.00
Retail Store	13.1	2.38	14.5	1.90

Errors come from a variety of sources: (1) objects that are too small to be detected, particularly in the store and PETS sequences D1C2 and D2C1 which have distant objects labeled; (2) in the intersection sequence several cars are in the scene at the beginning and ghosting effects mean that their tracks are not matched. (3) Failure to resolve occlusions correctly leads to multiple matches for some ground truth tracks.

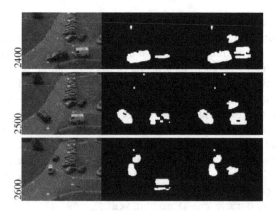

Fig. 4. Selected frames from a PETS2001 video sequence and corresponding foreground regions, demonstrating BGS adaptation without feedback from tracking (middle column) and with the feedback mechanisms (right column). Note the fragmentation (fr.2500) and ghosting (fr.2600) on the middle column. The central stopped car is lost on the middle, but maintained on the right.

Qualitative results are shown in Figure 4. This shows how the interaction between BGS and tracking prevents adaptation and fragmentation of the slowly moving and stopped vehicles, and prevents "ghosts" when they move away.

5 Conclusions

The two feedback mechanisms for handling slow moving and stopped objects work together to improve the results (in terms of underrepresentation, fragmentation and false negatives, with a small increase in false positives) on the tested sequences. The inhibition of background updates for tracked objects successfully prevents slow moving and stopped objects from being absorbed into the background. This inhibition interferes with existing healing mechanisms and requires the addition of the "active healing" controlled by the tracker. With the two mechanisms enabled together, the system shows significant improvement in tracking performance, particularly in the proportion of ground truth tracks that are detected and in reduced fragmentation.

References

1. Connell, J., Senior, A.W., Hampapur, A., Tian, Y.-L., Brown, L., Pankanti, S.: Detection and tracking in the IBM People-Vision system. In: IEEE ICME (June 2004)
2. Stauffer, C., Grimson, W.E.L.: Adaptive background mixture models for real-time tracking. In: Proc. IEEE Computer Society Conference on Computer Vision and Pattern Recognition (1999)
3. Toyama, K., Krumm, J., Brumitt, B., Meyers, B.: Wallflower: Principles and practice of background maintenance. In: Proc. IEEE International Conference on Computer Vision, vol. 1 (1999)
4. Boult, T., Micheals, R.J., Gao, X., Eckmann, M.: Into the woods: Visual surveillance of non-cooperative and camouflaged targets in compex outdoor settings. Proceedings of the IEEE 89(10), 1382–1402 (2001)
5. Tian, Y., Lu, M., Hampapur, A.: Robust and Efficien foreground Analysis for Real-time Video Surveillance. In: Computer Vision and Pattern Recognition (2005)
6. Javed, O., Shafique, K., Shah, M.: A hierarchical approach to robust background subtraction using color and gradient information. In: IEEE Workshop on Motion and Video Computing (2002)
7. Abbott, R., Williams, L.: Multiple target tracking with lazy background subtraction and connected components analysis. Tech. Rep., University of New Mexico (June 2005)
8. Harville, M.: A framework for high-level feedback to adaptive, per-pixel, mixture-of-gaussian background models. In: Heyden, A., Sparr, G., Nielsen, M., Johansen, P. (eds.) ECCV 2002. LNCS, vol. 2352, pp. 543–560. Springer, Heidelberg (2002)
9. Pnevmatikakis, A., Polymenakos, L.: Kalman tracking with target feedback on adaptive background learning. In: Workshop on Multimodal Interaction and Related Machine Learning Algorithms (2006)
10. Cheung, S.-C., Kamath, C.: Robust background subtraction with foreground validation for urban traffic video. EURASIP Journal of Applied Signal Processing, Special Issue on Advances in Intelligent Vision Systems (2005)
11. Wang, J.X., Bebis, G.N., Miller, R.: Robust video-based surveillance by integrating target detection with tracking. In: Computer Vision and Pattern Recognition (2006)

12. Senior, A.: Tracking with probabilistic appearance models. In: Third International Workshop on Performance Evaluation of Tracking and Surveillance systems (June 2002)
13. Venetianer, P., Zhang, Z., Yin, W., Lipton, A.: Stationary target detection using the objectvideo surveillance system. Advanced Video and Signal-based Surveillance (2007)
14. Yao, J., Odobez, J.-M.: Multi-layer background subtraction based on color and texture. In: Proc. IEEE Conference on Visual Surveillance (2007)
15. Taycher, L., Fisher III, J.W., Darrell, T.: Incorporating object tracking feedback into background maintenance framework. In: IEEE Workshop on Motion and Video Computing (2005)
16. PETS 2001 Benchmark Data (2001), http://www.cvg.rdg.ac.uk/PETS2001/
17. Collins, R., Lipton, A., Fujiyoshi, H., Kanade, T.: Algorithms for cooperative multisensor surveillance. Proceedings of the IEEE 89(10) (2001)

Frontal Face Generation from Multiple Low-Resolution Non-frontal Faces for Face Recognition

Yuki Kono[1], Tomokazu Takahashi[1,2], Daisuke Deguchi[1], Ichiro Ide[1], and Hiroshi Murase[1]

[1] Graduate School of Information Science, Nagoya Univesity,
Furo-cho, Chikusa-ku, Nagoya, 464-0856, Japan
[2] Faculty of Economics and Information, Gifu Shotoku Gakuen University, Japan

Abstract. We propose a method of frontal face generation from multiple low-resolution non-frontal faces for face recognition. The proposed method achieves an image-based face pose transformation by using the information obtained from multiple input face images without considering three-dimensional face structure. To achieve this, we employ a patch-wise image transformation strategy that calculates small image patches in the output frontal face from patches in the multiple input non-frontal faces by using a face image dataset. The dataset contains faces of a large number of individuals other than the input one. Using frontal face images actually transformed from low-resolution non-frontal face images, two kinds of experiments were conducted. The experimental results demonstrates that increasing the number of input images improves the RMSEs and the recognition rates for low-resolution face images.

1 Introduction

Quality of face images captured by surveillance cameras tends to be poor. Because of restrictions on the installation position and the number of cameras, most of them would be in low-resolution and would not capture a face from desirable angles for person identification. These conditions make it difficult for both humans and computers to identify a person from the obtained face images. Aiming to overcome the difficulty due to low resolution, Baker and Kanade [1] have proposed a "Face Hallucination" method that obtains a high-resolution face image from a low-resolution face image. Since then, many studies related to super-resolution of face images have been reported [2,3,4,5].

Even if we could obtain high-resolution face images from the cameras, another problem still remains that poses of a face in the images are not necessarily desirable. Methods which utilize a 3D face model directly or indirectly to transform face poses have been reported [6,7]. However, it is difficult to apply them to low-resolution face images because they require a large number of accurate point correspondences between a face image and a face model to fit the image to the model. Approaches based on image-based pose transformation have also been reported. A view-transition model (VTM) proposed by Utsumi and Tetsutani [8]

R. Koch et al. (Eds.): ACCV 2010 Workshops, Part I, LNCS 6468, pp. 175–183, 2011.

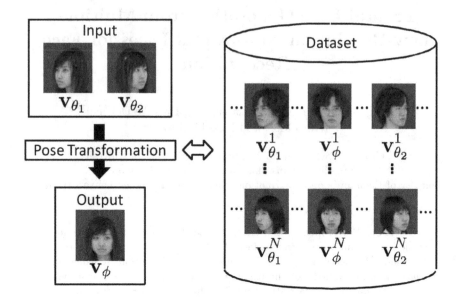

Fig. 1. Frontal face generation from two non-frontal faces

transforms views of an object between different postures by linear transformation of pixel values in images. For each pair of postures, a transformation matrix is calculated from image pairs of the postures of a large number of objects. Another work also took a similar approach [2]. Chai et al. [9] have proposed a Locally Linear Regression (LLR) method for pose-invariant face recognition, which generates a virtual frontal view from a single relatively high-resolution non-frontal face image by applying a patch-wise image transformation method.

Aiming to improve the accuracy of pose transformation and face recognition for low-resolution face images by using the information obtained from multiple input images, we propose a method for frontal face generation from multiple low-resolution non-frontal faces. The proposed method transforms multiple non-frontal input face images to a frontal face as shown in Figure 1. To achieve this, the proposed method uses a general image dataset consisting of faces of a large number of individuals viewed from various angles other than the input individual. The face pose transformation is achieved not by considering its three-dimensional structure, but by synthesizing a face image with a different pose from partial face image patches calculated from a large number of general individual's faces. This patch-wise image transformation, which we name Local VTM (LVTM), is achieved based on the VTM method.

2   Frontal Face Generation from Multiple Non-frontal Faces

The proposed method can be applied to frontal face recognition by using not only two input images but also one or any number of input images. However, in

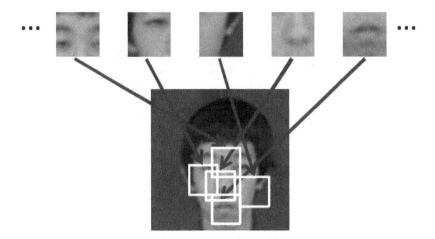

Fig. 2. Synthesis of face patches

the interest of simplicity, we describe a frontal face generation algorithm for two non-frontal face input images.

As shown in Figure 1, the proposed method transforms two input non-frontal face images $\mathbf{v}_{\theta_1}$ and $\mathbf{v}_{\theta_2}$ to an output frontal face image $\mathbf{v}_{\phi}$. Here, θ_1 and θ_2 represent poses of the input faces, and ϕ represents the frontal pose of the output face. To achieve this, the proposed method uses a dataset that contains $\mathbf{v}_{\theta_1}^n$, $\mathbf{v}_{\theta_2}^n$ and $\mathbf{v}_{\phi}^n$ ($n = 1, ..., N$). Here, n represents an individual in the dataset.

Faces of two persons have similar parts although these faces are not totally similar. Transforming the input face image using the information of the entire face image of other individuals might degrade the characteristics of the input individual's face. Therefore, instead of directly transforming the entire face image, we transform face patches that are partial images of a face image for each location in the face image. Then, as shown in Figure 2, an output frontal face image is synthesized from the transformed patches. The proposed method is summarized as follows.

1. Establish the correspondence of each patch position between frontal and non-frontal poses.
2. Transform each pair of images in two non-frontal patches to the corresponding frontal patch using LVTM.
3. Synthesize the transformed patches to obtain the entire frontal face image as shown in Figure 2.

We describe each step in detail below.

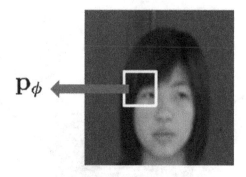

Fig. 3. Patch $\mathbf{p}_\phi$ in $\mathbf{v}_\phi$

2.1   Face Patch Correspondence

The proposed method transforms an entire face by patch-wise image transformation. Therefore we have to find the correspondence of similar facial texture patches between different poses. There are some algorithm to achieve this. For example, the LLR method [9] employs a cylinder face model for finding the correspondence between different poses. In our case, we employ a simple affine model calculated from three points that are the centers of the eyeballs and the lower tip of the nose. In other words, patch correspondences between different poses are found by transforming facial images so that the three points are located at the same positions.

2.2   Patch-Wise Face Image Transformation

A face patch is represented as $\mathbf{p}_\phi$, which is located in the output frontal image $\mathbf{v}_\phi$(Figure 3). Similarly, we represent the face patches corresponding to $\mathbf{p}_\phi$ in input non-frontal images $\mathbf{v}_{\theta_1}$ and $\mathbf{v}_{\theta_2}$ as $\mathbf{p}_{\theta_1}$ and $\mathbf{p}_{\theta_2}$, respectively. On the other hand, we represent face patches in $\mathbf{v}_{\theta_1}^n$, $\mathbf{v}_{\theta_2}^n$ and $\mathbf{v}_\phi^n$ in the dataset as $\mathbf{q}_{\theta_1}^n$, $\mathbf{q}_{\theta_2}^n$, and $\mathbf{q}_\phi^n$ $(n = 1, ..., N)$, respectively. The size of each patch is $W \times H$ pixels. We summarize below the symbols for the patches used in the algorithm.

Patch	Symbols
Input	$\mathbf{p}_{\theta_1}$, $\mathbf{p}_{\theta_2}$
Output	$\mathbf{p}_\phi$
Dataset	$\mathbf{q}_{\theta_1}^n$, $\mathbf{q}_{\theta_2}^n$, $\mathbf{q}_\phi^n$ $(n = 1, ..., N)$

This method transforms $\mathbf{p}_{\theta_1}$ and $\mathbf{p}_{\theta_2}$ to $\mathbf{p}_\phi$ using a transformation matrix $\mathbf{T}$ based on the VTM method [8]. The LVTM method proposed here, transforms each local area of an image while the VTM method transforms the entire area of an image.

The LVTM method calculates $\mathbf{p}_\phi$ as follows:

$$\mathbf{p}_\phi = \mathbf{T} \begin{bmatrix} \mathbf{p}_{\theta_1} \\ \mathbf{p}_{\theta_2} \end{bmatrix}, \tag{1}$$

where $\mathbf{p}$ also represents the vector form of the patch image, which is a column vector that has pixel values of the image as its elements. $\mathbf{T}$ is a $WH \times 2WH$ matrix which transforms $\mathbf{p}_{\theta_1}$ and $\mathbf{p}_{\theta_2}$ to $\mathbf{p}_\phi$. We calculate beforehand $\mathbf{T}$ using the dataset patches for each face patch position by solving the following equation:

$$\begin{bmatrix} \mathbf{q}_\phi^1 \cdots \mathbf{q}_\phi^N \end{bmatrix} = \mathbf{T} \begin{bmatrix} \mathbf{q}_{\theta_1}^1 \cdots \mathbf{q}_{\theta_1}^N \\ \mathbf{q}_{\theta_2}^1 \cdots \mathbf{q}_{\theta_2}^N \end{bmatrix}, \tag{2}$$

in the same manner as in [8].

2.3 Synthesis of Frontal Face

For each face patch position, the proposed method calculates $\mathbf{p}_\phi$ using the input patches and the dataset patches. After this, the proposed method synthesizes $\mathbf{v}_\phi$ from all $\mathbf{p}_\phi$. The pixel values of regions where face patches are overlapped are calculated by averaging the pixel values of the overlapped patches.

3 Experiment

To demonstrate the effectiveness of the proposed method, two experiments were conducted. In the first experiment, we transformed non-frontal face images to a frontal face image. Then we calculated the RMSE between each transformed face and the ground-truth frontal face while changing the number of input images and the input poses. In the second experiment, we input the transformed images to a system that recognizes individuals from the frontal face images. From the results, we evaluated the effectiveness of using multiple input images for frontal face generation from low-resolution non-frontal faces.

3.1 Frontal Face Generation

We used a face image dataset provided by SOFTPIA JAPAN [10], which contains face images of 300 individuals taken from horizontal angles of 0 (front), ±10, ±20, ±30 and ±40 degrees. We transformed all images by affine transformation with three points (the centers of the eyeballs and the lower tip of the nose) to find the patch correspondences between different poses. Figure 4 shows samples of images used for frontal face generation. The image size was 32×32 pixels and the face patch size was set to 16×16 pixels. Face images of 150 individuals were used for LVTM's training set and others were transformed. For the quantitative evaluation, we calculated the average RMSE to the ground-truth frontal faces in the cases of 1, 2 and 3 inputs.

3.2 Face Recognition

We input the transformed images to a recognition system that recognizes a person from a frontal face. The eigenspace method [11] was employed as the recognition strategy. We constructed an eigenspace from 150 actual frontal face images of the same individuals used for the input images. For the purpose of comparison, we calculated the average recognition rate in the cases of 1, 2 and 3 inputs.

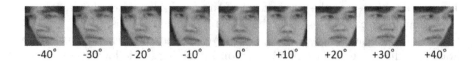

Fig. 4. Samples of images used for frontal face generation

Table 1. Overall average of RMSEs and recognition rates

Number of inputs	1	2	3
RMSE	19.1	17.7	16.7
Recognition rate [%]	61	73	77

4 Result and Discussion

Table 1 shows the average RMSEs and recognition rates. These averages were calculated from 150 individuals and all combinations of the input degrees for the cases of 1, 2 and 3 inputs. These results show that the more input images were used, the better RMSE and recognition rate were obtained. From this, we confirmed the effectiveness of the proposed method that uses multiple input images for frontal face generation. Figure 5 shows samples of the transformed

Table 2. RMSEs and recognition rates for each set of input degrees

(a)RMSE

Input angle[°]	−40	−30	−20	−10	+10	+ 20	+30	+40
−40	21.0	19.1	17.9	17.3	17.1	17.3	17.7	18.6
−30	-	19.5	18.0	17.2	17.0	17.2	17.8	18.0
−20	-	-	18.6	17.0	16.2	16.5	17.0	17.4
−10	-	-	-	17.6	15.9	16.2	16.7	16.8
+10	-	-	-	-	18.0	17.0	17.0	17.0
+20	-	-	-	-	-	19.0	18.0	17.9
+30	-	-	-	-	-	-	19.0	18.7
+40	-	-	-	-	-	-	-	20.2

(b)Recognition rate [%]

Input angle[°]	−40	−30	−20	−10	+10	+20	+30	+40
−40	45	55	69	73	73	71	67	65
−30	-	55	67	76	77	74	70	67
−20	-	-	68	79	83	80	75	69
−10	-	-	-	72	82	80	81	77
+10	-	-	-	-	70	80	80	72
+20	-	-	-	-	-	70	70	67
+30	-	-	-	-	-	-	60	59
+40	-	-	-	-	-	-	-	46

The number of input	1	2	3	Ground truth

RMSE	15.1	13.6	13.5	0.0

(a)Results with input angles of +30 for 1 input, ±30 for 2 inputs, and ±30 and −20 for 3 inputs

Ground truth				
Transformed face				
RMSE	19.6	21.5	14.8	17.2

(b)Results for various persons in the 2 input case with input angles of ±30

Input degree[]	+10	+20	+30	+40	Ground truth

RMSE	14.8	16.3	21.5	17.5	0.0

(c)Results for various input angles in the 1 input case

Fig. 5. Examples of transformed facial images

frontal face images and their RMSEs. The samples of the results for each of the 1, 2 and 3 input cases are shown in Figure 5 (a). On the other hand, Figure 5 (b) and (c) show the samples of the results for various individuals and various input angles, respectively. Additionally, Table 2 shows the average RMSEs and face recognition rates for each set of input degrees in the cases of 1 and 2 inputs. In the tables, the values in the fields where two input angles are the same are the 1 input case. These results show that the inputs with angles close to the front tend to achieve small RMSEs and high recognition rates. Increasing the number of input images increases the information of the input individual for frontal face generation. However, some cases where adding another input image degraded the accuracy were observed. For example, the recognition rate of the input degrees 30 and 40 was 59%, while that of the input degree 30 was 60%. Such situations were also observed in the case of 3 inputs. This is because appropriate patch correspondences would be difficult to obtain by the simple method when two poses are distant. The inappropriate patch correspondences might degrade the accuracy of the frontal face generation.

5   Conclusion

We proposed a method for frontal face generation from multiple low-resolution non-frontal faces for face recognition. The proposed method achieves the image-based face pose transformation by using the information obtained from multiple input face images without considering three-dimensional face structure. Using frontal face images actually transformed from low-resolution non-frontal face images, two kinds of experiments were conducted. The experimental results demonstrated that increasing the number of input images generally improves the RMSEs and the recognition rates for low-resolution face images.

Future work includes making use of the knowledge on the movement of face parts according to face pose change in order to obtain appropriate correspondences between face patches.

Acknowledgement

This work was supported by "R&D Program for Implementation of Anti-Crime and Anti-Terrorism Technologies for a Safe and Secure Society," Special Coordination Fund for Promoting Science and Technology of the Ministry of Education, Culture, Sports, Science and Technology, the Japanese Government.

References

1. Baker, S., Kanade, T.: Hallucinating faces. In: Proc. International Conference on Automatic Face and Gesture Recognition, pp. 83–88 (2000)
2. Jia, K., Gong, S.: Hallucinating multiple occluded face images of different resolutions. Pattern Recognition Letters 27, 1768–1775 (2006)

3. Dedeoğlu, G., Baker, S., Kanade, T.: Resolution-aware fitting of active appearance models to low resolution images. In: Leonardis, A., Bischof, H., Pinz, A. (eds.) ECCV 2006. LNCS, vol. 3952, pp. 83–97. Springer, Heidelberg (2006)
4. Lan, C., Hu, R., Lu, T., Luo, D., Han, Z.: Global face super resolution and contour region constraints. In: Zhang, L., Lu, B.-L., Kwok, J. (eds.) ISNN 2010. LNCS, vol. 6064, pp. 120–127. Springer, Heidelberg (2010)
5. Huang, H., He, H., Fan, X., Zhang, J.: Super-resolution of human face image using canonical correlation analysis. IEEE Trans. Pattern Recognition 43, 2532–2543 (2010)
6. Beymer, D., Poggio, T.: Face recognition from one example view. In: Proc. 5th International Conference on Computer Vision, pp. 500–507 (1995)
7. Blanz, V.G., Phillips, P.J., Vetter, T.: Face recognition based on frontal views generated from non-frontal images. In: Proc. International Conference on Computer Vision and Pattern Recognition, vol. 2, pp. 454–461 (2005)
8. Utsumi, A., Tetsutani, N.: Adaptation of appearance model for human tracking using geometrical pixel value distribution. In: Proc. 6th Asian Conference on Computer Vision, vol. 2, pp. 794–799 (2004)
9. Chai, X., Shan, S., Chen, X., Gao, W.: Locally linear regression for pose-invariant face recognition. IEEE Trans. Image Processing 16, 1716–1725 (2007)
10. SOFTPIA JAPAN.: HOIP-F (Face Image Dataset),
    http://www.softopia.or.jp/rd/facedb.html
11. Turk, M.A., Pentland, A.P.: Face recognition using eigenfaces. In: Proc. International Conference on Computer Vision and Pattern Recognition, vol. 2, pp. 586–591 (1991)

Probabilistic Index Histogram for Robust Object Tracking

Wei Li[1], Xiaoqin Zhang[2], Nianhua Xie[1], Weiming Hu[1],
Wenhan Luo[1], and Haibin Ling[3]

[1] [1]National Lab of Pattern Recognition, Institute of Automation,
CAS, Beijing, China
{weili,nhxie,wmhu,whluo}@nlpr.ia.ac.cn
[2] College of Mathematics & Information Science,
Wenzhou University, Zhejiang, China
[3] Dept. of Computer and Information Sciences Temple University
xqzhang@wzu.edu.cn, hbling@temple.edu

Abstract. Color histograms are widely used for visual tracking due to
their robustness against object deformations. However, traditional histogram representation often suffers from problems of partial occlusion,
background cluttering and other appearance corruptions. In this paper,
we propose a probabilistic index histogram to improve the discriminative power of the histogram representation. With this modeling, an input
frame is translated into an index map whose entries indicate indexes to a
separate bin. Based on the index map, we introduce spatial information
and the bin-ratio dissimilarity in histogram comparison. The proposed
probabilistic indexing technique, together with the two robust measurements, greatly increases the discriminative power of the histogram representation. Both qualitative and quantitative evaluations show the robustness of the proposed approach against partial occlusion, noisy and
clutter background.

1 Introduction

Appearance model is one of the most important issues in object tracking. Generally speaking, the appearance model can mainly be divided into two types: histogram [1] and non-parametric description [2],[3],[4],[5], [6],[7]. Histogram-based
models, which naturally capture the global statistic information of the target region, are one of the most popular models. This is due to their robustness against
target deformation and noises.

However, because the color histogram is a statistic description of the target
region, it loses the spatial information and not robust to background disturbance
and occlusion. Also the traditional ways obtain the histogram bin by equally dividing the color space. However this division can neither accurately nor efficiently
encode the color distribution of the target region.

The above drawbacks of histogram representation limit its application in visual tracking. In order to address the above issues, we propose a probabilistic

R. Koch et al. (Eds.): ACCV 2010 Workshops, Part I, LNCS 6468, pp. 184–194, 2011.

index histogram with spatial distance and cross bin-ratio dissimilarity measurement. The main contributions of the proposed algorithm are summarized as follows:

1. We propose a probabilistic index histogram as the appearance model. Instead of obtaining the histogram bins by equally dividing the color space, we define each bin adaptively as a palette. Using the palette indexing theory [8], each bin is considered as a color probabilistic distribution. An image is then translated into an index map whose entries are the bin number the pixels fall in.
2. For the probabilistic index histogram, we propose an efficient *spatial distance* between two bins. The spatial distance improves the matching accuracy by capturing spatial layout for the histogram.
3. Instead of using traditional distances (e.g., Bhattacharyya distance) for comparing two histograms, we use the cross *bin-ratio dissimilarity*, which is previously proposed for category and scene classification [9], to improve the robustness to background clutters and partial occlusion.

2   Related Work

Vast works have been done to increase the robustness of the histogram representation. In [10], oriented kernels are adopted to take the spatial information into consideration. Birchfiled et al. [11] introduce the spatial mean and covariance of the pixel positions of the given bin. In [12], Earth Mover's Distance (EMD) which is robust to illumination changes is employed to weighted the similarity of two histograms. In [13], object is represented by multiple image fragments and the histograms are compared with the corresponding image patch histogram.

Our method is different from the above histogram representations in both representation and similarity measurement. We model each bin as a palette and propose a new probabilistic histogram. Our probabilistic histogram code the color distribution more accurately than equally dividing the color space as histogram bin. With this modeling, an input frame is translated into an index map. Based on the index map, we introduce a spatial distance to compare spatial layout of the pixels falling in the same histogram bin. We also introduce the cross bin-ratio dissimilarity to compute the similarity of two histograms. This measurement together with the spatial distance enhances the robustness of histogram representation against occlusion, noisy and background cluttering.

The rest of this paper is structured as follows. In Section 3, the proposed algorithm is detailed. Experimental results are presented in Section 4, and Section 5 is devoted to conclusions.

3   Index Histogram

3.1   Palette Indexing

An efficient way to represent an image is to define it as an index matrix. Each entry in the matrix indicates index to a separate palette. The palettes are possible colors of the target region. By the definition of index matrix, image pixels corresponding to a palette share the same index. The image structure is better captured by carefully analyzing the index map. Let I be a $M \times N$ image and $\{L_s\}_{s=1}^m$ be the palettes. The index for each pixel $x_{i,j}$ is represented as $d_{i,j}$, where i, j is the location of the pixel. The palette L is a table of m color or feature. For the color image, $\{L_s\} = \mu_s$ can be an $[RGB]$ vector. The index $d_{i,j}$ of each pixel points to the palette the pixel belongs to.

Instead of including all the image color in the palette, each palette L_s is defined as a Gaussian distribution and the probability of a pixel $x_{i,j}$ belonging to a certain palette L_s is formulated by a Gaussian distribution:

$$p(x_{i,j}|L_s) = \phi(x_{i,j} : \mu_s, \Sigma_s) \tag{1}$$

where μ_s, Σ_s are the mean and covariance of Gaussian distribution $\phi(.)$. Through maximizing the probability each pixel belongs to all the palettes, each entry in the index map can be obtained.

3.2 Probabilistic Indexing Histogram

Following the idea of palette indexing, we model each histogram bin as a color palette. Let B_s be the sth histogram bin, $d_{i,j}$ indicate the bin the pixel $x_{i,j}$ falls in. Given an image I, the learning process aims to obtain the $d_{i,j}$ and B_d simultaneously. These two parameters can be obtained through maximizing the posterior probability $p(x|d, B)$. After treating the index variable d as hidden variables and bin B as the model parameters, $p(x|d, B)$ can be expressed as: $p(x|B) = \sum_d p(x, d|B)$. Unfortunately, this optimization is intractable, an approximate method is needed. The most popular approximation method is the variational method [14]. In the method, an alternative cost, *free energy* $\mathbb{E}$, is defined instead of directly maximizing $p(x|d, B)$:

$$\mathbb{E} = \sum_d q(d) \log \frac{q(d)}{p(x, d|B)} = \sum_d q(d) \log q(d) - \sum_d q(d) \log p(x, d|B) \tag{2}$$

where $q(d)$ can be an arbitrary distribution. If we define $q(d)$ as $p(x|B, d)$, $\mathbb{E}$ equals to $-log\ p(x|B)$. Using the Jensen's inequality, it can be shown that $\mathbb{E} \geq -log\ p(x|B)$. So the lower bound of $\mathbb{E}$ is the posterior probability $p(x|d, B)$ that we need to optimize. Using the variational method in [14], the free energy can be efficiently optimized using an iterative algorithm.

In order to minimize the free energy $\mathbb{E}$, we fix p and optimize q under the constraint $\sum_{i,j} q(d_{i,j}) = 1$. After minimizing the free energy $\mathbb{E}$, q is obtained as

$$q(d_{i,j}) \propto p(d_{i,j})p(x_{i,j}|d_{i,j}, B) \tag{3}$$

where $p(d_{i,j})$ is the prior distribution, and $p(x_{i,j}|d_{i,j}, B)$ is defined in Equ.(1). Then the bin parameters $B = \{\mu_s, \Sigma_s\}_{s=1}^m$ are estimated by minimizing the free energy $\mathbb{E}$ while keeping $q(d_{i,j})$ fixed:

Fig. 1. Probabilistic index histogram

$$\mu_s = \frac{\sum_{i,j} q(d_{i,j} = s)x_{i,j}}{\sum_{i,j} q(d_{i,j} = s)} \tag{4}$$

$$\Sigma_s = \frac{\sum_{i,j} q(d_{i,j} = s)[x_{i,j} - \mu_s][x_{i,j} - \mu_s]^T}{\sum_{i,j} q(d_{i,j} = s)} \tag{5}$$

These two steps are conducted iteratively until convergence. The results are probabilistic histogram whose bins are modeled as $B = \{\mu_s, \Sigma_s\}_{s=1}^m$. The index of the pixel $x_{i,j}$ can be obtained through minimizing the Mahalanobis distance between each pixel $x_{i,j}$ and the histogram bin:

$$d_{i,j} = \arg\min_s((x_{i,j} - \mu_s)\Sigma_s^{-1}(x_{i,j} - \mu_s)) \tag{6}$$

Fig.1 illustrates the result of probabilistic histogram and index map. Each color in the palette represents the mean of each histogram bin. For the clarity of illustration, each histogram bin is assigned a distinctive color instead of the original mean. Different color in the target region corresponds to different bins the pixels belong to. From Fig.1, the image can accurately be coded with the probabilistic index histogram.

3.3   Spatial Distance

The histogram representation provides rich statistic information at the cost of losing spatial layout of pixels falling into the same histogram bin. However, the index map of the target region obtained using Equ.(5)(6) captures the image structure. Specifically, the distribution of pixel position of the same index is an efficient way to represent the spatial layout of the histogram bin. Motivated by this observation, we model the spatial layout of the sth histogram bin using the spatial mean $\mu_{a,s}^T$ and covariance $\Sigma_{a,s}^T$ of the pixel position $a_{i,j}$ of index s,

$$\mu_{a,s}^T = \frac{\sum_{i,j} a_{i,j}\delta(d_{i,j} - s)}{\sum_{i,j} \delta(d_{i,j} - s)}$$

$$\Sigma_{a,s}^T = \frac{\sum_{i,j} [a_{i,j} - \mu_{a,s}^T]^T[a_{i,j} - \mu_{a,s}^T]\delta(d_{i,j} - s)}{\sum_{i,j} \delta(d_{i,j} - s)} \tag{7}$$

where δ is the Kronecker function such that $\delta(d_{i,j} - s) = 1$ if $d_{i,j} = s$ and $\delta(d_{i,j} - s) = 0$ otherwise.

The weight of each histogram bin contributes to the whole spatial distance is proportional to the number of pixels in the index:

$$\omega_s = \frac{\sum_{i,j} \delta(d_{i,j} - s)}{\sum_s \sum_{i,j} \delta(d_{i,j} - s)} \tag{8}$$

Given a candidate region, the spatial mean $\mu_{a,s}^C$ and covariance $\Sigma_{a,s}^C$ of the sth bin can be computed accordingly. The spatial distance between the target histograms H^T and a candidate histograms H^C is formulated as follows:

$$SD(H^T, H^C) = \sum_s \omega_s \exp\{-\frac{1}{2}[\mu_{a,s}^C - \mu_{a,s}^T]^T((\Sigma_{a,s}^T)^{-1} + (\Sigma_{a,s}^C)^{-1})[\mu_{a,s}^C - \mu_{a,s}^T]\} \tag{9}$$

3.4   Cross Bin-Ratio Dissimilarity

A widely used method to compare the target histogram H^T and candidate histogram H^C is the Bhattacharyya distance (e.g., in [1]):

$$\rho(H^T, H^C) = \sum_{u=1}^{m} \sqrt{h^T(u)h^C(u)} \tag{10}$$

However, this measurement only considers bin to bin information and loses the cross bin interaction. In addition, as the target region is usually represented with a rectangle, it is often corrupted by the background clutters and occlusion part that are irrelevant to the target. As shown in Fig.1, the histogram bin represented with blue is obviously the background and the pixels falling into this bin account for a large portion of the target region. Such background information and occluded part will introduce noises into histogram representation and which in turn brings the inaccurate matching. In order to overcome these drawbacks, we introduce a cross bin-ratio dissimilarity measurement.

Let h be an m-bin histogram. A ratio matrix W is defined to capture the cross bin relationship. Each element in the matrix is (h_u/h_v) which measure the relation between bin $h(u)$ and $h(v)$. The whole ratio matrix is written as follows:

$$W = \left(\frac{h_u}{h_v}\right)_{u,v} = \begin{bmatrix} \frac{h_1}{h_1} & \frac{h_2}{h_1} & \cdots & \frac{h_m}{h_1} \\ \frac{h_1}{h_2} & \frac{h_2}{h_2} & \cdots & \frac{h_m}{h_2} \\ \cdots & \cdots & \cdots & \cdots \\ \frac{h_1}{h_m} & \frac{h_2}{h_m} & \cdots & \frac{h_m}{h_m} \end{bmatrix} \tag{11}$$

With the definition of the ratio matrix, we compare the vth bin between two histogram H^T and H^C using dissimilarity M_v. M_v is defined as the sum of squared difference between the vth rows of corresponding ratio matrix:

$$M_v(H^T, H^C) = \sum_{u=1}^{m} (\frac{h_u^T}{h_v^T} - \frac{h_u^C}{h_v^C})/(\frac{1}{h_v^T} + \frac{1}{h_v^C}) \tag{12}$$

where $\frac{1}{h_v^T} + \frac{1}{h_v^C}$ is normalization term to avoid the instability problem when h_v^T and h_v^C close to zero. From the above definition, the influence of the clutter or occlusion part bin is weakened by the ratio operation. Thus this measurement is robust to background clutter and occlusion.

We simplify M_v using the $\mathbb{L}_2$ normalization $\sum_{k=1}^m h^2(k) = 1$ and formulate the cross bin-ratio dissimilarity M between histogram H^T and H^C as follows:

$$M(H^T, H^C) = \sum_{v=1}^m M_v(H^T, H^C) = \sum_{v=1}^m (1 - \frac{h_v^T h_v^C}{(h_v^T + h_v^C)^2} \|H^T + H^C\|_2^2) \quad (13)$$

3.5 Bayesian State Inference for Object Tracking

In this paper, the object is localized with a rectangular window and its state is represented using a six dimension affine parameter $X_t = (t_x, t_y, \theta, s, \alpha, \beta)$ where (t_x, t_y) denote the 2-D translation parameters and $(\theta, s, \alpha, \beta)$ are deforming parameters. Given the observation I_t, the goal of the tracking is to infer X_t. This inference can be cast as a Bayesian posterior probability inference process [15],

$$p(X_t|I_t) \propto p(I_t|X_t) \int p(X_t|X_{t-1}) p(X_{t-1}|I_{t-1}) dX_{t-1} \quad (14)$$

where $p(I_t|X_t)$ is the observation model and $p(X_t|X_{t-1})$ represents the dynamic model. A particle filter [15] is used to approximate the posterior probability with a set of weighted samples. We use a Gaussian distribution to model the state transition distribution. The observation model $p(I_t|X_t)$ reflects the similarity between the candidate histogram of state X_t and target histogram:

$$p(I_t|X_t) = \exp\{-\frac{1}{2\sigma^2}(1 - SD(H^T, H^C))\} * \exp\{-\frac{1}{2\sigma^2} * \alpha M(H^T, H^C)\} \quad (15)$$

where σ is the observation variance and α is a weighting factor to balance the influence of spatial distance and cross bin-ratio dissimilarity. If we draw particles from the state transition distribution, the weight $\mathbb{W}_t$ of each sample X_t can be evaluated by the observation likelihood $p(I_t|X_t)$. Then we use a *maximum a posterior* (MAP) estimate to obtain the state of the object at each frame.

4   Experiments

In order to validate the effectiveness of our proposed method, we perform a number of experiments on various sequences. Comparisons with other algorithms are also presented to further show the superiority of our approach. To give a fair comparison with the mean shift algorithm which can only deal with scale changes, we only consider the scale change of the affine parameter. The parameters are set to $\Sigma = \text{diag}(5^2, 5^2, 0.01^2, 0, 0.001^2, 0)$, and 420 particles are used.

Experiment 1. The first experiment is conducted to test the influence of the spatial distance and cross bin-ratio dissimilarity respectively. Also comparison

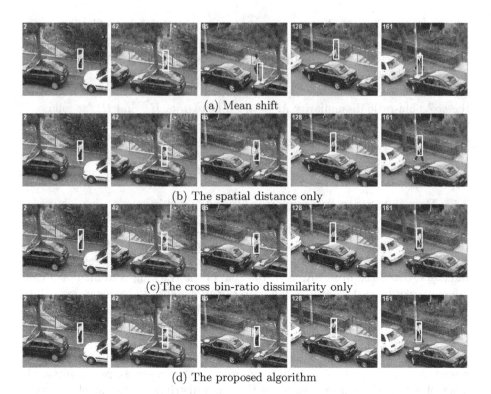

(a) Mean shift

(b) The spatial distance only

(c) The cross bin-ratio dissimilarity only

(d) The proposed algorithm

Fig. 2. The tracking results of Experiment 1

with the color histogram based mean shift algorithm [1] is presented. The sequence is a woman partially occluded by cars. The cars and some background are similar in appearance to the woman. Fig.2 (a) shows the tracking results of mean shift. Obviously the mean shift algorithm is distracted by the cars and similar background and can not deal with partial occlusion well. The results in Fig.2 (b) and Fig.2 (c) illustrate that both spatial distance and cross bin-ratio dissimilarity improve the tracking results. However, only one term can not always provide satisfying results. From the tracking results in Fig.2 (d), our proposed algorithm which combines the spatial distance and cross bin-ratio dissimilarity successfully tracks all the frames and provides accurate results.

A quantitative evaluation of four algorithms is presented in Table.1. We compute the RMSE (root mean square error) between the estimated position and the groundtruth. Here the groundtruth is marked by hand. The results in Table.1

Table 1. Quantitative results for Experiment 1

Tracking approach	Mean shift	Spatial only	Cross bin-ratio only	Our approach
RMSE of Position	14.7018	7.9759	6.6581	3.2451

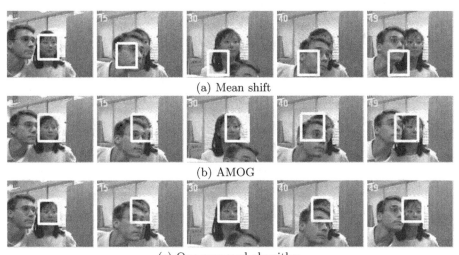

(a) Mean shift

(b) AMOG

(c) Our proposed algorithm

Fig. 3. The tracking results of the Experiment 2

validate that our algorithm with spatial distance and cross bin-ratio dissimilarity achieve the most accurate results.

Experiment 2. In the second experiment, we test the performance of our algorithm in handling partial occlusion and background distraction. We compare our algorithm with other two algorithms. One is the color histogram based mean shift [1] algorithm which only consider the statistical information of object. The other one is a popular parametric description algorithm [5], which adopts an *adaptive mixture of Gaussians* (AMOG) as the appearance model. From Fig.3(a), the mean shift algorithm is distracted away by another face with similar color and can not recover anymore. The AMOG also can not provide good results. On the contrary, our algorithm is capable of tracking the object through all the frames even though the face endures severely occlusion and background distraction.

Experiment 3. The third experiment aims to test the robustness of our algorithm against clutter background and scene blurring. As shown from Fig.4(a), the mean shift algorithm quickly drifts away and can not capture the object any more. This is mainly because the nearby background and the object share the

Table 2. Quantitative results for last three sequences

Tracking approach	Mean shift		AMOG		Our approach	
Evaluation method	RMSE	STF	RMSE	STF	RMSE	STF
Second sequence	20.8207	14/56	6.4717	43/56	2.5356	56/56
Third sequence	85.8135	1/100	92.2919	4/100	5.5866	100/100
Fourth sequence	15.6012	3/183	18.5673	27/183	3.7865	183/183

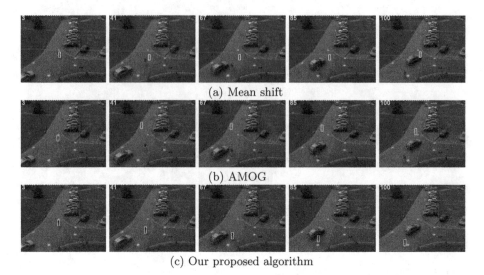

(a) Mean shift

(b) AMOG

(c) Our proposed algorithm

Fig. 4. The tracking results of Experiment 3

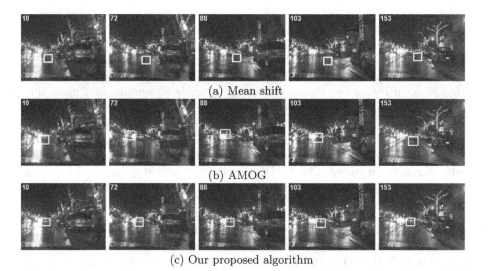

(a) Mean shift

(b) AMOG

(c) Our proposed algorithm

Fig. 5. The tracking results of Experiment 4

similar color histogram statistics. The tracking results in Fig.4(b) show that the AMOG also can not tackle the clutter background. However the good tracking results in Fig.4(c) illustrate that our algorithm is robust against clutter background and scene blurring.

Experiment 4. In the last experiment, we test our algorithm on a more challenging sequence. In this sequence, a car moves in a noisy background.

The nearby background is so noisy that the car can not easily be located even by eyes. Fig.5 presents the tracking results of three algorithms. As shown in Fig.5(c), the noisy background poses no challenges for our algorithms. However both mean shift and AMOG encounter troubles in the extremely noisy background.

A quantitative evaluations of the last three sequences are also given in Table 2 to further demonstrate the superiority of our algorithm. The evaluation is comprised of the following two aspects: RMSE, STF(the number of successfully tracked frames and the tracking is defined as failure if the center of the window is not in the object). From the results in Table 2, we make the following conclusions: (1) The mean shift and the AMOG algorithm are only suitable for the tracking when the appearance of the object is different from the background. Both these two algorithms can not deal with occlusion, noisy and clutter background well; (2) The spatial distance and the cross bin-ratio dissimilarity based on the probabilistic index histogram make our approach robust to occlusion, noisy and clutter background. As a result, our proposed approach is an effective way to improve the discriminative power of the traditional histogram representation.

5   Conclusions

In this paper, we propose a probabilistic index histogram to increase the robustness of the color histogram representation. Our new histogram representation, together with spatial distance and cross bin-ratio dissimilarity, greatly increase the discriminative power of the histogram representation. In experiments on several challenging sequences validate the claimed contributions.

Acknowledgement. This work is partly supported by NSFC (Grant No. 60825204 and 60935002) and the National 863 High-Tech R&D Program of China (Grant No. 2009AA01Z318).

References

1. Comaniciu, D., Ramesh, V., Meer, P.: Kernel-based object tracking (2003)
2. Black, M., Jepson, A.: Eigentracking: Robust matching and tracking of articulated objects using view-based representation. In: Proc. ICCV, pp. 329–342 (1995)
3. Jepson, A., Fleet, D., El-Maraghi, T.: Robust online appearance models for visual tracking. IEEE Transactions on Pattern Analysis and Machine Intelligence 25, 1296–1311 (2003)
4. Black, M., Fleet, D., Yacoob, Y.: A framework for modeling appearance change in image sequence. In: Proc. ICCV, pp. 660–667 (1998)
5. Zhou, S., Chellappa, R., Moghaddam, B.: Visual tracking and recongnition using appearance-adaptive models in particles filters. IEEE Transaction on Image Processing 13, 1491–1506 (2004)
6. Lim, J., Ross, D., Lin, R., Yang, M.: Incremental learning for visual tracking. In: NIPS, pp. 793–800 (2005)

7. Chen, S., Li, Y., Guan, Q., Xiao, G.: Real-time three-dimensional surface measurement by color encoded light projection. Applied Physics Letters 89, 111108 (2009)
8. Jojic, N., Caspi, Y.: Capturing image structure with probabilistic index maps (2004)
9. Xie, N., Ling, H., Hu, W., Zhang, Z.: Use bin-ratio information for category and scene classification (2010)
10. Georgescu, B., Meer, P.: Point matching under large image deformations and illumination changes. IEEE Transactions on Pattern Analysis and Machine Intelligence 26, 674–689 (2004)
11. Georgescu, B., Meer, P.: Spatiograms vs. histograms for region based tracking. In: IEEE Conf. on Computer Vision and Pattern Recognition (2005)
12. Zhao, Q., Brennan, S., Tao, H.: Differential emd tracking. In: Proc. ICCV (2007)
13. Adam, A., Rivlin, E., Shimshoni, I.: Robust fragmentsbased tracking using the integral histogram. In: IEEE Conf. Computer Vision and Pattern Recognition (2006)
14. Jordan, M., Ghahramani, Z., Jaakkola, T., Saul, L.: An introduction to variational methods for graphical models. In: Jordan, M.I. (ed.) Learning in Graphical Models. Kluwer Academic Publishers, Dordrecht (1998)
15. Isard, M., Blake, A.: Contour tracking by stochastic propagation of conditional density. In: Buxton, B.F., Cipolla, R. (eds.) ECCV 1996. LNCS, vol. 1065, pp. 343–356. Springer, Heidelberg (1996)

Mobile Surveillance by 3D-Outlier Analysis

Peter Holzer and Axel Pinz

Institute of Electrical Measurement and Measurement Signal Processing
Graz University of Technology, Austria
{peter.holzer,axel.pinz}@tugraz.at

Abstract. We present a novel online method to model independent foreground motion by using solely traditional structure and motion (S+M) algorithms. On the one hand, the visible static scene can be reconstructed and on the other hand, the position and orientation (pose) of the observer (mobile camera) are estimated. Additionally, we use 3D-outlier analysis for foreground motion detection and tracking. First, we cluster the available 3D-information such that, with high probability, each cluster corresponds to a moving object. Next, we establish a purely geometry-based object representation that can be used to reliably estimate each object's pose. Finally, we extend the purely geometry-based object representation and add local descriptors to solve the loop closing problem for the underlying S+M algorithm. Experimental results on single and multi-object video data demonstrate the viability of this method. Major results include the computation of a stable representation of moving foreground objects, basic recognition possibilities due to descriptors, and motion trajectories that can be used for motion analysis of objects. Our novel multibody structure and motion (MSaM) approach runs online and can be used to control active surveillance systems in terms of dynamic scenes, observer pose, and observer-to-object pose estimation, or to enrich available information in existing appearance- and shape-based object categorization.

1 Introduction

Multibody Structure and Motion (MSaM) extends existing Structure and Motion (S+M) or Simultaneous Localization and Mapping (SLAM) algorithms, because MSaM provides (i) the detection and tracking of rigid or sparsely rigid objects by spatial-temporal trajectories, (ii) the reconstruction of the (unknown) scene structure, and (iii) the pose estimation of the moving camera (observer). Hence, MSaM enables mobile and active surveillance systems.

This paper presents in detail our novel MSaM algorithm, which uses conventional S+M outlier information to model rigid and sparsely non-rigid object foreground motion, as well as experimental results. The main goal of our work is to provide both observer pose and observer-to-object(s) pose in real-time, using affordable hardware, e.g. a tablet or handheld PC with low CPU power.

3-D outlier information, gathered by an S+M algorithm, is the initial per-frame input of our online algorithm. Meanshift-clustering separates the outliers

R. Koch et al. (Eds.): ACCV 2010 Workshops, Part I, LNCS 6468, pp. 195–204, 2011.

into sets of moving objects. At each step, the current clustering information has an impact on future clustering, which prevents point features to change randomly between nearby objects. Thus, the interface between the S+M algorithm and the clustering procedure can be seen as feedback control system.

A stable object centered representation is computed per object, which constitutes the core of our algorithm. Based on a stable reference point, the object centered representation allows motion analysis and enables motion prediction based on position, velocity, and acceleration (Kalman Filter). For each object, rotation and translation information is gained over the tracked time. Finally, to solve loop-closing, an update routine for previously lost and re-appeared point features (e.g. after occlusion or self-rotation) is implemented.

Online MSaM is needed in mobile surveillance, augmented reality, or navigation, and it can enable active surveillance, replacing the human-in-the-loop.

2   Related Work

S+M and SLAM algorithms rely on point features and can simultaneously reconstruct 3-D scene information and observer motion. Both S+M (in computer vision) and SLAM (in robotics terminology) are general approaches because they are purely geometry-based, but SLAM requires real-time performance. Both approaches do not need prior model information, but their range of applications is limited to stationary scenes only. S+M/SLAM fails or produces erroneous reconstruction results in case of (dominant) independent foreground motion in the scene. S+M/SLAM algorithms can roughly be categorized into continuous tracking approaches [1, 2, 3] and keyframe-based approaches [4, 5, 6]. Latest work by Newcombe et al. [7] introduces the combination of state-of-the-art components to solve real-time monocular dense reconstruction of cluttered natural scenes. By combining SLAM with variational optical flow, accurate depth maps are generated. However, this approach is computationally expensive, as it needs a Desktop PC with a GPU.

In MSaM, Schindler et al. [8] distinguish between algebraic methods including factorization-based algorithms (e.g. [9, 10, 11]), and non-algebraic methods that combine rigid S+M with segmentation. Non factorization-based methods handling multi-view perspective sequences in dynamic scenes are addressed by [12, 13, 8, 14]. But most existing MSaM methods are computationally expensive and thus not applicable in real-time. Online MSaM systems, such as Leibe et al. [15] and Ess et al. [16] differ from basic S+M because their approaches are not purely geometry-based and require quite elaborated object detection algorithms. Furthermore, they are restricted to the processing of certain classes of objects only (cars and people).

Compared to the above non-factorization and factorization-based algorithms, our algorithm is applicable in real-time. Our work processes 3-D information, Schindler et al. [8] and Ozden et al. [14] cluster in the image plane. In contrast to Leibe et al. [15] and Ess et al. [16], our approach is not restricted to a certain class of object. Instead, it works for any rigid or sparsely rigid object.

3   Algorithm Overview

Our algorithm consists of three parts: (i) S+M information gathering briefly discussed in section 3.1, (ii) online clustering of outlier data as introduced in section 3.2, and (iii) an object-centered representation explained in section 3.3. In part (i) and (ii), state-of-the-art components are used. The main contribution of this paper - the rigid object representation for online tracking - is introduced in part (iii) and discussed in-depth in section 4.

3.1   Information Gathering by S+M

We start by reconstructing the scene and estimating the observer pose using an S+M algorithm that requires point correspondences. Typically, S+M reconstructs the static scene from an inlier point set, whereas instable points and noise are considered as outliers. Conventional S+M algorithms can reconstruct the static scene and the observer pose in case of up to 50% outliers. Our contribution is to analyse these outliers as candidates for independent foreground motion.

Several current S+M algorithms might be considered. Newcombe and Davison [7] use monocular S+M for scene reconstruction and observer pose estimation, but they neglect the outlier information. The most prominent contribution is certainly by Klein et al. [6,5,4], but their concept relies on keyframes and is inherently unable to handle significant changes in the scene (as posed by significant foreground motion). We decided to build on the S+M algorithm for general cameras by Schweighofer et al. [3]. The available implementation uses a calibrated stereo-rig, which eases the initialization process significantly (initial experiments with Klein's algorithm [6] failed on our video sequences, due to limited camera motion). But we want to emphasize, that the main focus of our paper is not on the S+M algorithm. We can use any algorithm that reliably reconstructs the observer motion, and provides access to 3D-inlier and outlier information gathered over time.

3.2   Motion Clustering

In conventional stable structure scenarios, outliers are mostly related to noise. In scenes with foreground motion, groups of consistently moving outliers can provide hypotheses on independently moving foreground objects. For each outlier, we build online a 3D-trajectory according to its 3D-coordinate in the scene. Having enough information (in our case: a trajectory with a length of minimum five frames), the trajectory's coordinates are passed to a clustering table. One column of the clustering table has the form $[X_t, X_{t-1}, \ldots, X_{t-4}]^T$ where X is the 3D-coordinate and t is the current frame. X_{t-1} is not necessarily the previous frame, it addresses the frame where the outlier was visible the last time. This also applies for $X_{t-2} \ldots X_{t-4}$. Figure 1(a) shows examples of 3D-outlier trajectories gathered by S+M at two different time steps in experiment 1 (cf. fig. 2). At first, only the cup (moving from left to right) contributes outliers, later, the

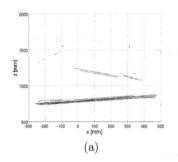
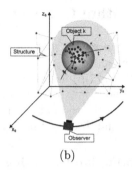

(a)   (b)

Fig. 1. (a) Example x/z-plots of 3D-outlier-trajectories. (b) Scene representration: Global scene coordinate system X_s attached to the static background structure; Local coordinate systems X_k attached to each independently moving cluster of outliers; One moving observer.

cow (moving from right to left) comes into view. For better legibility only the x/z-coordinates of the 3D-trajectories are plotted. Once the clustering table is set up, it is passed to a meanshift-clustering-algorithm [17,18]. To gain hypotheses for moving objects, we cluster the passed 3D-information online by position and by trajectory behavior. We do not use mean-corrected coordinates, as we want to preserve the trajectories' positions in the scene.

3.3   Maintaining the Object Centered Representation

For each previously gathered cluster, we establish a local object coordinate system, which can move independently with respect to the global scene coordinate system that is attached rigidly to the static background structure (see fig. 1(b)). For all moving objects, the origins of their object coordinate systems are selected as reference points. Thus, we link the object's representation and motion to one single reference point per object (in contrast to point cloud matching) without neglecting any object information. The core task of our algorithm is to stably maintain this reference point, as described in detail below in section 4.

4   Online Rigid Object Representation

In this section, we introduce a stable, purely geometry-based object representation that enables us to model the object behavior online, without prior knowledge of the scene. We obtain the object coordinate system by establishing difference vectors of available neighboring point features per object. The point features per object are determined by the online clustering-procedure described in section 3.2. When a cluster contains enough point features, a reference point is computed. Motion, i.e. rotation and translation of the object, is estimated according to each object's reference point. Finally, this geometry-based representation is extended by salient point descriptors to solve the loop-closing problem.

4.1   Initialization of the Object Centered Representation

Once $t \geq 4$ (at least four points are needed for pose estimation) point features are available on an object, the reference point is computed as the mean values of the available point coordinates. Thus, it coincides with a first rough estimation of the object's center of gravity. We define this reference point as the object's coordinate center and store its position in scene coordinates. The coordinates of all point features on the object are stored in object centered coordinates, i.e. the difference vectors Δd_i from point feature i to the reference point.

In case of temporary invisibility of an object, a 9-state Kalman Filter (KF) provides prediction for the reference point based on its position, velocity, and acceleration.

4.2   Update

Once the initialization process has been successfully finished, the updating procedure is continuously performed. The actual update process consists of three parts described below.

(i) In every subsequently processed frame, each point feature on an object provides one hypothesis[1] for the reference point. We require a feature set containing the same point features in two successive frames for pose estimation. Required point features can disappear over time (e.g. self-occlusion due to object rotation). If all point features of an object disappear, prediction by the KF is the only possibility, as no point feature provides a hypothesis for the reference point. However, if a subset of the object disappears temporarily, computing the reference point is still possible. The remaining $r \geq 1$ point features provide valid hypotheses for the reference point. Additionally, at every new frame j, available new point features provide new hypotheses for the object's reference point.

(ii) The self-rotation ΔR between two successive frames has to be estimated for each object. The difference vectors Δd_i created in the initialization process do not provide information on rotation. To estimate the rotation ΔR between the current frame j and the previous frame $j-1$, we proceed as follows:

1. In frame $j-1$, the reference has already been established by $n \geq 4$ neighboring point features. Thus, for each point feature $i \in n$, a difference vector Δd_i exists, indicating the position of the point feature w.r.t. reference point.
2. In frame j, the position of the reference point is assumed unchanged. Additionally, we know the scene coordinates of the same n point features $i_1 \ldots i_n$ found in frame $j-1$. In case of object motion between frames $j-1$ and j, the scene coordinates of the point features will be slightly different. We can compute a rotation matrix ΔR from these 3D point correspondences between frames $j-1$ and j.
3. The rotation with matrix ΔR is applied to all point features' difference vectors to obtain the relative position to the reference point in frame j. We update both "active" and "inactive" point features as well as all lost point

---

[1] A hypothesis is the difference vector from a point to the reference point.

features. To be robust against imprecise rotation estimations, the current position of point feature i is computed as mean of all past difference vectors.

4. Now, each point feature of the object provides a hypothesis for the new position of the reference point considering the inter-frame rotation. The mean of all hypotheses per object is used as new reference point, i.e. in our case the origin of the object-centered coordinate system.

(iii) To provide a stable reference point, we need a confidence measure that can distinguish between "active" and inactive" point features. We allow "active" point features to provide hypotheses for a reference point, whereas "inactive" point features must not. For each object, we generate the confidence measure by computing the median in x, y, and z direction of all hypotheses (i.e. median of 3D-coordinates of visible point features). Then, a certain range around the median values is chosen. In our case, this is 2 times the standard deviation. All point features within this adjusted range in all three directions are set "active". All other point features are set "inactive". If no hypothesis is available (i.e. no point feature is within the selected range) we increase the range stepwise, until a valid hypothesis emerges. Once the new reference point has been computed, the difference vectors from all point features are updated.

4.3   Re-mapping of Re-appeared Point Features

To handle the loop-closing problem of the underlying S+M algorithm, we extend this purely geometry-based algorithm by descriptors that are generated for each point feature on an object. Furthermore, we keep track of the visibility of all point features. In case of invisibility, continuous difference vector update can not be performed. Instead, a position estimation routine is used. While our focus is on loop-closing for object motion, the same method can be applied to stable background features to perform loop-closing for the static scene.

S+M performs continuous tracking, so that a temporarily lost point feature is not recognized on re-appearance. Providing (i) a stable reference point, (ii) a reliable object coordinate system, and (iii) descriptive information, re-mapping is possible. Upon re-mapping, descriptor and difference vector of a point feature are updated, as both are similar, but not equal.

Currently, we use the Pyramid Histogram of Oriented Gradients (PHOG) descriptors[2], because of their sufficient descriptive power.

5   Experimental Results

We present three selected experiments. All sequences were captured with two μeye 1220C USB cameras with 6.5 mm Cosmicar lenses mounted on a stereo rig and a baseline of approximately 30 cm. We used a constant frame rate of 20 Hz. The algorithm has been tested on an Intel Core 2 Quad PC with 2.8

---

[2] A. Bosch and A. Zissermann: Pyramid Histogram of Oriented Gradients (PHOG); available at http://www.robots.ox.ac.uk/∼vgg/research/caltech/phog.html

Fig. 2. Experiment 1 (row 1): 3D-output back-projected to the left image of the stereo-rig. Bounding boxes of each cluster (yellow); point features (colored circles): supporting a hypothesis (red), not supporting a hypothesis (magenta), lost (cyan); 2D projection of reference point (yellow), and of KF (green). **(a)** Output at frame 61. **(b)** Frame 109. First detection of the cow. **(c)** Frame 122. Not enough point features on the cow, only KF estimation is possible. **(d)** Frame 145. The cow is clustered correctly again. **Experiment 2 (row 2):** 3D-output back-projected to the left image of the stereo-rig at frames 48 **(a)**, 60 **(b)**, 87 **(c)**, and 93 **(d)**. **Experiment 3 (row 3):** 3D-output back-projected to the left image of the stereo-rig at frames 30 **(a)**, 95 **(b)**, 121 **(c)**, and 150 **(d)**.

GHz and 1 GB RAM using Matlab 7.6 on a 32 bit version of Ubuntu 9.10. However, Matlab was run on one core only. Online processing depends on the number of objects (i.e. clusters) and point features (descriptor generation) in the scene. The purely geometry-based model is very efficient at approximately 2 and 4 frames/second for 2 and 1 objects, respectively. Adding the Matlab implementation of the PHOG descriptor decreases this performance depending on the number of point features. However, this performance decrease is relevant to frames with descriptor generation only. In fact, the Matlab implementation of the PHOG descriptor is quite slow. A native C/C++ implementation would accelerate the online processing considerably.

In contrast to Ozden et al. [14] who require 1 minute/frame, our approach is applicable in real-time. However, their approach provides higher accuracy in object events like splitting or merging due to the high amount of hypotheses they are maintaining for each frame.

Experiment 1: The scene consists of 180 frames and shows static, textured background and two moving objects (a toy cow and a coffee cup that slide on a table by pulling them on a string) in the foreground. Figure 2 (row 1) shows the output of our algorithm at frames 61, 109, 122, and 145. In this scene, estimation works well in most parts. This is due to the low noise level in this scene. Fig. 3 shows the resulting motion trajectories of the reference points, in a 2D x/z plot (a) and in 3D (b). The KF prediction and its lag are shown as red trajectory.

In contrast to experiment 1 which was set up in our lab, experiments 2 and 3 are highly relevant to mobile surveillance. Both experiments show a moving person along with static background and a moving observer.

Experiment 2: The scene consists of 99 frames and shows static, textured background and the upper part of a walking person. Fig. 2 (row 2) shows the output of our algorithm at frames 48, 60, 87, and 93. Fig. 4 presents a 2D plot of the motion trajectory of the reference point.

Experiment 3: The scene consists of 161 frames and shows static, textured background and one moving person. Fig. 2 (row 3) shows the output of our

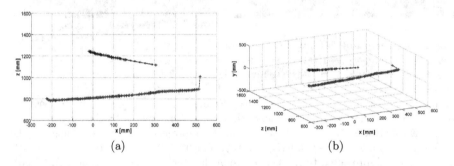

(a)   (b)

Fig. 3. Experiment 1: Computed reference point of cup over all frames (blue), cow (black), and the static structure (yellow). The KF output smooths the estimation (red). The motion is relative to the scene coordinate system initialized at the first observer view. **(a)** x/z-plot of 3D motion analysis. **(b)** 3D-plot of the same output.

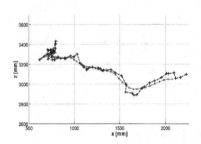

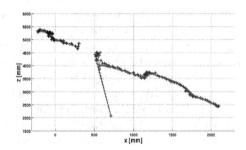

Fig. 4. Experiment 2: x/z-plot of the 3D-trajectory of the person. The motion is relative to the scene coordinate system initialized at the first frame.

Fig. 5. Experiment 3: the algorithm re-detects the person after loss. Too many points are lost, re-detection and re-mapping of point features is not feasible, resulting in two (black, blue) trajectories for one person.

Table 1. Quantitive evaluation of the the algorithm

	Frames	Features	Outliers	Valid Outliers	Objects	Points/Obj.	Detection in %
Exp. 1	180	820	200	131	2	12.6/6.2	91.5/54.5
Exp. 2	99	615	107	87	1	5.7	64.6
Exp. 3	161	564	128	108	1	7.8	79.4

algorithm at frames 30, 95, 121, and 150. Fig. 5 presents the motion trajectories of the reference points.

Table 1 shows a quantitative evaluation of our algorithm. Besides the number of frames per experiment, it contains the total number of point features (inliers and outliers). Furthermore, the number of outliers is listed separately. *Valid outliers* indicates outliers that lie on a moving object with high probability. The column *Points/Obj.* lists the average number of point features on the detected object(s). Each experiment was run three times, the average point amount has been taken. In experiment 1, the first number refers to the cup, the second to the cow. Finally, the detection rate for moving objects is shown in percent. Again, in experiment 1, the first number refers to the cup.

6   Conclusions

3-D reconstruction of dynamic scenes and tracking of independent foreground motion play an important role in application areas such as video surveillance, robotics, or augmented reality. In mobile surveillance, moving cameras substitute stationary ones, and pose estimation of the observing camera is an essential task for such kind of systems.

We have introduced a novel method to model foreground motion by extending an existing S+M environment [3] towards MSaM by 3D-outlier analysis. Our algorithm is purely geometry-based. The stable reference point per object and the positions of point features on the object w.r.t. the reference point provide strong information on the object pose and its motion behavior. A confidence measure in the update process improves the stability of the reference point. This object representation could be applied to any other multibody S+M or SLAM approach, such as [8]. Appearance- and shape-based object categorization algorithms can benefit from our geometry-based object model in conjunction with local descriptors. Both local descriptors and geometric information, including motion trajectories and motion patterns, can be very discriminative. So we combine both, available geometric information and local descriptors, to a higher level semantic descriptor.

Higher level semantic reasoning for stationary background point features could bring further improvement. Currently, more than 50% stable and reliable background is required, due to the underlying S+M algorithm [3]. Otherwise, the results of the proposed method deteriorate. A confidence measure introduced by higher level semantic reasoning could discriminate reliable and well detectable from weak background point features. Thus, the need for more than 50% background point features could be eliminated. As a first implementation of "strong background" selection, that could be seen as a kind of "good features to track" in the spirit of Shi and Tomasi [19], we consider a similar approach as for the confidence measure introduced in the reference point update routine.

References

1. Nistér, D., Narodistky, O., Bergen, J.: Visual odometry. In: CVPR, pp. 652–659 (2004)
2. Davison, A.J., Reid, I., Molton, N., Stasse, O.: Monoslam: Real-time single camera slam. PAMI 29, 1052–1067 (2007)
3. Schweighofer, G., Segvic, S., Pinz, A.: Online/realtime structure and motion for general camera models. In: IEEE WACV (2008)
4. Klein, G., Murray, D.: Parallel tracking and mapping for small AR workspaces. In: ISMAR (2007)
5. Williams, B., Klein, G., Reid, I.: Real-time slam relocalisation. In: ICCV (2007)
6. Klein, G., Murray, D.: Improving the agility of keyframe-based SLAM. In: Forsyth, D., Torr, P., Zisserman, A. (eds.) ECCV 2008, Part II. LNCS, vol. 5303, pp. 802–815. Springer, Heidelberg (2008)
7. Newcombe, R.A., Davison, A.J.: Live dense reconstruction with a single moving camera. In: CVPR (2010)
8. Schindler, K., Suter, D., Wang, H.: A model-selection framework for multibody structure-and-motion of image sequences. IJCV 79, 159–177 (2008)
9. Costeira, J., Kanade, T.: A multi-body factorization method for motion analysis. In: ICCV, pp. 1071–1076 (1995)
10. Costeira, J.P., Kanade, T.: A multibody factorization method for independently moving objects. IJCV 29, 159–179 (1998)
11. Yan, J., Pollefeys, M.: A general framework for motion segmentation: Independent, articulated, rigid, non-rigid, degenerate and non-degenerate. In: Leonardis, A., Bischof, H., Pinz, A. (eds.) ECCV 2006. LNCS, vol. 3954, pp. 94–106. Springer, Heidelberg (2006)
12. Fitzgibbon, A.W., Zisserman, A.: Multibody structure and motion: 3-D reconstruction of independently moving objects. In: Vernon, D. (ed.) ECCV 2000. LNCS, vol. 1842, pp. 891–906. Springer, Heidelberg (2000)
13. Li, T., Kallem, V., Singaraju, D., Vidal, R.: Projective factorization of multiple rigid-body motions. In: CVPR (2007)
14. Ozden, K., Schindler, K., Gool, L.V.: Multibody structure-from-motion in practice. PAMI 32, 1134–1141 (2010)
15. Leibe, B., Schindler, K., Cornelis, N., Gool, L.V.: Coupled object detection and tracking from static cameras and moving vehicles. PAMI 30, 1683–1698 (2008)
16. Ess, A., Leibe, B., Schindler, K., Gool, L.V.: A mobile vision system for robust multi-person tracking. In: CVPR (2008)
17. Fukunaga, K., Hostetler, L.D.: The estimation of the gradient of a density function, with applications in pattern recognition. IEEE Trans. on Information Theory 21, 32–40 (1975)
18. Comaniciu, D., Meer, P.: Mean shift: A robust approach toward feature space analysis. PAMI 24, 603–619 (2002)
19. Shi, J., Tomasi, C.: Good features to track. In: CVPR, pp. 593–600 (1994)

Person Re-identification Based on Global Color Context

Yinghao Cai and Matti Pietikäinen

Machine Vision Group, Department of Electrical and Information Engineering
University of Oulu, Finland
{yinghao.cai,mkp}@ee.oulu.fi

Abstract. In this paper, we present a new solution to the problem of person re-identification. Person re-identification means to match observations of the same person across different time and possibly different cameras. The appearance based person re-identification must deal with several challenges such as variations of illumination conditions, poses and occlusions. Our proposed method inspires from the spirit of self-similarity. Self-similarity is an attractive property in visual recognition. Instead of comparing image descriptors between two images directly, the self-similarity measures how similar they are to a neighborhood of themselves. The self-similarities of image patterns within the image are modeled in two different ways in the proposed Global Color Context (GCC) method. The spatial distributions of self-similarities w.r.t. color words are combined to characterize the appearance of pedestrians. Promising results are obtained in the public ETHZ database compared with state-of-art performances.

1 Introduction

Object Recognition has received tremendous interests in the communities of computer vision and pattern recognition. The general object recognition refers to categorization of objects that belong to the same class. Different from object recognition, object identification [1] aims to distinguish visually very similar objects from one class. In this paper, we fix the category of object identification to pedestrians and consider the problem of matching observations of the same person across different time and possibly different cameras. Identifying people separated in time and locations is known as person re-identification in [2, 3] which is of great interest in applications such as long term activity analysis [4] and continuously tracking across cameras [5].

Person re-identification is a difficult problem. Since the observations of people may come from different cameras, no spatial continuity information can be exploited in person re-identification. The appearance based person re-identification must deal with several challenges such as variations of illumination conditions, poses and occlusions across time and cameras. In addition, different people may dress quite similar. For example, one can hardly tell two people dressed in homogenous black apart solely by color information. Thus, a successful person

R. Koch et al. (Eds.): ACCV 2010 Workshops, Part I, LNCS 6468, pp. 205–215, 2011.

re-identification algorithm should be able to discriminate visually very similar objects while preserving invariance across different time and cameras.

Self-similarity is an attractive property in visual recognition [6, 7]. Instead of comparing image descriptors between two images directly, the self-similarity measures how similar they are to a neighborhood of themselves despite that the image patterns generating those self-similarities may be dramatically different across images [7]. The spirit of self-similarity is desirable in person re-identification since image patterns of the same person across time and cameras appear differently at pixel level. Many work have exploited the spirit of self-similarity in applications such as texture classification [8], image matching [7] and activity recognition [6]. In this paper, we mainly exploit the spatial distributions of self-similarities of features w.r.t. visual words to represent the appearance of pedestrians. The self-similarities of image patterns within the image are modeled in two different ways in the proposed Global Color Context (GCC) method. Experimental results on public benchmark dataset ETHZ [3, 9] demonstrate the effectiveness of the proposed method.

The rest of the paper is organized as follows. An overview of related work is in Section 2. We briefly introduce our Global Color Context (GCC) method in Section 3. Experimental results and conclusions are given in Section 4 and Section 5, respectively.

2 Related Work

Many methods have been put forward to address the problem of person re-identification [3, 9, 10, 11]. Color cue is widely used in person re-identification since the color of clothing provides information about the identity of the individual. Farenzenna et al. [3] combined HSV histogram, Maximally Stable Color Regions and recurrent patches together to get a description inside the silhouette of individuals. Those color features are weighted by their distances to the y-axis of symmetry of torso and legs. However, color based features are subject to variations of illumination conditions. To this end, various color invariants were proposed in [12, 13, 14]. The invariance properties of color descriptors depend on the types of illumination and the dataset used. An alternative solution to compensate illumination variations is by finding a transformation matrix [10] or a mapping function [11] which maps the appearance of one object to its appearance under another view. However, either transformation matrix or mapping function may not be unique in uncontrolled illumination conditions.

On the other hand, texture and edge features are exploited as complementary information to solely color information. Two families of texture filters, Schmid and Gabor, were explored in Gray and Tao [2]. Edge information was captured by histograms of oriented gradients (HOG) in Schwartz and Davis [9]. One image per person is required in [9] to obtain a high-dimensional feature vector composed of texture, gradient and color information for partial least square reduction [15]. Takala et al. [16] employed adaptive boosting on a wide collection of image features (shape, pose, color, texture, etc) to construct appearance models for tracked objects. It is shown that the overall performance of person re-identification is largely improved when combining multiple cues together.

As can be inferred from the name of our proposed method, GCC only considers the global layouts of self-similarities w.r.t. color words in the visual codebook. Any other color descriptors [12] and texture descriptors [8] can be easily plugged in our framework to further improve the performance. We employ color invariants [13] as features in this paper to handle the illumination variations in person re-identification. The self-similarities of image patterns are derived through computing their distances to color words in the codebook. Promising results are obtained in the public ETHZ dataset compared with state-of-art performances [3, 9].

3   Global Color Context

An overview of the proposed Global Color Context method can be seen from Figure 1. We first group visually similar color features to obtain a color codebook. The color codebook is obtained by k-means clustering at densely sampled image locations where color features are computed in a 3×3 neighborhood. Then, given a new image, the color features (Section 3.1) from the new image are assigned to color codebook (Section 3.2). Finally, the spatial occurrence distributions of self-similarities w.r.t. color words are learned and combined to characterize the appearance of pedestrians (Section 3.3).

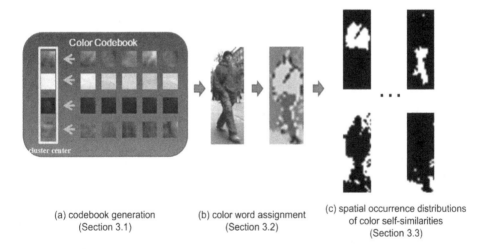

(a) codebook generation
(Section 3.1)

(b) color word assignment
(Section 3.2)

(c) spatial occurrence distributions
of color self-similarities
(Section 3.3)

Fig. 1. An overview of the proposed method. Local color features (Section 3.1) are extracted densely and clustered to form a color codebook. Then, the assignments of color features to color words in the codebook are explored in Section 3.2. Color features from the same visual color word are marked with the same color in Figure 1(b). For each color word in the codebook, the spatial occurrence distributions of color self-similarities are learned in Section 3.3(Figure 1(c)).

3.1 Color Descriptors

A wide range of color descriptors have been proposed in [12, 13, 14]. The invariance properties of color descriptors are summarized in van de et al. [12]. It is shown that the distinctiveness of color descriptors and their invariance properties are data-specific. In this section, we briefly review two color descriptors, hue histogram and opponent histogram from Van de Weijer and Schmid [13]. The two descriptors are chosen due to their superior performances on the ETHZ dataset we used.

Hue Histogram. In HSV color space, hue is proven to be both lighting geometry and specular invariant [13, 12]. However, hue becomes unstable near the grey axis. To this end, Van de Weijer and Schmid [13] applied an error analysis to the hue. The error analysis is based on the fact that the certainty of hue is inversely proportional to the saturation. Small values of saturation bring uncertainties in the computation of hue. Therefore, hue with small value of saturation should count less in histogram. In the construction of hue histogram, each sample of hue is weighted by its saturation [13, 12]. Hue and saturation can be computed from opponent colors [13, 12]:

$$hue = \arctan(\frac{O_1}{O_2}) = \arctan(\frac{\sqrt{3}(R-G)}{R+G-2B}) \tag{1}$$

$$saturation = \sqrt{O_1^2 + O_2^2} = \sqrt{\frac{2}{3}(R^2+G^2+B^2-RG-RB-GB)} \tag{2}$$

where O_1 and O_2 are two components from opponent color space:

$$O_1 = \frac{1}{\sqrt{2}}(R-G) \tag{3}$$

$$O_2 = \frac{1}{\sqrt{6}}(R+G-2B) \tag{4}$$

Finally, the hue histogram is divided into 36 bins according to Van de Weijer and Schmid [13].

Opponent Histogram. According to Van de Weijer and Schmid [13], the opponent angle ang_x^O in opponent color space is supposed to be specular invariant. The opponent angle ang_x^O is defined as:

$$ang_x^O = \arctan(\frac{O_{1x}}{O_{2x}}) \tag{5}$$

where O_{1x} denotes the first order derivative of O_1, etc. Similar to the error analysis of hue histogram, Van de Weijer and Schmid [13] also applied an error analysis to the opponent angle. Here, ∂ang_x^O is defined as the weight for the opponent angle:

$$\partial ang_x^O = \frac{1}{\sqrt{O_{1x}^2 + O_{2x}^2}} \tag{6}$$

The opponent histogram is also quantized to 36 bins.

3.2   Color Word Assignment

After extracting local color features at densely sampled image locations, we then group color features together to obtain color words (prototypes) of local appearances by k-means clustering. K-means clustering is a popular method in the Bag-of-Features framework due to its computational simplicity. A histogram of the visual words is usually obtained to characterize the appearance of an image in the Bag-of-Features framework. In this paper, we mainly exploit the spatial distributions of self-similarities of color features w.r.t. color words to represent the appearance of pedestrians.

It is known that the main deficiency of k-means clustering lies in the user needs to specify the number of clusters in advance. However, the number of clusters affects the performance of final person re-identification. Some color features may lie in-between several cluster centers which results in ambiguity in color word assignment. We mainly discuss two methods for color word assignment, hard assignment and soft assignment in this section where self-similarities are modeled in two different ways.

Hard Assignment. In hard assignment, each color feature is assigned to exactly one color word in the codebook learned by k-means clustering. Hard assignment explicitly models the self-similarities of image patterns w.r.t. one visual word to binary. It is assumed in hard assignment that two image patterns are similar to each other only if they are assigned to the same visual word. In the hard assignment, the occurrence frequency of each color word is computed as [17]:

$$\text{Count}(w) = \sum_{i=1}^{N} \begin{cases} 1 & \text{if } w = \underset{v \in V}{\arg\min}(D(f_i, v)) \\ 0 & \text{otherwise} \end{cases} \tag{7}$$

where w is the color word in the codebook. N is the number of local image regions. f_i is the color feature computed in image region. $D(f_i, v)$ is the Euclidean distance between color word v in codebook V and color feature f_i. Since the assignments of color words can be done once and for all, we do not need to compute the pairwise sum of squared differences (SSD) as in Schechtman and Irani [7]. Thus computing self-similarities of image patterns based on their visual words is computationally more efficient.

Soft Assignment. In the hard assignment, we assume that each color feature can be well represented by one single word in the codebook. However, it is often the case that a color feature has multiple candidates in the visual codebook which gives rise to visual word ambiguity [17]. In addition, as we mentioned before, the classification performance is closely related to the size of the codebook. While larger values of k bring rich representations over a wide variety of colors, they lead to overfitting in k-means clustering. On the other hand, small numbers of visual words are generally not representative of all local features. Soft assignment of visual words provides a tradeoff for this problem.

Furthermore, in computing the self-similarities of image patterns, the assumption in the hard assignment that two color features are similar to each other only if they are assigned to the same word provides strict constraints for matching pedestrians while soft assignment allows for appearance variations within the image to be compared. The soft assignment method assigns color words according to:

$$\text{Count}(w) = \sum_{i=1}^{N} \exp\left(-\frac{D(f_i, w)}{\sigma}\right) \tag{8}$$

where σ is a parameter controlling the smoothness of the self-similarities w.r.t. color word w. Figure 2 shows an example of hard assignment and soft assignment. Here, we compute the self-similarities of the clothing of the pedestrian. It can be seen from Figure 2 that hard assignment models the self-similarities w.r.t. the word occurred to binary while soft assignment provides a more smooth spatial distribution of the occurred visual word.

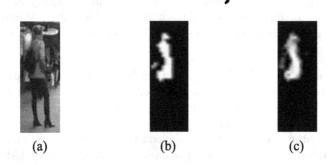

(a)   (b)   (c)

Fig. 2. An example of hard assignment and soft assignment. (a) Original pedestrian image. (b) The self-similarities w.r.t. color word occurred by hard assignment. (c) The self-similarities w.r.t. color word occurred by soft assignment.

3.3   Global Color Context

In previous section, we have explored two different ways of modeling self-similarities of image patterns within the image. In this section, we will learn how these self-similarities are distributed in the spatial domain. For each visual word in the codebook, we compute its occurrence frequency in a log-polar grid. The log-polar grid is partitioned into 32 bins (8 angles and 4 radial intervals) centered at the image center. The log-polar representation accounts for pose variations across images. To alleviate the influence of background clutters, each pixel is weighted by a Gaussian function in Figure 3(b) where pixels near the image center count more. The spatial distribution of each color word is normalized to one to characterize the appearance of pedestrians.

We name our method Global Color Context (GCC) since our method captures the self-similarities of image patterns w.r.t. color words in the entire image. Each pedestrian image has k color contexts in total where k equals to the number of color words in the codebook. Each color context records the spatial distribution of self-similarities w.r.t. the specific word in 4×8 bins. The similarity between two images is computed as the mean Chi-Square distance of k color contexts. Finally,

(a)   (b)

Fig. 3. The occurrence frequency of color words are computed in a log-polar grid in (a). To alleviate the influence of background clutters, each pixel is weighted by a Gaussian function in (b) where pixels near the image center count more.

the correspondences between pedestrian images are determined according to nearest neighbor classifier.

4   Experimental Results and Analysis

We evaluate our proposed GCC method on the public ETHZ dataset [18]. ETHZ dataset [18] was originally used for human detection. Schwartz and Davis [9] cropped pedestrian images by the ground truth locations of people in videos for person re-identification. The cropped ETHZ dataset contains three video sequences. Information about the cropped dataset is summarized in Table 1. The number of images per person varies from a few to hundreds. The main challenges of ETHZ dataset lie in variations in pedestrian's appearances and occlusions. Some sample images of ETHZ dataset are shown in Figure 4. Schwartz and Davis [9] carried out experiments on the ETHZ dataset to test their Partial Least Squares (PLS) method [15]. Recently, Farenzenna et al. [3] also tested their algorithms on the ETHZ dataset. We follow the evaluation methods of Farenzenna et al. [3] to validate the effectiveness of the proposed method.

According to Farenzenna et al. [3], the problem of person re-identification can be divided into two cases, single-shot case and multiple-shot case. The first situation matches people across time and locations based on one single image while multiple-shot case employs sequences of images for identification. In the single-shot case, we randomly select one image for each pedestrian as the gallery image while another randomly selected image forms the probe set. The procedure is repeated 10 times according to Farenzenna et al. [3]. The multiple-shot

Table 1. The ETHZ Dataset

	SEQ 1	SEQ 2	SEQ 3
Num of People	83	35	28
Total Num of Images	4857	1936	1762

Fig. 4. Sample images from ETHZ dataset

case is carried out on $N = 2, 5$ for multiple-shot vs single-shot(MvsS) with 100 independent times [3] where $N = 2, 5$ numbers of images are used as gallery set and one image forms the probe set. In this paper, the averaged cumulative matching characteristic curve (CMC) [2,3] is used to evaluate the performance of person re-identification. In CMC curve, rank i performance is the rate that the correct person is in the top i of the retrieved list.

In learning the color codebook, we carry out k-means clustering on VIPeR dataset [2]. The CMC curves of multiple choices of $k, k = 30, 50, 80$ on ETHZ dataset by hue histogram method are shown in Figure 5. Only the results of one-shot case are reported in Figure 5. In Figure 5, Hue_{80Hard} denotes the performance of hard assignment of $k = 80$ by hue histogram, etc. In soft assignment, the parameter σ is set to 0.02 in all experiments. We can see from Figure 5 that assigning color words by soft assignment method generally performs better than hard assignment method under various choices of k. For simplicity, we fix the size of codebook k to 30 and only consider the performances of soft assignment in the following experiments.

The CMC curves of hue histogram and opponent histogram by soft assignment method are shown in Figure 6 and Figure 7, respectively. We compare our proposed method with the PLS method in Schwartz and Davis [15] and the SDALF method in Farenzenna et al. [3]. The results of the PLS method and

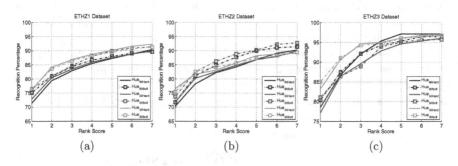

Fig. 5. The CMC curves of multiple choices of $k, k = 30, 50, 80$ on ETHZ dataset of hue histogram

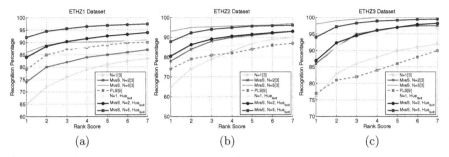

Fig. 6. The CMC curves of hue histogram on ETHZ dataset

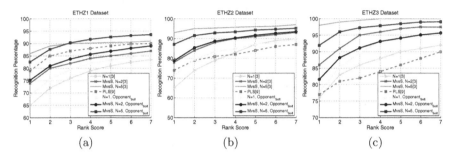

Fig. 7. The CMC curves of opponent histogram on ETHZ dataset

SDALF method are taken directly from [9] and [3], respectively. In Figure 6 and Figure 7, $N = 1$ denotes single-shot case in Farenzenna et al. [3]. MvsS, $N = 2$ and MvsS, $N = 5$ are different choices of N in multiple-shot case. We can see from Figure 6 and Figure 7 that our proposed GCC method achieves promising results in most cases. One possible reason for our success is that the pose variation is relatively small in ETHZ dataset. Our proposed GCC method captures the spatial distributions of color self-similarities well. Furthermore, the influence of background clutters is minimized through Gaussian weighting while Schwartz and Davis [9] exploited all foreground and background information in their PLS method.

5   Conclusions

In this paper, we have presented an approach to person re-identification inspired from the spirit of self-similarity. Experimental results on the public ETHZ dataset demonstrate the effectiveness of the proposed method. Our proposed method only considered the self-similarities w.r.t. color words. Future work will focus on exploring more texture descriptors in the current framework to further improve the performance of person re-identification.

Acknowledgement. This work was partly supported by the Academy of Finland.

References

1. Ferencz, A., Learned-Miller, E.G., Malik, J.: Learning to locate informative features for visual identification. International Journal of Computer Vision 77, 3–24 (2008)
2. Gray, D., Tao, H.: Viewpoint invariant pedestrian recognition with an ensemble of localized features. In: Forsyth, D., Torr, P., Zisserman, A. (eds.) ECCV 2008, Part I. LNCS, vol. 5302, pp. 262–275. Springer, Heidelberg (2008)
3. Farenzena, M., Bazzani, L., Perina, A., Murino, V., Cristani, M.: Person re-identification by symmetry-driven accumulation of local features. In: Proceedings of Computer Vision and Pattern Recognition (2010)
4. Shet, V.D., Harwood, D., Davis, L.S.: Multivalued default logic for identity maintenance in visual surveillance. In: Leonardis, A., Bischof, H., Pinz, A. (eds.) ECCV 2006. LNCS, vol. 3954, pp. 119–132. Springer, Heidelberg (2006)
5. Javed, O., Rasheed, Z., Shafique, K., Shah, M.: Tracking across multiple cameras with disjoint views. In: Proceedings of International Conference on Computer Vision, pp. 952–957 (2003)
6. Junejo, I.N., Dexter, E., Laptev, I., Perez, P.: View-independent action recognition from temporal self-similarities. IEEE Transactions on Pattern Analysis and Machine Intelligence 99 (2010)
7. Shechtman, E., Irani, M.: Matching local self-similarities across images and videos. In: Proceedings of Computer Vision and Pattern Recognition, pp. 1–8 (2007)
8. Ojala, T., Pietikäinen, M., Mäenpää, T.: Multiresolution gray-scale and rotation invariant texture classification with local binary patterns. IEEE Transactions on Pattern Analysis and Machine Intelligence, 971–987 (2002)
9. Schwartz, W.R., Davis, L.S.: Learning discriminative appearance-based models using partial least squares. In: Proceedings of the XXII Brazilian Symposium on Computer Graphics and Image Processing (2009)
10. Gilbert, A., Bowden, R.: Tracking objects across cameras by incrementally learning inter-camera colour calibration and patterns of activity. In: Leonardis, A., Bischof, H., Pinz, A. (eds.) ECCV 2006. LNCS, vol. 3952, pp. 125–136. Springer, Heidelberg (2006)
11. Javed, O., Shafique, K., Shah, M.: Appearance modeling for tracking in multiple non-overlapping cameras. In: Proceedings of Computer Vision and Pattern Recognition, pp. 26–33 (2005)
12. van de Sande, K.E.A., Gevers, T., Snoek, C.G.M.: Evaluating color descriptors for object and scene recognition. IEEE Transactions on Pattern Analysis and Machine Intelligence 32 (2010)
13. van de Weijer, J., Schmid, C.: Coloring local feature extraction. In: Leonardis, A., Bischof, H., Pinz, A. (eds.) ECCV 2006. LNCS, vol. 3952, pp. 334–348. Springer, Heidelberg (2006)
14. Burghouts, G.J., Geusebroek, J.M.: Performance evaluation of local colour invariants. Computer Vision and Image Understanding 113, 48–62 (2009)

15. Schwartz, W., Kembhavi, A., Harwood, D., Davis, L.: Human detection using partial least squares analysis. In: Proceedings of International Conference on Computer Vision (2009)
16. Takala, V., Cai, Y., Pietikäinen, M.: Boosting clusters of samples for sequence matching in camera networks. In: Proceedings of International Conference on Pattern Recognition (2010)
17. van Gemert, J.C., Veenman, C.J., Smeulders, A.W.M., Geusebroek, J.M.: Visual word ambiguity. IEEE Transactions on Pattern Analysis and Machine Intelligence 32, 1271–1283 (2010)
18. Ess, A., Leibe, B., Schindler, K., Gool, L.V.: A mobile vision system for robust multi-person tracking. In: Proceedings of Computer Vision and Pattern Recognition, pp. 1–8 (2008)

Visual Object Tracking via One-Class SVM

Li Li, Zhenjun Han, Qixiang Ye, and Jianbin Jiao*

Graduate University of Chinese Academy of Sciences, Beijing, China
Fax: +86-10-88256278
lily_sdl@gucas.ac.cn, hanzhenjun06@mails.gucas.ac.cn,
{qxye,jiaojb}@gucas.ac.cn

Abstract. In this paper, we propose a new visual object tracking approach via one-class SVM (OC-SVM), inspired by the fact that OC-SVM's support vectors can form a hyper-sphere, whose center can be regarded as a robust object estimation from samples. In the tracking approach, a set of tracking samples are constructed in a predefined searching window of a video frame. And then a threshold strategy is proposed to select examples from the tracking sample set. Selected examples are used to train an OC-SVM model which estimates a hyper-sphere encircling most of the examples. Finally, we locate the center of the hyper sphere as the tracked object in the current frame. Extensive experiments demonstrate the effectiveness and robustness of the proposed approach in complex background.

Keywords: Object tracking, One-class SVM, Tracking sample set.

1 Introduction

Object tracking has been becoming one of the most popular research topics and playing an important role in many video applications, such as video-based human computer interaction systems [3], automatic driving systems [14], intelligent video surveillance [6] and robotics [16]. However, owing to the difficulties arising from the object motion state variation, the appearance variation of either object or background and the occlusions, the performance of tracking algorithm remains to be improved.

Object tracking is to automatically find the same object in adjacent video frames after the object's location is initialized. That is to accurately calculate the object's location(u, v) and scale s in a new video frame, even when the object's motion state varies or its appearance is affected by noise, occlusion or clutter background during the tracking process. In the existing object tracking researches, lots of methods and strategies have been proposed to deal with the noise, occlusion, and background clutter problems. In these methods, SVM (support vector machine) based method is attracting more and more attentions [2, 20, 7, 17, 18, 10].

---

* Corresponding author.

R. Koch et al. (Eds.): ACCV 2010 Workshops, Part I, LNCS 6468, pp. 216–225, 2011.
© Springer-Verlag Berlin Heidelberg 2011

There are totally three types of SVM, namely support vector classification (SVC), support vector regression (SVR) and distribution estimation (one-class SVM), which have been exploited for object tracking in previous literatures [2, 20, 7, 17, 18, 10]. In [2], Avidan proposes Support Vector Tracking method, which integrates SVM classifier with an optic-flow based tracker by maximizing the SVM classification score. In [20] [7], the authors combine SVM with different types of tracking filters, particle filter and adaptive Kalman filter for efficient visual tracking, while in [17, 18] the authors treat object tracking as a foreground/background classification problem, and use SVM as a classifier to distinguish an object from its background. And in [10] Kabaoglu applied SVR to multiple targets tracking, where combining SVR with particle filters, they can obtain an effective probability distribution of multiple targets for tracking.

As Zhou et al. has proposed [19], "Many Could Be Better Then All" is a proved theory learning algorithms, and can also be extended to tracking problems. Considering the basic issue of tracking is to find the object in video frames, we can reasonably make a hypothesis that the object can be well estimated by "many" object examples inside the searching window. One-class SVM (OC-SVM) is proposed to find a tighter hyper-sphere encircling "many" given examples [15], so we can represent the tracked object with "many" selected examples and the center of the hyper-sphere can be regarded as a robust location estimation of tracked object.

As shown in Fig.1, the solid dots and small boxes respectively represent examples and tracking template. After the OC-SVM training process performed on these solid dots, we can obtain the center (hollow circle) and the hyper-sphere encircling "many" of those dots, with the noise examples excluded by the hyper-sphere. The center of the hyper-sphere can be regarded as a robust object estimation for tracking.

In this paper, inspired by the property of OC-SVM and the characteristics of tracking, we propose an approach to model object tracking problem by finding the center of the hyper-sphere of OC-SVM. In our tracking system, a set of tracking samples are constructed in a predefined searching window of a new

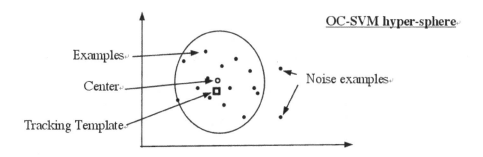

Fig. 1. Illustration of OC-SVM hyper-sphere. The solid black dots, small boxes, and hollow circles respectively represent the examples, tracking template, and center of hyper-sphere.

video frame, and then tracking examples are selected from the set using threshold method. Because of the limited size of the searching window and our set construction strategy, a relatively small number of tracking examples will be left to represent the object. Different from the previous work [2,20,7,17,18,10], OC-SVM in our method is directly used to estimate the distribution in feature space and then to locate object for tracking, instead of transferring part of tracking problem into SVM classification or SVR regression problem.

The rest of this paper is organized as follows. Section 2 describes the proposed OC-SVM based tracking approach in details. Experiments are presented in section 3, and conclusions and future work are presented in section 4.

2 Object Tracking Based on OC-SVM

The flow chart of the proposed tracking approach is shown in Fig.2. In the following sections, we will present the details of the steps in the flow chart.

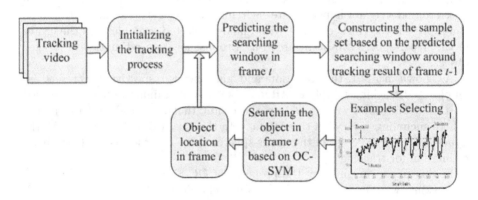

Fig. 2. Flow chart of our tracking approach based on OC-SVM

2.1 Tracking Sample Set Construction

In the current video frame, a sample set is constructed by extracting image blocks (sub-windows) from a searching window, which is a rectangle of size $W \times H$ surrounding the previous tracking results in the new tracking frame (the black rectangle in Fig.3a). This window can be determined with a Kalman filter method with constant velocity motion model. And each sample in the set is defined as a sub-window of the searching window (the rectangle) in Fig.3b). A sample rectangle is specified by $r = (x, y, s, \alpha)$ with $0 < x < W, 0 < y < H, s > 0, 0° \leq \alpha \leq 360°$. This sample set is (almost) infinitely large. For practical reasons, it is reduced as follows:

1. The (x, y) varies with the step of n pixels in horizontal and vertical orientations;

2. The s is uniformly range from 0.8 to 1.2 times of the tracked object's size;
3. The α is set as $0°$ in our approach, for rotation of the samples is not considered.

These restrictions lead to a reasonable number of samples in the set. Suppose that totally K samples are extracted to construct a set $\left\{ \overrightarrow{S_t^i} : i = 1, \cdots, K \right\}$ for the tth video frame, it can be seen that most of the samples are associated with the background, and only a few of them are parts or the whole of object (shown in Fig.3c).

To represent the object and samples, we extract a composite feature set including location (x, y), scale s and HOGC [9], which is a 120 dimension vector including both color and gradient histograms, to represent each sample $\overrightarrow{S_t^i}$ in the set. HOGC can capture both the color and local contour characteristics. Then we can obtain a set $A_t = \left\{ A_t^i : i = 1, \cdots, K \right\}$ for all the samples at frame t, in the composite feature space.

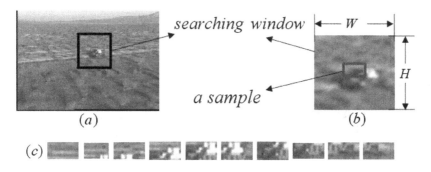

Fig. 3. (a) The searching window in which sample set is constructed. (b) A sample. (c) Some samples in the sample set.

2.2   Examples Selection for OC-SVM

As the input of OC-SVM, most of the tracking examples should be from one class. The samples which are quite different with the tracking object should be discarded. Therefore, we propose a scheme to select tracking examples by calculating the similarity between the instantaneous tracking samples in the constructed set and the tracked object. The similarity is determined by the Bhattacharyya similarity coefficient [11] of the feature as follows:

$$Bhat(F, A_t^i) = \sum_{j=1}^{M} \sqrt{F(j) A_t^i(j)} \tag{1}$$

where F, A_t^j are respectively the feature of the tracked object and the ith sample $\overrightarrow{S_t^i}$ in the composite feature space, and M is the number of feature dimension. The similarity between the tracked object with each sample in the set constructed at frame t is shown in Fig.4.

After the similarity between the instantaneous tracking samples and the previous tracked object is calculated, a threshold method is used to select the tracking examples from the whole sample set. We use some dynamic value between the maximum and minimum as a threshold, which is calculated as follow:

$$Threshold = \varphi^* \max + (1 - \varphi)^* \min \qquad (2)$$

where φ is a ratio parameter between 0 and 1, and we choose $\varphi = 0.2$ or 20% as division point following the eighty twenty rule, which can include most of the information, and prove to be best choice in our experiments. Then, we obtain the tracking examples $\overrightarrow{S_t^i}$ in frame as follows:

$$\begin{cases} Bhat \geq Threshold; & sample \in \{examples\} \\ Bhat < Threshold; & sample \notin \{examples\} \end{cases} \qquad (3)$$

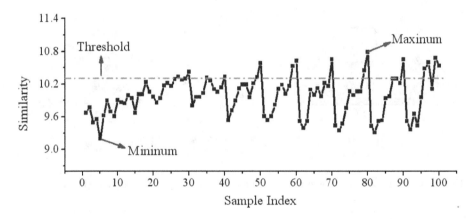

Fig. 4. The similarity between the tracked object template and 100 samples in the sample set at frame t. The maximum, minimum, and their threshold calculated using formula (2) can be found and marked in the graph. The samples above the threshold line are chosen as examples to construct the OC-SVM.

2.3   Object Tracking Based on OC-SVM

After we obtain the examples, OC-SVM is employed to estimate a hyper-plane encircling the most of a given examples without any class information, to locate object for tracking.

Given the tracking example set $\left\{ \overrightarrow{S_t^i} : i = 1, \cdots, K \right\}$ and their corresponding $A_t = \left\{ A_t^i : i = 1, \cdots, K \right\}$, the hyper-plane (R, c) that encircles the most of examples can be given by solving the following quadratic optimization procedure [15]:

$$\min_{R\in,\varepsilon\in,c\in,} R^2 + \frac{1}{vK}\sum_i \varepsilon_i \tag{4}$$

$$\textbf{Subject to } \left\|pos(\overrightarrow{S_t^i}^{*}) - c\right\|^2 \le R^2 + \varepsilon_i, \varepsilon_i \ge 0 \tag{5}$$

where R is the radius of the hyper-plane, $[0,1]$ acts as the proportion of the examples inside the hyper-plane to all examples, $pos(\overrightarrow{S_t^i}^{*})$ is the position of the ith selected example, c is the center of the hyper-plane or object position in current frame, and ε_i represents the distance of example $\overrightarrow{S_t^i}$ apart from the hyper-plane, and equals to zero when the example is inside the hyper-plane.

From (4) and (5), it can be seen that a hyper-plane (R,c) calculated by OC-SVM is used to encircle most of examples $\left\{\overrightarrow{S_t^i} : i = 1, \cdots, K\right\}$ as much as possible. After solving the quadratic optimization problem (4) and (5), we will obtain the object location c in current frame.

3 Experiments and Results

We carried out a variety of experiments with widely used videos from VIVID [4], CAVIAR [8] and SDL data set [1] to validate the proposed approach. The test videos consist of a variety of cases, including occlusions, lighting changes, scale variations, object rotations and complex backgrounds. Some of the videos are captured on moving platforms. The objects include moving humans and vehicles. Experimental results on three video clips of them are shown in Fig.5. For all sequences, video objects in the first frame are manually labeled and then initialized for tracking.

As is shown in Fig.5, the larger images on the left are the tracking scenes (the first frame that we track), and the smaller ones on the right are the tracking results using our approach (top row) and classical mean-shift approach [13] (bottom row). Since all the target regions on the right are small compared to the whole scene, only image patches that contain target regions are shown. Red rectangle is the result of tracking experiments, while the manually labeled blue ellipse is the actual location of the tracked object.

In the first video clip from VIVID data set, the car loops around on a runway, and the appearance varies largely from the initial state because of vehicle rotations ($405^{th}, 448^{th}, 479^{th}$ frames). Fig.5a shows that the stable tracking results based on our proposed approach for object rotations compared with mean-shift algorithm.

The second video clip from VIVID data set is quite challenging. Two groups of similar vehicles travel in opposite directions and then intersect with each other. Because of the similarity of the tracked vehicle and partial occlusions of the other vehicles, tracking errors happen in this case. Fig.5b presents the tracking results based on our proposed approach and mean-shift algorithm, which shows the robustness of our method in object rotations, mimic objects and similar background.

(a) 387th frame   405th frame   448th frame   479th frame

(b) 161th frame   314th frame   349th frame   533th frame

(c) 71th frame   94th frame   685th frame   762th frame

Fig. 5. Three examples of tracking experiments. The larger images on the left are the tracking scenes (the first frame that we track), and the smaller ones on the right are the tracking results using our approach *toprow* and classical mean-shift approach [1] *bottomrow*. Since the target regions on the right are small compared to the whole scene, only image patches that contain target regions are shown. Red rectangle is the result of tracking experiments, while the manually labeled blue ellipse is the actual location of the tracked object.

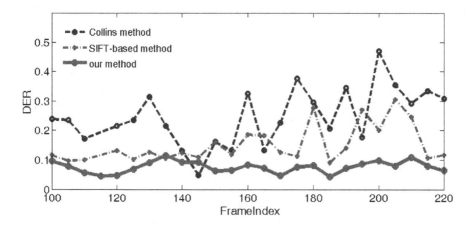

Fig. 6. Average DER of three methods

The third video clip in Fig. 5c from the SDL data set is very challenging. There are serious occlusions on the object, and the background has the similar color to the object (a person in small size) and there are some small trees which are quite similar to the object in shape. Our approach can track the object correctly. The tracking results of this video show that the proposed approach can effectively deal with partial occlusions, appearance variations and similar color and contour.

Besides contrast experiments, tracking efficiency is another evaluation criterion for a real-time tracking application. In the experiments, we find that our proposed tracking approach can work almost real time on a computer with Core(TM) 2 Duo CPU (2.53GHz) and 3GB memory. To quantitatively evaluate the proposed approach, relative displacement error rates (DER) is defined as follows.

$$\text{DER} = \frac{\text{Displacement error betweern tracked object position and groundtruth}}{\text{Size of the object}} \tag{6}$$

In the experiments we use the average DER of 10 video clips from the above 3 data sets to reflect the performance of each method. The lower the average DER is, the better the tracking performance is. We compare our method with other three representative ones, including single template tracking method with SIFT feature [12] and adaptive single template tracking method of Collins [5]. The results of three methods are shown in Fig.6. It can be seen from the figures that the average DER of our method (about 0.03 to 0.12) is smaller than that of the other two methods in almost the whole tracking process.

4 Conclusions and Future Works

In this paper, we propose a new object tracking approach via OC-SVM. The basic idea of this paper is that object tracking is to automatically find the same object in adjacent video frames. Based on this idea, the tracking sample set is firstly constructed, and then tracking examples are chosen from the tracking sample set as input of OC-SVM. We can obtain object location in each frame from previous frame information using OC-SVM for tracking. Experiments validate the effectiveness of the tracking approach in complex backgrounds. Comparisons indicate that object tracking with OC-SVM outperforms some of the representative tracking methods. The proposed approach can be extended to multiple object tracking in the future work.

References

1. Sdl data set, `http://www.coe.gucas.ac.cn/SDL-HomePage`
2. Avidan, S.: Support vector tracking. IEEE Transactions on Pattern Analysis and Machine Intelligence 26(8), 1064–1072 (2004)
3. Betke, M., Gips, J., Fleming, P.: The camera mouse: Visual tracking of body features to provide computer access for people with severe disabilities. IEEE Transactions on Neural Systems and Rehabilitation Engineering 10(1), 1–10 (2002)
4. Collins, R., Zhou, X., Teh, S.K.: An open source tracking testbed and evaluation web site. In: IEEE International Workshop on Performance Evaluation of Tracking and Surveillance (2005)
5. Collins, R.T., Liu, Y., Leordeanu, M.: Online selection of discriminative tracking features. IEEE Transactions on Pattern Analysis and Machine Intelligence, 1631–1643 (2005)
6. Cucchiara, R., Grana, C., Prati, A., Vezzani, R.: Computer vision system for in-house video surveillance, vol. 152, pp. 242–249 (2005)
7. Dai, H., Chen, M., Zhou, S.: Support vector machine based adaptive kalman filtering for target tracking. In: General System and Control System, pp. 114–118 (2007)
8. Fisher, R., Santos-Victor, J., Crowley, J.: Caviar test case scenarios, `http://www.homepages.inf.ed.ac.uk/rbf/CAVIAR`
9. Han, Z., Ye, Q., Jiao, J.: Online feature evaluation for object tracking using kalman filter. In: IEEE International Conference on Pattern Recognition, pp. 3105–3108 (2008)
10. Kabaoglu, N.: Target tracking using particle filters with support vector regression. IEEE Transactions on Vehicular Technology 58(5), 2569–2573 (2009)
11. Kailath, T.: The divergence and bhattacharyya distance measures in signal selection(divergence and bhattacharyya distance measures in signal selection). IEEE Transactions on Communication Technology 15, 52–60 (1967)
12. Li, Y., Yang, J., Wu, R., Gong, F.: Efficient object tracking based on local invariant features. In: International Symposium on Communications and Information Technologies, pp. 697–700 (2006)
13. Liang, D., Huang, Q., Jiang, S., Yao, H., Gao, W.: Mean-shift blob tracking with adaptive feature selection and scale adaptation. In: IEEE International Conference on Image Processing, vol. 3, pp. 1497–1500 (2007)
14. McCall, J.C., Trivedi, M.M.: Video-based lane estimation and tracking for driver assistance: survey, system, and evaluation. IEEE Transactions on Intelligent Transportation Systems 7(1), 20–37 (2006)

15. Schölkopf, B., Platt, J.C., Shawe-Taylor, J., Smola, A.J., Williamson, R.C.: Estimating the support of a high-dimensional distribution. Neural computation 13(7), 1443–1471 (2001)
16. Schulz, D., Burgard, W., Fox, D., Cremers, A.B.: People tracking with mobile robots using sample-based joint probabilistic data association filters. International Journal of Robotics Research 22(2), 99–116 (2003)
17. Tang, F., Brennan, S., Zhao, Q., Tao, H.: Co-tracking using semi-supervised support vector machines. In: IEEE International Conference on Computer Vision, pp. 992–999 (2007)
18. Zhang, J., Chen, D., Tang, M.: Combining discriminative and descriptive models for tracking. In: Computer Vision CACCV 2009, pp. 113–122 (2010)
19. Zhou, Z.H., Wu, J., Tang, W.: Ensembling neural networks: Many could be better than all. Artificial intelligence 137(1-2), 239–263 (2002)
20. Zhu, G., Liang, D., Liu, Y., Huang, Q., Gao, W.: Improving particle filter with support vector regression for efficient visual tracking. In: IEEE International Conference on Image Processing, pp. 1501–1504 (2005)

Attenuated Sequential Importance Resampling (A-SIR) Algorithm for Object Tracking

Md. Zahidul Islam, Chi-Min Oh, and Chil-Woo Lee

Chonnam National University, Gwangju 500-757, South Korea

Abstract. This paper presents a newly developed attenuating resampling algorithm for particle filtering that can be applied to object tracking. In any filtering algorithm adopting concept of particles, especially in visual tracking, re-sampling is a vital process that determines the algorithm's performance and accuracy in the implementation step.It is usually a linear function of the weight of the particles, which decide the number of particles copied. If we use many particles to prevent sample impoverishment, however, the system becomes computationally too expensive. For better real-time performance with high accuracy, we introduce a steep Attenuated Sequential Importance Re-sample (A-SIR) algorithm that can require fewer highly weighted particles by introducing a nonlinear function into the resampling method. Using our proposed algorithm, we have obtained very impressive results for visual tracking with only a few particles instead of many. Dynamic parameter setting increases the steepness of resampling and reduces computational time without degrading performance. Since resampling is not dependent on any particular application, the A-SIR analysis is appropriate for any type of particle filtering algorithm that adopts a resampling procedure. We show that the A-SIR algorithm can improve the performance of a complex visual tracking algorithm using only a few particles compared with a traditional SIR-based particle filter.

1 Introduction

Particle filter is the combination of two main elements: sequential importance sampling (SIS) [1,2] and resampling. This combination of SIS and resampling is called sequential importance resampling (SIR). In SIS algorithm, after some iterations, only very few particles have non-zero importance weights. This phenomenon is often represented as weight degeneracy or sample impoverishment. An intuitive solution is to multiply the particles with high normalized importance weights, and discard those with low normalized importance weights, which can be done in the resampling step. In practical situation, however, present resampling algorithm can not really prevent the weight degeneracy problem, it just saves further calculations time by discarding the particles associated with insignificant weights. In this proposed A-SIR algorithm, we change the conventional resampling principle of SIR by using a nonlinear function that attenuates particles and uses fewer more effective and higher-weighted particles. The conventional resampling method in SIR replaces the high important weights with

R. Koch et al. (Eds.): ACCV 2010 Workshops, Part I, LNCS 6468, pp. 226–235, 2011.

many replicates of particles, thus introducing high correlation between particles. the attenuating parameter in A-SIR, can control the number of best particles based on weight. So in A-SIR based system, using only few numbers of effective particles are able to give impressive result than conventional SIR. Resampling usually (but not necessarily) occurs between two importance sampling steps. It can be taken at every step or only taken if regarded necessary. In our proposed A-SIR algorithm, resampling schedule has been used as deterministic instead of dynamic way. In deterministic framework, resampling is taken at every k time step (usually $k = 1$). In a dynamic schedule, a sequence of thresholds (constant of time varying) is set up and the variance of the importance weight is monitored; resampling is taken only when the variance is over the threshold. The strength of resampling step in SIS algorithm has been verified by many researchers as described in [7], since resampling step also bring some extra variation, some special schemes are needed.

Further more, the performance of a tracking system depends highly on the target object representation and the similarity measurement between the target and the reference object which can be called the measurement model or the observation model. Most of the proposed tracking algorithms are mainly application dependent [4,5]. Many of them rely on a single cue as for example color, which can be chosen according to the application context. Color based tracker has some advantages, but in some cases, there are some disadvantages to having an object in plain color. An efficient color based target representation can be made with multiple regions of the color histogram by multiple integral image [3], which can be termed as Multi-Part Histogram (MPH) method and it is very helpful to deal with occlusions. In this paper our A-SIR based object tracking method is driven by MPH based measurement technique. The most weighted particles are found in the central region of the target by a weighting function, because the other areas of the target are not as important as the center. To calculate the similarity of the multi-region histogram the Bhattacharyya coefficient [6] has been used as the metric.

The rest of paper is organized as follows: Section 2 describes about the related work of our current study of this paper. Section 3 presents the brief about particle filter and resampling algorithm. Sections 4 introduce the proposed A-SIR algorithm. Section 5 presents the proposed human body descriptor used for tracking with MPH. The experimental results using various real time videos with severe occlusions are discussed in Section 6. An evaluation and comparison study are presented in this part. The concluding remarks are addressed in Section 7.

2   Related Work

There are several approaches have been done to improve resampling strategy in visual tracking. In [8], they propose systematic resampling with adaptive template for visual tracking. The systematic resampling already was established in [1] and this is still linear type function. In [9], they propose a sampling strategy aiming at reducing computational complexity in particle filtering framework.

This strategy combines the particle filtering with the transition prior and the unscented Kalman filter. Our approach is different to them which is non-linear type and ideally suited for visual tracking for real time performance with high accuracy. In this article, we use a non-linear function to change this resampling algorithm for choosing only few number of the best particle with high weight by reducing search area. All high weight particles are concentrated gently on the tracked object and reduce the failure possibility and enhancing performance significantly.

3   Particle Filter

In particle filtering, we want to compute the filtered estimates of x_t that is, $p(x_t|y_t)$ based on the set of all the available measurements up to time t. According to the Bayesian estimation, it recursively computes $p(x_t|y_t)$, that is, in terms of the posterior density at previous time step $p(x_{t-1}|y_{t-1})$. Particle filter algorithm uses a set of weighted samples drawn from the posterior distribution to approximate integrals as discrete sums. Given a set of S random samples $\{x_{1:t-1}^i, w_{1:t-1}^i\}_{i=1,2,...,N}$, where, $w_{1:t-1}^i$ is corresponding weights and $y_{1:t-1}$ are available measurements up to time t. According to SIS strategy, the posterior distribution can be computed as

$$p(x_t|y_{1:t}) \approx \sum_{i=1}^{N} w_t^i \delta(x_t - x_t^i), \tag{1}$$

where $\delta(.)$ is the Driac delta function in Equ. 1. It is not usually not possible to sample from the posterior distribution directly. This matter can be resolved by drawing samples from a proposal distribution $q(x_{1:t-1}|y_{1:t-1})$. It is a significant step by choosing proper proposal distribution when using importance sampling algorithm. The most popular choice of proposal distribution is the prior distribution because of the convenience in calculating it. The proposal distribution can be expressed as

$$p(x_t|x_{t-1}) = q(x_t|x_{t-1}, y_t)_{i=1,2,...,N}. \tag{2}$$

Selecting the prior distribution as proposal distribution, the importance weight calculation can be simply expressed as

$$w_t^i = w_{t-1}^i \frac{p(y_t|x_t^i)p(x_t|x_{t-1})}{q(x_t|x_{1:t-1}^i, y_{1:t})}. \tag{3}$$

The mean state of an object is estimated at each time step by

$$\hat{E}[x_t] = \sum_{i=1}^{N} w_t^i x_t^i. \tag{4}$$

But this straightforward algorithm creates some problem which is called weight degeneration. Resampling algorithm has been applied to overcome this problem.

4   Non-linear Resampling Algorithm

We basically modified SIR based particle filter by changing the resampling function as non-linear function. In our real time visual tracking case, the SIR filter works well, but the effective particle sorting with higher weight in every iteration is computationally expensive. Mean while the tracking failure possibility increases. Our ultimate goal can be divided in two parts. First is, we want to use less number of the best weighted particle and second one is, by reducing particle number we want to get the best tracking output. Our proposed method saves more calculation time and use the lowest number of the highest weight particle by attenuating function. The attenuating parameter also can control the best used particle number as our desire which is mainly application dependent. The traditional resampling algorithm is linear mapping function to copy or replace particle with high weight which can be expressed as

$$\mathbb{N} = w_t^i . n, \tag{5}$$

where, w in associated weight and n is the particle number. We can copy the more effective particle by discarding the particles associated with insignificant weights by the following equation

$$\mathbb{W} = a(\exp(b(w_t^i))) + c, \tag{6}$$

where, b is the attenuating factor and a and c is the arbitrary constant (a and $b \neq 0$). The number of particle copies for resampling can be controlled by the parameter b as shown in figure 1. We can see from this figure, that, this non-linear mapping helps to attenuate particles by discarding low weight particle which is better than linear mapping in conventional resampling. To normalize Equ. 6, we can write the equation as

$$\mathbf{W}_t^i = \frac{\mathbb{W}}{\sum_{i=1}^{N} w_t^i}. \tag{7}$$

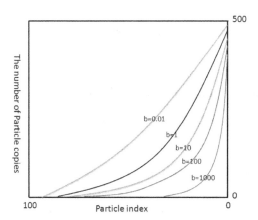

Fig. 1. Effect of attenuating parameter b in resampling step based on weight

Finally the Equ. 5 can be re-written with the help of Equ. 7 as

$$\mathbb{N} = round(\mathbf{W}_t^i.n). \tag{8}$$

5   Multi-Part Histogram (MPH) Based Measurement

5.1   Object Feature Descriptor

In this paper the tracked human body is considered to be a composition of some regions, as shown in figure 2. We introduce the Multi-Part Histogram (MPH) using integral image based representation [3] which characterizes the human body using its detailed spatial information. As we see from figure 2, the shadow (a) region is the most important part during tracking; it almost keeps the same structure with low variance. The remaining (b) region of figure 2 has a high variance during the tracking period. That means the centre region histogram of the bounded rectangle is the most weighted and as the distance increases the whole rectangle center can be assigned smaller weights by employing the weighting function

$$w_i^t(d) = \begin{cases} 1 - d^2 & d < 1 \\ 0 & otherwise \end{cases} \tag{9}$$

where, d is the distance from center to rectangle boundary. Now we denote the human body parts as

$$\mathcal{R} = \{r_i, p, w, h\}_{i=1,2,\ldots,L} \tag{10}$$

where, r, p, w and h are the single region, the position, the width, and the height of the whole rectangle, respectively. All the parts of the rectangle are formed with a color integral histogram from the target intensity image I and the reference template image T, which can be denoted as $m_I(r_i)$ and $m_T(r_i)$ respectively. The distance between the reference template image T and the target intensity image I over time t is given by:

$$\Phi(m) = [\sum_{i=1}^{L} (m_T(r_i) - m_I^t(r_i))^2]^{\frac{1}{2}} \tag{11}$$

5.2   Color Measurement Model

We adopt the Gaussian density for the likelihood function of the measured color histogram as follows:

$$p(q_t|x_t) \propto N(D_t; 0, \sigma^2) = \frac{1}{\sqrt{2\pi}\sigma} \exp\{\frac{-D_t^2}{2\sigma^2}\} \tag{12}$$

where $D_t = dist[p, q_t]$ is the distance between the reference histogram p of the objects to be tracked and the histogram q_t computed from image z_t in the region defined by the state vector x_t. If we consider $p = \{p(u)\}_{u=1,2\ldots,m}$ and

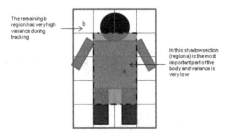

Fig. 2. By using the multiple region histogram of the target by integral image-based representation, we can get accurate spatial information specific to human tracking. The integral image base representation is computationally less expensive than the direct pixel method.

$q_t = \{q_t(u)\}_{u=1,2...,m}$ are the two histograms calculated over m bins, then we adopt the distance D_t between two histograms from the Bhattacharyya similarity coefficient [6] as:

$$D_t = \sqrt{1 - \sum_{u=1}^{m} \sqrt{p(u)q_t(u)}} \qquad (13)$$

We have compared the similarity of the histogram for each corresponding region between the reference and the target image. If we have N regions, then we calculate the similarity by the Bhattacharyya coefficient N times for N regions; each region has $\{(bin_1), (bin_2), ..., (bin_i)\}$ number of integral images. Finally the Bhattacharyya similarity coefficient calculation is given by:

$$D_T = \frac{\sum_{j=1}^{N} D_{tj}}{N} \qquad (14)$$

6   Experiments and Results

In this section, we show the single object tracking performance by our new proposed resampling based algorithm in particle filtering framework. The performance is verified using our own video sequences, in which we aim to track a pre-selected moving person. In this sequence, *circleocc* (500 frames), there are two persons are walking towards each other from opposite sides. They meet, shake hands, circle each other and our subject is completely occluded more than three times. Figure 3 shows the comparison performance of *circleocc* between our proposed re-sample based algorithm with traditional algorithm using 100 particles. For this comparison we use same frame number 71, 99, 109, 181, 296 and 480 in all testing purpose. Our proposed system even works well with only 10 particles because of its effectiveness, which is shown in figure 3(c). The tracking failure possibilities drastically reduced using only 10 particles. The overall performance can be verified by the red bars in the all frames which are shown

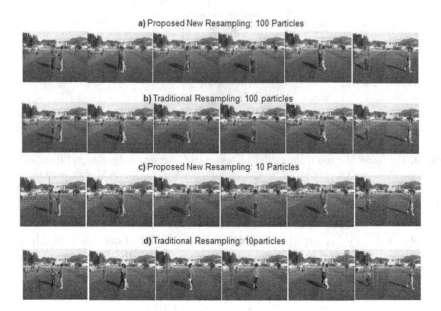

Fig. 3. Tracking result of *circleocc* with traditional and proposed resamlping using 100 and 10 particles. The overall performance can be verified by the red bars in the all frames which are shown horizontally and vertically. These red bars represent as probability densities of the estimated state.

horizontally and vertically. These red bars represent as probability densities of the estimated state. From figure 3(b), we can see that, the probability densities become scattered to find the best weighted particle for the next state estimation. This works well sometimes but using only large number of particles. However the tracking failure possibility still remains due to the many real time challenges in visual tracking.

6.1   The Error Metric the Performance Evaluation

Basically the evaluation of our proposed system is measured qualitatively. Also, the root mean squared error (RMSE) method in the state space has been used to evaluate the performance of our developed algorithm. The RMSE can be formulated by:

$$RMSE(t) = \sqrt{0.5((g_t - \hat{g}_t)^2 + (h_t - \hat{h}_t)^2)} \qquad (15)$$

where, $(\hat{g}_t, \hat{h}_t)$ stands for the upper-left corner coordinates of the tracking box determined by the central position, corresponding to the state estimated by the particle filter in the frame . The ground truth states (g_t, h_t) correspond to the true positions of the object and have been generated by manually creating the tracking box surrounding the object in the test videos. We evaluate our proposed

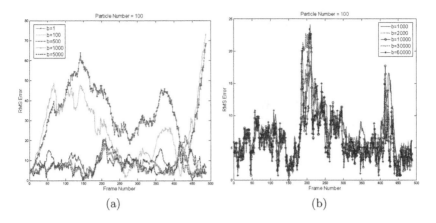

(a)   (b)

Fig. 4. Performance analysis with different attenuating parameter b using 100 particle. (a) attenuating paramneter from $b = 1$ to 5000, (b) attenuating parameter $b = 1000$ to 60000.

system with 100 particles using different attenuating parameter to observe the tracking output. The RMS error graph of this video stream at different attenuating parameter is shown in figure 4(a) and 4(b). From this figure we can see, that the range of b between 1,000 to 10,000, the tracking performance are almost remain same and it is better than below $b = 1,000$. The more numeric analysis we can find from table 1. From this table 1 we can see that, $b = 1,000$ to 10,000 is better region than others. In the mean time we compare our result with conventional SIR filter. Our proposed A-SIR is much better than SIR. For example, when $b = 1000$, the maximum RMS error and average error are 17.73 and 7.2 respectively. On the other hand in conventional SIR, the maximum RMS error and average error are 24.95 and 8.15 respectively. Also we can see from the last row of this table as the number of used best particle (**P**) at different attenuating parameter (att. par.) b.

The more we increase b the used particle will decrease. It saves our further calculation time and our system become more faster without any tracking performance degradation. We chose our perfect attenuating factor based on the tracking environment and desire. The graph as shown in figure 5 shows the

Table 1. Performance evaluation with differrent attenuating parameter b using 100 particles btween proposed resampling and SIR filter

Att.Par. b	1	100	500	1000	2000	5000	10000	30000	60000	Trad. SIR
Max RMSE	68.43	73.25	22.47	17.73	18.5	20.5	22.6	24.1	24.2	24.95
Min RMSE	4.43	1.41	2.82	1	0.7	0	0	0.7	0	0
Avg. RMSE	34.25	25.3	10.3	7.2	6.8	6.5	6.4	6.5	6.37	8.15
P	100	99	95	87	75	57	43	22	12	N/A

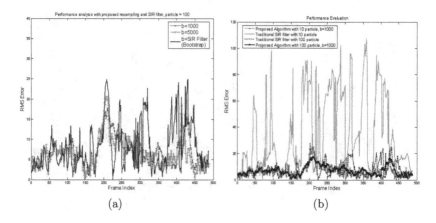

(a)   (b)

Fig. 5. (a)Performance comparison between proposed A-SIR and conventional SIR filter, (b) performance comparison between proposed A-SIR and SIR with 100 and 10 particles

performance analysis between proposed A-SIR and conventional SIR filter. Our proposed algorithm works well with only 10 particles, where, SIR based particle filter totally fail to track the object with 10 particles. Also the graph as shown in figure 5(b) represent the performance analysis with 10 and 100 particles between proposed A-SIR and SIR based particle filter. Table 2 summarizes the RMS error at different attenuating parameter with only 10 particles. The last column of this table also shows the correspondence tracking performance with conventional SIR based particle filter.

Table 2. Performance evaluation with different attenuating parameter b using 10 particles between proposed resampling and SIR filter

Att.Par. b	1	100	500	1000	2000	5000	Trad. SIR
Max RMSE	78.93	23.1	24.7	27.6	29.2	27.7	107.9
Min RMSE	1.6	0	0	0	0.7	0	0.7
Avg. RMSE	36.15	8.2	6.87	6.82	7.0	6.78	36.2
P	10	9	6	4	3	2	N/A

7   Conclusion

A new re-sample based A-SIR algorithm has been proposed in this paper in the particle filtering framework. This proposed re-sample design issues related with efficient control of best weighted particle in attenuating form. The attenuating factor can control the high weight particle as our desire and also it saves the further calculation time during tracking and only few particles can give the satisfactory tracking result. This A-SIR algorithm boost up the object tracking performance than conventional SIR based filter. The proposed non-linear type resampling can find out the most important particle and attenuate the other particle in very efficient way. Also, from the all results and tables, we can conclude that this proposed algorithm minimizes the real time performance degradation, and their complexity is reduced remarkably.

Acknowledgement. This research was supported by the MKE(The Ministry of Knowledge Economy), Korea, under the ITRC(Information Technology Research Center) support program supervised by the NIPA(National IT Industry Promotion Agency) (NIPA-2010-C1090-1011-0008).

References

1. Sanjeev, A.M., Simon, M., Neil, G., Tim, C.: A tutorial on particle filters for online nonlinear/non-gaussian Bayesian tracking. IEEE Transactions on Signal Processing 50(2), 174–188 (2002)
2. Douc, R., Cappe, O.: Comparison of resampling schemes for particle filtering. In: 4th International Symposium on in Image and Signal Processing and Analysis, pp. 64–69 (2005)
3. Viola, P., Jones, M.: Rapid object detection using a boosted cascade of simple features. In: IEEE Conference on Computer Vision and Pattern Recognition, vol. 1, pp. 511–518 (2001)
4. Yunqiang, C., Yong, R.: Real time object tracking in video sequences. In: Signals and Communications Technologies, Interactive Video, vol. II, pp. 67–88. Springer, Heidelberg (2006)
5. Artuar, L., Lyudmila, M., David, B.: Structural Similarity-based Object Tracking in Multimodality Surveillance Videos. Machine Vision and Applications 20(2), 71–83 (2009)
6. Aherne, F.J., Thacker, N.A., Rockett, P.I.: The Bhattacharyya Metric as an Absolute Similarity Measure for Frequency Coded Data. Kybernetika 32(4), 1–7 (1997)
7. Liu, J.S., Chen, R.: Blind deconvolution via sequential imputation. Journal of American Statistical Association 90, 567–576 (1995)
8. Wu, G., Tang, Z.: A new resampling strategy about particle filter algorithm applied in Monte Carlo framework. In: Second International Conference on Intelligent Computation Technology and Automation, pp. 507–510 (2009)
9. Wang, F., Lin, Y.: Improving Particle Filter with A New Sampling Strategy. In: 4th International Conference on Computer Science and Education, pp. 408–412 (2009)

An Appearance-Based Approach to Assistive Identity Inference Using LBP and Colour Histograms

Sareh Abolahrari Shirazi[1,2], Farhad Dadgostar[1,2], and Brian C. Lovell[1,2]

[1] NICTA, PO Box 6020, St Lucia, QLD 4067, Australia
[2] The University of Queensland, School of ITEE, QLD 4072, Australia

Abstract. Robust identity inference is one of the biggest challenges in current visual surveillance systems. Although, face is an important biometric for generic identity inference, it is not always accessible in video-based surveillance systems due to the poor quality of the video or ineffective viewpoints where the captured face is not clearly visible. Hence, taking advantage of additional features to increase the accuracy and reliability of these systems is an increasing need. Appearance and clothing are potentially suitable for visual identification and tracking suspects. In this research we present a novel approach for recognition of upper body clothing, using local binary patterns (LBP) and colour information, as an assistive tool for identity inference.

Keywords: Local Binary Patterns, Colour Histogram, Object Recognition, Ensemble-Learning.

1 Introduction

Vision based surveillance systems are widely used in public spaces in many metropolitan cities. The increasing number of cameras and pervasiveness of CCTV networks have created new possibilities of automated or assistive security monitoring. While some of the tasks of the security personnel such as tracking an individual when in a multi-camera network may seem trivial, it is an exhaustive task. Therefore any assistive or automated method that can reduce the search will save resources and more importantly may save time.

The London bombings on July 7th 2005, were a series of coordinated suicide attacks to the city's public transport. There was initially a great deal of confusing information from police sources about the origin, method, and even timings of the explosions. But London's network of closed-circuit TV (CCTV) cameras helped investigators track down those responsible for the terrorist bomb attacks within 24 hours. Police examined about 2,500 items of CCTV footage and forensic evidence from the scenes of the attacks and finally investigators identified four men whom alleged had been the suicide bombers.

It is obvious that performing the above task in a timely manner is a matter of national security, and any delay in decision making can have huge impacts on the society. The main challenge in such scenarios always is allocating a sufficient number of expert people to perform the task. It is a common knowledge that the huge volume of recorded data always makes "human resource" the main bottleneck in

R. Koch et al. (Eds.): ACCV 2010 Workshops, Part I, LNCS 6468, pp. 236–245, 2011.

performing a pervasive search on wide-area surveillance video. Therefore any assistive or automated method that can reduce the search space will save resources and time.

2 Research Background

Tracking is one of the first steps in active visual surveillance. An active visual surveillance system must be able to handle challenges like occlusion, splitting and merging which are the results of moving objects in the scene [1, 2]. Although, face can be considered as a reliable attribute for people tracking, face recognition on video surveillance is an extremely challenging task. The quality of video footage and the distance of the subject from the camera are instances which make 'face' poorly suited to identity inference. An example is shown in Fig. 1.

Fig. 1. Face extracted from video surveillance

Upper body clothing is a good assistive feature for tracking individuals in surveillance footage for security personnel. Upper body is considered a more reliable clue, in comparison to lower body. Firstly it is more observable through surveillance footage, and secondly it has more variety of colours and textures which makes it more discriminative for recognition and tracking.

In order to overcome the many complicated challenges in a tracking environment, one should take advantage of multiple image properties, such as texture, colour, etc [1]. Several content-based methods use visual information as suitable features for retrieval of similar images to a given query image [3]. In this paper we are investigating the usability of Local Binary Pattern (LBP) features for describing human's clothing textural properties.

One important property of the LBP operator is its robustness to illumination variation. Moreover, its computational simplicity makes it usable for real-time applications [4].

The LBP operator was first introduced in 1996 by Ojala [5]. The basic LBP operator is based on the comparison of the eight neighbours of each pixel, presented as a binary sequence, or a binary pattern. The histogram of the binary patterns represents the local spatial structure of an image [6]. Since transformation of an image from RGB to greyscale has no effect on patterns, LBP texture operator is defined as a greyscale invariant texture measure [7]. In 2002 an extended version of LBP was introduced by Ojala. To reduce the size of the binary pattern and the computational costs, two other extensions of LBP were proposed subsequently. The first one was "uniform LBP" based on the fact that some features occur more frequently than others and the second one was "Rotation Invariant LBP". To remove the effect of rotation, a unique identifier is assigned to each rotation invariant local binary pattern [8]. In 2004 two further extensions of LBP were introduced [9, 10].

Since the introduction of LBP it has made a significant contribution to texture analysis [4, 7]. It has also been a powerful feature in some computer vision applications which might not be considered as texture problems such as face analysis. In addition to face and facial expression recognition, LBP has been used for iris and finger print recognition [11-14].

Another visual attribute can be used for tracking is "Colour" which is an intrinsic attribute of an image. Hence, it could be helpful for this purpose. The most well known colour descriptor is the colour histogram which has been used for tracking in many situations [15-17]. In these methods, similarity is measured by the distance between matching bins of histogram [18]. Although histogram-based methods have been widely used, they have a serious limitation since they do not include spatial information. There has been some approaches based on Colour Coherent Vector (CCV) to represent spatial information with colours but they are computationally complex [19]. The proper choice of colour space would be important in this context. Using proper colour spaces may help in achieving invariance against illumination changes. In HSI (Hue Saturation Intensity Colour Space) chromaticity and intensity information are separated which provides intensity invariant chromaticity measures [20].

Since none of these attributes are invariant to different imaging conditions and each of them has pros and cons, using a collection of properties can potentially enhance the performance of the tracking system [1]. Clothing can be roughly categorised into plain and complex textures, and be classified using a texture analyser in combination with a colour descriptor is very efficient.

Paschos et.al proposed a monitoring system based on colour and texture information. Their approach was based on extending greyscale algorithms to colour images where each colour band is processed separately by applying greyscale texture analysis techniques. It has been used in colour texture segmentation and classification [21]. In this context, colour and texture can be processed jointly or separately. In separate use of colour and texture, textural information is derived from the luminance plane along with pure colour features. This is particularly useful for segmentation [22]. In some of these works texture features were extracted from greyscale images and then combined with colour histograms and moments. Some other methods have investigated a combination of colour bands to extract better features [23].

In this paper we demonstrate a novel approach in applying LBP features and colour information compared to the existing methods for upper body recognition. To achieve more robustness against some of the tracking challenges such as partial occlusion, pose variation and so on, an ensemble of colour and LBP classifiers are employed. Appearance-based descriptors like LBP, SIFT and many others are specifically designed for grey level images. One way to extend these descriptors to colour images is applying them over all colour channels. Though this method has been successful in some applications, no specific framework has been proposed to address this problem. In this paper we introduced a rule-based machine to overcome the aforementioned problem.This paper is organised as follows: In section 3 our approach is described.

The experiments and results are presented in section 4. Finally, conclusion, limitations of the proposed method, and future work are described in section 5.

3   Proposed Method

Intuitively colour and local appearance are important factors for clothing recognition. Several local descriptors exist such as LBP and SIFT. LBP gives us useful textural information which suits our application. Unlike LBP, SIFT is not an appropriate choice as a local descriptor. Fig.2 shows some examples of SIFT applied on clothing images. As it is shown, descriptors work well for clothing with specific trademarks (image on the right) and not for plain ones (image in the middle).

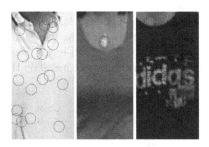

Fig. 2. Samples for SIFT descriptor

In our recognition model, clothing in still images is to be recognized. Fig.3 gives a brief description of our approach.

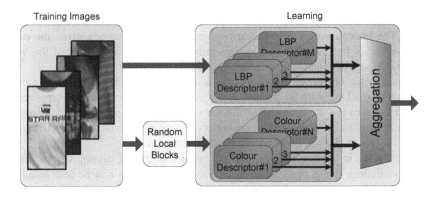

Fig. 3. The proposed clothing recognition model

Based on the study in [24], HSI colour space is chosen as the colour model here. HSI colour space decouples the intensity component from colour carrying information in a colour image, making Hue and Saturation components less sensitive to intensity changes. A well-known fact about HSI space is the instability of the Hue channel near

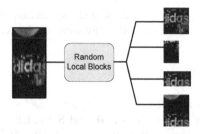

Fig. 4. Random local blocks description

the grey axis [25]. Hence, in our classifier, when intensity is low and saturation is close to zero, Hue information is removed from the descriptor. To avoid the curse of dimensionality and reduce the computational load of the algorithm, colours are quantized accordingly. When intensity is used, the values are divided into six levels, and when Hue is used values are divided into 36 levels. Based on this allocation, we have 42 distinct colours bins which are computed based on HSI values (Fig. 5).

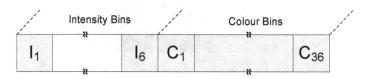

Fig. 5. Colour based feature vector

After colour quantization, clothing image are divided into 10 vertical and 10 horizontal stripes to use informative patterns in both directions and building the colour histogram for the colour quantized image based on the 42 existing colour indices. The concatenation of resulted histograms from 20 stripes gives the colour feature vector of the image. Using stripes rather than using histogram of the whole image helped us to include spatial information in the classification method.

As one of our objectives is to provide a more robust algorithm to the existing challenges in tracking such as occlusion, we computed colour descriptors for several random blocks in the clothing image. This method overcomes partial occlusion and would be helpful in identification, in presence of specific patterns on the clothing. The result is a series of classifiers based on colour descriptors.

The last stage is the aggregation. The notion of Ensemble Learning is to aggregate several predictions using multiple learners. Ensemble learning can be also intuitively defined as solving a hard classification problem through breaking it down into smaller ones. Computer vision applications usually encounter high-dimensional feature vectors with only a small number of training samples (compared to the data dimensionality). As a result, it is extremely difficult, if not impossible, to construct an efficient single classification rule. To solve such issues, ensemble learning techniques have become very popular over the last few years in computer vision applications. In fact, several studies even showed that ensemble of classifiers significantly outperform their single base counterparts. Since the outputs of the base classifiers are just class-labels, the simplest way to combine classifiers is the majority voting. In majority voting, votes for each class over the input classifiers are counted and the class with the maximum number of votes is selected as the winner.

It is important to be mentioned that to do the similarity measurement for two feature vectors $X = (X_1, ..., X_N)$, $Y = (Y_1, ..., Y_N)$ we used the following metric:

$$Similarity = \sum_{i=1}^{N} |X_i - Y_i|$$

In addition, we evaluated each classifier by computing the percentage of correct recognitions (True Positives) out of total number of recognitions (True Positives + False Positives).

4 Experiments

To evaluate the proposed classification framework, we designed an experimental data collection setting to collect the required data samples. For this purpose we recorded 12 subjects in four locations, when walking through a gate. In each location, three different, synchronized cameras were recording the subjects. Camera set up is shown in Fig. 6.

1. Camera #1 (25 fps): Height: 3m; Distance to the door: 1.9m
2. Camera #2 (18 fps): Height: 1.8m; Distance to the door: 1.7m
3. Camera #3 (30 fps): Height: 1.8m; Distance to the door: 1.8m

In order to have some illumination variation in the sequences, the recordings were conducted in three different locations to allow some changes in lighting conditions. Samples are shown in Fig. 7. The recorded data was then used for experimentation.

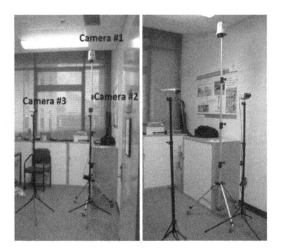

Fig. 6. Camera set up

The first step in the experiments was applying LBP on the clothing of the subjects' images. The clothing segment was estimated from manually marked eye locations. To compute LBP feature vectors, we divided the image into several overlapping blocks of the same size and then we applied LBP on each block. The resulted histogram represents block's feature vector. The LBP feature vector is resulted by concatenation of each block's feature vector.

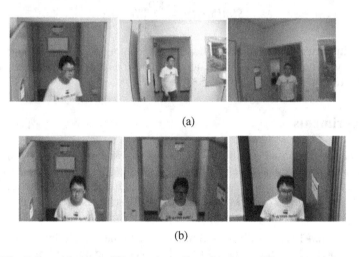

(a)

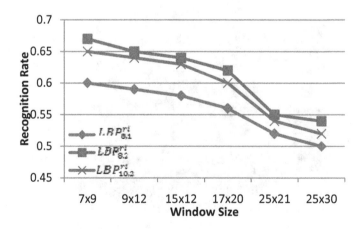

(b)

Fig. 7. Row (a) shows different views, Row (b) shows different locations

In this experiment we investigated the effects of different parameters on the final result; including block sizes, number of neighbours and the radius of the circle on which neighbours are located and mapping mode (Rotation Invariant (RI), Uniform2 (U2), Rotation Invariant and Uniform2 (RIU2)). Multiple value selection for these

Fig. 8. LBP based recognition rates

parameters would give us multiple classifiers with different recognition rates. The following figures present the recognition rates of LBP classifiers based on multiple parameter values. These values are selected, since it is a good trade-off between recognition performance and feature vector length.

After running random block sampling, Fig.9 shows the average value and standard deviation of the resulted recognition rates for colour based classifiers.

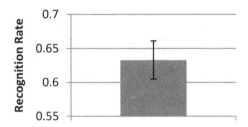

Fig. 9. Average and standard deviation value of recognition rate for our proposed colour based method

As the last step, we used "Majority Voting" approach for the aggregation stage and combined all the classifiers' outputs. In this context, a comparable work to our approach is Choi *et al.* [24]. In their method unlike ours, they only used colour descriptor as a classification and in their feature extraction process, they have considered three vertical and horizontal stripes from the input image. The achieved result for their algorithm was 0.51. Table.1 demonstrates the recognition accuracy of the proposed methods against some benchmark algorithms. RGB and HSI histogram are holistic methods, i.e. the histogram was computed for the whole image. For LBP the best obtained result is shown. The last two rows show the recognition accuracy of the proposed colour descriptor and the ensemble of colour and LBP classifiers. Table.1 reveals that the proposed algorithm outperforms the other studied methods. It comes as no surprise that holistic approaches do not perform satisfactory as the studied dataset presents substantial illumination and pose variations. Comparing to Choi's method that is specifically designed for clothing recognition, our ensemble approach shows 50% improvement.

Table 1. Comparison of recognition rates

Method	Maximum recognition rate
RGB histogram [25]	0.28
HSI histogram	0.34
LBP [5]	0.67
Choi's method [24]	0.51
Proposed Colour descriptor	0.70
Proposed ensemble method	**0.77**

5 Conclusion

In this paper, we proposed a novel approach to recognise clothing from images captured by CCTV cameras. Appearance-based descriptors like LBP, SIFT and many others are specifically designed for grey level images. The normal way to extend these descriptors for colour images is to apply them over all colour channels. Though some ad hoc methods are successful in specific applications, no generic framework has been proposed to address the general problem; extending appearance-based descriptors for colour images. In this paper we introduced a rule-based machine to incorporate colour information along with LBP descriptors. To make the approach more robust to tracking challenges such as partial occlusion, pose variation and so on which affect CCTV footage, an ensemble of colour and LBP classifiers are employed.

Experimental results on a new dataset showed that the proposed approach improves the recognition accuracy significantly compared to existing benchmark methods. We plan to investigate the performance of our algorithm on other popular tracking datasets in the future.

Acknowledgement

This project is financially supported by the Australian Government through the National Security Science and Technology Branch within the Department of the Prime Minister and Cabinet. This support does not represent and endorsement of the contents or conclusions of the project. NICTA is funded by the Australian Government's Backing Australia's Ability initiative, in part through the Australian Research Council.

References

1. Takala, V., Pietikainen, M.: Multi-Object Tracking Using Color, Texture and Motion. In: IEEE Conference on Computer Vision and Pattern Recognition, CVPR 2007 (2007)
2. Ning, J.: Robust Object Tracking using Joint Color-Texture Histogram. International Journal of Pattern Recognition 23(7), 1245–1263 (2009)
3. Iqbal, Q., Aggarwal, J.: CIRES: A system for content-based retrieval in digital image libraries. In: International Conference on Control, Automation, Robotics and Vision, pp. 205–210 (2002)
4. Mäenpää, T., Pietikäinen, M.: Texture analysis with local binary patterns. In: Handbook of Pattern Recognition and Computer Vision, vol. 3, pp. 197–216.
5. Ojala, T., Pietikäinen, M., Harwood, D.: A comparative study of texture measures with classification based on featured distributions. Pattern Recognition 29(1), 51–59 (1996)
6. Marcel, S., Rodriguez, Y., Heusch, G.: On the recent use of local binary patterns for face authentication. International Journal on Image and Video Processing Special Issue on Facial Image Processing, 06–34 (2007)
7. Pietikäinen, M.: Image analysis with local binary patterns. In: Kalviainen, H., Parkkinen, J., Kaarna, A. (eds.) SCIA 2005. LNCS, vol. 3540, pp. 115–118. Springer, Heidelberg (2005)

8. Ojala, T., Pietikäinen, M., Mäenpää, T., Viertola, J.: Multiresolution gray-scale and rotation invariant texture classification with loval binary patterns. Pattern Analysis and Machine Intelligence 24, 971–987 (2002)
9. Jin, H.: Face detection using improved LBP under bayesian framework. In: International Conference on Image and Graphics, pp. 306–309 (2004)
10. Huang, X., Li, S., Wang, Y.: Shape localization based on statistical method using extended local binary pattern. In: International Conference on Image and Graphics (2004)
11. Feng, X., Pietikainen, M., Hadid, A.: Facial expression recognition with local binary patterns and linear programming. Pattern Recognition And Image Analysis C/C of Raspoznavaniye Obrazov I Analiz Izobrazhenii 15(2), 546 (2005)
12. Sun, Z., Tan, T., Qiu, X.: Graph matching iris image blocks with local binary pattern. In: Zhang, D., Jain, A.K. (eds.) ICB 2005. LNCS, vol. 3832, pp. 366–372. Springer, Heidelberg (2005)
13. Nanni, L., Lumini, A.: Local binary patterns for a hybrid fingerprint matcher. Pattern Recognition 41(11), 3461–3466 (2008)
14. Ahonen, T., Hadid, A., Pietikainen, M.: Face description with local binary patterns: Application to face recognition. IEEE Transactions on Pattern Analysis and Machine Intelligence 28(12), 2037–2041 (2006)
15. Bradski, G.: Computer vision face tracking for use in a perceptual user interface. Intel Technology Journal 2(2), 12–21 (1998)
16. Collins, R., Liu, Y., Leordeanu, M.: Online selection of discriminative tracking features. IEEE Transactions on Pattern Analysis and Machine Intelligence 27(10), 1631–1643 (2005)
17. Yang, T., et al.: Real-time multiple objects tracking with occlusion handling in dynamic scenes, vol. 1: pp. 970–975 (2005)
18. Swain, M., Ballard, D.: Color indexing. International Journal of Computer Vision 7(1), 11–32 (1991)
19. Hongli, X., De, X., Yong, G.: Region-based image retrieval using color coherence region vectors. In: International Conference on Signal Processing, vol. 1, pp. 761–764 (2004)
20. Mäenpää, T., Pietikäinen, M.: Classification with color and texture: jointly or separately? Pattern Recognition 37(8), 1629–1640 (2004)
21. Thai, B., Healey, G.: Modeling and classifying symmetries using a multiscale opponent color representation. IEEE Transactions on Pattern Analysis and Machine Intelligence 20(11), 12 (1998)
22. Paschos, G., Valavanis, K.P.: A color texture based visual monitoring system for automated surveillance. IEEE Transactions on Systems 29(2), 10 (1999)
23. Akhloufi, M.A.: A new color-texture approach for industrial products inspection. Multimedia 3, 8 (2008)
24. Choi, Y., et al.: Retrieval of Identical Clothing Images Based on Local Color Histograms. In: International Conference on Convergence and Hybrid Information Technology, pp. 818–823 (2008)
25. van de Sande, K., Gevers, T., Snoek, C.: Evaluating color descriptors for object and scene recognition. IEEE Transactions On Pattern Analysis And Machine Intelligence 23(9), 1582–1596 (2009)

Vehicle Class Recognition Using Multiple Video Cameras

Dongjin Han[1], Jae Hwang[2], Hern-soo Hahn[1], and David B. Cooper[3]

[1] Soongsil University, Seoul, Korea
[2] The George Washington University, DC, USA
[3] Brown University, RI, USA
{dongjin.d.han,xinuya}@gmail.com,hahn@ssu.ac.kr,cooper@lems.brown.edu

Abstract. We present an approach to 3D vehicle class recognition (which of SUV, mini-van, sedan, pickup truck) with one or more fixed video-cameras in arbitrary positions with respect to a road. The vehicle motion is assumed to be straight. We propose an efficient method of Structure from Motion (SfM) for camera calibration and 3D reconstruction. 3D geometry such as vehicle and cabin length, width, height, and functions of these are computed and become features for use in a classifier. Classification is done by a minimum probability of error recognizer. Finally, when additional video clips taken elsewhere are available, we design classifiers based on two or more video clips, and this results in significant classification-error reduction.

1 Introduction

Vehicle class recognition requires the measurement of features used in a classifier. The images in the frames of a video clip are quite noisy and may contain considerable clutter that can change from frame to frame. As such, highly reliable recognition requires multiple images to extract features. 3D-based recognition makes it possible to relate and incorporate information from multiple frames in a video clip coherently. It may facilitate a 3D model for a vehicle and a 3D estimation of the vehicle position in each image frame. This paper presents a new 3D-based system for recognizing vehicle classes from video clips that are taken by multiple cameras under varying conditions. Recognition utilizes 3D geometrical information via a newly-developed Structure from Motion (SfM) method with cross ratio invariance. When multiple video clips are combined, the experiment shows better recognition performance.

1.1 Vehicle Recognition Literature

There are plenty of vehicle recognition related works and most of them use view-based methods. Ikeuchi developed an system based on local-feature configuration, which is a generalization of the Eigen-window method [1]. This system works on the training images made from 3D computer graphic (CG) model. Ozcanli developed contour and appearance based method [2]. They construct

R. Koch et al. (Eds.): ACCV 2010 Workshops, Part I, LNCS 6468, pp. 246–255, 2011.

a dense correspondence between the interior regions of two shapes using a mutual information paradigm. They need good vehicle boundary segmentation for
recognition. Few authors have dealt with the problem of detecting and recognizing 3D objects in images primarily from their shape information. Jolly and Jain
used 2D geometric models by deformable parametric vehicle template [3] [4].
They showed 91.9% recognition rate. All of the above view-based methods require a fixed camera viewpoint since viewpoint change will result in appearance
variation in 2D image.

Vehicle Recognition using 3D. 2D or 3D geometric vehicle model has been
used for the tracking [5] or the recognition [6], [7]. Ferryman fitted a 3D wire
model onto single image and showed 92.2% recognition rate on training data and
much lower recognition rate on test set [8]. Koller [7] projected a 3D polyhedral
model on video sequence and recursively estimate the parameters to recognize
vehicle class. A generic vehicle model, represented by a 3D polyhedral model
described by 12 length parameters, was used to cover the different shapes of
road vehicles (Fig. 1). This work is closely related to our approach since video
sequence is used and camera parameter is roughly used for the alignment of the
image projection of the 3D model (For comparison, our method can be regarded
as bottom-up process while Koller's method is top-down). However, Koller did
not show enough experiment results to show the capability of recognition [7]. 3D
based methods for vehicle recognition have not shown comparable results to the
view-based methods, yet.

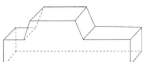

Fig. 1. 3D model for Sedan   **Fig. 2.** Vehicle examples used in the experiments: (a)-
with 24 line segments   (d) show cropped examples

1.2   System Overview

Our solution to the problem of vehicle classification based on a single video clip
is: 1. Compute the sequence of approximate silhouette using Grimson's background modeling algorithm [9] and morphological filter (usually the silhouette
is in considerable error). 2. Estimate epipole. 3. Estimate the position of the vehicle, on this straight-line trajectory, using robust invariant method (New SfM
using cross ratio invariance). 4. With calibration of N-virtual cameras, triangu-
late 3D locations on the vehicle at apparent contours. 5. Use the features from 4.
in a Bayesian Recognition. If two or more video clips are available with cameras
in arbitrary positions and maybe not viewing common backgrounds, then: 6.
Combine the results of processing the video clips individually to obtain a more
accurate classification.

2   Structure from Motion Using Cross Ratio Invariance

2.1   Epipolar Geometry for Our Problem

When the camera is fixed and the object is moving, it can be thought of that the object is fixed and the camera is moving as in Fig. 3. The moving camera at the positions corresponding to the image frames are called 'Virtual Cameras'.

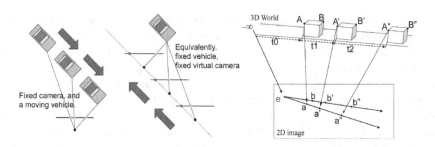

Fig. 3. Virtual cameras and illustration of cross ratio invariance

For the short intervals, the assumption of straight-line trajectory is good, i.e., covers essentially all of the 200 clips recorded. For the long intervals, straight-line trajectory is usually a good approximation since the system automatically segment the sequence of vehicle locations into sub-intervals in each of which the vehicle trajectory is straight, and has been assumed in this project.

2.2   Cross Ratio Invariance

We assume camera internal parameters are estimated from known 3D points in the scene. In Fig. 3, a box is traveling along straight line and observed at three locations $t = t_0, t_1, t_2$, denote the position of the fixed point A on the box at location $t = t_0 = 0$, A' at $t = t_1$, and A'' at t_2. a, a', a'' are mappings of A, A', A'', and b, b', b'' are mapping of B, B', B'', respectively.

Homogeneous Equations which relate real 3D coordinate system and image projection in 1D are

$$\begin{pmatrix} a \\ 1 \end{pmatrix} = \begin{pmatrix} p_1 & p_2 \\ p_3 & 1 \end{pmatrix} \begin{pmatrix} 0 \\ 1 \end{pmatrix} \Rightarrow a = B \tag{1}$$

$$\begin{pmatrix} a' \\ 1 \end{pmatrix} = \begin{pmatrix} p_1 & p_2 \\ p_3 & 1 \end{pmatrix} \begin{pmatrix} t_1 \\ 1 \end{pmatrix} \Rightarrow a' = \frac{p_1 t_1 + p_2}{p_3 t_1 + 1} \tag{2}$$

$$\begin{pmatrix} a'' \\ 1 \end{pmatrix} = \begin{pmatrix} p_1 & p_2 \\ p_3 & 1 \end{pmatrix} \begin{pmatrix} t_2 \\ 1 \end{pmatrix} \Rightarrow a'' = \frac{p_1 t_2 + p_2}{p_3 t_2 + 1} \tag{3}$$

where the projection matrix $\begin{pmatrix} p_1 & p_2 \\ p_3 & 1 \end{pmatrix}$ is already known.

From the definition of cross ratio invariance, we have

$$r(cross\ ratio = const.) = \frac{a'-a}{a''-a}\frac{a''}{a'} = \frac{b'-b}{b''-b}\frac{b''}{b'} \tag{4}$$

Solving Eq. (1)-(4) for the five unknown A_0, B_0, C_0, t_1, t_2 results in

$$t_1 = rt_2, \tag{5}$$

where r is the cross ratio and t_1 and t_2 are the translation distances multiplied by an unknown scale constant. This is true for all the points on a moving object.

2.3 Histogram Method for Cross Ratio Invariance

The cross ratio is estimated by computing histogram of cross ratios based on all edges inside of approximate silhouettes shown red as in Fig. 5.

Let E_i, E_j, E_k be edge maps of any 3 frames indexed i, j, k among all N frames and e is the estimated vanishing point. Also, let the points $p_1 \in E_i^\theta$, $p_2 \in E_j^\theta$ and $p_3 \in E_k^\theta$, then let d_1, d_2, d_3 be the distances between p_1, p_2, p_3 and the epipole e, respectively. Then Eq. (4) just becomes $r = \frac{d2-d1}{d3-d1} \cdot \frac{d3}{d2}$. For every combination of edge points p_1, p_2, p_3, r value is computed. Fig. 6 shows the examples of histogram. The peak of each histogram corresponds to the estimation of cross ratio $\hat{r}$ for each group of three frames.

Fig. 4. Estimate of epipole using approximate contours and epipolar line l_θ

Fig. 5. Edges inside of approximate contour (red) are used

2.4 N frames and Scale Factor

With N frames, there are $N-1$ translations between N frames $(t_{1,2}, ..t_{i,i+1}, ..t_{N-1,N})$ and we can find $M = \binom{N}{3} = \frac{N!}{3!(N-3)!} = \frac{N(N-1)(N-2)}{6}$ cross ratios among N frames taken 3 at a time, where $t_{i,j}$ is the translation distance between frames N_i and N_j. To determine all of the translations $t_{i,j}$ for the entire sequence N frames, we have to solve M linear equations. Generally, there will be more homogeneous equations than unknown parameters and will be computed in the least square sense. Each $\hat{t}_{i,i+1}$ of the estimation ($\hat{T} = \hat{t}_{1,2}, ...\hat{t}_{i,i+1}, \hat{t}_{i+1,i+2}, ...\hat{t}_{N-1,N}$) describes the translation ratio between adjacent 2 frames of N frames.

$$P^i = \begin{pmatrix} p_{11}^0 & p_{12}^0 & p_{13}^0 & p_{14}^0 + S \cdot e(0)\sum_{k=1}^{i-1}\hat{t}_{k,k+1} \\ p_{21}^0 & p_{22}^0 & p_{23}^0 & p_{24}^0 + S \cdot e(1)\sum_{k=1}^{i-1}\hat{t}_{k,k+1} \\ p_{31}^0 & p_{32}^0 & p_{33}^0 & p_{34}^0 + S\sum_{k=1}^{i-1}\hat{t}_{k,k+1} \end{pmatrix}, \tag{6}$$

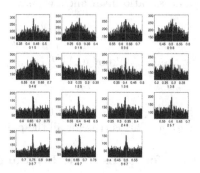

Fig. 6. All of the cross ratio histograms used for a video clip calibration

where $e(0), e(1)$ are x,y coordinate of the epipole, and $\hat{t}_{k,k+1}$ is the translation ratio between $k, k+1$ frames. With all $\hat{t}_{i,j}$'s, Eq. (6) is the final result of our camera calibration. The epipole e determines the direction of $\hat{T}$ via $\hat{T} \propto K^{-1}e$, where K^{-1} is the inverse of intrinsic camera matrix.

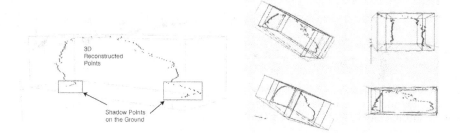

Fig. 7. points cluster on the ground   **Fig. 8.** Example: 3D reconstructed points

Without knowledge of scale constant (i.e., with arbitrary scale), 3D points are reconstructed. The clustered points on the ground determine the scale constant S. Given initial camera matrix P^0, the center of camera is null vector of P^0, i.e., $P^0 C = 0$ ([10]) and the height(z) of the center is $C[2]/C[3]$. The equation $S = cam\_height/(cam\_height - min\_z)$ gives the scale constant S where min_z represents the clustered shadow cast points on the ground ($Z = 0$) (Fig. 7). Even on a cloudy day where shadows are not clearly defined, this scheme of finding scale shows successful results in our experiments.

Dense 3D points on the vehicle contour generator are estimated. Given a line through the vanishing point (epipolar line), the sequence of intersections of that line with the sequence of apparent contours is taken as a sequence of observations of the same 3D point on the vehicle surface.

2.5 Recognition Using 3D Information: Reconstruction Error

Regardless of the benefit using 3D for recognition, triangulation error appear due to uncertainty of the image correspondence as shown in Fig. 9. The error of stereo vision system [11] and the accuracy of N-ocular vision system [12] are studied to analyze such errors. In general, the stereo localization error is described by an error ellipsoid in 3D space bounded by the down range and cross range errors of ΔR and ΔC. (Reconstruction using N-views will be similar to those of stereo case.) The down range error increases as the square of the range while the cross range error increases linearly. As the range increases, the down range error along the line of sight is dominant. Such analyses explain the elliptical shape of points spread in SfM method shown as in Fig. 10. The estimated camera positions (black asterisks) are also shown. Virtual cameras are about 60 feet (19 m) away from a vehicle and the maximum camera baseline is about 10 feet (3 m). Red and blue dots are triangulated points from all the pairs of 7 cameras. Outliers are removed (blue dots) and red points (14 points) are used for the point estimation ([-20.83, -1.91, 2.98]). Its covariance (using only red ones) is $\begin{bmatrix} 0.2183 & -0.0545 & 0.0255 \\ -0.0545 & 0.0136 & -0.0064 \\ 0.0255 & -0.0064 & 0.0030 \end{bmatrix}$. The clustered 14 red points lie within 1.8 feet range and are elliptically shaped. In Fig. 8, all of the points are reconstructed by such procedure. 3D reconstruction results are manually compared to manufacturers'

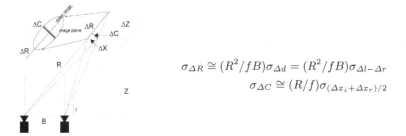

$$\sigma_{\Delta R} \cong (R^2/fB)\sigma_{\Delta d} = (R^2/fB)\sigma_{\Delta l - \Delta r}$$
$$\sigma_{\Delta C} \cong (R/f)\sigma_{(\Delta x_i + \Delta x_r)/2}$$

Fig. 9. Stereo localization. ΔR and ΔC define the error ellipsoid with $\sigma_{\Delta R}$ and $\sigma_{\Delta C}$.

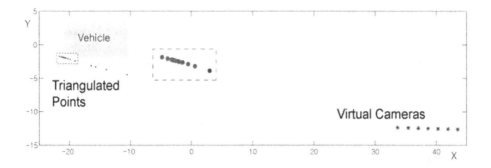

Fig. 10. Example of a point reconstruction P=[-20.83, -1.91, 2.98]

specification. The error in height was small (less than one foot) but the length of a vehicle can vary 3-5 feet. There exists non-negligible amount of error in reconstruction. 68 cases of 3D reconstructions on sedans are investigated. The length of 68 sedans has the mean 14.7 (feet) and the standard deviation 3.39. The height has the mean 4.5 and the std 0.39.

Features from Estimated 3D Points. Features and a probe used are shown in Fig. 11. Features are extracted from 3D points on the vehicle apparent contour through use of 3D probes. Probes are the 3D templates for estimating feature positions (Shown in top right corner).

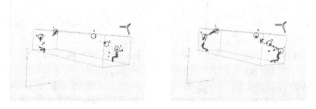

Fig. 11. Examples of 3D points and detected probes (features) using 3D probe model

3   Bayesian Recognition

The discriminant for vehicle class j (ω_j) is as follow. The class conditional probability given a video clip can be represented as $P(\omega_j | \bar{\mathbf{I}})$ for vehicle class j (ω_j) and a full Bayesian classifier should compute

$$
\begin{aligned}
P(\omega_j | \bar{\mathbf{I}}) &= P(\omega_j) P(\bar{\mathbf{I}} | \omega_j) / P(\bar{\mathbf{I}}) \\
&\propto P(\omega_j) P(\bar{\mathbf{I}} | \omega_j) = P(\omega_j) P(\bar{\mathbf{F}} | \omega_j)
\end{aligned}
\tag{7}
$$

where $P(\omega_j)$ is relative frequency of occurrence of vehicle class j and regarded as equal for all 4 categories (*i.e.*, $= 1/4$). $\bar{\mathbf{I}} = \{I_1, I_2, ..., I_N\}$ represents N images frames in a video clip. $\bar{\mathbf{F}} = \{f_1, f_2, ..., f_M\}$ is M-primitive or compound features, in our experiment $M = 3$).

Prior probability $P(\bar{\mathbf{F}} | \bar{\mathbf{I}}; \omega_j)$ ($= P(\bar{\mathbf{F}} | \omega_j)$) is the probability of features detected from 3D reconstruction of a vehicle given class j. The vector $\bar{\mathbf{F}}$ denotes the primitive and compound features and $P(\bar{\mathbf{F}} | \omega_j)$ is trained from learning samples.

One set of three features are selected based on their apparent discriminatory power together, which are: Height, (Rear of vehicle - Rear of cabin), (Cabin Center- Vehicle Center)). Those 3 dimensional feature vectors cluster in the feature space and its 2D subspace projections are shown in Fig. 12 and 13. Note that SUV's and mini-vans are overlapped (not separable).

When two video clips are combined, the discriminant used for vehicle class j is (independence assumption)

$$
P(\bar{\mathbf{F}} | \omega_j) = P(\bar{\mathbf{F}}_1, \bar{\mathbf{F}}_2 | \omega_j) = P(\bar{\mathbf{F}}_1 | \omega_j) P(\bar{\mathbf{F}}_2 | \omega_j)
\tag{8}
$$

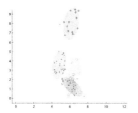 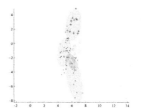

Fig. 12. Learning samples for the features "Height" (x-axis) and "Rear of vehicle - Rear of cabin" (y-axis) with a Gaussian fit

Fig. 13. Learning samples for the features "Height" (x-axis) and "Cabin Center- Vehicle Center" (y-axis) with a Gaussian fit

- legend: blue - sedan, red - SUV, green - minivan, black cross - pickup truck.

where $\bar{\mathbf{F}}_1, \bar{\mathbf{F}}_2$ are feature vectors from 3D point reconstructions based in 2 video clips and $\bar{\mathbf{F}} = [\bar{\mathbf{F}}_1, \bar{\mathbf{F}}_2]$. For L video clips, $P(\bar{\mathbf{F}}|\omega_j) = \prod_{i=1}^{L} P(\bar{\mathbf{F}}_i|\omega_j)$.

4   Recognition Experiments

200 single video clips were collected from the street at various times (summer or winter, day or dusk, cloudy, sunny, or snowy days). Among the 200 video clips, there are 66 pairs of video clips where each camera in a pair captures the same vehicle (a total of 132 video clips with 68 single clips remaining). Two different tests were conducted.

Test 1: Learn and test data are separated. 15 training samples for each class are selected (a total of 60 learn data). The recognizer was run on 140 test sets.

Test 2: 15 training sets as in Test 1, then the recognizer was run on all 200 video clips (learn+test data).

Each test consists of the following sub-tests: (a) Classify each single video clip individually. (b) Treat the clips in sub-test (c) as individual single clips and classify each individual clip. (c) Classify each pair of video clips together. The results are shown in Table 1-6 as confusion matrices.

4.1   Recognition Using Pairwise Video Clips

Camera positions for a pairwise video clips and 3D reconstruction's of them are shown in Fig. 15. Image frames by such a video camera set-up are also shown in Fig. 14. Table 3 and 6 show pairwise video clips recognition confusion matrix. For table 2 and 3, 32 pairs are used. For *Test 1* (b) (table 2), the 32 pairs are treated as independent 64 single video clips and the recognition was performed. Those 32 pairs were tested combined in *Test 1* (c) (table 3). It shows the increment of correct recognition rate from 89% to 94%. Similarly, *Test 2* (b) and *Test 2* (c) show the correct recognition rate from 86% to 91%. *Both Test 1 and 2 show the 5% increment in the correct recognition rate when the information of pairwise video clips is combined.*

Table 1. *Test 1* (a): 140 single test samples. **81%**.

	classification			
	C0	C1	C2	C3
C0	47	1		5
C1	2	48	6	1
C2		11	9	
C3	1			9

Table 2. *Test 1*(b): 64 singles in 32 pairs. **89%**.

	classification			
	C0	C1	C2	C3
C0	26	1		3
C1	1	26	1	
C2		1	1	
C3				4

Table 3. *Test 1*(c): 32 pairs. **94%**.

	classification			
	C0	C1	C2	C3
C0	28			2
C1		26	2	
C2			2	
C3				4

In *Test 1*, 140 test video clips consist of 32 pairwise video clips (64) + 76 single video clips

Table 4. *Test 2*(a): 200 all samples. **82%**.

	classification			
	C0	C1	C2	C3
C0	61	1		6
C1	3	59	8	2
C2		15	20	
C3	1	1		23

Table 5. *Test 2*(b): 132 singles in 66 pairs. **86%**.

	classification			
	C0	C1	C2	C3
C0	41	1		4
C1	3	36	4	1
C2		5	15	
C3	1			21

Table 6. *Test 2*(c): 66 pairs. **91%**.

	classification			
	C0	C1	C2	C3
C0	44			2
C1	2	36	6	
C2		2	18	
C3				22

In *Test 2*, 200 test video clips consist of 66 pairwise video clips (132) + 68 single video clips

† C0- sedan, C1- SUV, C2- minivan, C3- pickup truck.
‡ 60 learning samples (15 for each class).

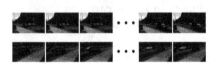

Fig. 14. Two video clips from camera 0 and 1 in Fig. 15

Fig. 15. Camera positions and 3D reconstruction from two video clips in Fig. 14

5 Conclusion

A new method using cross ratio invariance for Structure from Motion is developed and can be used as the input of Bundle Adjustment.

When SUV's and mini-vans are treated as one class, the recognition rate increases from 82% to 97% for the 200 single video clips test while reaching to 98% for the pairwise clip test. The experiment result combining two video clips shows an significant improvement on the recognition rate (from 86% to 91% for test 2, and from 89% to 94% for test 1). The recognition rate will increase with more video cameras.

In overall, our classification rate is comparable to view-based methods [1], [3], although a comprehensive comparison to other systems is not performed due to the differences of individual system set-up.

References

1. Mohottala, S., Masataka Kagesawa, K.I.: Vehicle class recognition using 3d cg. In. In: Proc. of 2003 ITS World Congress (2003)
2. Ozcanli, O., Tamrakar, A., Kimia, B., Mundy, J.: Augmenting shape with appearance in vehicle category recognition. In: IEEE Proceedings of the 2006 IEEE Computer Society Conference on Computer Vision and Pattern Recognition, CVPR 2006 (2006)
3. Jolly, M., Lakshmanan, S., Jain, A.: Vehicle segmentation and classification using deformable templates. IEEE Transactions on Pattern Analysis and Machine Intelligence 18 (1996)
4. Jain, A., Zhong, Y., Dubuisson-Jolly, M.: Deformable template models: A review. Signal Processing 71(2), 109–129 (1998)
5. Lou, J., Tan, T., Hu, W., Yang, H., Maybank, S.: 3-d model-based vehicle tracking. IEEE Transactions on Image Processing 14, 1561 (2005)
6. Han, D., Leotta, M., Cooper, D., Mundy, J.: Vehicle class recognition from video-based on 3d curve probes. In: 2nd Joint IEEE International Workshop on Visual Surveillance and Performance Evaluation of Tracking and Surveillance 2005, pp. 285–292 (2005)
7. Koller, D.: Moving object recognition and classification based on recursive shape parameter estimation. In: Proc. 12th Israel Conf. Artificial Intelligence, Computer Vision, pp. 27–28 (1993)
8. Ferryman, J., Worrall, A., Sullivan, G., Baker, K.: A generic deformable model for vehicle recognition. In: Proceedings of the 1995 British conference on Machine vision, vol. 1 table of contents, pp. 127–136. BMVA Press Surrey, UK (1995)
9. Stauffer, C., Grimson, W.: Adaptive background mixture models for real-time tracking, II: 246–252 (1999)
10. Hartley, R., Zisserman, A.: Multiple View Geometry in Computer Vision. Cambridge University Press, Cambridge (2000)
11. Kim, W., Ansar, A., Steele, R., Steinke, R.: Performance analysis and validation of a stereo vision system. In: IEEE International Conference on Systems, Man and Cybernetics (2005)
12. Firoozfam, P., Negahdaripour, S.: Theoretical accuracy analysis of N-ocular vision systems for scene reconstruction, motion estimation, and positioning. In: 2nd Internat. Symp. on 3D Data Processing, Visualization, and Transmission (3DPVT 2004). Citeseer (2004)

Efficient Head Tracking Using an Integral Histogram Constructing Based on Sparse Matrix Technology

Jia-Tao Qiu, Yu-Shan Li, and Xiu-Qin Chu

NO. 2 South Taibai Road, Xidian University, Xi'an, 710071, China

Abstract. In this paper, a sparse matrix technology-based integral histogram constructing is applied to a particle filter for efficient head tracking, which can significantly enhance the performance of the particle filter of large number of particles in terms of speed. Also, by exploiting the integral histogram constructing, a novel orientation histogram matching-based proposal is proposed for head tracking based on a circular shift orientation histogram matching, which is robust to in-plane rotation. The proposed head tracking is validated on S.Birchfields image sequences.

1 Introduction

Detecting and tracking of objects using their appearances play an important role in numerous computer vision applications such as video surveillance and human computer interaction. Fast appearance feature extraction is very important for the real-time requirement of object tracking[1].

Recently, the integral image representation is generalized to the integral histogram representation [1] to speed up search algorithms which are based on histogram-comparison. At a frame, the integral histograms are computed and stored beforehand. The histogram for any rectangular region is extracted with the pre-computed integral values at its four corner points. Therefore, the integral histogram technique has gained popularity in many domains for its high efficiency. For a large image, however, integral histogram extraction still demands significant computational resources because it is required to operate on the large number of pixels of the entire image. In this paper, the sparse matrix technology-based integral histogram constructing is applied to head tracking in a particle filter framework. For readability, this integral histogram constructing is referred to as SMIHC in the rest of the paper. As can be seen in Section 4, the performance of the particle filter tracker in terms of execution time is greatly enhanced by exploiting the SMIHC for calculation of box type filters during the particle sample phase. In this phase, a detection proposal is devised to generate particles based on matching gradient orientation histograms between the reference target and each candidate target in the current frame. Incorporating the up to date detection observation in the proposal distribution allows for stable head tracking with small number of particles. The proposed head tracking has

R. Koch et al. (Eds.): ACCV 2010 Workshops, Part I, LNCS 6468, pp. 256–265, 2011.

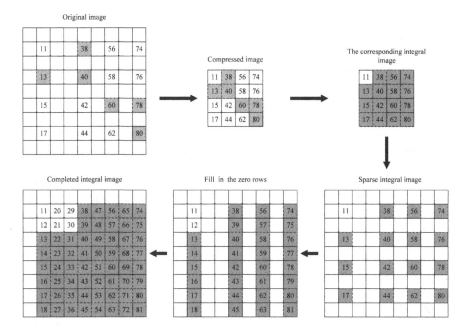

Fig. 1. Illustration of the efficient integral image constructing with sparse matrices

been tested on S.Birchfields image sequences and some promising results have been achieved.

The remainder of this paper is organized as follows. The use of the SMIHC for extracting orientation histograms will be described in Section 2. In Section 3, the orientation histogram matching-based head tracking using particle filters will be discussed. Some experimental results will be reported in Section 4. Finally, a brief summary will be given in Section 5.

2   Efficient Orientation Histogram Extraction

Integral histogram representation can speed up search algorithms which are based on histogram-comparison[1]. We follow the idea to use integral images for speeding up the extraction of gradient orientation histograms for fast head tracking.

In general, a gradient image is first thresholded to suppress noises. Then, the gradient orientation histograms are constructed with this threshold gradient image. Therefore, the gradient image used is very sparse and the gradient image of each orientation bin is even sparser. This sparseness property can be exploited to make the integral histogram constructing efficient, thus resulting in an improvement of orientation histogram extraction efficiency.The efficient integral image constructing is described as follows. For an orientation bin, its full gradient magnitude image is first converted to sparse form by squeezing out any zero rows and columns. These rows and columns contain only zeros and do not have any contributions to the integral image. But operating on them will consume significant computational resources. Therefore, the integral image for the orientation bin can be constructed based on this compressed image with small size. In Fig.1, the basic steps for constructing the integral image are illustrated. We start the pixel number from the upper left corner and count down to the bottom of the first column, and then start counting again at the top of the next column. The white pixels have zero values. The color pixels contain the non-zero values. The non-zero gradient magnitude pixels are labeled in sky blue. The non-zero integral pixels are labeled in green. The original image contains the magnitudes of gradients whose orientations falling into the bin. There are five typical steps in the process. First, extracting the compressed image by removing the zero rows and columns of the original image; second, computing its integral image with the traditional image integrating; third, substituting the original magnitude values in the original image with their integral values yielding the sparse integral image ; fourth, duplicating the integrated rows to the down zero rows; finally, operating on the last obtained matrix and duplicating the integrated columns to the right zero columns. In this process, the zeros in the zero rows are not involved in the row accumulating and the zeros in the zero columns are also not involved in the column accumulating.At the first step, the original image and its indices of the nonzero rows and columns are obtained with the sparse matrix technology. In the integrating of the compressed image, however, the row and column accumulating are implemented with the full matrix, for the accumulators may not be sparse vectors any more after the first several accumulations. It is not that all the zero elements in the matrix are removed but that the zero rows and columns are removed, which is different from the sparse matrix technology. At current frame, the integral histogram is calculated only for once and used throughout the histogram matching in the object searching. The histogram of any rectangular region in the frame can be constructed by indexing and arithmetic intersecting the integral histogram values at its four corner points. Please see [1] for more details.

3   Proposed Head Tracking

3.1   Head Detecting with a Circular Shift Orientation Histogram Matching

In this paper, we use detection responses to guide the tracker to the most likely region for further tracking. The detection response map is constructed by comparing the reference orientation histogram with that of each candidate target in a brute force search manner. In the search, only the candidate regions centered on a subset of nodes of a regular grid (see Fig.2 (d)) are considered. The brute force search is made practical by the integral histogram representation. We compare these candidate histograms with the reference histogram with a circular shift orientation histogram matching. Then, the candidate with the highest similarity score can serve as the crude detecting result.

We extract the reference model from the tracking head from early frames. But the tracking head has different orientations in the frames as rotating continually, as shown in Fig.2. To remove the effect of rotation, therefore, each candidate region should be rotated back to the reference patch before matching their orientation histograms.

To this end, image warping may demand significant computational resources. In fact, after extracting the histogram, all we need to do is to permute it with a circular shift histogram matching, which yields the same result as having rotated

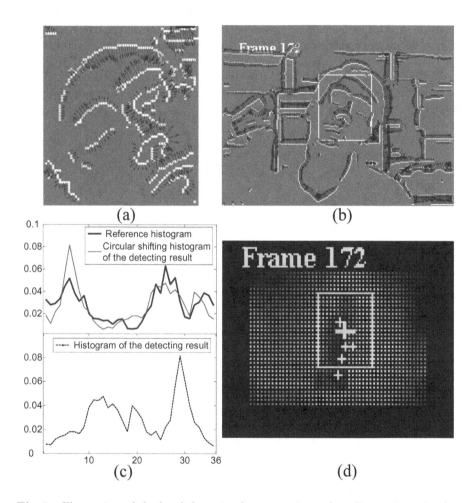

Fig. 2. Illustration of the head detection by comparison of gradient orientation histograms. (a) The reference gradient image (from frame 160). (b) The current gradient image (frame 172) and detecting result. (c) Illustration of the circular shift orientation histogram matching.(d)The detection response map and particles.

the image patch by the width of the associated number of angular bins (see Fig.2(c)).

Therefore, the histogram similarity between each circular shifted version of the original candidate histogram and reference histogram must be first measured. Then, the minimum over all the distances is chosen as the final distance between the candidate and reference regions. Let $\mathbf{h}_{ref} = \{h_{u,ref}\}_{u=1,\ldots,B}$ denote the reference orientation histogram , with B denoting the number of histogram bins. Within a candidate region $X^i = \{x^i, y^i, w^i, h^i\}$ in some later frames, with $(x^i, y^i), w^i$, and h^i denoting the location, width, and height respectively, the orientation histogram is denoted by $\mathbf{h}_{X^i} = \{h_{u,X^i}\}_{u=1,\ldots,B}$. Let $\mathbf{h}_X^s$ denote a circular shifted version of the original candidate histogram $\mathbf{h}_X$, with s denoting the number of bins that is shifted. For readability, we have omitted the superscript of X^i. The frequency contained in the new bin u_n in $\mathbf{h}_X^s$ is updated by that of its corresponding old bin u_o in $\mathbf{h}_X$, which can be computed by $u_o = mod(u_n - s + B - 1, B) + 1$, with the modulus function $mod\{A, D\}$ returning the residue after A divided by D . We borrow the distance metric on histogram models defined in [2] to compute the distance between the $\mathbf{h}_X^s$ and $\mathbf{h}_{ref}$, as

$$D(\mathbf{h}_{ref}, \mathbf{h}_X^s) = \sqrt{1 - \sum_{u=1}^{B} \sqrt{h_{u,ref}, h_{u,X}^s}} \qquad (1)$$

Based on this distance, we finally define a circular shift distance, as

$$d(\mathbf{h}_{ref}, \mathbf{h}_X) = \min_{0 \leq s < B} D(\mathbf{h}_{ref}, \mathbf{h}_X^s) \qquad (2)$$

Where $d(\mathbf{h}_{ref}, \mathbf{h}_X)$ denotes the circular shift distance between the $\mathbf{h}_{ref}$ and $\mathbf{h}_X$. In general, there are many distance functions to compare orientation histograms, the L2-norm distance [3], for instance. According to [4], the symmetric KL-divergence can also be well suited for the comparison of orientation histograms. In fact, the circular shifted orientation histogram is still the orientation histogram of the associated rotated patch. As shown in Fig.2(c), the proposed distance metric performs well in comparing the reference histogram with that of the in-plane rotated patches. Figure 2 illustrates an example of orientation histogram matching between the reference histogram and shifted versions of each candidate histogram by using Eq.(2). In this case, the sequence seq-sb is used. The reference gradient image of the tracking head (from frame 160) and current gradient image (from frame 172) are shown in Fig.2 (a) and Fig.2 (b), respectively. The solid blue graph in Fig.2 (c) shows the reference histogram extracted from the reference gradient image. We use 36-bin orientation histograms throughout the paper. The histograms of rectangular candidate regions are extracted from the current gradient image. The rectangular candidates are centered on the nodes of the regular grid (see Fig.2 (d)). All of them have the fixed size of the estimate state of previous frame. Figure 2 (d) shows the detection response map which is constructed via computing histogram distances using Eq.(2). The detecting result is labeled with rectangles.

The small crosses around the large one represent the particles. The dash graph at the bottom of Fig.2 (c) shows the original histogram of the detecting result, which is extracted from the rectangular region in Fig.2 (b). In Fig.2 (c), the solid red graph shows its shifted version that is the most similar to the reference histogram. Obviously, the rotated head can be accurately detected via performing orientation histogram comparisons in this case.

3.2 Head Tracking Using the Orientation Histogram Matching-Based Proposal

In this paper, the proposed head tracking follows the particle filter[2,5] framework. The basic steps of particle filtering described in [2] are followed. We define the hidden state as $X^i = \{x^i, y^i, w^i, h^i\}$, which corresponds to its associated candidate region. Particle filters can be described as follows. Starting with a weighted set of samples $\left\{X_{t-1}^{(i)}, \pi_{t-1}^{(i)}\right\}_{i=1,\ldots,N}$ approximately distributed according to $p\left(\mathbf{x}_{t-1} \mid \mathbf{y}_{1:t-1}\right)$, new samples are generated from a suitably designed importance function (also called proposal). The new importance weights are computed with a recursive equation , as

$$\pi_t^{(i)} \propto \pi_{t-1}^{(i)} \frac{p\left(\mathbf{y}_t \mid \mathbf{x}_t^{(i)}\right) p\left(\mathbf{x}_t^{(i)} \mid \mathbf{x}_{t-1}^{(i)}\right)}{q(\mathbf{x}_t^{(i)} \mid \mathbf{x}_{t-1}^{(i)}, \mathbf{y}_t)} \qquad \sum_i \pi_t^{(i)} = 1 \qquad (3)$$

The new particle set $\left\{X_t^{(i)}, \pi_t^{(i)}\right\}_{i=1,\ldots,N}$ is then approximately distributed according to $p\left(\mathbf{x}_t \mid \mathbf{y}_{1:t}\right)$. Monte Carlo techniques can then be used to obtain approximations to the desired point estimates . Color histograms are used for the construction of the color likelihood model that we use. The orientation histogram matching described in Section 3.1 is implemented for the construction of a proposal distribution to utilize the current observation data, enhancing the tracker's robustness to peaked likelihoods. The proposal distribution is constructed by evaluating the histogram similarity measure on a subset of locations over the image. Firstly, the orientation histograms of the candidate regions centered on the subset of nodes of the regular grid (see Fig.2(d)) are extracted. Secondly, each candidate orientation histogram is circularly shifted back to the reference histogram and the distance between them is computed by using Eq.(2), resulting in a detection response map (see Fig.2(d)). Thirdly, the response values are thresholded according to $d(\mathbf{h}_{ref}, \mathbf{h}_X) < \tau$, which is followed by computing non-minimum suppression. In this way the extrema of high similarity scores are detected . Based on locations of these extrema, denoted by $\mathbf{p}_j = \{x_j, y_j\}$,with $j = 1, \ldots, N_p$, we define an orientation histogram matching-based proposal for the head location, as:

$$q(x_t, y_t \mid x_{t-1}, y_{t-1}, \mathbf{y}_t) = \xi N((x_t, y_t)|(x_{t-1}, y_{t-1}), (\sigma_x^2, \sigma_y^2))$$

$$+ \frac{(1-\xi)}{N_p} \sum_{j=1}^{N_p} N((x_t, y_t)|\mathbf{p}_j, (\sigma_x^2, \sigma_y^2)) \qquad (4)$$

where the first term at the right-hand side of Eq.(4) is the prior density. The second term is a mixture-of-Gaussians containing the current observation information. As a result, this proposal allows for jumping in the state space to regions of high similarity scores. In this paper, the mixture coefficient ξ is set to be 0.2. An example of generating particles by using the proposed proposal is illustrated in Fig.2 (d). The reference histograms are extracted from the tracking head from early frames. The reliabilities of the reference histograms are judged by aligning them with a few pre-selected key frame templates, which are carefully extracted from the head under different poses without being polluted by the background pixels. The key templates are also used as reference models in themselves. This will help when the head undergoes large rotations and new parts of it come into view.

4   Experimental Results

4.1   Evaluation of the Proposed Head Tracking

In this Section, some experimental results are reported. The effectiveness of the proposed method had been tested on several image sequences provided by S.Birchfield [6]. Both the sequence seq-mb and seq-sb contain 500 frames. The sequence seq-dt has 150 frames. All the sequences have the resolution of 128x96. To verify the efficiency improvements brought by the proposed method, we had resized them to 512x384.

The robustness of the circular shift histogram matching to in-plane rotation was tested with seq-mb, as illustrated in Fig.3. The head in frame 290 had an

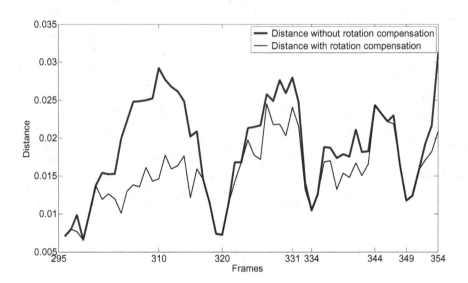

Fig. 3. Illustration of the performance of the circular shift distance metric

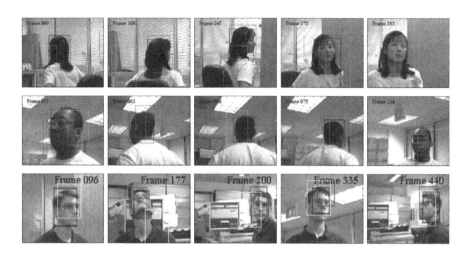

Fig. 4. Comparison of the rectangular results of the detecting and tracking methods. The Detecting:red rectangles. the mean-shift:yellow rectangles. the proposed:blue rectangles. Top row: seq-mb. Middle row: seq-dt. Bottom row: seq-sb.

upright posture and its orientation histogram served as the reference template for detecting head during frames 295-354. The distances between the head histograms of these frames and the reference histogram are shown. The thick blue graph shows the distances measured with the conventional distance metric without compensating the in-plane rotation. The thin black graph shows the distances measured using the proposed distance metric for compensating the in-plane rotation. As can be seen, the head at frames 320,334 and 349 had the histograms which are most similar to the reference histogram when taking upright postures. The head in the frames 310,331 and 344 showed the most divergence when rotating large angles from the upright posture. Clearly, the proposed distance metric on orientation histogram models had a better performance than the conventional one when the head underwent in-plane rotations.

Fig.4 shows some samples of tracking results. As can be seen, the proposed algorithm had a better performance than the traditional mean-shift algorithm in some challenging conditions such as head rotation, large appearance changes, and illuminating changes. The crude detecting results and the final tracking results also conformed to each other at most frames. This implies that the circular shift histogram matching-based proposal had a good performance. The performance of the proposed method was also evaluated by comparing the tracking results with the ground truth locations provided by S.Birchfield [6]. This is shown in Fig.5. The errors of the crude detecting results and the tracking errors of the traditional mean-shift are also shown. As can been seen from Fig.5(a), the traditional color mean-shift was failure as the girl rotated her head causing large appearance changes. From Fig.5(b), we can see that it also did not perform well for sequence seq-sb at frame 177 when the head experienced illumination

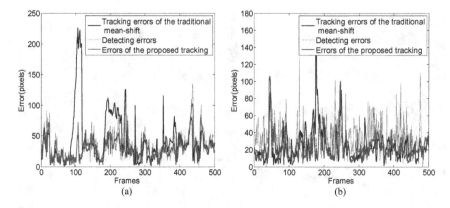

Fig. 5. Comparison between the tracking results and the ground truth and errors of different methods. (a) The errors for seq-mb. (b) The errors for seq-sb.

changes. The proposed head tracking, however, could lock on the head throughout the image sequence. The proposed tracking could also have large error values at some frames. But the head was in general well tracked, for the average head size in the image sequence is 140x180. It can also be seen that the crude detecting results basically conformed to the final tracing results.

4.2   Running Time Analysis

The simulation was accomplished with 2.0 GHz CPU, without any hardware acceleration.We list the processing time for five algorithm steps.

For integrating the gradient images for the integral orientation histogram, the traditional integrating consumed around 407.38 ms while the SMIHC required 196.41 ms. Furthermore, the SMIHC method provided just the same result. The processing time comparison between the SMIHC-based head tracking and the one with the straightforward histogram extracting was also performed, as shown in Table 1. It is impractical to implement the brute force search step with the conventional histogram extracting. In our experiments both the two used the integral histogram technical to perform the brute force search. To evaluate the performance of the proposed tracker in terms of speed when large number of particles were used, we had tested it by using 25, 50,100,200 particles. The color likelihood computation exploited the conventional histogram extraction method. If a larger number of particles were used, computing the color likelihood with integral histogram representation would significantly improve the efficiency of the tracker . It can be seen that the particle sample and brute force search steps only made up a small portion of the overall computation time by exploiting the SMIHC. The SMIHC-based particle filter tracker yielded a speedup of 1.2X over the one with the straightforward histogram extracting when using 200 particles. If a larger number of particles were used the speedup would be more satisfactory.

Table 1. Efficiency comparison between the proposed integral histogram-based head tracking and the one with the straightforward histogram extracting(unit:ms)

Stage	25 par.	50 par.	100 par.	200 par.
Sample(conventional)	46.92	89.38	176.91	353.16
Sample(proposed)	1.11	1.76	3.44	8.24
Likelihood	47.97	91.83	180.97	358.79
Gradient		71.80		
Search		11.54		

5   Conclusions

In this paper, the SMIHC is applied to particle filters for efficient head tracking. The SMIHC is around two times faster than the traditional orientation histogram integrating. Therefore, a brute force search can be implemented using the SMIHC for the construction of the orientation histogram matching-based proposal. Experimental results showed that the proposed head tracking had a good performance when the head underwent rotations and experienced illumination changes compared with the traditional color mean-shift algorithm.

References

1. Porikli, F.: Integral histogram: A fast way to extract histogram features. In: IEEE computer Society Conference on Computer Vision and Pattern Recognition (CVPR) (2005)
2. Perez, P., Vermaak, J., Blake, A.: Data Fusion for Visual Tracking with Particles. In: IEEE Proceedings Issue on State Estimation (2004)
3. Lowe, D.: Distinctive Image Features from Scale-invariant Key Points. International Journal of Computer Vision 60, 91–110 (2004)
4. Takacs, G., Chandrasekhar, V., Chen, H., Chen, D., Tsai, S., Grzeszczuk, R., Girod, B.: Permutable Descriptors for Orientation-Invariant Image Matching. In: IEEE International Conference on Computer Vision and Pattern Recognition, CVPR (2010)
5. Isard, M., Blake, A.: Condensation: Conditional Density Propagation for Visual Tracking. International Journal of Computer Vision 29, 5–28 (1998)
6. Birchfeild, S.: Elliptical Head Tracking Using Intensity Gradients and Color Histograms. In: IEEE International Conference on Computer Vision and Pattern Recognition, CVPR (1998)

Analyzing Diving: A Dataset for Judging Action Quality

Kamil Wnuk and Stefano Soatto

University of California, Los Angeles, CA, 90095
{kwnuk,soatto}@cs.ucla.edu

Abstract. This work presents a unique new dataset and objectives for action analysis. The data presents 3 key challenges: tracking, classification, and judging action quality. The last of these, to our knowledge, has not yet been attempted in the vision literature as applied to sports where technique is scored.

This work performs an initial analysis of the dataset with classification experiments, confirming that temporal information is more useful than holistic bag-of-features style analysis in distinguishing dives. Our investigation lays a groundwork of effective tools for working with this type of sports data for future investigations into judging the quality of actions.

1 Introduction

For sports as well as rehabilitation, quality control, security, and interfaces, the ability to make a critical judgment about how a particular action is performed can be imperative. With a significant portion of recent work focused on classification and detection of action categories, we instead propose focusing on analysis in domains where small details of temporal phenomena are the key informative elements. To this end, we have collected footage from a diving competition in which each dive is scored for technique irrespective of dive type or difficulty.

The convenience of working with a sport like diving is that it naturally avoids some of the early action dataset perils such as forced or exaggerated actions or loosely defined categories like "dance". All dive types are strictly defined and recorded in competition. Although the dive types vary, their quality is evaluated independently of type, meaning there are subtle universal details that are important and must be discovered.

In this work we evaluate representations for the diving data on the more familiar classification task to build our intuitions about the dataset. In particular, if a representation is not useful in discriminating between different dives, we can hardly expect it to capture the minutiae that might have a significant influence on a score for technique.

In the process we make incremental modifications to background subtraction and setup a feature extraction pipeline suited for individual performance sports captured at relatively high frame rates. Our methods are able to overcome noticeably compressed video with significant motion blur, a highly deforming subject, and changing illumination.

R. Koch et al. (Eds.): ACCV 2010 Workshops, Part I, LNCS 6468, pp. 266–276, 2011.

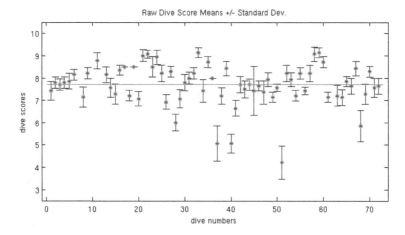

Fig. 1. Mean raw scores for all dives, shown with one standard deviation. The overall average of mean raw scores is shown in green. It is interesting to note that the most variance in scores took place during the middle of the competition: this is typically when divers take the most risks to try to get ahead.

1.1 Related Work

The most notable recent foray into sports footage in the literature was a broadcast sports dataset collected by [1]. However the dataset was a mixture of many different sports captured at highly variable angles and the task was limited to a categorization exercise. People have also demonstrated detection on particular ice skating or ballet moves, but to our knowledge none of the sports datasets available to date have provided a corresponding set of performance scores describing the quality with which the actors executed their moves.

In this preliminary work, our investigation focuses on classification experiments to test our intuition about suitable features and inference techniques. Notably we draw upon the success of descriptors based on gradient orientation histograms [2,3,4] in person detection, and adopt a variant thereof to encode pose.

We explore classification, drawing from both the holistic philosophy of treating the entire video as a bag of features [5,6,7] (which in our case encode poses), and from the temporal analysis perspective [8,9,10] of comparing multidimensional time series or their features.

2 The FINA09 Diving Dataset

We introduce a new dataset gathered from the finals of the men's 3 meter springboard diving event at the 13th FINA World Championships which took place in Rome, Italy on July 23, 2009. The dataset is gathered with the intent of motivating a new action analysis application of *judging action quality*, however, it presents a challenge for tracking and classification as well.

The dataset includes 12 different divers, each performing 6 unique dives. Each dive is recorded with two synchronized high speed cameras from orthogonal viewpoints. The first camera captures the dive head-on while the second records video from the side. Unfortunately 4 videos were partially omitted in the available footage, thus we only have 68 total samples. Example frames are shown in Fig. 2. The dive types and their distribution of scores are also shown in Table 1 and Fig. 1.

2.1   How Diving Is Scored

In the diving competition, each dive was given a raw score for technique by 7 independent judges. The scores are in the range from 1 to 10, and must be expressed in increments of 0.5 (ex: 6.0 and 6.5 are acceptable scores, 6.2 is not). The 2 highest and 2 lowest scores are discarded. The total raw score explaining how well the dive was executed is obtained by summing the remaining 3 scores. To obtain the final dive score this sum is then multiplied by the difficulty of the dive. Dive difficulties are constants assigned to each dive type agreed upon by FINA (the organization overseeing the competition).

As an end goal we are interested in evaluating the quality of a dive, irrespective of its difficulty class. However, in this work we explore the issue of representation in the context of the more familiar task of classification.

3   Background Subtraction and Tracking

When considering representations our intuition steers us towards a person-centric frame of reference, since factors such as symmetry and pose typically have high significance in perception of technique. We do however, take the time to compare classification performance with a representation requiring minimal preprocessing, which represents a video as a collection of spatial-temporal interest points with histogram of oriented gradients (HoG) and histogram of optical flow (HoF)

Table 1. Distribution of Dives by Type

Dive Type ID	Num Samples	Description
107B	11	forward 3.5 somersaults
205B	11	back 2.5 somersaults
307C	11	reverse 3.5 somersaults
405B	3	inward 2.5 somersaults
407C	9	inward 3.5 somersaults
5154B	9	forward 2.5 somersaults 2 twists
5156B	1	forward 2.5 somersaults 3 twists
5253B	2	back 2.5 somersaults 1.5 twists
5353B	10	reverse 2.5 somersaults 1.5 twists
5355B	1	reverse 2.5 somersaults 2.5 twists

Fig. 2. Sample synchronized frame pairs from various dives, exemplifying the type of color and shape irregularity, as well as effects of motion blur, compression, and regions of background with similar statistics to the foreground object

descriptors [11] (see Table 2). A person-centric representation implies tracking over the entire duration of the video. Throughout each video a diver undergoes significant deformations and passes through zones with varying shadows and lighting, while the background motions at high frame rate remain relatively small. Therefore we choose to model the background instead of the appearance of the diver for tracking purposes. In this section we explain our pipeline and underscore any adaptations made to existing works due to the unique challenges and properties of our data.

3.1 Robust Registration

In order to fix the small variation in scale and viewpoint between frames, we register sequential frames with an affine transform. We have found that simply applying RANSAC [12] to matching SIFT keypoints [2] works satisfactory in our sequences if we add correspondences with the frame at time $t - 5$ to the constraints in the least-squares estimation. Sample results are shown in Fig. 3.

3.2 Background Subtraction

In addition to stabilizing the video and allowing us to express events in a canonical reference frame, registration enables the construction of an initial background model (from mosaics Fig. 3). Since the videos in our dataset are recorded at "Internet streaming quality" they are plagued with compression artifacts that create constant jitter. These quantization artifacts are an additional nuisance on top of the already existing small background motions in the audience, and motion blur

Fig. 3. Typical median mosaic results from the sequential affine registration

due to fast motion of the camera or actor. To maximize our robustness to such nuisances we select the background subtraction method of [13], which models the background with a color histogram at each pixel taking into account a spatial and temporal neighborhood. We use a neighborhood radius of 4 (9×9 window), with color quantized via k-means to 32 bins. An update rate of $\alpha = 0.1$ is used to keep the background current. Local pixel histograms are compared with the Bhattacharyya Coefficient measure of similarity.

Improving foreground coherence. In the above background subtraction approach a threshold is typically selected on the Bhattacharyya Coefficient to separate foreground pixels from background pixels (0.76 in our case, determined emprically). An artifact of thresholding is that regions belonging to a solid object often get fragmented. This is most evident when the foreground object passes over an area where the color distribution in the background model is very similar to a region on the object. To encourage better label consistency we augment the above approach by constructing a Random Field on the pixel lattice V, E:

$$\hat{c} = \arg\min_{\mathbf{c}} \sum_{x_i \in V} \Psi(c_i|x_i) + \lambda \sum_{x_i, x_j \in E} \Phi(c_i, c_j|x_i, x_j), \tag{1}$$

where c_i is the binary label assigned to a given pixel, and each pixel $x_i \in V$ is connected to its 4 vertical and horizontal neighbors via the edges, E, represented as tuples x_i, x_j.

The unary term is simply the Bhattacharyya Coefficient resulting from the background subtraction approach, which ranges between 0 and 1. For the pairwise potential we use a function of the Euclidean distance between pixel color values in LUV space. This way the pairwise cost is highest when neighboring pixels have different labels but are similar in color, and low if different labels are assigned to neighboring pixels with a large difference in color:

$$\Psi(c_i|x_i) = BC_i, \qquad \Phi(c_i, c_j|x_i, x_j) = \frac{1}{1 + \|x_i - x_j\|}[c_i \neq c_j]. \qquad (2)$$

3.3 Foreground Object Tracking

Once we have established a background subtraction method to apply sequentially to each frame, we then have to find correspondences between foreground regions to obtain time series of region properties.

Naive Correspondence. We first try the naive correspondence approach by matching the nearest region with similar area in consecutive frames. However this approach breaks down when background subtraction fails to produce a single coherent region. Once a mismatch occurs in a noisy background subtraction result, the correspondence may continue to drift and produce nonsensical results.

Kalman Filtering in 3D. We leverage the orthogonal viewpoints in our dataset to implement a Kalman Filter in three dimensions to track the center of mass of the diver. The state includes the location, velocity, and acceleration of the center of mass and uses a random walk as the motion model. The observations are the naive correspondences described above.

We have manually annotated the time of the frames at which the diver begins receiving the final lift from the springboard and the time of impact with the water. In our experiments we do not use information before the diver starts being lifted by the springboard and stop tracking at the time of impact with the water to capture the angle of entry and splash. In classification experiments, the post-entry information is also discarded, and only the time spent in the air is considered.

4 Representation

4.1 Describing Pose with Gradient Orientation Histograms

The works of [2,3] and many others have shown variations on gradient orientation histograms to be highly successful in object recognition as well as person detection. Taking a cue from the above works, we encode diver poses by computing a SIFT descriptor [2,14] at a fixed scale and orientation centered at the

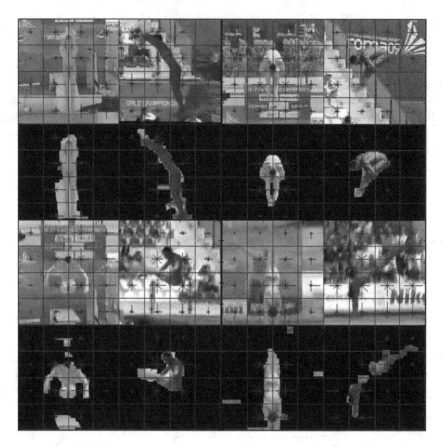

Fig. 4. Our diver-centered unsmoothed SIFT feature in various frames, computed without and with background subtraction for comparison. Notice in the descriptors using background subtraction that even frames with discontinuities and false foreground patches produce seemingly little noise in responses.

diver's tracked location. However, instead of smoothing the image to the scale of the descriptor before computing gradients as typically done in SIFT, we take the suggestion of [3] and skip the smoothing to capture finer level gradient information. The scale is selected so that it spans a square window size of 250×250 pixels. We have found empirically that the default 4×4 spatial bin configuration with 8 bins for orientation works sufficiently well for this dataset.

4.2   Leveraging Background Subtraction

We choose to incorporate background information into the descriptor by masking the gradient responses before descriptor computation. Since the descriptor encoding is designed with insensitivity to noise in mind, it overcomes frames where the background subtraction result was less than ideal. Notice in the examples in

Fig. 4 that even frames with discontinuities and false foreground patches produce seemingly little noise in responses. We compare the performance gained by incorporating background subtraction in Table 2 by running classification experiments both with descriptors computed before (1-SIFT) and after (1-SIFT-fg) applying the foreground mask.

5   Classifying Time Series

By extracting a diver centered descriptor, as described above, at each frame in both viewpoints, we obtain a 256 dimensional time series encoding the sequence of the diver's poses. We approach the classification task in 2 ways: without regard for temporal ordering of poses using a bag method, and with a dynamic time warping based kernel that keeps the sequence in tact during comparison.

5.1   Bag of Poses

The first approach, disregarding temporal ordering, uses the familiar bag-of-features pipeline. Since our representation has one feature per frame centered around the diver, this method essentially represents each dive as an unordered collection of poses.

To construct the pose dictionary we randomly sample features from dives in the training set (15 per dive sample) and compute a k-means clustering of a predetermined size. For each dive, we then project the descriptor in every frame to a given dictionary and represent the full dive sequence as a single histogram of pose occurrences. To discriminate between histograms we train a χ^2-SVM. We also experiment with the RBF-χ^2 kernel, but find that the former performs better in our case. Table 2 shows our results over a range of parameters.

5.2   Dynamic Time Alignment Kernel (DTAK)

To preserve the sequential nature of the dives while comparing time series and still get the benefit of using SVMs we can use the Dynamic Time Alignment Kernel (DTAK) proposed by Shimodaira et al. [15,16]. The DTAK kernel between two time series, $\mathbf{X} = (\mathbf{x}_1, \ldots, \mathbf{x}_n)$ and $\mathbf{Y} = (\mathbf{y}_1, \ldots, \mathbf{y}_m)$, is defined as:

$$K_{DTAK}(\mathbf{X}, \mathbf{Y}) = \frac{p_{n,m}}{n+m}, \tag{3}$$

$$p_{i,j} = \max \begin{cases} p_{i-1,j} + k(\mathbf{x}_i, \mathbf{y}_j) \\ p_{i-1,j-1} + 2k(\mathbf{x}_i, \mathbf{y}_j) \\ p_{i,j-1} + k(\mathbf{x}_i, \mathbf{y}_j) \end{cases} \tag{4}$$

where $k(\mathbf{x}_i, \mathbf{y}_j) = e^{-\frac{1}{2\sigma^2} \|\mathbf{x}_i - \mathbf{y}_j\|^2}$.

When using the DTAK kernel, we bypass the dictionary construction stage used in the bag of poses framework and compare descriptors directly. That is, $\mathbf{X}, \mathbf{Y}$ are sequences of descriptors taken from each frame of two different dives.

To add a level of robustness when comparing pose descriptors, similar to that offered by the dictionary creation stage in the bag pipeline, we do a reduced dimensionality approximation of the descriptors in each frame via PCA. Descriptors for PCA analysis were selected randomly from the training data in the same way as for k-means clustering in the bag approach. In our experiments we chose to represent each frame with 25 principal components, which capture approximately 43% of the variance of the data.

We apply the DTAK kernel on sequences of these reduced dimensionality descriptors and feed the computed kernel into an SVM.

6   Classification Experiments and Results

For our classification experiments we select the set of dive types to consist of those of which we have at least 6 samples. Referring back to Table 1, this leaves us with 6 unique dive types. We perform our tests by leave-one-out cross validation, where 1 dive sample is left out each round as the test sample, while all the other dives serve as training data. Dictionaries and principal components are always recomputed at each iteration to avoid training and testing on the same data.

Table 2 summarizes our results using various descriptors, parameters, and classification approaches. The top section of Table 2 demonstrates that without incorporating background subtraction, a person centric representation which requires tracking provides little gain if the classification is performed without regard for temporal order. We do notice a boost in performance when we classify the person-centric representation as a sequence. For this comparison we used the binaries provided by [11] to extract features from our videos.

Comparing results computed without incorporating background subtraction into the descriptor (1-SIFT) to those where the descriptor was masked to keep

Table 2. Mean Classification Accuracies for All Experiments

Tracking	Representation	Classifier	Mean Accuracy
none	[11]:HoG+HoF bag (k=500)	χ^2-SVM	58.67
3d-KF	1-SIFT bag (k=200)	χ^2-SVM	47.73
3d-KF	1-SIFT bag (k=500)	χ^2-SVM	58.13
3d-KF	1-SIFT PCA (npc=25)	dtak-SVM ($\sigma = 500$)	62.19
3d-KF	1-SIFT-fg bag (k=10)	χ^2-SVM	51.06
3d-KF	1-SIFT-fg bag (k=50)	χ^2-SVM	64.01
3d-KF	1-SIFT-fg bag (k=100)	χ^2-SVM	75.10
3d-KF	1-SIFT-fg bag (k=200)	χ^2-SVM	74.26
3d-KF	1-SIFT-fg bag (k=500)	χ^2-SVM	80.81
3d-KF	1-SIFT-fg bag (k=100)	RBF-χ^2-SVM	73.59
3d-KF	1-SIFT-fg bag (k=500)	RBF-χ^2-SVM	75.77
3d-KF	1-SIFT-fg PCA (npc=25)	dtak-SVM ($\sigma = 500$)	**91.75**
3d-KF	1-SIFT-fg PCA (npc=25)	dtak-SVM ($\sigma = 750$)	**91.75**
3d-KF	1-SIFT-fg PCA (npc=25)	dtak-SVM ($\sigma = 250$)	81.84

foreground (1-SIFT-fg) we see the benefit of this procedure. It boosts performance by almost 30% for the best performing classifier. Also, notice that DTAK kernel based classification shows significantly better performance and appears fairly stable with respect to changes in its σ parameter. The bag performance could be improved with temporal binning to include some sequential information, but the point was to compare both methods at their basic level. We also noticed that in our case there was no benefit to be had from using the RBF-χ^2 kernel.

7   Conclusion

We have presented a preliminary analysis of a new diving dataset and constructed an effective pipeline of tools for processing video of individual technique-based sports. Our results confirm that temporally constrained analysis is preferable for distinguishing dives, and that a gradient orientation histogram based pose representation is effective in the classification task. The question left open for our future investigations is whether these techniques will also hold true for dive quality score estimation. The diving dataset will be made publicly available on the website of the authors.

Acknowledgements. We would like to acknowledge the support of ONR 67F-1080868/N00014-08-1-0414, ARO56765-CI and AFOSR FA9550-09-1-0427.

References

1. Rodriguez, M., Ahmed, J., Shah, M.: Action mach: A spatio-temporal maximum average correlation height filter for action recognition. In: Proc. CVPR (2008)
2. Lowe, D.G.: Distinctive image features from scale-invariant keypoints. IJCV 2, 91–110 (2004)
3. Dalal, N., Triggs, B.: Histogram of oriented gradients for human detection. In: Proc. CVPR (2005)
4. Felzenszwalb, P., Girshick, D., McAllester, D., Ramanan, D.: Object detection with discriminatively trained part based models. PAMI (2010)
5. Efros, A.A., Berg, A.C., Mori, G., Malik, J.: Recognizing action at a distance. In: Proc. ICCV (2003)
6. Schuldt, C., Laptev, I., Caputo, B.: Recognizing human actions: A local svm approach. In: Proc. ICPR (2004)
7. Yeffet, L., Wolf, L.: Local trinary patterns for human action recognition. In: Proc. ICCV (2009)
8. Wilson, A., Bobick, A.: Parametric hidden markov models for gesture recognition. PAMI 21 (1999)
9. Raptis, M., Wnuk, K., Soatto, S.: Flexible dictionaries for action recognition. In: Proceedings of the 1st International Workshop on Machine Learning for Vision-based Motion Analysis, in conjunction with ECCV (2008)
10. Saad, A., Arslan, B., Mubarak, S.: Chaotic invariants for human action recognition. In: IEEE 11th International Conference on Computer Vision, pp. 1–8 (2007)

11. Laptev, I., Marszalek, M., Schmid, C., Rozenfeld, B.: Learning realistic human actions from movies. In: Proc. CVPR (2008)
12. Fischler, M., Bolles, R.: Random sample consensus: A paradigm for model fitting with applications to image analysis and automated cartography. Comm. of the ACM. 24, 381–395 (1981)
13. Ko, T., Soatto, S., Estrin, D.: Background subtraction on distributions. In: Forsyth, D., Torr, P., Zisserman, A. (eds.) ECCV 2008, Part III. LNCS, vol. 5304, pp. 276–289. Springer, Heidelberg (2008)
14. Vedaldi, A., Fulkerson, B.: VLFeat: An open and portable library of computer vision algorithms (2008), http://www.vlfeat.org/
15. Shimodaira, H., Noma, K.I., Nakai, M., Sagayama, S.: Dynamic time alignment kernel in support vector machine. In: Proc. NIPS (2002)
16. Zhou, F., De la Torre, F., Hodgins, J.K.: Aligned cluster analysis for temporal segmentation of human motion. In: IEEE Conference on Automatic Face and Gestures Recognition (2008)

Appearance-Based Smile Intensity Estimation by Cascaded Support Vector Machines

Keiji Shimada[1], Tetsu Matsukawa[2], Yoshihiro Noguchi[1], and Takio Kurita[3]

[1] Human Technology Research Institute, Advanced Industrial Science and Technology, Tsukuba, Japan
[2] Graduate School of Systems and Information Engineering, University of Tsukuba, Tsukuba, Japan
[3] Faculty of Engineering, Hiroshima University, Higashi-Hiroshima, Japan

Abstract. Facial expression recognition is one of the most challenging research area in the image recognition field and has been studied actively for a long time. Especially, we think that smile is important facial expression to communicate well between human beings and also between human and machines. Therefore, if we can detect smile and also estimate its intensity at low calculation cost and high accuracy, it will raise the possibility of inviting many new applications in the future. In this paper, we focus on smile in facial expressions and study feature extraction methods to detect a smile and estimate its intensity only by facial appearance information (Facial parts detection, not required). We use Local Intensity Histogram (LIH), Center-Symmetric Local Binary Pattern (CS-LBP) or features concatenated LIH and CS-LBP to train Support Vector Machine (SVM) for smile detection. Moreover, we construct SVM smile detector as a cascaded structure both to keep the performance and reduce the calculation cost, and estimate the smile intensity by posterior probability. As a consequence, we achieved both low calculation cost and high performance with practical images and we also implemented the proposed methods to the PC demonstration system.

1 Introduction

The visual information plays a very important role in our everyday life. Especially, in regard to communication between human beings, we can come to understand deeply and smoothly each other to pay attention to behaviors and facial expressions as well as languages. Facial expression analysis has been approached by several research fields, for example in psychology[1], brain science, etc. In engineering[2] too, many researchers have tried to analyze and estimate facial expressions and human emotions by face images, by voice signals, by biosignals, etc. for a long time. But, it is still difficult to recognize facial expressions only by face images automatically, because there are many problems such as inconsistencies in individuals, lack of criterion to judge facial expressions, disparities between simulation data and practical data and a mismatch between the expressions and the emotions. Therefore, there is no critical solution to work well under the practical environment and active research is still much in progress.

R. Koch et al. (Eds.): ACCV 2010 Workshops, Part I, LNCS 6468, pp. 277–286, 2011.

In particular, smile (In a wide sense, facial expressions, which are observed when human beings derive pleasure) is one of the most important facial expression used to communicate well between human beings and also between human and machines. If we can automatically detect smile on real-time and at high accuracy, it will serve a useful function to existing applications like digital still camera and HMI (Human Machine Interface), and also raise the possibility of inviting new applications like rehabilitation and welfare in the near future. Furthermore, we think that such applications, which contain a camera should work on-the-fly, because of privacy issues.

In general, there are two major approaches to detect smile. One is feature-based method[3] and the other is appearance-based method[4]. Feature-based method has the robustness for the variation of face positions and angles, because it can normalize those and analyze more detailed information around facial parts. But it generally requires to find some facial parts such as eyes, mouth, etc. So, if it does not find those facial parts, it can't provide the result. On the other hand, appearance-based method does not need to find facial parts and can provides the result of smile detection without facial parts detection. As a result, although it is susceptible to the position of facial parts and the variation of face angle, it has low calculation cost.

In this paper, we study the method to detect smile and estimate its intensity using only facial appearance information on real-time and high performance, which is robust to the position gap of facial parts and face angle within approximately ±30 degrees of frontal. We have also implemented our proposed methods to on-the-fly PC demonstration system.

2   Smile Detection and Intensity Estimation

We try to detect smile and estimate smile intensity using 256 gray values, where the size of face is 40 x 40 pixels. That means it is not necessary to identify facial parts in our method. Fig. 1 shows the process flow of our smile detection and smile intensity estimation. In this paper, we study three feature extractions, namely such as Local Intensity Histogram (LIH), Center-Symmetric Local Binary Pattern (CS-LBP)[5] and LIH+CS-LBP, which combines the above two features as facial appearance information. In addition, we consist SVM smile detector as cascaded structure like face detector proposed in [6]. It consists of sub detector, which has a small number of support vectors by applying Reduced Set Method (RSM)[7] and main detector, which consists of all support vectors.

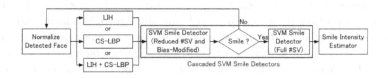

Fig. 1. Smile detection and intensity estimation flow

This cascaded structure has the ability to keep high performance, while reducing the calculation cost.

At the end, we estimate smile intensity based on the posterior probability estimated by the output from SVM smile detector.

2.1   Feature Extraction

Generally, face detector does not insure the accuracy of the detected face positions, means that positions of facial parts such as eyes, mouth and etc., are not always corresponding for each detected face. Therefore, the robust features for facial parts positions and face angles are necessary to detect smile only by appearance information, accurately. In this paper, we divide the face image into some grid cells and extract local features for each cell, after that we build the final feature by concatenating all local features. We use Local Intensity Histogram (LIH) and Center-Symmetric Local Binary Pattern (CS-LBP) as local feature and describe how to extract those features in the following subsections.

Local Intensity Histogram (LIH). LIH is build by concatenating the intensity histograms in local regions and the extraction steps are as follows:

1. Divide face image into M x N cells
2. Build an intensity histogram with L bins for each cell
3. Normalize the histogram for each cell
4. Build the final feature by concatenating the normalized intensity histograms of all cells to form a (M x N x L) dimensional vector

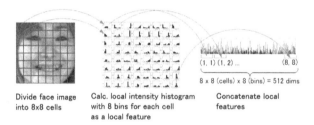

Divide face image   Calc. local intensity histogram   Concatenate local
into 8x8 cells   with 8 bins for each cell   features
                        as a local feature

Fig. 2. Example of feature extraction by LIH (8 x 8 cells, 8 bins)

Fig. 2 shows the processing example, where face image is divided into 8 x 8 cells and 8 bins.

Center-Symmetric Local Binary Pattern (CS-LBP). CS-LBP is a simple method and it is also has the ability to extract features, which has robustness for illumination changes. Additionally, it can also represent a texture information as more compact binary patterns. CS-LBP is calculated by,

$$CS-LBP_{R,N,T}(x,y) = \sum_{i=0}^{(N/2)-1} s(n_i - n_{i+(N/2)})2^i, \quad s(x) = \begin{cases} 1 & x > T \\ 0 & otherwise \end{cases} \quad (1)$$

where T is an encoding threshold, n_i and $n_{i+(N/2)}$ correspond to the gray values of center symmetric pairs of pixels of N equally spaced pixels on a circle of radius R. In this paper, N is fixed to 8 and R is fixed to 1. The following steps show the process to extract CS-LBP feature:

1. Divide face image into M x N cells
2. Calculate a CS-LBP for each cell and build a CS-LBP histogram
3. Buid the final feature by concatenating the CS-LBP histograms of all cells to form a (M x N x 16) dimensional vector.

Fig. 3 shows the processing example, where face image is divided into 5 x 5 cells.

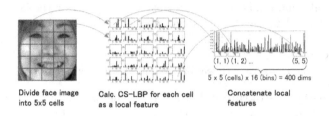

Divide face image   Calc. CS-LBP for each cell   Concatenate local
into 5x5 cells   as a local feature   features

Fig. 3. Example of feature extraction process by CS-LBP (5 x 5 cells)

2.2 Detection and Intensity Estimation

In this paper, we use SVM with RBF kernel function. RBF kernel function is defined as,

$$K(\mathbf{x}_i, \mathbf{x}_j) = \exp(-\frac{\|\mathbf{x}_i - \mathbf{x}_j\|}{2\sigma^2}) \tag{2}$$

where σ is a kernel parameter. Then the decision function with kernel is given as,

$$y = \sum_{i \in S} \alpha_i K(\mathbf{x}_i, \mathbf{x}) - h \tag{3}$$

where S and $\mathbf{x}_i$ means a set of support vectors and support vector, K is a kernel function, $\mathbf{x}$ is the input vector. Here, α_i shows weight for support vector and h represent the bias term. In smile detection case, If $y \geq 0$ implies smile and non-smile, otherwise.

In addition, we estimate smile intensity to evaluate the posterior probability of SVM outputs. Using sigmoid function, smile intensity is defined as,

$$si = \frac{1}{1 + \exp(-\lambda y)} \tag{4}$$

where si is smile intensity that ranges from 0 to 1 and λ is the gain. In this paper, we fix λ to 5.

Furthermore, we construct cascaded smile detectors described in Sec. 2. It consists of sub smile detector, which is reduced the number of support vectors constructed by RSM and main smile detector, which has all the set of support vectors.

3   Data Preparation

We constructed the original image database, which consisted of sports games TV programs to train and test our smile detector. In such TV programs, the spectators are often shooted and their expressions and emotions vary with the outcome of games. It is good for us, because the most spectators show same facial expressions with the outcome of their supporting athletes or teams. Therefore, we can collect smile and non-smile images with a high degree of efficiency. Moreover, face direction, gender and age of spectators are so various that we can evaluate the practical performance of our proposed method. As a result, our original smile/non-smile database contained 6,460 faces, which consisted of 2,730 smile samples and 3,730 non-smile samples.

In addition, we also used the pubic databases, "Facial Expression and Emotion Database (FEED)"[8], to evaluate the smile intensity estimation method.

4   Experiments

In this paper, we tested our system by 5-fold cross validation, which was one of several approaches commonly used for evaluation purpose. We compared each performance by Area Under the Curve (AUC), which was obtained by Receiver Operating Characteristics (ROC) Analysis.

4.1   Performance Evaluation by LIH

With respect to LIH, we investigated the optimal number of cells and bins. We first compared the performance according to the number of cells with the fixed number of bins, which is 8 (See Fig. 4 (Left)). Here, we used our original database and smile detector, which consisted of all support vectors (not cascaded). When increasing the number of cells, AUC was gradually improved. So, 8 x 8 cells showed the best performance. Next, we compared AUC by varying the number of bins, while keeping the number of cells constant as 8 x 8 (See Fig. 4 (Right)).

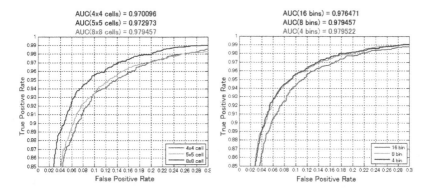

Fig. 4. Comparison of performance by LIH

4 and 8 bins showed almost the same better performance, but 4 bins provided
the best. It means that we need just 4 gray values to detect smile in this experiment. As a result of these experiments, with respect to LIH, the optimal
parameters were 8 x 8 cells and 4 bins (That is a 256 dimensional vector) and
that performance provided 0.979522 for AUC.

4.2   Performance Evaluation by CS-LBP

With respect to CS-LBP, we investigated the optimal number of cells and the
encoding threshold. We first compared the performance according to the number
of cells with constant encoding threshold equal to 0.00 (See Fig. 5 (Left)). Here,
we used our original database and smile detector, which consisted of all support
vectors (not cascaded). As increasing the number of cells, AUC was higher, but
too match cause to degraded the performance. In this experiment, 5 x 5 cells
gave the best. Next, we compared AUC according to the encoding threshold
with constant 5 x 5 cells (See Fig. 5 (Right)). Almost the same performances
were shown, but in this experiment, the encoding threshold of 0.02 provided
the best AUC. As a result of these experiments, with respect to CS-LBP, the
optimal parameters were 5 x 5 cells and encoding threshold of 0.02 (That is a
400 dimensional vector) and it provided 0.979423 for AUC.

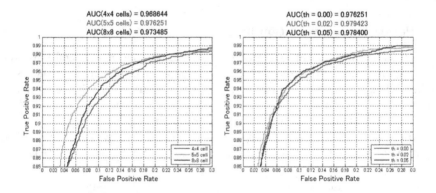

Fig. 5.   Comparison of performance by CS-LBP

4.3   Performance Evaluation by LIH+CS-LBP

In this section, we describe the experiments with LIH+CS-LBP feature, which
is combined LIH and CS-LBP. Here, the parameters of LIH were set to 8 x 8
cells and 4 bins and the parameters of CS-LBP were set to 5 x 5 cells and the
encoding threshold of 0.02 from the above experimental results. LIH+CS-LBP,
which was a 656 (= 256 (LIH) + 400 (CS-LBP)) dimensional vector, improved
the performance. Here, it provided 0.982320 for AUC and it was better than
using only LIH or CS-LBP (See Fig. 6).

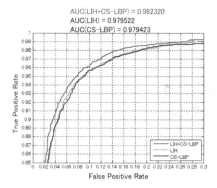

Fig. 6. Performance comparison of all the three features

4.4   Performance of Cascaded SVM Smile Detector

In this section, we describe the comparison of the performance and calculation cost, with our cascaded SVM smile detectors. Here, we used LIH+CS-LBP features and the parameters of LIH were set to 8 x 8 cells and 4 bins and the parameters of CS-LBP were set to 5 x 5 cells and the encoding threshold of 0.02 according to the results of Sec 4.1∼ 4.3. The number of support vectors of sub smile detector were reduced either to 32, 64, 128, 256, 512 or 1024 by RSM. And we also adjusted a bias term (h in Equ. 3) of sub smile detector lower to suppress the miss rejection cases at the sub smile detector. In this paper, we adjusted a bias to achieve True Positive Rate as well as main smile detector's one in advance. Fig. 7 shows the performance according to the several cascaded SVMs. When the number of support vectors of sub smile detector decreased, the performance degraded. But it kept to provide over 0.98 for AUC with 1024 support vectors in this experiment.

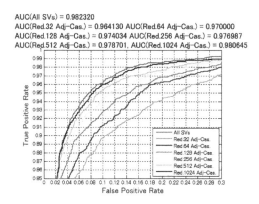

Fig. 7. Performance of cascaded SVMs (feature extraction is LIH+CS-LBP)

Table 1. Comparison of the number of SVs, AUC and calculation speed (Matlab@3.0GHz Core 2 Quad)

Classifier	# SVs	AUC (LIH+CS-LBP)	CPU Time (msec)
Normal SVM (with all SVs)	2481	0.982320	9.6418
Cascaded SVMs	32 (& 2481)	0.964130	3.9463
	64 (& 2481)	0.970000	4.0364
	128 (& 2481)	0.974034	4.1089
	256 (& 2481)	0.976987	4.6670
	512 (& 2481)	0.978701	5.4277
	1024 (& 2481)	0.980645	7.5097

on Matlab @ 3.0GHz Core 2 Quad

At the end of this section, we showed a comparison of AUC and calculation speed for each smile detector structure (See Table 1).

Cascaded SVMs with 1024 support vectors in sub smile detector achieved comparable in performance ($AUC > 0.98$) to normal SVM (non-cascaded), while reducing over 20% in the calculation cost. As a result of these experiments, our proposed cascaded SVM smile detectors could reduce calculation cost with a little performance degradation.

4.5   Smile Intensity Estimation

The FEED database is suitable to evaluate a change of a certain facial expression, because it has $100 \sim 150$ image sequences, which contain the variation from neutral face to a certain facial expression for each subject. Fig. 8 showed the results of our smile intensity estimator to "happy" expression of Subject #0001. This result proved that our smile intensity estimator could track the transition from neutral to smile well. Especially, it could represent the subtle facial expression changes as shown in red rectangle area in Fig. 8.

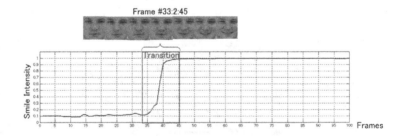

Fig. 8. Smile intensity estimation result of "happy" sequence #1 of Subject #0001

5   Demonstration System

In this section, we introduce the PC demonstration system, which was implemented according to the methods proposed in this paper. Fig. 9 (Top) shows

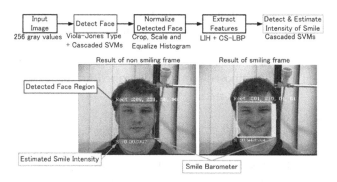

Fig. 9. Process flow of the PC demonstration system (Top) and examples of detection results (Bottom)

the process flow. At first, we detect face from the input image by face detector and crop, scale and normalize the influence of illumination changes by the histogram equalization. After that, we extract LIH+CS-LBP features as facial appearance information and detect smile and estimate its intensity by cascaded SVMs. Fig. 9 (Bottom) shows examples of detection results by our demonstration system. Our demonstration system roughly spends $48ms$ for face detection and $8ms$ for smile intensity estimation per face on the average (on a Core 2 Quad at 3GHz). It shows that our system can detect face and estimate its smile intensity from input image on semi real-time.

6   Conclusion

In this paper, we studied how to detect smile and estimate smile intensity only by facial appearance information, and proved the validity of our proposed system through several experiments. We also built the semi real-time PC demonstration system, which was implemented according to our proposed smile intensity estimation methods.

We constructed original smile/non-smile practical database from sports games TV programs, which contained various spectators in both indoor and outdoor. We investigated the optimal parameters for LIH and CS-LBP to detect smile with the above database and compared the performance of smile detection using LIH, CS-LBP and LIH+CS-LBP. In our result, with respect to LIH, we achieved 0.979522 for AUC with 8 x 8 cells and 4 bins and for CS-LBP, achieved 0.979423 with 5 x 5 cells and the encoding threshold of 0.02. Combined feature, LIH+CS-LBP worked better among all the three features. That produced a AUC value of 0.982320 as the best performance. This result indicates that our proposed system is robust and works well even under the practical environment.

In addition, we constructed cascaded SVMs for smile detector, which was composed of a sub detector, consisted of small number of support vectors and a main detector consisted of all support vectors. As a result, we could keep AUC higher than 0.98, while delivering about 20% reduction in the calculation cost.

With respect to smile intensity estimation, we showed that our estimator could track the subtle expression changes from neutral to smile using the FEED database.

In the future, we plan to detect the other facial expressions and estimate those intensities based on the proposed methods described in this paper.

References

1. Ekman, P., Friesen, W.V.: Unmasking the face. A guide to recognizing emotions from facial cues. Prentice-Hall, Englewood Cliffs (1975)
2. Fasel, B., Luettin, J.: Automatic facial expression analysis: A survey. Pattern Recognition 36, 259–275 (1999)
3. Whitehill, J., Littlewort, G., Fasel, I., Bartlett, M., Movellan, J.: Toward practical smile detection. IEEE Transactions on Pattern Analysis and Machine Intelligence 31, 2106–2111 (2009)
4. Deniz, O., Castrillon, M., Lorenzo, J., Anton, L., Bueno, G.: Smile detection for user interfaces. In: Bebis, G., Boyle, R., Parvin, B., Koracin, D., Remagnino, P., Porikli, F., Peters, J., Klosowski, J., Arns, L., Chun, Y.K., Rhyne, T.-M., Monroe, L. (eds.) ISVC 2008, Part II. LNCS, vol. 5359, pp. 602–611. Springer, Heidelberg (2008)
5. Heikkilä, M., Pietikäinen, M., Schmid, C.: Description of interest regions with local binary patterns. Pattern Recogn. 42(3), 425–436 (2009)
6. Shimada, K., Noguchi, Y., Sasahara, H., Yamamoto, M., Tamegai, H.: Detection of driverfs face orientation for safety driving assistance. Transactions of JSAE 41, 775–780 (2010)
7. Scholkopf, B., Burges, C.J.C., Smola, A.J.: Advances in Kernel Methods. The MIT Press, Cambridge (1998)
8. Wallhoff, F.: Facial expressions and emotion database (2006),
   http://www.mmk.ei.tum.de/~waf/fgnet/feedtum.html

Detecting Frequent Patterns in Video Using Partly Locality Sensitive Hashing

Koichi Ogawara, Yasufumi Tanabe, Ryo Kurazume, and Tsutomu Hasegawa

Kyushu University

Abstract. Frequent patterns in video are useful clues to learn previously unknown events in an unsupervised way. This paper presents a novel method for detecting relatively long variable-length frequent patterns in video efficiently. The major contribution of the paper is that Partly Locality Sensitive Hashing (PLSH) is proposed as a sparse sampling method to detect frequent patterns faster than the conventional method with LSH. The proposed method was evaluated by detecting frequent everyday whole body motions in video.

1 Introduction

Detection of previously learned human actions in video [1,2,3] or detection of irregular, i.e. not learned, human actions in video [4,5] have been an active research topic because they can be applied to various tasks including content-based retrieval from video, surveillance, etc.

For this, a human action database is usually built in advance in a supervised way by manually annotating a large collection of videos. To accelerate this process, several methods have been proposed to detect frequent human actions in video based on frequent data mining techniques [6,7,8,9]. This line of research can be grouped into 2 approaches. In the first approach, a large number of spatio-temporal patches are extracted from video and are classified into different actions [10,11], however it lacks the ability to detect relatively long actions. In the second approach, dynamic programming is employed to detect relatively long actions [12], however the computational complexity becomes $O(N^2)$ where N is the length of a video, thus it is not appropriate to deal with a long video.

To decrease the computational time, Meng et al. proposed a method that finds data similar to the data at t from the entire input time series by Locality Sensitive Hashing (LSH) [13] for all t and connects them along the time axis so as to detect frequent actions from a motion capture data in $O(N^{1+1/\alpha})$ [14]. The problem is that there is a large overlap between the found data at successive times, thus the search is redundant.

In this paper, we propose a novel method for detecting frequent patterns in video efficiently where the redundancy in [14] is resolved. In the proposed method, detected nearby data in the first frame are maintained by a linked list and are updated along the time axis. However, dynamic change of nearby data is not handled correctly, thus the linked list is modified based on a small number of

R. Koch et al. (Eds.): ACCV 2010 Workshops, Part I, LNCS 6468, pp. 287–296, 2011.

data sparsely sampled in each frame, whereas nearby data are densely sampled in every frame in [14] which results in a large computational time.

As a sparse sampling method, Partly Locality Sensitive Hashing (PLSH) is proposed which is an extension to LSH. Experimental results show that the proposed method can detect frequent patterns in video much faster than the conventional method with LSH.

2   Overview of the Proposed Method

The purpose of this study is to detect unknown frequent patterns appeared in a d-dimensional time series. Frequent patterns are a set of subsequences of a time series where minor variation of shape and length is allowed.

Fig. 1 shows a 2D slice of a d-dimensional time series. If a data point $o(t)$ observed at t is on a frequent pattern, many similar shaped patterns exist at around $o(t)$. So, if a sequence of data has many other data in its neighborhood, it can be considered as a good candidate for a frequent pattern.

Here, terminology is defined as follows. "Neighborhood" is defined as the inside of a hyper sphere of radius R in d-dimensional space. "Segment" is defined as a subsequence of a time series bounded by a hyper sphere. "Data density" is defined as the total length of segments in a hyper sphere. Then, the data density at t is calculated as

$$D(t) = \sum_{i \in S(t)} \|o(i) - o(i+1)\| \tag{1}$$

$$\text{where } S(t) = \{i; \|o(i) - o(t)\| \leq R\}.$$

Data density is used to evaluate the existence of frequent patterns in each frame. However, it takes quadratic time in total to calculate data density exactly in all frames even if a tree-like data structure is used.

To overcome this problem, data density is calculated efficiently by an algorithm outlined in Table 1 using the proposed approximate nearest neighbor search scheme named PLSH.

At $t = 1$, a linked list is initialized by checking all N data so that each element in the linked list holds the endpoints, i.e. the start time (st) and the end time (ed), of all the segments as in Fig. 1 (a).

From $t = 2$ to N, the linked list is updated along the time axis. Firstly, the endpoints of the segments found in the previous frame are shifted so that they lie on the boundary of the current hyper sphere as in Fig. 1(b). The amount of shift between neighboring frames is usually very small, so the time taken to update the linked list can be considered as a constant when the total number of elements in the linked list is limited. When a segment goes out of the sphere, it is removed from the linked list. When 2 segments are connected together, the corresponding elements in the linked list are merged as in Fig. 1 (c).

However, newly appeared segments cannot be detected immediately by the update method above, so they have to be found and added to the linked list

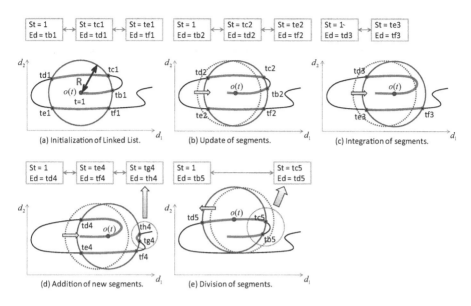

Fig. 1. Calculation of data density

separately as in Fig. 1 (d). For this, instead of checking all N data which results in quadratic time in total, nearby data are sparsely sampled in the proposed method. If a sample is inside the hyper sphere and is not included in the linked list yet, the segment including that sample are added to the linked list. A new segment is not necessarily detected at the time it enters into the hyper sphere, because it will be detected in the subsequent frames if it is sufficiently similar to the segment at t. When a new segment is detected later, the data density in the past can be modified at that time.

Then, it is important to select nearby data sparsely and efficiently. To achieve this, Partly Locality Sensitive Hashing (PLSH) is proposed in Section 3.

Another problem is that divided segments cannot be detected immediately by the update method above, so they also have to be found separately as in Fig. 1 (e). For this, nearby data at around $t - T_{delay}$ are sparsely sampled in the

Table 1. Algorithm to find frequent patterns

---

1. At $t = 1$:
    Initialize a linked list that maintains segments in the hyper sphere. (Fig. 1(a))
2. From $t = 2$ to N:
    Update the linked list. (Fig. 1(b),(c))
    Detect new segments by PLSH and modify the linked list. (Fig. 1(d))
    Detect divided segments by PLSH and modify the linked list. (Fig. 1(e))
3. From $t = 1$ to N:
    Detect frequent patterns by a global optimization method.

---

proposed method. If a sample is outside the current hyper sphere and is included in the linked list, the segment including that sample are divided. In this case, a divided segment is not necessarily detected at the time the hyper sphere divides it, because it will be detected in the subsequent frames if division is persistent. When a divided segment is detected later, the data density in the past can be modified at that time.

Finally, frequent patterns are detected by a global optimization method using Dynamic Programming as explained in Section 4.

3   Partly Locality Sensitive Hashing

Partly Locality Sensitive Hashing (PLSH) is an extension to Locality Sensitive Hashing (LSH) [13] which is an approximate nearest neighbor search algorithm.

3.1   Combining Locality Sensitive and Insensitive Hash Functions

In PLSH, a set of hash functions $g_l(\boldsymbol{p})$ $(1 \le l \le L)$ are defined as

$$g_l(\boldsymbol{p}) = < hs_{l,1}(\boldsymbol{p}), \ldots, hs_{l,K_s}(\boldsymbol{p}), hi_{l,1}(\boldsymbol{p}), \ldots, hi_{l,K_i}(\boldsymbol{p}) > .$$

where $hs_{l,k}(\boldsymbol{p})$ is an arbitrary locality sensitive hash function $hs_{l,k} : R^d \to U$, while $hi_{l,k}(\boldsymbol{p})$ is an arbitrary locality insensitive hash function $hi_{l,k} : R^d \to U$. In a projection-based scheme as in Fig. 2, these functions can be defined as

$$hs_{l,k}(\boldsymbol{p}) = \lfloor (\boldsymbol{a}_{sl,k} \cdot \boldsymbol{p} + b_{sl,k})/w_{sl,k} \rfloor,$$
$$hi_{l,k}(\boldsymbol{p}) = \lfloor (\boldsymbol{a}_{il,k} \cdot \boldsymbol{p} + b_{il,k}) \rfloor \bmod w_{il,k},$$

where $\boldsymbol{a}, b$ are randomly chosen to satisfy $\boldsymbol{a} \in R^d, ||\boldsymbol{a}|| = 1, 0 \le b < w$ for each hash function.

Approximate nearest neighbor search proceeds as follows. Firstly, L hash values are calculated for all N data by applying L hash functions and they are stored in the corresponding hash buckets in L hash spaces. Given an input query $\boldsymbol{p}$, L hash values are calculated in the same way and the data in the corresponding hash buckets in L hash spaces are examined.

3.2   Sparse Sampling Using PLSH

LSH is not appropriate to select nearby data inside a hyper sphere sparsely and efficiently because the data in a hash bucket are densely examined as in Fig. 3(a) [14].

PLSH is useful in this case. In PLSH, an input data point is defined as $d + 1$ dimensional vector $(p_1, \ldots, p_d, t)^T$ where the first d elements represent an original data point and the last 1 element represents the time when the data point is observed.

A hash function $g_l(\boldsymbol{p})$ is composed of K_s locality sensitive hash functions and a single locality insensitive hash function. As for the locality sensitive hash

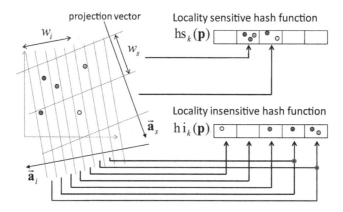

Fig. 2. Partly locality sensitive hash functions

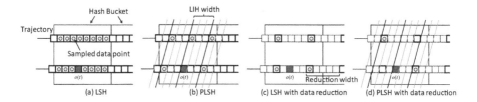

Fig. 3. Difference between LSH and PLSH

functions, the last value of a_s is always 0 and w_s is R. As for the locality insensitive hash function, the first d values of a_i is 0 and the last value is 1 so that data points in neighboring frames never collide in a same hash bucket. This means the data are scattered to w_i (= LIH width) hash spaces.

In this way, data are selected from different and independent hash spaces in neighboring frames as in Fig. 3(b). The number of data in a hash bucket is reduced by $\frac{1}{w_i}$ and the computational time is reduced similarly without affecting the probability of detecting new segments or divided segments because all data are examined anyway.

In LSH, similar reduction rate of $\frac{1}{w_r}$ can be achieved by sampling a time series in w_r intervals (= Reduction width) beforehand as in Fig. 3 (c). However, reduction rate of $\frac{1}{w_i \cdot w_r}$ is achieved in PLSH by combining data reduction ($\frac{1}{w_r}$) and data scattering ($\frac{1}{w_i}$) as in Fig. 3 (d).

4   Detection of Frequent Patterns

Given a time series $O = (o(1), \cdots, o(N))$, frequent patterns are detected in 2 steps:

1. Detection of frequent patterns using data density
2. Classification of frequent patterns

4.1   Detection of Frequent Patterns Using Data Density

Firstly, frequent patterns are detected from O irrespective of types of patterns. This problem is formulated as a combinatorial optimization problem in that each frame is assigned binary labels as $X = (x_1, \cdots, x_N)$ where $x_t \in \{1 = $ frequent pattern, $0 = $ not frequent pattern$\}$. This problem is re-formulated as to find X that minimizes the energy function defined as

$$E(O, X) = E_v(O, X) + E_d(O, X) + E_s(X). \tag{2}$$

The energy function is composed of 3 terms: velocity term, data density term and smoothing term.

Velocity term $E_v(O, X)$ penalizes data points with small velocity and is defined as

$$E_v(O, X) = \sum_t - \log(1 - \exp(-\frac{|\dot{o}_{x_t}(t)|}{< |\dot{o}_{x_t}(t)| >}))$$

where $< |\dot{o}_{x_t}(t)| >$ is the mean value of $|\dot{o}_{x_t}(t)|$.

Data density term $E_d(O, X)$ penalizes data points with small data density and is defined as

$$E_d(O, X) = \sum_t - \log(1 - \exp(-\frac{D_{x_t}(t)}{< D_{x_t}(t) >}))$$

where $< D_{x_t}(t) >$ is the mean value of $D_{x_t}(t)$.

Smoothing term $E_s(X)$ penalizes different neighboring labels to reject short patterns and is defined as

$$E_s(X) = \sum_t T(x_t \neq x_{t+1}) \cdot C_{\text{smooth}}$$

where C_{smooth} is a constant and $T(s)$ is defined as $T(\text{true}) = 1, T(\text{false}) = 0$.

Because all the terms in Eq. (2) satisfies the first order Markovian property, the energy function can be minimized analytically by Dynamic Programming.

4.2   Classification of Frequent Patterns

Because different types of actions are mixed in the detected frequent patterns, an agglomerative clustering technique is applied to classify them by iteratively grouping 2 patterns together whose average distance is smaller than the radius R of the hyper sphere.

5   Experimental Results

The proposed method was evaluated by detecting frequent whole body motions in video. To evaluate the sparse sampling methods for detecting new segments and divided segments, PLSH was compared with LSH. L is fixed to 8 and K_s is

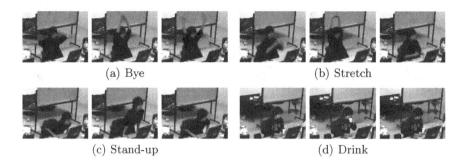

(a) Bye   (b) Stretch

(c) Stand-up   (d) Drink

Fig. 4. 4 whole body motions

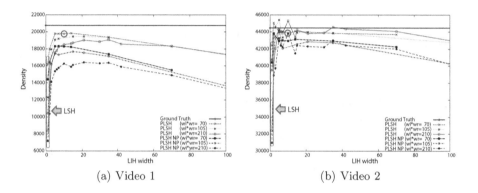

(a) Video 1   (b) Video 2

Fig. 5. Estimation of data density

fixed to 3 in both PLSH and LSH for all the experiments. All the experiments were performed on a Xeon 3.0GHz PC.

2 videos were prepared that contain several of 4 types of whole body motions as in Fig. 4. Video 1 contains 5 Byes, 6 Stretches and 5 Stand-ups. Video 2 contains 7 Byes, 7 Stretches, 8 Drinks and 4 noisy motions. Cubic Higher-order Local Auto Correlation (CHLAC) [15] is used as an image descriptor to represent patterns in video in a position invariant way.

5.1 Evaluation of Sampling Methods in Data Density Estimation

As described in Section 3.2, the sparse sampling method using PLSH achieves reduction rate of $\frac{1}{w_i \cdot w_r}$ by combining data reduction ($\frac{1}{w_r}$) and data scattering ($\frac{1}{w_i}$).

Because the computational time is roughly proportional to $\frac{1}{w_i \cdot w_r}$, the optimal w_i, w_r that does not degrade the detection rate of frequent patterns, i.e. accuracy in data density estimation, should be determined.

Fig. 5 shows the accumulated data density for different pairs of w_i, w_r. The vertical axis represents the accumulated data density and the horizontal axis

Table 2. Evaluation of video 1 [2700 frames]

Action	Bye	Stretch	Stand-up	False	False	Precision	Recall	Time
Presented #	5	6	5	Positive	Negative			[msec]
(1)LSH (w_r=1)	5.00	5.00	5.00	0.00	1.00	1.00	0.94	5807
(2)LSH (w_r=15)	5.00	5.00	5.00	0.00	1.00	1.00	0.94	512
(3)LSH (w_r=70)	5.00	4.90	0.00	0.00	6.10	1.00	0.62	198
(4)PLSH	5.00	5.00	5.00	0.00	1.00	1.00	0.94	224

Table 3. Evaluation of video 2 [3600 frames]

Action	Bye	Stretch	Drink	False	False	Precision	Recall	Time
Presented #	7	7	8	Positive	Negative			[msec]
(1)LSH (w_r=1)	7.00	6.00	8.00	1.00	1.00	0.95	0.95	13499
(2)LSH (w_r=15)	7.00	4.00	8.00	1.00	3.00	0.95	0.86	1134
(3)LSH (w_r=70)	4.00	0.00	7.20	0.30	10.80	0.97	0.51	338
(4)PLSH	7.00	3.00	8.00	1.00	4.00	0.95	0.82	482

represents w_i. The polygonal lines in the figure connect the pairs where $\frac{1}{w_i \cdot w_r}$ is the same. NP (Non Propagation) means data density in the past is not modified when new segments or divided segments are detected.

In Fig. 5 (b), data density exceeds the ground truth for some pairs. This is because divided segments were not detected well in those parameters.

The left most point in each polygonal line, i.e. w_i is 1 which means data are not scattered, represents the case where LSH is applied. From the figure, we can see that data density is estimated better when w_i increases from 1 when $\frac{1}{w_i \cdot w_r}$ is a constant, however data density moves away from the ground truth when w_i increases further. This is because data are not sampled sufficiently when one of w_i and w_r takes a large value that decreases detection rate of frequent patterns.

The red circles in Fig. 5 (a) and (b), i.e. w_i is 10 and w_r is 7, were chosen as the best parameters and were used in the following experiments.

5.2   Detection Rate of Frequent Patterns

4 different sampling methods were evaluated: (1) LSH ($w_r = 1$, no data reduction)(2) LSH ($w_r = 15$, experimentally chosen to achieve the same accumulated data density as PLSH)(3) LSH ($w_r = 70$, same reduction rate as PLSH)(4) PLSH ($w_i \cdot w_r = 70$). The detection rate of frequent patterns were summarized in Table 2 and Table 3. To evaluate the results, Precision and Recall were calculated as

$$\text{Precision} = \frac{TP}{TP + FP}, \quad \text{Recall} = \frac{TP}{TP + FN}$$

where TP means True PositiveFP means False Positive and FN means False Negative. Each experiment was evaluated 10 times and the average is shown because the parameters of hash functions were determined randomly,

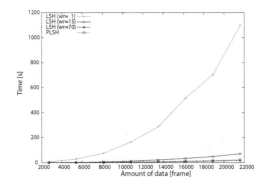

Fig. 6. Computational time v.s. amount of data

Similar detection rate to PLSH was achieved with (1) and (2), but the computational time became longer because the reduction rate was worse. On the other hand, the detection rate with (3) was considerably bad while the computational time was slightly better. From these results, we see that PLSH can be better in both computational time and detection rate than LSH.

5.3   Computational Time vs. Amount of Data

10 videos of different length were generated by simply concatenating the video in Table 2 with Gaussian noise added. The same 4 methods were applied to these videos and the results are shown in Fig. 6.

The figure shows the computational time is drastically reduced by data reduction and data scattering of PLSH because the number of data in a hash bucket is reduced.

6   Conclusions

This paper presents a method for detecting relatively long variable-length frequent patterns efficiently from video.

Partly Locality Sensitive Hashing (PLSH) is proposed by combining locality sensitive hash functions and locality insensitive hash functions. Data reduction and data scattering of PLSH enables faster and much accurate detection of frequent patterns than the conventional method with LSH.

Acknowledgement

This study was supported in part by Program for Improvement of Research Environment for Young Researchers from Special Coordination Funds for Promoting Science and Technology (SCF) commissioned by the MEXT of Japan, and in part by Grant-in-Aid for Young Scientists (B)(21700224).

References

1. Schuldt, C., Laptev, I., Caputo, B.: Recognizing human actions: a local svm approach. In: Proc. of ICPR (2004)
2. Ke, Y., Sukthankar, R., Hebert, M.: Efficient visual event detection using volumetric features. In: Proc. of ICCV (2005)
3. Niebles, J.C., Fei-Fei, L.: A hierarchical model of shape and appearance for human action classification. In: Proc. of IEEE Conference on Computer Vision and Pattern Recognition: CVPR, pp. 1–8 (2007)
4. Zhong, H., Shi, J., Visontai, M.: Detecting unusual activity in video. In: Proc. of IEEE Conference on Computer Vision and Pattern Recognition: CVPR, pp. 819–826 (2004)
5. Boiman, O., Irani, M.: Detecting irregularities in images and in video. International Journal of Computer Vision: IJCV 74, 17–31 (2007)
6. Staden, R.: Methods for discovering novel motifs in nucleic acid sequences. Computer Applications in the Biosciences 5, 293–298 (1989)
7. Lin, J., Keogh, E., Lonardi, S., Patel, P.: Finding motifs in time series. In: Proc. of the 2nd Workshop on Temporal Data Mining, pp. 53–68 (2002)
8. Yankov, D., Keogh, E., Medina, J., Chiu, B., Zordan, V.: Detecting time series motifs under uniform scaling. In: Proc. of the 13th ACM KDD Intl. Conf. on Knowledge Discovery and Data Mining, pp. 844–853 (2007)
9. Mueen, A., Keogh, E., Zhu, Q., Cash, S., Westover, B.: Exact discovery of time series motifs. In: Proc. of 2009 SIAM International Conference on Data Mining: SDM, pp. 1–12 (2009)
10. Niebles, J.C., Wang, H., Fei-Fei, L.: Unsupervised learning of human action categories using spatial-temporal words. In: Proc. of BMVC (2006)
11. Kitani, K.M., Okabe, T., Sato, Y., Sugimoto, A.: Discovering primitive action categories by leveraging relevant visual context. In: Proc. Int. Workshop on Visual Surveillance (in conjunction with, ECCV 2008) (2008)
12. Uchida, S., Mori, A., Kurazume, R., Taniguchi, R., Hasegawa, T.: Logical dp matching for detecting similar subsequence. In: Proc. of Asian Conference of Computer Vision (2007)
13. Datar, M., Immorlica, N., Indyk, P., Mirrokni, V.: Locality-sensitive hashing scheme based on p-stable distributions. In: Proc. of the twentieth annual Symposium on Computational Geometry, pp. 253–262 (2004)
14. Meng, J., Yuan, J., Hans, M., Wu, Y.: Mining motifs from human motion. In: Proc. of EUROGRAPHICS 2008 (2008)
15. Kobayashi, T., Otsu, N.: Action and simultaneous multiple-person identification using cubic higher-order local auto-correlation. In: Proc. Int. Conference on Pattern Recognition: ICPR, pp. 741–744 (2004)

Foot Contact Detection for Sprint Training

Robert Harle, Jonathan Cameron, and Joan Lasenby

University of Cambridge

Abstract. We introduce a new algorithm to automatically identify the time and pixel location of foot contact events in high speed video of sprinters. We use this information to autonomously synchronise and overlay multiple recorded performances to provide feedback to athletes and coaches during their training sessions.

The algorithm exploits the variation in speed of different parts of the body during sprinting. We use an array of foreground accumulators to identify short-term static pixels and a temporal analysis of the associated static regions to identify foot contacts.

We evaluated the technique using 13 videos of three sprinters. It successfully identifed 55 of the 56 contacts, with a mean localisation error of 1.39±1.05 pixels. Some videos were also seen to produce additional, spurious contacts. We present heuristics to help identify the true contacts.

1   Introduction

High speed video (100 fps or greater) is used extensively for detailed performance review in sports training. With high speed motion, the camera captures more detail than the human eye and allows more specific feedback to be given to the athlete. Unfortunately, the sheer volume of information captured by such video makes in-field video review cumbersome with the result that feedback is not available when it may be of most use.

We present a simple but robust machine vision technique that identifies foot contact events in high speed videos of sprinters. We use this information to autonomously synchronise and composite two videos of sprint repetitions to facilitate feedback.

2   Domain-Specific Constraints

Arbitrary event identification within video is a difficult and as-yet unsolved problem. In order to get robust event identification, we make a number of simplifying assumptions about the video being recorded as follows:

Static camera. The camera viewpoint does not change, either during or between repetitions. This is appropriate for sprint training, as different viewpoints would complicate comparison of two repetitions.

Sprinter moves across the frame. The sprinter must have a significant component of motion perpendicular to the camera axis. A view from the side of the track is a natural viewpoint for review of sprint technique.

R. Koch et al. (Eds.): ACCV 2010 Workshops, Part I, LNCS 6468, pp. 297–306, 2011.

Single athlete. Whilst the technique could be adapted for multiple simultaneous sprinters, here we concentrate here on creating a system for a single
sprinter.

Fast but not real time. The nature of sports training means that live feedback is rarely useful. Instead we concentrate on supporting in-field review
within 60 s of the end of a repetition (this is less time than it typically takes
for an athlete to return to their start).

3   Identifying Foot Contacts

To usefully composite two videos we need to synchronise them both temporally
and spatially. We analyse the video to identify the times and locations of foot
contacts and use one such event from each video to provide an estimate of the
temporal and spatial offsets between them.

3.1   Properties of a Contact Event

The sprinting movement can be viewed as a series of short ground contacts,
interspersed with longer periods of flight. During a contact, the foot is rooted to
the ground and used to propel the rest of the body forwards.

Sports scientists identify *foot-down* and *toe-off* events. Figure 1 illustrates a
foot contact. First the fore of the foot contacts the ground, forming the foot-
down event. Thereafter the foot flattens (the heel may never touch the ground),
before 'peeling' away from the floor. This peeling action means that the last
contact between foot and ground occurs at the toe, forming the toe-off event.
Unlike the foot-down event, the toe-off is very localised spatially, making it a
good candidate for spatiotemporal synchronisation.

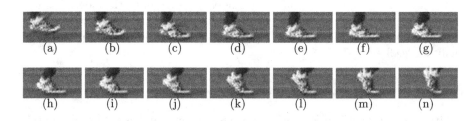

Fig. 1. A captured foot contact sequence. The *foot-on* event occurs in (c), the centre
of mass then passes over the foot, which is 'peeled' off the ground towards the *toe-off*
event (not shown).

3.2   Recognising the Toe-off Event

During a contact, the foot is pushing against the ground without slip and must
therefore be stationary. It is, in fact, the only stationary part, since the rest of the

body is propelled forward and is in constant motion. Our technique to identify the toe-off exploits this invariant by searching for foreground pixels whose values are not significantly changing. We divide the method into four phases:

1. Background subtraction;
2. Static foreground accumulation;
3. Create candidate toe-offs from temporal analysis of accumulations;
4. Select toe-offs from candidates.

We now address each in turn.

3.3   Background Subtraction

The technique we present relies on being able to segment the runner (or at least their feet) from the background. There have been many background subtraction algorithms proposed [1] and most are sufficient for this task.

Our dataset shows athletes training outdoors, with natural lighting and little background distraction. After some experimentation, we favour the background subtraction algorithm of [2]. Regardless of the background subtracter choice, we do not assume a noise-free estimate of the foreground. We do, however, assume that the runner's feet are identified as foreground throughout a contact.

3.4   Static Foreground Accumulation

Working with each foreground pixel, we accumulate the number of frames since it last changed value significantly. Any accumulator associated with a changing pixel or a background pixel is reset to zero. More formally, for the nth frame, I_n, we identify the set of foreground pixels, F_n, and we maintain an accumulator array A_n for each pixel (x, y) as follows:

$$A_{n,x,y} = \begin{cases} A_{n-1,x,y} + 1, & \text{if } (x,y) \in F_n \text{ AND compare } (I_{n,x,y}, I_{n-1,x,y}) \\ 0, & \text{otherwise} \end{cases} \quad (1)$$

where

$$\text{compare}(I_{n,x,y}, I_{n-1,x,y}) = \begin{cases} true & \text{if } \sum_{k=R,G,B}(I^k_{n,x,y} - I^k_{n-1,x,y})^2 < K^2 \\ false & \text{otherwise} \end{cases}$$

$$(2)$$

for threshold K, a Euclidean distance in RGB colour space. We have not found this threshold to be particularly sensitive — in our experiments we set $K = 20$.

Figure 2 shows the development of the accumulator values throughout the sprint cycle. Non-zero accumulations occur for foot contacts (since the foot is stationary); for background marked incorrectly as foreground (noise); and for *false accumulations*, which occur when a single-coloured object moves through the field of view. This is particularly observable at the top back of the upper leg in Figure 2.

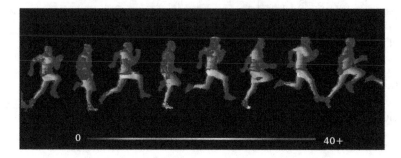

Fig. 2. Visualisation of the accumulator values during the sprint action. Accumulation values are in units of frames; this video was captured at 370 fps. Notice that the accumulators rarely reach 40 or higher, except at toe-off.

3.5 Identifying Candidate Toe-off Events

Let us temorarily assume ideal foreground extraction whilst considering the foot contact itself. Ideally the accumulation at the foot begins with the foot-on event and continues as the foot flattens mid-contact. Thereafter, the peeling motion described above means that the greatest accumulations are associated with pixels imaging the toe. Just before the toe-off event, only those few pixels remain static; just after the event the entire foot is moving and no significant accumulation should be observable.

A foot contact is thus associated with the appearance of a significant localised accumulation in the accumulation array that grows before shrinking to zero. We label a given pixel location as *static* if its accumulator value exceeds a threshold value, T i.e. a *static* pixel is one whose value has not significantly changed in T frames.

Sensitivity to T. The threshold T, above which we label a pixel static, is chosen to capture the foot pixels as static during a contact but not the moving body. The duration of a sprint contact lies between 90 ms and 150 ms. At a frame rate of 370 fps this corresponds to approximately 30 frames, during which the rest of the athlete's body should have moved significantly. This suggests a threshold of $T =30$. In fact, since the body moves so much in comparison to the foot, the threshold is not particularly sensitive.

Figure 3 shows the variation of $|S|$ throughout a foot contact for various choices of T. Small values of T introduce more false accumulations that can mask those associated with the foot. Values of T that significantly exceed the foot contact time result in an empty set S. We conclude that an appropriate threshold can be chosen on the basis of expected contact time and framerate and that it is not particularly sensitive.

Handling Background Subtraction Errors. As background subtraction is never perfect in practice, we perform temporal and spatial region growing on the set of static pixels for each frame. Each region of static pixels is monitored

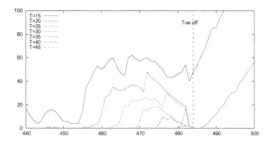

Fig. 3. Variation in the number of static pixels, $|S|$, with threshold, T, during a foot contact

individually, preventing foreground noise from masking the disappearance of a set of static pixels that might indicate a toe-off event.

Stationary pixels are first grouped in a frame using an efficient region grower [3]. The spatial regions are then merged through time using the criterion of spatial overlap between neighboring frames, resulting in a set of region histories. Each history consists of a list of sets of static pixels, one such set for each frame of the underlying video in which the region is present.

3.6   Toe-off Selection

Having applied this analysis, we are left with a set of candidate toe-off events. These candidates may be from genuine toe-offs, from false accumulations or even from foreground noise if the foreground filtering has failed. Here we present two filters that may assist in identifying the true toe-offs.

Sample Consensus. With a sprinter we expect toe-off events to occur with an approximately fixed period and in an approximately straight line in space, mapping under the assumption of a relatively long focal length lense (fish-eye lenses are of little use in sports training) to a straight line in the image.

This pair of constraints can be used to form a model suitable for application of the RANSAC algorithm [4]. In fact the small number of contacts detected (typically < 50) make complete evaluation of sample sets trivial.

Each pair of contact points provide estimates of the spatial line near which contacts should occur and the period of contacts.

Votes are tallied for each contact pair based on a spatial threshold on the distance of a contact from the line and a temporal threshold to the nearest multiple of the period. Multiple hits for a particular step (caused by two close contacts) are suppressed by selecting the one that agrees best with the predicted contact time. The set of inliers found for the pair with the highest number of votes are identified as the contact times.

Running Cycle Phase. As an additional hint, if we are able to estimate where in the running cycle the sprinter is, we can estimate when contacts might (and might not) be expected. This phase can be estimated from the foreground silhouette of the runner. With a side-on view as above, the silhouette expands and contracts as the arms and legs merge and diverge during the running action. Figure 4 illustrates the variation in the size of the foreground silhouette throughout the running cycle. A low-pass filter applied to these data produces a smooth signal with maxima during a flight period and minima during a contact. We can use this to exclude any candidate toe-off events occuring between a maximum and the subsequent minimum. Our algorithm uses the slope of this foreground size trend to decide whether we are approaching a maximum or a minimum and hence whether to allow toe-off events.

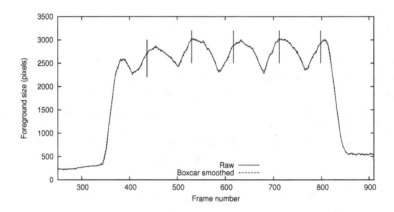

Fig. 4. The variation in the number of foreground pixels identified in each frame for a sample video. The short vertical lines indicate where toe-off events occur in the video.

4   Evaluation and Results

4.1   Video Capture System

We collected video data using custom software that captured the raw Bayer image from an AVT Pike IEEE1394b camera[1]. This camera is capable of 205 fps at a resolution of 640×480. We chose to window the camera output to 640×240 to achieve a faster framerate of 370 fps, which better represents what can be found in a typical training environment today. Unlike typical solutions of today, however, our system streams the data over a firewire connection rather than storing it locally on the camera for later download. This means it is immediately available for processing once the repetition is complete. The bandwidth limitations of the firewire bus mean that we can only support the lower resolution of 640×240. However, coaches have responded well to the data present even at the lower resolutions we have used here.

When processing the Bayer video frames [5], we used a custom implementation of the AHD demosaicing algorithm [6] running on an Nvidia GPU using the CUDA parallel processing framework. The AHD algorithm is the de-facto

---

[1] http://www.alliedvisiontec.com

demosaicing algorithm for Bayer images but is processor-intensive and the use of the parallel processing capabilities of the GPU allowed the demosaicing to occur fast enough to process the entire video within the allowed timeframe of 30 s.

4.2 Results

We collected a set of 13 videos containing 56 toe-off events from three different runners: two male sprinters and one female, all wearing spiked running shoes. The videos were collected at an outdoor running track in a variety of conditions including strong sun, wind, intermittent cloud cover and overcast skies. The background was empty of any major distractions, although many videos contained distant athletes and swaying vegetation.

The low resolution of the video meant it was not possible to reliably identify (by eye or automatically) the single nearest frame to the toe-off, although each toe-off could be bracketed to within three or four frames. Because of this relatively high error in our temporal ground truth, we present results based on the locations of the events.

Our core algorithm returns a set of candidate toe-off events. The set of all candidates contained 266 possible event, including 55 of the 56 toe-off events (as well as additional, unwanted events). For those 55 correctly identified toe-offs, the average spatial error was 1.39 ± 1.05 pixels.

Figure 5 provides an example set of candidate toe-offs, annotated for clarity. Note from that many of the noise candidates come from pixels that are not imaging the ground: these could be masked out if the camera position and orientation is known in advance, dramatically reducing the set of candidates. However, we trialed both the Sampling Consenses and the cycle phase filters without such masking. Figure 6 shows an example result from the Sampling Consensus, which produced consistently good results.

The cycle filter was also successful in discarding many incorrect events, although it depended on a good foreground extraction result. Figure 7 shows a failed background subtraction where skin tones of the lower leg are confused with the track surface, resulting in an unexpected foreground silhouette that adversely impacts the size trend. Although the broad trend remains, there is

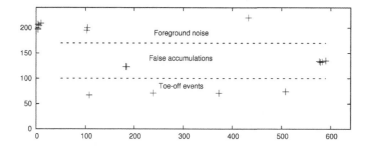

Fig. 5. The candidate events generated for single video

Fig. 6. Example showing the result of the Sample Consensus algorithm. Circles are drawn to highlight the single pixels that mark the locations. The spatial threshold was 10 pixels and the temporal 10 frames. A video of this dataset, rob.avi is included as supplementary material.

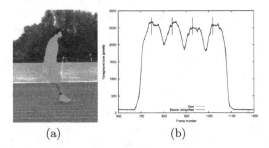

(a)   (b)

Fig. 7. (a) An example of background subtraction failure. (b) The resultant variation in the number of foreground pixels (c.f. Figure 4).

now a double peak between frame 700 and 800, even after smoothing. For this reason we favour the Sampling Consensus approach.

5   Related Work

Much of the existing literature in sports event detection is concerned with large scale classification of moves or activities, or easily detected events such as a ball crossing the goal line or identifying where a ball hit the ground.

The market leaders in the case of ball sports is Hawk-Eye[2]. Whilst it would be possible to apply a similar technique to that presented here, the typical ball tracking problem is actually considerably simpler. As long as the ball is traveling in the air it's trajectory is relatively predictable and it is distinctive and well localised assisting foreground segmentation. This allows contact points to be estimated from the intersection of the ball trajectory with the ground.

Similar problems occur in the classification / video understanding literature. This is a substantial field so here we will just pull out a few representative

---

[2] http://www.hawkeyeinnovations.co.uk/

Fig. 8. An example of a blended pair of videos done using toe off information. A video of this sequence is available in the supplementary material (blend.avi).

works. Most of these works use relatively low level features, applying classification techniques on the output. Compressed videos allow the use of available block motion vectors from which various features are extracted [7]. [8] has some elements in common with the method presented here (background subtraction, spatial segmentation and region tracking). However the region tracking has to be considerably more complex as the work is primarily concerned with moving regions whereas ours points of interest are stationary. [9] uses the evolution of a 2D shape over time to identify events. The foreground size approach used above is a very simple version of this. [10] use a derived template to perform motion recognition on aerobics exercise. A general review can be found in [11].

Most methods are directed at the handling of more general motion and as such cannot take advantage of the domain specific elements used in this paper.

6   Conclusions

We have presented a novel technique to identify the time and pixel locations of foot contacts within videos of sprint training. The algorithm was able to identify foot contacts within $640{\times}240$, 370 fps video to within $1.39{\pm}1.05$ pixels. An example of a resulting video overlay is shown in Figure 8.

Acknowledgments

The authors would like to acknowledge the support of the EPSRC through the SESAME WINES project and the sprinters who assisted in this work.

References

1. Parks, D.H., Fels, S.S.: Evaluation of background subtraction algorithms with post-processing. In: AVSS 2008: Proceedings of the 2008 IEEE Fifth International Conference on Advanced Video and Signal Based Surveillance, pp. 192–199. IEEE Computer Society, Washington, DC, USA (2008)
2. Cucchiara, R., Grana, C., Piccardi, M., Prati, A.: Detecting moving objects, ghosts, and shadows in video streams. IEEE Transactions on Pattern Analysis and Machine Intelligence 25, 1337–1342 (2003)

3. Tarjan, R.: Efficiency of a good but not linear set union algorithm. J. ACM 22, 215–225 (1975)
4. Fischler, M., Bolles, R.: Random sample consensus: a paradigm for model fitting with application to image analysis and automated cartography. Commun. Assoc. Comp. Mach. 24, 381–395 (1981)
5. Bayer, B.E.: Color imaging array. US Patent 3971065 (1976), http://patft.uspto.gov/netacgi/nph-Parser?patentnumber=3971065
6. Hirakawa, K., Parks, T.: Adaptive homogeneity-directed demosaicing algorithm. IEEE Transactions on Image Processing 14, 360–369 (2005)
7. Zhou, W., Vallaikal, A., Kuo, C.C.J.: Rule-based video classification system for baseketball video indexing. In: International Multimedia Conference, pp. 213–216. ACM, New York (2000)
8. Courtney, J.: Automatic video indexing via object motion analysis. Pattern Recognition 30, 607–625 (1997)
9. Gorelick, L., Shechtman, E., Irani, M., Basri, R.: Actions as space-time shapes. IEEE Transactions on Pattern Analysis and Machine Intelligence 29, 2247–2253 (2007)
10. Bobick, A., Davis, J.: The recognition of human movement using temporal templates. IEEE Transactions on Pattern Analysis and Machine Intelligence 23, 257–267 (2001)
11. Gavrila, D.: The visual analysis of human movement: A survey. Computer Vision and Image Understanding 73, 82–98 (1999)

Interpreting Dynamic Meanings by Integrating Gesture and Posture Recognition System

Omer Rashid Ahmed, Ayoub Al-Hamadi, and Bernd Michaelis

Institute for Electronics, Signal Processing and Communications (IESK)
Otto-von-Guericke-University Magdeburg, Germany
{Omer.Ahmad,Ayoub.Al-Hamadi}@ovgu.de

Abstract. Integration of information from different systems support enhanced functionality however it requires a rigorous pre-determined results for the fusion. This paper proposes a novel approach for determining the integration criteria using Particle filter for the fusion of hand gesture and posture recognition system at decision level. For decision level fusion, integration framework requires the classification of hand gesture and posture symbols in which HMM is used to classify the alphabets and numbers from hand gesture recognition system whereas ASL finger spelling signs (alphabets and numbers) are classified by posture recognition system using SVM. These classification results are input to integration framework to compute the contribution-weights. For this purpose, Condensation algorithm approximates the optimal a-posterior probability using a-prior probability and Gaussian based likelihood function thus making the weights independent of classification ambiguities. Considering the recognition as a problem of regular grammar, we have developed our production rules based on context free grammar (CFG) for the restaurant scenario. On the basis of contribution-weights, we mapped the recognized outcome over CFG rules and infer meaningful expressions. Experiments are conducted on 500 different combinations of restaurant orders with the overall 98.3% inference accuracy which proves the significance of proposed approach.

1 Introduction

Human Computer Interaction (HCI) is emerged as a new field which aims to bridge the communication gap between humans and computers. An intensive research has been done in computer vision to assist HCI particularly using gesture and posture recognition [1]. Many pioneering techniques have been proposed to address the issues of effective interaction-interface but a natural mean of interaction still remains and yet to solve. Gesture and posture recognition are the application areas in HCI to communicate with computers. In gesture recognition, Yoon et al. [2] developed a hand gesture system in which combination of location, angle and velocity is used for the recognition. Liu et al. [3] developed a system to recognize 26 alphabets by using different HMM topologies. Hunter et al. [4] used HMM for recognition where Zernike moments are used as image features for hand gesture sequences.

R. Koch et al. (Eds.): ACCV 2010 Workshops, Part I, LNCS 6468, pp. 307–317, 2011.
© Springer-Verlag Berlin Heidelberg 2011

In posture recognition, Adaptive Neuro-Fuzzy Inference Systems (ANFIS) model is used for the recognition of Arabic Sign Language [5]. However, the use of colored gloves avoids the segmentation problem and helps the system to obtain good features. Elliptic Fourier Descriptor (EFD) is used by Malassiotis and Strintzis [6] for 3D hand posture recognition. Similarly, Licsar and Sziranyi [7] used Fourier coefficients to represent hand shape in their system which enables them to analyze hand gestures for the recognition. Handouyahia et al. [8] presents a recognition system based on the shape description using size functions for International Sign Language (ISL). Freeman and Roth [9] used orientation histogram for the classification of gesture symbols, but huge training data is used to solve the orientation problem.

Integration of different systems is used to enhance the performance and results in better recognition of subjects under observation. In this context, integration of different modalities have been used to improve the recognition (i.e. identification of a human by combining face and voice traits [10]) in the field of biometrics. Similarly, for multi-modal biometric systems, fusion takes place at different levels which includes sample level, feature level, match score level and decision level fusion [11]. Particularly in computer vision, improvement in performance of face recognition system is proposed by Chang et al. [12] in which the fusion of 2D and 3D information of the face images is done. Kumar et al. [13] performed fusion at feature level and match score level to combine the palm prints and hand geometrical features. Similarly, Wu et al. [14] proposed a multi-model system to combine the gait recognition with face recognition system for the human recognition. However, it is observed that the main motivation of exploiting different modalities is to achieve better performance and to cop the limitations of uni-modal approach.

According to our knowledge, integration of gesture and posture recognition system is not addressed yet. In this paper, our principle objective is the integration of these systems which allows us to design an effective interaction-interface for HCI. Moreover, proposed integration framework enables us to extract multiple inferences from gesture and posture recognition systems at decision level by estimating contribution-weights using particle filter. These contribution-weights are exploited for the combination of extracted symbols which results in the interpretation of new meaningful *expressions* from the developed lexicon database.

2   Proposed System

The proposed framework is staged in several phases to integrate and infer from the gesture and posture recognition systems as shown in Fig. 1. First, we start with the image acquisition phase from Bumblebee2 camera and extract the objects of interest (i.e. hands and face) using the color and depth information. Second, gesture and posture feature vectors are computed by exploiting different properties of hand. In classification process, HMM recognizes the gesture symbols from alphabets and numbers whereas SVM is used for finger-spelling ASL signs in posture recognition. Third, a novel particle filter system is proposed for the

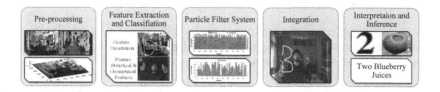

Fig. 1. Presents the process flow of the proposed framework

integration of gesture and posture system by computing the contribution-weights thus determining the integration-criteria . After computing the contribution-weights, the interpretation is performed by processing the Context Free Grammar production rules which results in the inference of meaningful expression.

3   Pre-processing

The image acquisition is done by Bumblebee2 camera which gives 2D images and depth images. The depth image sequences are exploited to select region of interest for segmentation of objects (i.e. hands and face) where the depth lies in range from 30 cm to 200 cm (i.e. in our experiments) as shown in Fig. 2(a). In this region, we extract the objects (i.e. hands and face) from skin color distribution which is modeled by normal Gaussian distribution characterized by mean and variance as shown in Fig. 2(b). We have used YC_bC_r color space because skin color lies in a small region of chrominance components where as the effect of brightness variation is reduced by ignoring the luminance channel. After that, skin color image is binarized and the contours are extracted by computing chain code representation for detection of hands and face as shown in Fig. 2(c).

Fig. 2. (a) Original Image with selected depth region (b) Results of Normal Gaussian distribution using the depth Information (c) Detected hands and face

4   Feature Extraction and Classification

In this section, feature extraction and classification approaches used for gesture and posture system are briefly described to build an understanding about the proposed approach. However, the details can be found in [15].

4.1 Feature Extraction

Feature extraction is an essential phase for the classification in which the selection of optimal features play a significant role. Hand gesture and posture features are extracted for the recognition which are described as follows.

Hand Gesture Features. Orientation is used as a main feature for gesture recognition and for its computation; centroid points of the hands are computed. The orientation is determined between two consecutive centroid points when drawing gesture path. The equation used to compute the orientation θ_t is:

$$\theta_t = \arctan\left(\frac{y_{t+1} - y_t}{x_{t+1} - x_t}\right) ; t = 1, 2, ..., T - 1 \qquad (1)$$

where T represents length of gesture path. x_t and y_t are centroid point at frame t. The computed angle θ_t is quantized in range from 1 to 18 by dividing it by 20 degrees. These quantized values give us discrete vector $F_{gesture} = \theta_t$ which is used in HMM to classify gesture symbols.

Hand Posture Features. Two types of feature vectors are computed for posture recognition namely statistical and geometrical feature vectors. For statistical feature vectors, Hu-Moments [16] are used which are derived from basic moments and describe the properties of objects shape statistically (i.e.area, mean, variance, covariance and skewness). Hu [16] derived a set of seven moments which are translation, orientation and scale invariant. Feature vectors of Hu-Moment are written as $F_{stat} = (\phi_1, \phi_2, \phi_3, \phi_4, \phi_5, \phi_6, \phi_7)^T$ in which ϕ_1 is the first Hu-Moment and so on. The geometrical feature vectors (i.e.circularity and rectangularity) are computed by exploiting the hand geometry from the standard shapes such as circle and rectangle. Geometrical feature vector $F_{geo} = (Cir, Rect)^T$ set varies from symbol to symbol and is useful to recognize the alphabets and numbers.

Statistical and geometrical feature vector set are combined together to form a set of feature set $F_{posture}$. It is denoted as:

$$F_{posture} = F_{stat} \wedge F_{geo}, F_{posture} = (\phi_1, \phi_2, \phi_3, \phi_4, \phi_5, \phi_6, \phi_7, Cir, Rect)^T \qquad (2)$$

4.2 Classification

Gesture and posture feature vectors are classified using HMM and SVM respectively. In gesture classification, HMM recognizes the gesture symbols from alphabets (A-Z) and numbers (0-9) whereas SVM classifies the finger-spelling ASL signs in the posture recognition. Baum-Welch algorithm (BW) is used for the training of HMM parameters by discrete vector θ_t. We have used Left-Right banded model with 9 states for hand motion recognition of gesture path. Classification of hand gesture path is done by selecting the maximal observation probability of the gesture model by Viterbi algorithm. SVM classifier learns from statistical and geometrical features of the hand using Radial Basis Function (RBF) as the Gaussian kernel. For the gesture and posture classification results, please refer this [15].

5   Integration

The basic idea in the proposed integration framework is to parallelly interpret multiple signs from gesture and posture recognition systems. The goal behind integration is to combine signs which are driven from different approaches (i.e. gesture and posture system) and results in the inference of new symbols at any instance of time. Integration ($Intgr$) of gesture and posture systems fused at decision level is formulated as:

$$Intgr = \alpha_{gstr} * Rec_{hmm} \wedge \alpha_{pstr} * Rec_{svm} \tag{3}$$

where Rec_{hmm} and Rec_{svm} are the classification results of gesture and posture system. α_{gstr} and α_{pstr} are the contribution-weights associated with gesture and posture system which acts as a reliability criteria for fusion of these systems. These weights are computed as follows.

5.1   Particle Filter System

Condensation algorithm [17] approximates the contribution-weights optimally for the fusion of gesture and posture systems. The key functionality of Condensation algorithm is to approximate a-posteriori probability (i.e. contribution-weight) by a set of random samples called particles to simulate the probabilistic model of the system. We have proposed a particle filter system which consists of two separate particle filters (i.e. for gesture and posture), explained as follows.

Initialization. The classification outcome of both gesture and posture system is input to separate particle filter. A vector $S(n)$ is constructed to initialize the Condensation process which is represented as follows:

$$S(n) = \{s_k^{(gstr)}, s_k^{(pstr)}\} \tag{4}$$

A set of N (i.e. 100) random points called particles x_k^n with weights w_k^n denotes the initial distribution of particles at time k for both gesture and posture systems. These particles are denoted as:

$$s_k^{(gstr|pstr)} = \{x_k^n, w_k^n\}_N^{n=1} \tag{5}$$

(Note. From this point on, the same notation is used for both the particle filters (i.e. gesture and posture), except when stated otherwise).

Prediction. The a-priori probability $p(x_k|z_{k-1})$ is computed from previous a-posteriori probability $p(x_{k-1}|z_{k-1})$ and the dynamic model $p(x_k|x_{k-1})$ as shown in Fig. 3. The formulation is represented as:

$$p^{(n)}(x_k|z_{k-1}) = p^{(n)}(x_k|x_{k-1})\,p^{(n)}(x_{k-1}|z_{k-1}) \tag{6}$$

Updation. The a-posteriori probability (i.e. contribution-weights) of the state is calculated from the a-priori probability $p(x_k|z_{k-1})$ and the likelihood function

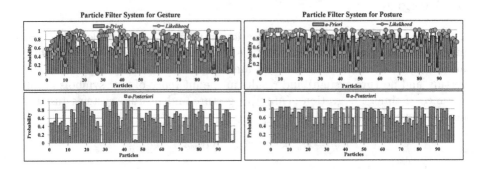

Fig. 3. Shows the proposed particle filter system for gesture and posture system from frame 42 in Fig.4(a). The top-graphs presents a-priori and Gaussian likelihood function of gesture and posture system whereas bottom-graphs show a-posteriori probability (i.e. contribution-weights) for fusion of these system.

$p(z_k|x_k)$ by incorporating the new measurement data z_k as shown in Fig. 3. Likelihood function is formulated as follows:

$$p^{(n)}\left(z_k|x_k\right) = \pi_k^{(n)} = e^{-((z_k - x_k^n)^2)/(2\sigma^2)} \tag{7}$$

where σ is the standard deviation of particle weights. The contribution-weights $\alpha_{gstr|pstr}$ or a-posteriori probability $p(x_k|z_k)$ for gesture and posture system is computed as follows:

$$p\left(x_k|z_k\right) = \frac{\sum_{n=1}^{N} p^{(n)}\left(z_k|x_k\right) p^{(n)}\left(x_k|z_{k-1}\right)}{\sum_{n=1}^{N} p^{(n)}\left(z_k|x_k\right)} \tag{8}$$

Using N values of $p(z_k|x_k)$, we have built a probability distribution for the whole space at any time instant. The conditional probability acts as a weighting factor for its corresponding state with successive iterations. The normalized weighting probabilities are calculated as follows:

$$\pi_k^{(n)} = \frac{p^{(n)}\left(z_k \mid x_k\right)}{\sum_{n=1}^{N} p^{(n)}\left(z_k \mid x_k\right)} \tag{9}$$

In this way, we obtain the contribution-weights which defines the integration-criteria for the fusion of these systems.

5.2   Lexicon and Regular Language

Before making the interpretation and inferences from contribution-weights computed above, first we describe the proposed structure of the language. We consider the recognition as a problem of regular langauge therefore, we mapped the recognition outcome over the context free grammar (CFG) rules. The grammar is defined as a 4-tuple in CFG and is written as $Grammar = (V, T, S, R)$ where V is the set of objects and contains non-terminals as well as terminals symbols,

T is the set of teminals, S is start symbol and it is a subset of V (i.e. $S \in V$), and R is the set of production rules. We have presented integration results in the form of CFG rules as follows:

$$V = \{S, PostureAlphabet, X, GestureAlphabet, Alphabet,$$

$$PostureDigit, Y, Digit, 0_p|1_p, ..., 9, a_g|b_g, ..., z_g, a_p|b_p, ..., z_p\}$$

$$T = \{0_p|1_p, ..., 9_p, a_g|b_g, ..., z_g, a_p|b_p, ..., z_p\}$$

$$S \rightarrow PostureAlphabet \ X$$

$$PostureAlphabet \rightarrow Alphabet \ PostureAlphabet \ | \ Alphabet$$

$$X \rightarrow PostureDigit \ Y$$

$$PostureDigit \rightarrow Digit \ PostureDigit \ | \ Digit$$

$$Y \rightarrow GestureAlphabet \ PostureDigit$$

$$Digit \rightarrow 0_p|1_p|2_p, ..., 9_p$$

$$Alphabet \rightarrow a_p|b_p|c_p, ..., z_p$$

$$GestureAlphabet \rightarrow a_g|b_g|c_g, ..., z_g$$

In the above CFG production rules, PostureAlphabet computes set of recognized posture alphabet signs, GestureAlphabet results in recognized gesture symbols and PostureDigit is the set of recognized numbers. Different symbols can be devised in integration process depending upon the lexicon as shown in Table 1.

Table 1. Lexicon of Gesture Symbols

Gesture ⇒ Order	Gesture ⇒ Order	Gesture ⇒ Order
A⇒Apple, Apricot	J⇒Jackfruit, Jambolan	S⇒Star Fruit, Strawberry
B⇒Blueberry, Banana	K⇒Kaffir Lime, Kiwi	T⇒Tangerine, Tart Cherry
C⇒Cherry, Cantaloupe	L⇒Lemon, Lychee	U⇒Ugli Fruit, Uniq Fruit
D⇒Date, Dewberry	M⇒Mango, Melon	V⇒Voavanga
E⇒Elderberry, Eggfruit	N⇒Nectarine	W⇒Watermelon, Wolfberry
F⇒Fig, Farkleberry	O⇒Orange, Oval Kumquat	X⇒Xigua
G⇒Grapes, Gooseberry	P⇒Pear, Peach	Y⇒Yunnan Hackberry
H⇒Honeymelon, Hackberry	Q⇒Quince	Z⇒Zinfandel Grapes
I⇒Imbe	R⇒Raspberry, Rambutan	

5.3 Interpretation and Inference

In this module, contribution-weights whose threshold is above 70% are selected for the fusion process and is written as:

$$(\alpha_{gstr}|\alpha_{pstr}) \geq T$$

To infer from gesture recognition system, HMM classifier and its states model recognizes the alphabets and numbers after processing some frames. However, posture recognition system recognizes the symbol at every frame because a single frame is sufficient to recognize ASL symbols in finger spelling domain. Besides, integration is carried out from contribution-weights of gesture and posture symbols at any time frame after passing the threshold criteria. In this regard, different approaches are proposed for the fusion of different systems which includes AND/OR combination, majority voting, behavior knowledge method and weighted voting method [18]. However, we have used AND/OR combination for gesture and posture recognition symbols. Integration ($Integ$) is formulated as:

$$Integ = (\alpha_{gstr} * Rec_{hmm} \wedge \alpha_{pstr} * Rec_{svm})$$
$$Integ = (\alpha_{gstr} * Rec_{hmm} \wedge \{\alpha_{pstr} * Rec_{svm}, \alpha_{pstr} * Rec_{svm}, ..., \alpha_{pstr} * Rec_{svm}\})$$

The combination of CFG rules yield us to the integration of gesture and posture recognition in which multiple posture symbols are combined with the gesture symbol. To make inferences of results from CFG, the possible derivation of posture results is *PostureAlphabet* followed by *PostureDigit* whereas *GestureAlphabet* yield to only one possible outcome in the integration. The inference which is derived from CFG rules is as follows:

$$S \rightarrow PostureAlphabet\ GestureAlphabet\ PostureDigit$$

Different interpretations can be devised for integration process which includes:

- The ideal case of integration, both gesture and posture systems recognize the symbol at any time frame.
- Gesture system does not classify any symbol because HMM is not activated when gesture drawing process starts. However, posture system recognizes a symbol based on classification results and weight computation above threshold.
- There are some predictions about gesture symbols dependent upon the inference from HMM states. In this case, gesture symbol is still incomplete and it gives a clue about user's intention while drawing the gesture symbol. Intentions are predicted if weight computation result is above the threshold.
- No match has occurred from gesture and posture systems. In this way, the symbols are not present in the lexicon.

6   Experimental Results

In the proposed approach, input sequence is captured by Bumblebee2 stereo camera with 240*320 pixels image resolution. The proposed concept of integration is tested on a real-time scenario which is in our case, the restaurant lexicon. We have chosen 45 different fruits for this choice as shown in Table. 1 and make different (i.e. currently our system supports 500 combinations) orders for it by combining gestures and postures. The integration concept is defined as the first and second alphabet of the fruit from gesture and ASL posture respectively and then combined it with another posture number, thus making an order.

Fig. 4(a and b) shows an interpretation based on fusion of gesture and posture recognition system. In this sequence Fig. 4(a), posture system firstly recognizes the alphabet "A". However, gesture recognition system did not recognize any

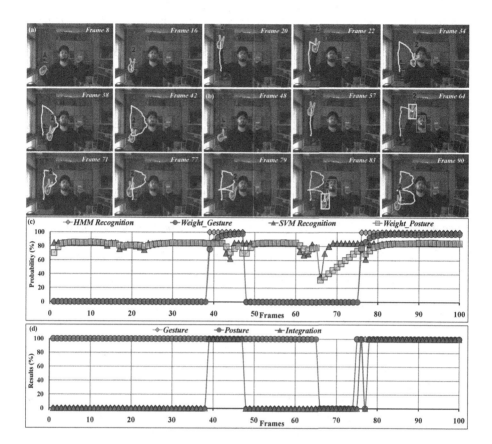

Fig. 4. (a) Shows the recognized gesture symbol "D" whereas the classified postures are alphabet "A" and Number "2". The meaningful expression "Two Date Juices" is inferred from this sequence. (b) Second meaningful expression is "Two Blueberries Juices" which results from recognized gesture symbol "B" and classified posture alphabet "L" and Number "2". (c) Graph shows the recognition rate along with contribution-weights from particle filter system. Graph(d) presents gesture and posture results along with their integration.

symbol during the initial frames. The next posture symbol recognized is "2" which indicates the quantity of order. From frames 38 to 48, gesture recognition system computes the probability of possible signs which the user can draw depending on HMM states and most likely candidates for the gesture recognition. Moreover, it selects the highest probability element and mark it the "best" element for recognition. At frame 48, first gesture ends and the recognized symbol is "D", thus completing the order (i.e. $Rec_{pstr} = $ "A", $Rec_{pstr} = $ "2", $Rec_{gstr} = $ "D"). The rise and decay of the contribution-weight for both systems for the integration is shown in Fig. 4 (c and d).

The next interpretation starts from frame 49 in which the user draws the posture symbol "L". The next posture recognized is the number which describes the quantity as "2" and finally the gesture symbol which has been recognized is the symbol "B" (i.e. $Rec_{pstr} = $ "L", $Rec_{pstr} = $ "2", $Rec_{gstr} = $ "B"). Gesture and posture recognition works optimally and recognizes the signs correctly.

Fig. 4(c) presents the classification and weight-contribution results of gesture and posture recognition for the whole sequence. Moreover, the recognition of gesture and posture system after applying the threshold is presented in Fig. 4(d) along with the integration of these systems. In this sequence, the recognized gesture elements for the first order is $Date = $ "D", $PostureAlphabet = $ "A" which means $Date$ and from the posture recognized symbol, it is "2". It means $TwoDateJuices$. The second order is $TwoBlueberryJuices$. By changing the lexicon, the proposed approach can be used in other scenarios.

We have tested our proposed approach on the restaurant lexicon database with the overall 98.3% inference accuracy. It is observed that the classification inaccuracies does not effect the performance due to particle filter based weight computation technique. One of the potential reasons is, the particle filter works on the principle of prediction and updation mechanism, therefore, the continuous inferencing of meaningful expression is achieved successfully.

7 Conclusion and Future Work

In this paper, a novel approach is proposed for the integration of gesture and posture recognition in which contribution-weights are computed using Particle filter. The proposed approach is tested on restaurant lexicon which successfully integrates both systems and enables to interpret multiple inferences at the same instance of time. The future research is focused on the words recognition for gesture and posture systems along with their integration.

Acknowledgement

This work is supported by Transregional Collaborative Research Centre SFB/TRR 62 Companion-Technology for Cognitive Technical Systems funded by DFG and Forschungspraemie (BMBF-Froederung, FKZ: 03FPB00213).

References

1. Jaimes, A., Sebe, N.: Multimodal human-computer interaction: A survey. In: Computer Vision and Image Understanding, pp. 116–134 (2007)
2. Yoon, H., Soh, J., Bae, Y., Yang, H.: Hand gesture recognition using combined features of location, angle and velocity. Pattern Recognition 34, 1491–1501 (2001)
3. Liu, N., Lovel, B., Kootsookos, P.: Evaluation of hmm training algorithms for letter hand gesture recognition. In: IEEE Int. Sym. on SPIT, pp. 648–651 (2003)
4. Hunter, E., Schlenzig, J., Jain, R.: Posture estimation in reduced-model gesture input systems. In: International Workshop on Automatic Face-and Gesture-Recognition, pp. 290–295 (1995)

5. Hussain, M.: Automatic recognition of sign language gestures. Master thesis, Jordan University of Science and Technology (1999)
6. Malassiotis, S., Strintzis, M.: Real-time hand posture recognition using range data. Image and Vision Computing 26, 1027–1037 (2008)
7. Licsar, A., Sziranyi, T.: Supervised training based hand gesture recognition system. In: International Conference on Pattern Recognition, pp. 999–1002 (2002)
8. Handouyahia, M., Ziou, D., Wang, S.: Sign language recognition using moment-based size functions. In: Int. Conference of Vision Interface, pp. 210–216 (1999)
9. Freeman, W., Roth, M.: Orientation histograms for hand gesture recognition. In: Int. Workshop on Automatic Face and Gesture Recognition, pp. 296–301 (1994)
10. Brunelli, R., Falavigna, D.: Person identification using multiple cues. IEEE Trans. on PAMI 17, 955–966 (1995)
11. Ross, A., Jain, A.: Multimodal biometrics: An overview. In: 12th Signal Processing Conference, pp. 1221–1224 (2004)
12. Chang, K., Bowyer, K.W., Flynn, P.J.: Face recognition using 2d and 3d facial data. In: ACM Workshop on Multimodal User Authentication, pp. 25–32 (2003)
13. Kumar, A., Wong, D., Shen, H., Jain, A.: Personal verification using palmprint and hand geometry biometric. In: 4th Int. Conf. on Audio and Video-based Biometric Person Authentication, pp. 668–678 (2003)
14. Wu, Q., Wang, L., Geng, X., Li, M., He, X.: Dynamic biometrics fusion at feature level for video-based human recognition, pp. 152–157 (2007)
15. Rashid, O., Hamadi, A., Michaelis, B.: A framework for integration of gesture and posture recognition using hmm and svm. In: IEEE ICIS, pp. 572–577 (2009)
16. Hu, M.: Visual pattern recognition by moment invariants. IRE Transaction on Information Theory 8, 179–187 (1962)
17. Isard, M., Blake, A.: Condensation - conditional density propagation for visual tracking. Int. Jour. of Computer Vision 29, 5–28 (1998)
18. Monwar, M., Gavrilova, M.: A robust authentication system using multiple biometrics. In: Computer and Information Science, pp. 189–201 (2008)

Learning from Mistakes: Object Movement Classification by the Boosted Features

Shigeyuki Odashima, Tomomasa Sato, and Taketoshi Mori

Graduate School of Information Science and Technology,
The University of Tokyo, 7-3-1 Hongo, Bunkyo-ku, Tokyo, Japan
{odashima,tsato,tmori}@ics.t.u-tokyo.ac.jp

Abstract. This paper proposes a robust object movement detection method via a classifier trained by mis-detection samples. The mis-detection are related to the environment, such as reflection on a display or small movement of a curtain, so learning the patterns of mis-detections will improve the detection precision. The mis-detections are expected to have several features, but selecting manually optimal features and thresholds is difficult. In order to acquire optimal classifier automatically, we employ a ensemble learning framework. The experiment shows the method can detect object movements sufficiently by constructing the classifier automatically by the proposed framework.

1 Introduction

Managing objects in the intelligent household environments can give information of "where the object is now" or "when the object used", and it enables the systems tell people where the lost objects are, and support people by observing human-object interactions [1]. To realize object management, we first need to know "where and when the object moved" - especially, "object placement" and "object removal". This paper deals with a object movement detection method in the indoor environments.

As object movement detection methods with fixed cameras in the indoor environments, methods with the background subtraction method are commonly used [2,3,4,5]. To achieve robust object movement detection, the method needs to handle static human (e.g. a person is sitting down and reading book) and non-object movements, such as small shift of furniture or background clutters (e.g. small shift of sofa, shadow). Our previous work [3] handles these problems by detecting "stable changes of the images" and setting some thresholds for detected object regions. However, the method needs to set various thresholds manually to classify the non-object movements, so it is very difficult to decide optimal thresholds and to add the features for classification.

In this paper, we propose a robust object detection method by classifying object movements and non-object movements via a strong classifier trained by mis-detection samples. Some non-object movements are related to the environment (e.g. small movement of curtain, reflection on a display), so by learning these mis-detections caused by non-object movements, the method can improve its

R. Koch et al. (Eds.): ACCV 2010 Workshops, Part I, LNCS 6468, pp. 318–327, 2011.

detection performance automatically. To handle automatically various features such as shape or color difference, the proposed method employs the framework of feature selection via ensemble learning.

The structure of this paper is organized as follows. The rest of this section discusses the related works. Section 2 describes an overview of the object movement detection method via the stable changes of the images. Section 3 provides the details of the ensemble learning framework and the method of classifying the object movement candidates into object movements and non-object movements. In section 4, the experimental results show that the proposed framework can detect object movements sufficiently via the object movement classifier constructed by the correct object movement detection results and mis-detected results. Finally, conclusions are presented in section 5.

1.1 Related Works

Handling small background clutters with unexpected small shift of objects is also big problem when extracting changes via background subtraction methods. The adaptive background subtraction method [4,5,6] handle the small clutters by constructing the background models dynamically. Though the adaptive background subtraction method can handle gradual changes, but the method update its background models automatically regardless of whether the changed regions are object or not, so it is not suitable for detecting objects.

The feature selection via ensemble learning has made great progress in several computer vision areas [7,8,9]. By classifying object movements and non-object movements via learning with stored mis-detections in the framework of ensemble learning, the proposed method detects object movements robustly.

2   Overview of the Proposed Method

In this work, we follow mainly our object movement detection method via the stable image changes [3]. Fig. 1 depicts an overview of the object movement detection method. The method has two major stages: attentive region detection and object detection.

First, the method extracts changed regions by a background subtraction method with energy optimization [10], and then tracks the extracted regions ("attentive region detection" stage in Fig. 1). In this stage, (1) the method extracts changed pixels by a background subtraction technique and categorizes them into "something inserted" state, called the foreground state, and "something removed" state, called removed-layer state. (2) The method then employs the blob detection algorithm to the pixels and extracts changed regions. After extracting the changed regions, (3) the method tracks the extracted regions.

Second, the method categorizes the extracted regions into non-objects and objects via their motion, and finally detects object placement and object removal. In this stage, (4) the method detects object movement candidate regions by discriminating between the non-object state and the object state via the regions'

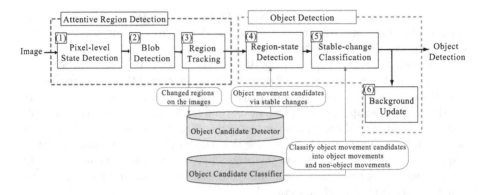

Fig. 1. Overview of the object movement detection method

motion detection result for past some frames (this motion detection results are called as "motion history"). (5) The method classifies the object movement candidates into object movements and non-object movements, and detects object placement and object removal. Finally, (6) the method updates its background model according to the object detection result.

The method detects object movements from stable changes extracted by a background subtraction method. Common background subtraction methods have only one background model, so these background subtraction methods have only "changed" state, called foreground state, and "not changed" state, called background state. To detect object movements, we need to classify "object placement" and "object removal" from the "changed" state, so the method adopt a multiple-layered background model [2,4] (called "layered background model"). Moreover, the method adopts an object placement / removal classification method based on edge subtraction, to handle "object removal which exists in the initial state", which the method only with the layered background model cannot handle properly.

When object movement detection is performed, non-objects such as humans also exist in the images. So the method needs to categorize objects and non-objects to detect objects robustly. The object movement detection method categorizes objects and non-objects via the stable image changes. The stable image change is the state that the region is changing from the recorded state but the change is settled. For example, when a book is placed on the sofa, the region of the book is changing the "sofa" region. The changed region caused by object movement don't move, so the "stable changed" regions are different from those caused by non-objects such as human. To detect the stable changes, the method extracts motion of the extracted regions for several frames, called motion history, and categorizes them into objects and non-objects by the state machine which is driven by the motion history.

The method via the stable changes works well for classifying non-objects with movement such as humans even if the human region is occluded. But, the method cannot handle properly the stable changes caused by non-object movements,

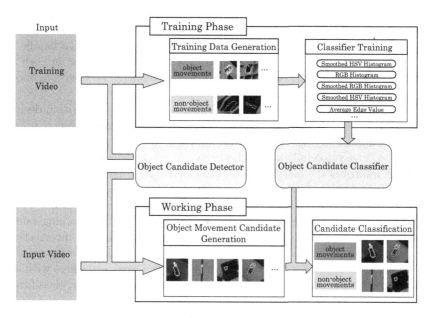

Fig. 2. Overview of the object movement classification. The object candidates extracted by the stable image changes are classified into object movements and non-object movements. In the training phase, the method trains the classifier via the labeled object movement detection results extracted by the stable image changes. In the working phase, the method classifies the object movement candidates by the learning results.

such as small shift of furniture or shadow, because the non-object movements also make the stable changes. In the next section, we provide a method to reject the stable changes without object movements via the classifier constructed by the mis-detections.

3   Rejection of the Stable Changes without Object Movements via Learning by Mis-detections

3.1   Classification Framework

The proposed method classifies the object movement candidate regions detected by the stable changes (Fig. 1 (4)) by the object candidate classifier. Fig. 2 depicts an overview of the object movement classification.

The proposed method has two phases: the training phase and the working phase. In the training phase, the method gathers the object movement candidates by the detection method by the stable changes, and give class labels (object movement or non-object movement). The class labels of the candidates are acquired manually. Then, the method trains the object candidate classifier by the labeled candidates. In the working phase, the method detects the object movements by the stable changes, and then classifies them into object movements and non-object movements by the classifier acquired in the training phase.

The non-object movements are expected to have several features in their appearance: small color difference between the input image and in the background image, long narrow shape, low contrast in the region boundary. To handle several features properly, we adopt the AdaBoost framework utilized in such as a face detector [7], and construct a strong object movement classifier by choosing effective features from the feature sets.

The original AdaBoost algorithm is a supervised learning algorithm designed to construct a strong binary classifier. The input of the algorithm is a set of training examples $(\mathbf{y}_n, z_n), n = 1, ..., N_{train}$, where each $\mathbf{y}_n$ is an example and z_n is an boolean value indicating whether $\mathbf{y}_n$ is a positive or negative example. AdaBoost improves the classification performance by combining a collection of a set of weak classifiers, which are associated with each feature. In each training step, the AdaBoost algorithm acquires an optimal weak classifier via the weighted training examples, and re-weights the weights of the training samples to increase the importance of the samples which were classified incorrectly by the previous weak classifier. The final strong classifier takes the forms of the Perceptron. The method assigns large weights to the weak classifiers with good classification performance whereas small weights to those with poor classification performance.

In our implementation, a weak classifier $h_i(\mathbf{y})$ has the form of threshold function.

$$h_i(\mathbf{y}) = \begin{cases} 1 & \text{if } p_i f_i(\mathbf{y}) < p_i \theta_i \\ 0 & \text{otherwise.} \end{cases} \tag{1}$$

where θ_i is a threshold and p_i is the value which represents the direction of the inequality (p_i is -1 or 1). The optimal values for θ_i and p_i are chosen by minimizing the sum of the weight of the misclassified training examples as shown Eq. (2).

$$(p_i, \theta_i) = \underset{(p_i, \theta_i)}{\operatorname{argmin}} \sum_{n=1}^{N_{train}} w_{i,n} |h_i(\mathbf{y}_n) - z_n| \tag{2}$$

where $w_{i,n}$ is the weight of nth training sample in the training step i. In our implementation, the same weak classifier may be repeatedly chosen in the different learning steps. The resulting algorithm is given by Table 1.

3.2   Feature Set of Weak Classifiers

In our experiments, the stable image changes caused by non-object movements has mainly three features: shape, color difference, contrast. The proposed method extracts the features shown in Table 2 from the object candidate region R.

As the HSV color histogram, we employ Pérez's method [11]. For the metric of color histograms, we use Bhattacharrya distance. Bhattacharrya distance $d_h(q(E_i), q(E_b))$ between a histogram $q(E_i)$ and a histogram $q(E_b)$ is calculated as follows.

Table 1. The AdaBoost algorithm

- Input : set of example $(\mathbf{y}_1, z_1), ..., (\mathbf{y}_{N_{train}}, z_{N_{train}})$
- Let m be the number of negative examples and l be the number of

  positive exmaples. Initialize weights $w_{1,n} = \dfrac{1}{2m}, \dfrac{1}{2l}$ depending on the value of z_n
- For $t = 1, ..., T$:

  1) Normailze the weights $w_{t,n}$ so that $\sum_{n=1}^{N_{train}} w_{t,n} = 1$.

  2) For each feature f_i, train a weak classifier h_i.

  3) The error ε_i of a classifier h_i is determined with respect to the

  weights $w_{t,1}, ..., w_{t,N_{train}}$:

  $$\varepsilon_i = \sum_n^{N_{train}} w_{t,n} |h_i(\mathbf{y}_n) - z_n|.$$

  4) Choose the classifier h_i with the lowest error ε_i and set $(h_t, \varepsilon_t) = (h_i, \varepsilon_i)$.

  5) Update the weights $w_{t+1,n} = w_{t,n} \beta_t^{1-e_n}$ where $\beta_t = \dfrac{\varepsilon_t}{1-\varepsilon_t}$

  and $e_n = 0$, if example $\mathbf{y}_n$ is classified correctly by h_t and 1, otherwise.
- The final strong classifier is given by :

  $$h(\mathbf{y}) = \begin{cases} 1 & \text{if } \sum_{t=1}^T \log\dfrac{1}{\beta_t} h_t(\mathbf{y}) \geq T_o \sum_{t=1}^T \log\dfrac{1}{\beta_t} \\ 0 & \text{otherwise.} \end{cases}$$

$$d_h(q(E_i), q(E_b)) = \sqrt{1 - \sum_{k=1}^N \sqrt{q(E_i; k) q(E_b; k)}} \qquad (3)$$

where $q(E; k)$ is the kth bin of the histogram $q(E)$.

When using simple color histograms, noise affects largely if the number of the sampled points is small. The object regions on the input images are sometimes small, so the method must be robust even if the object candidate region is small. We also employ the smoothed color histograms (feature type 5, 7 in Table 2) as well as the simple color histograms (feature type 4, 6 in Table 2). The smoothed color histogram $q(R)$ of the object candidate region R is calculated as follows [12]:

$$q(E; k) = \frac{N_B(k, R) + C}{N_T(R) + K \times C} \qquad (4)$$

where $N_B(k, R)$ is the histogram bin value of the kth bin, $N_T(R)$ is the number of pixels in the R, K is the number of bins and C is a constant value (in our implementation, $C = 1$). The smoothed histogram is robust for noise even if the sampled points are few because of the smoothing term C. At the same time, because of the smoothing term, the difference between the smoothed histograms is small when the number of the sampled points is small, so the smoothed histogram is affected by the shape (especially, the size of R) as well as color difference.

Table 2. The feature set

	ID	Feature description
Shape	1	The size of R.
	2	The ratio of major axis and minor axis of R when R is approximated to a ellipse.
	3	The average width of R.
Color Difference	4	The Bhattacharrya distance between the RGB color histogram of the input image and the one of the background image in R.
	5	The Bhattacharrya distance between the smoothed RGB color histogram of the input image and the one of the background image in R.
	6	The Bhattacharrya distance between the HSV color histogram of the input image and the one of the background image in R.
	7	The Bhattacharrya distance between the smoothed HSV color histogram of the input image and the one of the background image in R.
Contrast	8	The average edge value on the contour of R. To calculate edge value, the input image and the background image are used when R is classified as object placement and object removal, respectively.
	9	The average edge difference of the input image and the background image on the contour of R.

4 Experiments

We evaluate the object movement detection performance of proposed method with 6 video sequences which are captured in 4 viewpoints (total 24 different video sequences, total 7516 frames). These evaluation video sequences consist of the images of 320×240 resolution recorded at 7.5 fps. The evaluation video sequences contain 69 object placement and 41 object removal (total 110 events).

In the experiment, the object candidate classifier is trained by 2 video sequences which are captured in 4 viewpoints (total 8 different video sequences, total 2808 frames). The classifier is trained by 42 object movements and 43 non-object movements (total 83 object movement candidate samples) extracted by the object candidate detector. The method does not have classifiers specified in the each viewpoint, but the method has only one classifier trained by the all samples, to avoid to be affected by the distance between the objects and the cameras. The number of weak classifiers was 50.

The object detection method was implemented on a PC with an Intel Core 2 Duo 2.5 GHz processor. The method ran with single-thread processing.

In this experiment, we implement false positive and recall as performance evaluation measures, as defined below:

$$\text{false positive} = 1 - \frac{\text{correctly detected object movements}}{\text{total detected object movements}} \tag{5}$$

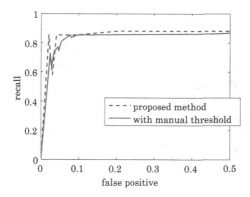

Fig. 3. ROC curves of the proposed method. The broken line: by the proposed method. The solid line: by manual thresholds with our previous method [3].

$$\text{recall} = \frac{\text{correctly detected object movements}}{\text{total object movements in the images}} \tag{6}$$

In this experiment, we calculated performance of the proposed method under variant threshold parameter T_o of AdaBoost. We compare the proposed method with our previous object movement detection method with manual thresholds [3].

Fig. 3 shows the resulting detection performance in various parameters. As can be seen from the graph, the performance of the proposed method is almost the same of the result by manually determined threshold or is slightly improved. The proposed method has only 1 threshold parameter, though our previous method has 5 threshold parameters, so the proposed method can be tuned easily.

Fig. 4 shows some sample detection results (these results are taken with $T_o = 0.5$ when false positive $= 0.04$ and recall $= 0.85$). In the images, rectangles are overlaid on the object candidate regions and are given different colors according to the classification result (blue: object placement, red: object removal and yellow: non-object movement). (a) and (b) in Fig. 4 are when the object candidate detector detects correct object movements, and (c) and (d) are when the object candidate detector wrongly detects non-object movements. As can been seen from Fig. 4, the object candidate detector sometimes wrongly detects non-object movements, but the proposed object candidate classifier rejects non-object movements.

The average calculation time was roughly 120[ms/frame]. The calculation time is almost same by our previous method with manual thresholds [3], so the proposed method works in sufficient frame rates.

4.1   Discussion

Selected features in the object candidate classifier. The object candidate classifier acquired in the experiment selected the first three features from the average edge value (feature type 8, with sum of weights $= 0.30$) ,the Bhattacharrya distance of the smoothed HSV color histograms (feature type 7, with sum

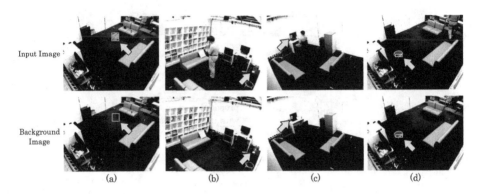

Input Image

Background
Image

(a)   (b)   (c)   (d)

Fig. 4. Examples of detection results. The top row and the bottom row are the input images and the background images when object movement candidates are detected, respectively. Rectangles are overlaid on the object candidate regions of the images (blue: object placement, red: object removal, yellow: non-object movement). (a) is when placement of a box is detected, (b) is when removal of a phone is detected, and (c) and (d) are when non-object movements are wrongly detected by the object candidate detector. The object candidate detector sometimes wrongly detects non-object movements, but the proposed method classifies the object movements correctly.

of weights $= 0.28$) and the average width (feature type 3, with sum of weights $= 0.13$) while the classifier gave no weights to the features of the size and the ratio of major axis and minor axis (feature type 1, 2). One reason of this feature selection is considered that the features has same elements (the smoothed HSV histogram has some elements of the size and the average width and the ratio of major axis and minor axis have the element of "thickness").

Limitations. The proposed framework works well for long-term small background clatters, but has several limitations. First, the proposed framework obviously cannot handle the large changes for the whole image with strong illumination changes. To handle strong illumination changes, the method need to be improved in the primary process, especially in the background subtraction methods. Second, the proposed method cannot handle large shift of furniture (e.g. opening or closing of a door, large movement of a sofa). However, handling the large shift would be beyond the scope of this paper, because detecting these large shift has benefits in some applications, so handling the large shift of furniture should be done by detecting the state of "large shift".

5   Conclusion

This paper proposed an robust object movement detection method via the classifier trained by the mis-detection results. The classifier automatically selects optimal features of shape, color difference and contrast. Our experiment shows the proposed method has similar or improved detection performance compared to the classifier with manual thresholds.

One of the future tasks is constructing a system which manages "where what object is now" and "what the person is doing with the managed objects" by integrating object recognition and behavior recognition.

References

1. Gupta, A., Kembhavi, A., Davis, L.: Observing human-object interactions: using spatial and functional compatibility for recognition. IEEE Transactions on Pattern Analysis and Machine Intelligence 31, 1775–1789 (2009)
2. Maki, K., Shirai, N., Shirai, Y.: Interactive inquiry of indoor scene transition with awareness and automatic correction of mis-understanding. In: The 1st international workshop on video event categorization, tagging and retrieval, VECTaR 2009 (2009)
3. Odashima, S., Mori, T., Shimosaka, M., Noguchi, H., Sato, T.: Event understanding of human-object interaction: Object movement detection via stable changes. In: Zhang, J., Shao, L., Zhang, L., Jones, G.A. (eds.) Intelligent Video Event Analysis and Understanding. Studies in Computational Intelligence, vol. 332, pp. 195–210. Springer, Heidelberg (2011)
4. Kim, K., Chalidabhongse, T., Harwood, D., Davis, L.: Real-time foreground–background segmentation using codebook model. Real-Time Imaging 11, 172–185 (2005)
5. Tian, Y., Lu, M., Hampapur, A.: Robust and efficient foreground analysis for real-time video surveillance. In: CVPR (2005)
6. Stauffer, C., Grimson, W.: Adaptive background mixture models for real-time tracking. In: CVPR (1999)
7. Viola, P., Jones, M.J.: Robust real-time face detection. IJCV 57, 137–154 (2004)
8. Li, Y., Huang, C., Nevatia, R.: Learning to associate: Hybridboosted multi-target tracker for crowded scene. In: CVPR (2009)
9. Gehler, P., Nowozin, S.: On feature combination for multiclass object classification. In: ICCV (2009)
10. Boykov, Y., Kolmogorov, V.: An experimental comparison of min-cut/max-flow algorithms for energy minimization in vision. IEEE Transactions on Pattern Analysis and Machine Intelligence 26, 1124–1137 (2004)
11. Pérez, P., Hue, C., Vermaak, J., Gangnet, M.: Color-based probabilistic tracking. In: Heyden, A., Sparr, G., Nielsen, M., Johansen, P. (eds.) ECCV 2002. LNCS, vol. 2350, pp. 661–675. Springer, Heidelberg (2002)
12. Özuysal, M., Calonder, M., Lepetit, V., Fua, P.: Fast keypoint recognition using random ferns. IEEE Transactions on Pattern Analysis and Machine Intelligence 32, 448–461 (2009)

Modeling Multi-Object Activities in Phase Space

Ricky J. Sethi and Amit K. Roy-Chowdhury*

UC Riverside

Abstract. Modeling and recognition of complex activities involving multiple, interacting objects in video is a significant problem in computer vision. In this paper, we examine activities using relative distances in phase space via pairwise analysis of all objects. This allows us to characterize simple interactions directly by modeling multi-object activities with the Multiple Objects, Pairwise Analysis (MOPA) feature vector, which is based upon physical models of multiple interactions in phase space. In this initial formulation, we model paired motion as a damped oscillator in phase space. Experimental validation of the theory is provided on the standard VIVID and UCR Videoweb datasets capturing a variety of problem settings.

1 Introduction

Motion underlies all activities; human activities, in fact, are defined by motion. The rigorous study of motion has been the cornerstone of physics for the last 450 years, over which physicists have unlocked its deep, underlying structure. In this paper, we exploit the physics of motion to understand interactions between objects by utilizing relative distance in phase space with a pairwise analysis of all the interacting objects. In this initial formulation, we model paired motion as a damped oscillator in phase space.

2 Related Work and Contributions

We build liberally upon theoretical thrusts from several different disciplines, including Analytical Hamiltonian Mechanics and human activity recognition [1], especially for multiple activities [2]. Human activities, in fact, can be categorized into four classes: kinesics, haptics, proxemics, and chronemics. Most work in activity recognition [1] deals with actions of individuals (walking, running, waving, hugging, shaking hands, etc.) that, even for multiple activities [3,2], fall in the domain of kinesics and haptics. Our approach, on the other hand, rests in the domain of proxemics and chronemics since we model the spatio-temporal relationships between multiple, interacting objects. In particular, we capture the proxemics and chronemics by examining relative distances between objects in phase space with respect to time.

Others, such as [4], have utilized relative distances within a coupled HMM but our methodology does not require an external stochastic framework and can characterize motion directly from tracks without requiring training or classifiers. In addition, they

* Both authors were supported by the DARPA VIRAT program at UCR. The first author was also partially supported by NSF Award 1019343/Sub Award CIF-B-17 at UCLA.

R. Koch et al. (Eds.): ACCV 2010 Workshops, Part I, LNCS 6468, pp. 328–337, 2011.

interpret pedestrian actions only and create prior models of human behaviour by using synthetic agents that encapsulate their assumptions for simple actions that are atomic in nature and only look at single interaction detections. Our approach, on the other hand, looks at complex activities betwen multiple, interacting objects of any variety without the need for synthetic agents or prior models of the objects. Also, unlike the heuristic examination of simple activities in a single domain using relative distances in [5], we create a consistent framework to derive and unify different representations of motion for activity recognition. All these previous approaches also rely on a classifier (coupled HMM in [4] and a simple hypothesis testing framework based on two-class nearest neighbor classification with extensive parameterizations and thresholds in [5]) whereas our approach uses our physics-based models to do the recognition directly.

In recent years, there has been some work in leveraging results from physics to the analysis of videos. For example, Energy-Based Models (EBMs) [6] capture dependencies between variables for *image recognition* by associating a scalar energy to each configuration of the variables. Others [7] take local and global optical flow approaches and compute confidence measures while in [8], they recognize single-person activities using a pseudo-Hamiltonian, which is a scalar or multi-dimensional time-series that represents the motion of an object over the course of an activity. In this paper, however, we formalize the idea of using scalar energies and pseudo-Hamiltonians to a generalized phase space analysis by applying the physics-based methodology to modeling multi-object activities in phase space. The *phase space* of a system consists of all possible values of the coordinates, which is usually represented as the space of position vs momentum (x, p) or position vs velocity (x, v) but can be any set of coordinates like the relative distance vs time, (r, t) or the generalized coordinates and Hamiltonian values discussed in detail below.

Our approach also draws inspiration from the method employed in [9], which detects global motion patterns by constructing super tracks using flow vectors for tracking high-density crowd flows in low-resolution. Our methodology in this paper, on the other hand, works in both high- and low-resolution and for densely- and sparsely-distributed objects since all it requires is the (x, y, t) tracks for the various objects' motion analysis in phase space, as shown in Figure 1.

Specifically, we develop physical models of complex interactions in phase space by doing a **Multiple Objects Pairwise Analysis (MOPA)** in which we model paired motion as a damped oscillator in phase space using relative distances. MOPA thus contains phase space features for paired activities, with physical models of complex interactions in phase space.

Our method starts with the input containing multiple, interacting objects. We then use a detector and tracker, as discussed in Section 4, to find objects and their tracks.

Fig. 1. From Tracks to Phase Space: the phase space of a system consists of all possible values of the coordinates, which can be (q,p) or (q,p,t), for example; we may also look at modified phase plots of (H,t), (H,q,p), etc.

We then analyze all individual objects pairwise and iteratively do a pairwise analysis of all pairs and individuals. Finally, we recognize different activities based on the above model for all tracks in the video, as shown in Table 2.

3   Modeling Multi-Object Activities

Hamiltonian Dynamics is an elegant and powerful alternative formulation of classical mechanics that not only gives the equations of motion for a system but, more importantly, provides greater, and often more abstract, insight about the system. It provides a framework based upon the Principle of Least Action that can be extended to other laws of physics; in fact, almost all fundamental laws of physics can be expressed in terms of a least action principle. Hamilton's equations, using the Hamiltonian H, are equivalent to the Euler-Lagrange equations; Hamilton's equations are primarily of interest in establishing basic theoretical results, rather than determining the motions of particular systems. The Hamiltonian is usually stated most compactly, in generalized coordinates, as [10]:

$$H(q, p, t) = \sum_i p_i \dot{q}_i - L(q, \dot{q}, t) \tag{1}$$

where H is the Hamiltonian, p is the generalized momentum, and $\dot{q}$ is the time derivative of the generalized coordinates, q, and L is the Lagrangian. This defines the dynamics on the system's phase space, in which the q_i and p_i are regarded as functions of time [11,12]. The **phase space** of a system consists of all possible values of the generalized coordinate variables q_i and the generalized momenta variables p_i. If the Hamiltonian is time-independent, then phase space is 2-dimensional, (q,p); if the Hamiltonian is time-dependent, then phase space is 3-dimensional, (q,p,t) [8]. In general, the phase space of a system consists of all possible values of the coordinates, which can be (q,p) or (q,p,t), for example; we may also look at modified phase plots of (H,t), (H,q,p), etc.

In this section, we do a pairwise analysis of multiple objects. In particular, we develop physical models of complex interactions in phase space and do a **Multiple Objects Pairwise Analysis (MOPA)** in which we model paired motion as a damped oscillator in phase space using relative distances. MOPA thus contains phase space features for paired activities, with physical models of complex interactions in phase space.

We start with the problem of trying to categorize the motion of two objects in video and to see if their motion is correlated (this is taken as a simple example and can be generalized further). We hypothesize that the motion of two objects can be modeled as an oscillation with the envelope of that oscillation, as seen in the under-damped oscillator in Figure 2, being the average distance over time between the two. We thus calculate the relative distance with respect to time between these two objects and use that as the envelope for the oscillation.

To further elucidate our method, let us consider two people walking. In order to model the motion of two people walking, we consider the three possibilities: they can walk towards each other, they can walk away from each other, or they can walk parallel to each other [4]. These three situations are shown in Figure 3 where we model all three types of motion as an oscillator. In Figure 3a, two people walking towards each other

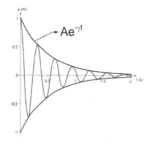

Fig. 2. An Under-damped Oscillation Envelope

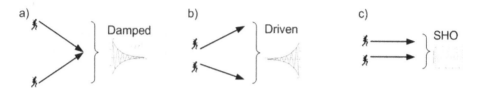

Fig. 3. Modeling Paired Motion: a) Two people walking towards each other is modeled as a Damped Oscillator; b) Two people walking away from each other is modeled as a Resonant Driven Oscillator; c) Two people walking parallel to each other is modeled as an Un-driven SHO

is modeled as a Damped Oscillator; in Figure 3b, two people walking away from each other is modeled as a Resonant Driven Oscillator; in Figure 3c, two people walking parallel to each other is modeled as an Un-driven Simple Harmonic Oscillator (SHO).

A damped oscillator is described by the following second order differential equation of motion, which can represent all three cases and can model mass and damping; in addition, we set the spring constant to zero since we do not model it here but it should be considered in the more general case. Thus, setting the restoring force to zero yields the following second order differential equation:

$$m\ddot{x} + c\dot{x} = 0 \qquad (2)$$

where c is the *damping constant*. This leads to

$$x(t) = A_1 + A_2 e^{-\gamma t} \qquad (3)$$

where $\gamma = \frac{c}{m}$ is the *damping factor* with mass m and A_1 and A_2 are the coefficients. For an under-damped oscillator, this gives $x(t) = Ae^{-\gamma t}$ with amplitude A.

Thus, the damping is determined by γ, which is determined by the coefficients, and, as in systems theory, x need not only be the position. This damping is pictured in Figure 2 along with the envelope, which is given by $Ae^{-\gamma t}$. This gives the Hamiltonian for the damped oscillator as [13]:

$$H(x,p) = \frac{p^2}{2m} + \frac{1}{2}m\omega_0^2 x^2 = \frac{1}{2}m\dot{x}^2 + \frac{1}{2}mx^2 \qquad (4)$$

The change in energy is then given as:

$$\frac{dH}{dt} = -c\dot{x}^2 \tag{5}$$

The under-damped oscillator is also shown in the (x, v), (H, t), and $(\frac{dH}{dt}, t)$ modified phase space plots in Figure 4.

The *damping ratio*, ζ, determines whether the damping is critical, under-, or over-damping. *Logarithmic decrement*, δ, is used to find the damping ratio of an under-damped system in the time domain. The logarithmic decrement is the natural log of the amplitudes of any two successive peaks:

$$\delta = \frac{1}{n} \ln \frac{x_0}{x_n} \tag{6}$$

where x_0 is the greater of the two amplitudes and x_n is the amplitude of a peak n periods away. The damping ratio is then found from the logarithmic decrement as:

$$\zeta = \frac{1}{\sqrt{1 + (\frac{2\pi}{\delta})^2}} \tag{7}$$

When two people are walking as in Figure 3, we can thus estimate the kind of oscillation via the damping factor, γ; in particular, we do an exponential fit to the average distance between the two people with respect to time in order to determine the damping factor. We then conclude SHO if $\gamma = 0$, driven resonant oscillator if $\gamma < 0$, and damping if $\gamma > 0$. To further qualify the damping, we utilize the *quality factor*, Q. We could use the *damping time*, $\tau = \frac{m}{\gamma}$, to define $Q = \omega_0 \tau$ provided $\omega_0 \tau \gg 2\pi$. Q is also defined as:

$$Q = \frac{1}{2\zeta} \tag{8}$$

We can then use (8) to determine the kind of damping as critical damping when $Q = \frac{1}{2}$; over-damping when $Q < \frac{1}{2}$; and under-damping when $Q > \frac{1}{2}$.

3.1 Advantages

The advantages of our approach over a simple linear fit are manifold. One of the main is the robustness in tracking: short-term tracking errors would not affect our method since we fit to a model. In addition, the utilization of the Q-factor lets us determine the extent of the motion (if an object is headed for another, this lets us characterize if they head directly there or meander and go back and forth, instead). Finally, the formalism afforded by our method provides a framework that is extensible with more complex models to an even wider variety of situations and domains.

3.2 Application to Activity Modeling

For example, we model two people walking towards each other, as in Figure 3a, as an under-damped oscillator. We thus use the logarithmic decrement, δ, to estimate the damping ratio, ζ, by estimating n in (6). We use this damping ratio to compute the quality factor, Q, and determine the specific kind of damping. We can also use the damping ratio to get the angular frequency, ω, and then plot x vs ω or use ω directly. Finally, we can use average distance and ω to get average velocity since $v = r\omega$. In fact, it is also possible to estimate Q from the (x, v) phase space plot, as in Figure 4a (under-damped oscillations spiral in SHO is an ellipse, e.g.), or to use an exponential fit on the (H, t) or $(\frac{dH}{dt}, t)$ phase space plot, as in Figure 4b.

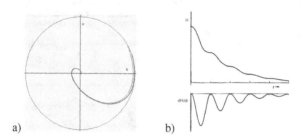

Fig. 4. Modified phase space plots of an under-damped oscillator in a) (x, v) and b) (H, t) and $(\frac{dH}{dt}, t)$ phase space plots

3.3 Generalization

The proposed method is also generalizable to many other cases of arbitrary motion interactions. For example, it can be applied to deal with intersections since intersections imply a transition from over-damped to under-damped or critically damped. In fact, it can be generalized to more than two objects by considering pairwise combinations (as seen in Figure 5 and Figure 7) and future work can consider more efficient methods than this combinatorial approach.

This generalization can also be extended to more complex interactions, as shown in Figure 5a. Here, we see interactions that are not atomic, direct interactions; instead, we might observe interactions like people milling together, where they alternately approach and recede from each other. Similarly, they might interact with a static object, like a car or a building, by meandering around it, rather than approaching it directly and then becoming static in its vicinity. In video of activities in the "wild", as in the UCR Videoweb dataset, these are exactly the kinds of activities observed, as shown in Figure 5b, where we see how convoluted the trajectories of individual objects can seem. When two objects interact, there's a temporal overlap to their trajectories, as shown in Figure 5c, where we see the overlap of frames between all the objects in a scene.

Thus, in these complex situations, people don't generally exhibit the simple, direct motions examined in [4,5]. Instead, there are multiple *turning points* of their motion as they might alternate between approaching each other or moving away from each other; these turning points are characterized as the extrema of the (r, t) plot between two objects, as shown in Figure 5d. Thus, in order to model these behaviours, we do a pathwise MOPA analysis between all the turning points by using the extrema of the path to indicate the turning points for each segment.

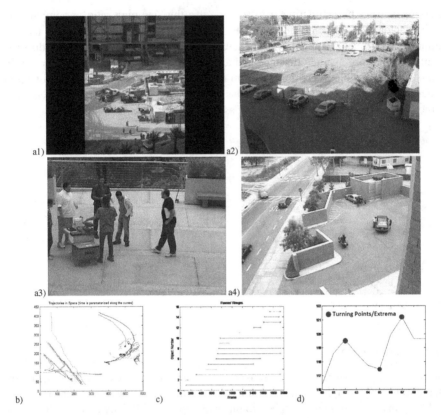

Fig. 5. Multi-Object Activity Modeling. a) Sample frames representing complex interactions captured by our approach: construction site, courtyard, and parking lot; (b) Sample trajectories in space (time is parameterized along the curve) for the courtyard and parking lot; (c) Temporal overlap of objects' trajectories; (d) Turning points for a single trajectory.

4   MOPA Experimental Results

We experimented with videos consisting of people, vehicles, and buildings, which encompasses a large class of possible activities. We used a 15-minute combined dataset of high-resolution and low-resolution video from standard datasets like the UCR Videoweb (http://vwdata.ee.ucr.edu/) and VIVID (https://www.sdms.afrl.af.mil/request/data_request.php) datasets.

For object detection and tracking, we utilized the methodology we have developed in an alternate paper [14]; in this approach, we analyze the statistical properties of tracklets (short-term tracks) and develop associations between them to come up with longer tracks via a stochastic optimization step that considers the statistical properties of individual tracklets, as well as the statistics of the targets along each proposed longterm track. We then used these (x, y, t) tracks to compute the relative distance trajectories, (r, t), with which we subsequently compute the MOPA.

In this section, we show the results of modeling a paired activity from the UCR Videoweb dataset. In Figure 6a, we see three representative samples from a video of a person walking to their car. We model this as a paired activity where the stationary track of the car finally intersects the dynamic track of the person. In Figure 6b, we plot the average distance between the two tracks with respect to time and then do an exponential fit to that curve. The results are analyzed further in Table 1, where we see the analysis of Section 3 applied to the video and chart represented in Figure 6. We find a value for γ of 0.007, which indicates Damping; subsequent analysis yields a Q-factor of 37.878, thus indicating Under-Damping and showing that the two tracks eventually converge.

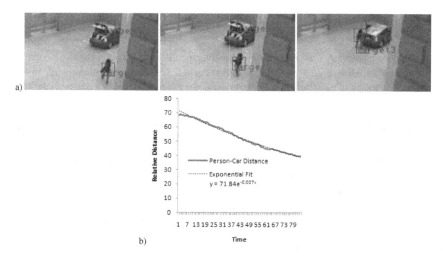

Fig. 6. Person-Car Paired Activity Modeling. In a) we see three representative samples from a video of a person walking to their car. In b) we plot the average distance between the two tracks with respect to time and then do an exponential fit.

Table 1. Person-Car Paired Activity Values. We find a value for γ of 0.007, which indicates Damping; subsequent analysis yields a Q-factor of 37.878, thus indicating Under-Damping and showing that the two tracks eventually converge.

Factor	Value	Result
γ	0.007	Damping
δ	0.083	
ζ	0.013	
Q-Factor	37.878	Under-Damping

In addition, we apply our method to a video of two people running away from each other and towards a car. By doing a pairwise plot of the average distance between the tracks for each pair with respect to time and then doing an exponential fit to each curve, we find its value for γ. As can be seen in Figure 7a, we do an exponential fit to the curve of the two people running away from each other to see a driven oscillator, exactly as expected for such a case; while in Figure 7b and Figure 7c, we see the motion of each person running towards the car as being a damped oscillator, where subsequent analysis yields a Q-factor greater than $\frac{1}{2}$, thus indicating Under-Damping and showing that the two tracks eventually converge.

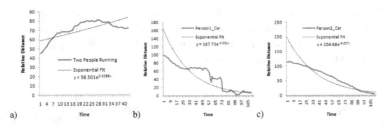

a)   b)   c)

Fig. 7. Two People and A Car: in a) we plot the average distance between the two people with respect to time and then do an exponential fit to that curve to see a driven oscillator while in b) and c) we see the motion of each person running towards the car is modeled as a damped oscillator

We also apply our modelling methodology to activity recognition by testing it within a query-based retrieval framework. We use a combined database from the UCR Videoweb and VIVID datasets and the results are shown in Table 2, where we see the precision/recall values for this experiment of database query and retrieval. As shown in the table, the detection rate for the two-object activity "People Walking Together" is lower because, as can be seen in the video, the participants tend to walk parallel to each other and the start and end frames for the multi-object activities are occasionally ambiguous.

Robustness of Tracking: Short-term tracking errors would not affect our method since we fit to a model. This is the advantage of using an exponential fit over a simple linear fit. Long-term tracking errors also would not affect the transtions (the formations and dispersals of the pairs). Although tracking and object detection fail when objects are close together, our method works with both.

Table 2. Precision/Recall Values for DB query and retrieval using the combined VIVID and UCR Videoweb database

Activity	Precision	Recall	Total Fetched	True Positive	Ground Truth
Person Entering Building	1	1	9	9	9
Person Exiting Building	1	1	7	7	7
Person Entering Vehicle	0.9	0.9	11	10	10
Person Exiting Vehicle	1	1	6	6	6
People Walking Together	1	0.71	5	5	7
People Coming Together	0.86	0.86	7	6	6
People Going Apart	0.8	1	5	4	5

5   Conclusions and Future Work

Our formulation takes an altogether novel approach whereby we attempt to create a theoretical framework rooted in physics to gain insight into the problem of activity recognition in video. The framework we present provides a structured approach for activity recognition that only requires tracks for the motion; it can be generalized across different application domains and even applied to coupled systems, interactions between sparse objects, and other systems without requiring separate heuristics for each. Future work will study how to obtain robust physics-based features, develop more complex physics models (e.g., determine driving forces for the driven oscillator, use field theory, etc.), and use shape or learning algorithms to determine mass and potentials.

References

1. Turaga, P., Chellappa, R., Subrahmanian, V., Udrea, O.: Machine recognition of human activities: A survey. In: CSVT (2008)
2. Ryoo, M., Aggarwal, J.: Spatio-temporal relationship match: Video structure comparison for recognition of complex human activities. In: ICCV (2009)
3. Duchenne, O., Laptev, I., Sivic, J., Bach, F., Ponce, J.: Automatic annotation of human actions in video. In: ICCV (2009)
4. Oliver, N., Rosario, B., Pentland, A.: A bayesian computer vision system for modeling human interactions. In: ICVS (1999)
5. Gaur, U., Song, B., Roy-Chowdhury, A.: Query-based retrieval of complex activities using strings of motion-words. In: WMVC (2009)
6. LeCun, Y., Chopra, S., Ranzato, M., Huang, F.: Energy-based models in document recognition and computer vision. In: ICDAR (2007)
7. Bruhn, A., Weickert, J., Schnorr, C.: Lucas/kanade meets horn/schunck: combining local and global optic flow methods. In: IJCV, pp. 211–231 (2005)
8. Sethi, R., Roy-Chowdhury, A., Ali, S.: Activity recognition by integrating the physics of motion with a neuromorphic model of perception. In: WMVC (2009)
9. Hu, M., Ali, S., Shah, M.: Detecting global motion patterns in complex videos. In: ICPR (2008)
10. Goldstein, H.: Classical Mechanics, 2nd edn. Addison-Wesley, Reading (1980)
11. Landau, L., Lifshitz, E.: Course of Theoretical Physics: Mechanics, 3rd edn (1976)
12. Marion, J., Thornton, S.: Classical Dynamics of Particles and Systems, 4th edn. Saunders, Philadelphia (1995)
13. Fowles, G., Cassiday, G.: Analytical Mechanics, 6th edn. Brooks Cole, Pacific Grove (2004)
14. Anonymous: A stochastic optimization framework for stable multi-target tracking (2010)

Sparse Motion Segmentation Using Multiple Six-Point Consistencies*

Vasileios Zografos, Klas Nordberg, and Liam Ellis

Computer Vision Laboratory, Linköping University, Sweden
{zografos,klas,liam}@isy.liu.se

Abstract. We present a method for segmenting an arbitrary number of
moving objects in image sequences using the geometry of 6 points in 2D
to infer motion consistency. The method has been evaluated on the Hopkins 155 database and surpasses current state-of-the-art methods such as
SSC, both in terms of overall performance on two and three motions but
also in terms of maximum errors. The method works by finding initial
clusters in the spatial domain, and then classifying each remaining point
as belonging to the cluster that minimizes a motion consistency score.
In contrast to most other motion segmentation methods that are based
on an affine camera model, the proposed method is fully projective.

1 Introduction

Motion segmentation can be defined as the task of separating a sequence of
images into different regions, each corresponding to a distinct rigid motion. There
are several strategies for solving the motion segmentation problem, some of which
are based on first producing a dense motion field, using optical flow techniques,
and then analyzing this field. Examples of this approach are [1] where the optic
flow is given as a parametric model and the parameters are determined for each
distinct object, or the normalised graph cuts by [2].

Other approaches are instead applied to a sparse set of points, typically interest points that are tracked over time, and their trajectories analysed in the
image. A common simplifying assumption is that only small depth variations
occur and an affine camera model may be used. The problem can then be solved
using the factorization method by [3]. This approach has attracted a large interest in recent literature, with the two current state-of-the-art methods, relative
to standard datasets such as Hopkins 155 [4], being Sparse Subspace Clustering
(SSC) [5] and Spectral Clustering of linear subspaces (SC) [6].

Other common methods in the literature are based on Spectral Curvature
Clustering (SCC) [7], penalised MAP estimation of mixtures of subspaces using
linear programming (LP) [8], Normalised Subspace Inclusion (NSI) [9], Non-
negative Matrix Factorisation (NNMF) [10], Multi-Stage unsupervised Learning (MSL) [11], Local Subspace Affinity (LSA), Connected Component Search

---

* This work has been supported by ELLITT, the strateigc area for ICI research funded
by the Swedish goverment.

R. Koch et al. (Eds.): ACCV 2010 Workshops, Part I, LNCS 6468, pp. 338–348, 2011.

(CCS) [12], unsupervised manifold clustering using LLE (LLMC) [13], Agglomerative Lossy Compression (ALC) [14], Generalised Principal Component Analysis
(GPCA) [15], or on RANdom SAmple Consensus (RANSAC) [4].

In this paper we describe a motion segmentation method for sparse point trajectories, which is based on the previous work on six point consistency (SPC) [16],
but with the additional novelties and improvements: (i) an alternative method
for estimating the vector **s** (Sec. 2.2), (ii) a new matching score (Sec. 2.3), and
(iii) a modified classification algorithm (Sec. 3).

2 Mathematical Background

Our proposed method uses the consistent motion in the image plane generated by
6 points located on a rigid 3D object. The mathematical foundation of this theory
was formulated by Quan [17] and later extended by other authors [18,19,20]. A
similar idea was presented in [21], and later used for motion segmentation in [16].
[21] shows that the consistency test can be formulated as a constraint directly
on the image coordinates of the 6 points and that, similarly to epipolar lines
emerging from the epipolar constraint, this 6-point constraint generates 6 lines
that each must intersect its corresponding point.

More formally, we consider a set of six 3D points, with homogeneous coordinates $\mathbf{x}_k$, projected onto an image according to the pinhole camera model:

$$\mathbf{y}_k \sim \mathbf{C\,T\,x}_k, \quad k = 1, \ldots, 6, \tag{1}$$

where $\mathbf{y}_k$ are the corresponding homogeneous image coordinates, $\mathbf{C}$ is the 3×4
camera matrix, and $\sim$ denotes equality up to a scalar multiplication. $\mathbf{T}$ is a
4×4 time dependent transformation matrix that rotates and translates the set
of 3D points from some reference configuration to the specific observation that
produces $\mathbf{y}_k$. This implies that also $\mathbf{y}_k$ is time dependent. The problem addressed
here is how we can determine if an observed set of image points $\mathbf{y}_k$ really is given
by (1) for a particular set of 3D points $\mathbf{x}_k$ but with $\mathbf{C}$ and $\mathbf{T}$ unknown.

In general, the homogeneous coordinates of the 3D points can be transformed
by a suitable 3D homography $\mathbf{H_x}$ to *canonical* homogeneous 3D coordinates
$\mathbf{x}'=\mathbf{H_x\,x}$, and similarly, for a particular observation of the image points we
can transform them to canonical homogeneous 2D coordinates $\mathbf{y}'_k=\mathbf{H_y\,y}_k$. The
canonical coordinates are given by:

$$(\mathbf{x}'_1\,\mathbf{x}'_2\,\mathbf{x}'_3\,\mathbf{x}'_4\,\mathbf{x}'_5\,\mathbf{x}'_6) \sim\sim \begin{pmatrix} 1\,0\,0\,0\,1\,X \\ 0\,1\,0\,0\,1\,Y \\ 0\,0\,1\,0\,1\,Z \\ 0\,0\,0\,1\,1\,T \end{pmatrix}, (\mathbf{y}'_1\,\mathbf{y}'_2\,\mathbf{y}'_3\,\mathbf{y}'_4\,\mathbf{y}'_5\,\mathbf{y}'_6) \sim\sim \begin{pmatrix} 1\,0\,0\,1\,u_5\,u_6 \\ 0\,1\,0\,1\,v_5\,v_6 \\ 0\,0\,1\,1\,w_5\,w_6 \end{pmatrix}.$$

Here $\sim\sim$ denotes equality up to an individual scalar multiplication on each col-
umn. $\mathbf{H_x}$ and $\mathbf{H_y}$ depend on the 3D points $\mathbf{x}_1,...,\mathbf{x}_5$ and on the image points
$\mathbf{y}_1,...,\mathbf{y}_4$, respectively, and after these transformation are made the relation between 3D points and image points is given by $\mathbf{y}'_k\sim\mathbf{H_y\,C\,T\,H_x^{-1}\,x}'_k$. One of the

main results in [17] is that from these transformed coordinates we can compute a set of five *relative invariants* of the image points, denoted i_k, and of the 3D points, denoted $\tilde{I}_k$, according to:

$$\mathbf{z} = \begin{pmatrix} i_1 \\ i_2 \\ i_3 \\ i_4 \\ i_5 \end{pmatrix} = \begin{pmatrix} w_6(u_5 - v_5) \\ v_6(w_5 - u_5) \\ u_5(v_6 - w_6) \\ u_6(v_5 - w_5) \\ v_5(w_6 - u_6) \end{pmatrix}, \qquad \mathbf{s} = \begin{pmatrix} \tilde{I}_1 \\ \tilde{I}_2 \\ \tilde{I}_3 \\ \tilde{I}_4 \\ \tilde{I}_5 \end{pmatrix} = \begin{pmatrix} XY - ZT \\ XZ - ZT \\ XT - ZT \\ YZ - ZT \\ YT - ZT \end{pmatrix} \qquad (2)$$

such that they satisfy the constraint $\mathbf{z} \cdot \mathbf{s} = i_1\,\tilde{I}_1 + i_2\,\tilde{I}_2 + i_3\,\tilde{I}_3 + i_4\,\tilde{I}_4 + i_5\,\tilde{I}_5 = 0$.

To realize what this means, we notice that this constraint includes scalars derived from the reference 3D coordinates $\mathbf{x}_k$ (before they are transformed) and observed image points $\mathbf{y}_k$ (after the transformation $\mathbf{T}$ is made), but neither $\mathbf{C}$ nor $\mathbf{T}$ are explicitly included. Therefore, the constraint is satisfied regardless of how we transform the 3D points (or move the camera), as long as they are all transformed by the same $\mathbf{T}$. As long as the observed image coordinates are consistent with (1), the corresponding relative image invariants $\mathbf{z}$ must satisfy the constraint for a fixed $\mathbf{s}$ computed from the 3D reference points. The canonical transformations $\mathbf{H_x}$ and $\mathbf{H_y}$ can conveniently be included into the unknowns $\mathbf{C}$ and $\mathbf{T}$. In short, the above constraint is necessary but not sufficient for the matching between the observed image points and the 3D reference points.

2.1   The 6-Point Matching Constraint

The matching constraint is expressed in terms of the relative invariants $\mathbf{z}$ and $\mathbf{s}$ that have been derived by transforming image and 3D coordinates. In particular, this means that it cannot be applied directly onto the image coordinates, similar to the epipolar constraint. The transformation $\mathbf{H_y}$ is *not* a linear transformation on the homogeneous image coordinates since it also depends on these coordinates (see the Appendix of [17]). If however, we make an explicit derivation of how $\mathbf{z}$ depends on the 6 image points, it turns out that it has a relatively simply and also useful form:

$$\mathbf{z} = \alpha \begin{pmatrix} D_{126}D_{354} \\ D_{136}D_{245} \\ D_{146}D_{253} \\ D_{145}D_{263} \\ D_{135}D_{246} \end{pmatrix}, \qquad \alpha = \frac{D_{123}}{D_{124}D_{234}D_{314}}, \qquad (3)$$

$$D_{ijk} = (\mathbf{y}_i \times \mathbf{y}_j) \cdot \mathbf{y}_k = \det\left(\mathbf{y}_i\ \mathbf{y}_j\ \mathbf{y}_k\right).$$

Since $\mathbf{z}$ can be represented as a projective element, the scalar α can be omitted in the computation of $\mathbf{z}$. An important feature of this formulation is that each element of $\mathbf{z}$ is computed as a multi-linear expression in the 6 image coordinates. This can be seen from the fact that each point appears exactly once in the computations of the two determinants in each element of $\mathbf{z}$.

This formulation of $\mathbf{z}$ allows us to rewrite the constraint as $\mathbf{z} \cdot \mathbf{s} = l_1 \cdot \mathbf{y}_1 = 0$ with

$$l_1 = l_{26}D_{354}\tilde{I}_1 + l_{36}D_{245}\tilde{I}_2 + l_{46}D_{253}\tilde{I}_3 + l_{45}D_{263}\tilde{I}_4 + l_{35}D_{246}\tilde{I}_5 \qquad (4)$$

where $l_{ij}=\mathbf{y}_i\times\mathbf{y}_j$. $\mathbf{l}_1$ depends on the five image points $\mathbf{y}_2,...,\mathbf{y}_6$ and on the elements of $\mathbf{s}$. A similar exercise can be made for the other five image points and in general we can write the matching constraint as $\mathbf{z}\cdot\mathbf{s}=\mathbf{l}_k\cdot\mathbf{y}_k=0$ where $\mathbf{l}_k$ depends on $\mathbf{s}$ and five image points: $\{\mathbf{y}_i, i\neq k\}$. With this description of the matching constraint it makes sense to interpret $\mathbf{l}_k$ as the dual homogeneous coordinates of a line in the image plane. To each of the 6 image points, $\mathbf{y}_k$, there is a corresponding line, $\mathbf{l}_k$, and the constraint is satisfied if any of the 6 lines intersects its corresponding image point. The existence of the lines allows us to quantify the matching constraint in terms of the Euclidean distance in the image between a point and its corresponding line. Assuming that $\mathbf{y}_k$ and $\mathbf{l}_k$ have been suitably normalized, their distance is given simply as

$$d_k = |\mathbf{y}_k \cdot \mathbf{l}_k| \qquad (5)$$

2.2   Estimation of s

$\mathbf{s}$ can be computed from (2), given that 3D positions are available, but it can also be estimated from observations of the 6 image points based on the constraint. For example, from only three observations of the 5-dimensional vector $\mathbf{z}$, $\mathbf{s}$ can be restricted to a 2-dimensional subspace of $\mathbb{R}^5$. From this subspace, $\mathbf{s}$ can be determined using the internal constraint [17]. This gives in general three solutions for $\mathbf{s}$, that satisfy the internal constraint and are unique except for degenerate cases. This approach was used in [16].

Alternatively, for $B \geq 4$ observations of $\mathbf{z}$ a simple linear method finds $\mathbf{s}$ as a total least squares solution of minimizing $\|\mathbf{Z}\,\mathbf{s}\|$ for $\|\mathbf{s}\|=1$, where $\mathbf{Z}$ is a $B \times 5$ matrix consisting of the observed vectors $\mathbf{z}$ in its rows. $\mathbf{z}$ is then given by the right singular vector of $\mathbf{Z}$ corresponding to the smallest singular value. This approach has the advantage of producing a single solution for $\mathbf{s}$ which, on the other hand, may not satisfy the internal constraint. However, this can be compensated for by including a large number of observations, B, in the estimation of $\mathbf{s}$. This is the estimation strategy we use in this paper and it works well, provided that there are enough images in each sequence.

2.3   Matching Score

In the case of motion segmentation we want to be able to consider a set of 6 points, estimate $\mathbf{s}$, and then see how well this $\mathbf{s}$ matches to the their trajectories. The matching between $\mathbf{s}$ and observations of the 6 points over time is measured as follows. For each observation (at time t) of the 6 points $\mathbf{y}_1(t),...,\mathbf{y}_6(t)$ we use $\mathbf{s}$ to compute the 6 corresponding lines, $\mathbf{l}_1(t),...,\mathbf{l}_6(t)$, and then compute the distances d_k from (5). Finally, we compute a matching score $\tilde{E}$ of the 6 point trajectories:

$$\tilde{E}(P_1,\ldots,P_6) = \underset{t}{\mathrm{median}}\left[d_1^2(t) + \ldots + d_6^2(t)\right]^{1/2}, \qquad (6)$$

where P_k denotes image point k, but without reference to a particular image position in a particular frame. The median operation is used here in order to effectively reduce the influence of possible outliers.

```
Create spatial clusters using k-means;
foreach point P_k do
    foreach cluster C_j do
        Select 6 points {P_k, P_2^j, ..., P_6^j}. ;
        Calculate score E(P_k, C_j) from (6). ;
    end
    Assign P_i to cluster with min(E(P_k, C_j)). ;
end
Reject inconsistent clusters. ;
Initial NBC merging. ;
Final refinement merging. ;
```

Algorithm 1. Motion segmentation pseudocode

Fig. 1. A K-means initialisation example on the left. On the centre the classification result before the merging, and the final merged results on the left.

3   A Motion Segmentation Algorithm

In this section we describe a simple yet effective algorithm that can be used for the segmentation of multiple moving rigid 3D objects in a scene. The input data is the number of motion segments and a set of N point trajectories over a set of images in an image sequence. Our approach includes: a *spatial initialisation* step for establishing the initial motion hypotheses (or *seed* clusters), from which the segmentation will evolve; a *classification* stage, whereby each tracked point P_k, is assigned to the appropriate motion cluster; and a *merging* step, that combines clusters based on their similarity, to form the final number of moving objects in the scene.

Initialisation: The first step is the generation of initial 6-point clusters, each representing a 3D motion hypothesis. For this we use spatial K-means clustering in the image domain (see Fig. 1). The initial clustering is carried out in an arbitrary frame from each sequence (usually the first or the last). We define a seed cluster $C_j=\{P_1^j, ..., P_I^j\}$ as the I points at minimum distance to each K-means center. From the subsequent computations it is required that $I \geq 5$, and we use $I = 6$.

Point classification: Following the initialisation step, we assign the remaining points to the appropriate seed cluster. For each of the unclassified points P_k and for each seed cluster C_j, we estimate **s** according to Sec. 2.2 and compute a point-to-cluster score E from (6) as $E(P_k, C_j) = \tilde{E}(P_k, P_2^j, ..., P_6^j)$. This gives

$M(N\text{-}6M)$ score calculations in total, and produces an $M \times (N\text{-}6M)$ matrix $\mathbf{A}=[a_{ik}]$, with column k referring to particular point P_k and element a_{ik} as the index of the cluster that has the i-th smallest score relative to P_k. We employ a "winner takes all" approach with P_k assigned to the cluster that produces the lowest score, i.e., to the cluster index a_{1k}. This implies that the clusters will grow during the classification step, however, it should be noted that the scores for a particular point are always computed relative to the seed clusters. Note also that there is no threshold associated with the actual classification stage. A typical classification result can be seen in Fig. 1. The growth of the clusters is independent of the order that the points are classified, so the latter may be considered in parallel, leading to a very efficient and fast implementation.

Cluster merging and rejection: This is the final stage of our method, and results in the generation of motion consistent clusters each associated with a unique moving object in the scene. This stage consists of a quick *cluster rejection* step; an *initial merging* step using redundant classification information; and a final merging or *refinement* step where intermediate clusters are combined using agglomerative clustering based on some similarity measure.

-*Cluster rejection:* Any clusters that contain very few points (e.g. ≤ 7) are indicative of seed initialisation between motion boundaries, and represent unique and erroneous motion hypotheses. Therefore, any such clusters are promptly removed and their points re-classified with the remaining clusters.

-*Initial merging:* A direct result of the classification in Sec. 3 is the matrix $\mathbf{A}$, where so far we have only used the top row in order to classify points. However, $\mathbf{A}$ provides also information on cluster similarity, which we can exploit to infer initial merge pairings. We call this "Next-Best Classification" (NBC) merging and we now look at the cluster with the second best score for each point, since it contains enough discriminative power to accurately merge clusters. NBC merging involves generating the zero-diagonal sparse symmetric $M \times M$ matrix $\mathbf{L}=[l_{ij}]$ that contains the merging similarity between the clusters. Its elements are defined as:

$$l_{ij} = \sum_{k=1}^{N-6M} \left[\frac{1(k,i,j)}{E(P_k,C_j)} + \frac{1(k,j,i)}{E(P_k,C_i)} \right], \qquad (7)$$

where the summation is made over the $N - 6M$ points not included in the seed clusters. $1(k,i,j)$ is an indicator function that takes the value 1 when $a_{1k}=i$ and $a_{2k}=j$ and 0 otherwise. In other words, this function is $=1$ iff P_k is assigned to cluster i and has cluster j as second best option.

The matrix $\mathbf{L}$ describes all the consistent pairings inferred by the NBC merging. However, since usually inconsistent clusters will generate non-zero entries in $\mathbf{L}$ we need to threshold out low response entries due to noise. Using a threshold τ we obtain the sparser *adjacency matrix* $\mathbf{L}^*$. From $\mathbf{L}^*$ we can then construct an undirected graph G which contains the intermediate clusters as disconnected sub-graphs. If $\mathbf{L}^*$ is insufficient to provide the final motion clusters, due to for example noisy data, then a final refinement step may be required. The result of the cluster rejection and initial merging steps is a set of $\tilde{M} \leq M$ clusters $\tilde{C}_1,...,\tilde{C}_{\tilde{M}}$.

Table 1. 2 motion results

	GPCA	LSA	RANSAC	MSL	ALC	SSC	SCC	SPC	SC	LP	NNMF	NSI	LLMC	CCS	MSPC
Checkerboard: 78 sequences															
Mean:	6.09	2.57	6.52	4.46	1.55	1.12	1.77	4.49	0.85	3.21	-	3.75	4.37	16.37	**0.41**
Median:	1.03	0.27	1.75	0.00	0.29	0.00	0.00	3.69	0.00	0.11	-	-	0.00	10.62	0.00
Traffic: 31 sequences															
Mean:	1.41	5.43	2.55	2.23	1.59	**0.02**	0.63	0.22	0.90	0.33	0.1-	1.69	0.84	5.27	0.09
Median:	0.00	1.48	0.21	0.00	1.17	0.00	0.14	0.00	0.00	0.00	0.-	-	0.00	0.00	0.00
Articulated: 11 sequences															
Mean:	2.88	4.10	7.25	7.23	10.70	**0.62**	4.02	2.18	1.71	4.06	10.-	8.05	6.16	17.58	0.95
Median:	0.00	1.22	2.64	0.00	0.95	0.00	2.13	0.00	0.00	0.00	2.6-	-	1.37	7.07	0.00
All: 120 sequences															
Mean:	4.59	3.45	5.56	4.14	2.40	0.82	1.68	3.18	0.94	2.20	-	-	3.62	12.16	**0.37**
Median:	0.38	0.59	1.18	0.00	0.43	0.00	0.07	1.08	0.00	0.00	-	-	0.00	0.00	0.00

-*Refinement merging:* The last step involves the merging of the intermediate clusters, (resulting from the NBC merging), into the final clusters each representing a distinct motion hypothesis. This is achieved by pairwise agglomerative clustering and a maximum similarity measure between clusters. Assume that we wish to merge two clusters, say $\tilde{C}_1$ and $\tilde{C}_2$. We can generate K 6-point mixture clusters $\tilde{C}'$ by randomly selecting 3 points each from $\tilde{C}_1$ and $\tilde{C}_2$. If $\tilde{C}_1$ and $\tilde{C}_2$ belong to the same motion-consistent object and there is little noise present, we expect the scores $\tilde{E}$ calculated for each selection of C' to be grouped near zero, with little variation and few outliers. Conversely, if $\tilde{C}_1$ and $\tilde{C}_2$ come from different objects, $\tilde{E}$ should exhibit a larger dispersion and be grouped further away from zero. Instead of defining the similarity based on location and dispersion of sample statistics, we fit a parametric model to the sample data (using Maximum Likelihood Estimation) and compute the statistics from the model parameters. This allows for a much smaller number of samples and a more accurate estimate than what can be obtained from sample statistics (e.g. mean and variance). Given therefore that the scores in (6) should generally group around a median value with a few extremal outliers and assuming that the distances d_k in (5) are i.i.d., then the score distribution may be well approximated by a Generalised Extreme Value (GEV) distribution [?]. A robust indication of average location in a data sample with outliers is the mode, which for the GEV model can be computed by:

$$\tilde{m} = \mu + \sigma \left[(1 + \xi)^{-\xi} - 1 \right] / \xi \quad \text{for } \xi \neq 0, \tag{8}$$

where μ, σ and ξ are the location, scale and shape parameters respectively recovered by the MLE. Using this as a similarity metric we can merge two clusters when (8) is small or reject them when it is large. The clustering proceeds until we reach the pre-defined number of motions in the scene. The overall method is included in pseudocode in Algorithm 1.

Table 2. 3 motion results

	GPCA	LSA	RANSAC	MSL	ALC	SSC	SCC	SPC	SC	LP	NNMF	NSI	LLMC	CCS	MSPC
Checkerboard: 26 sequences															
Mean:	31.95	5.80	25.78	10.38	5.20	2.97	6.23	10.71	2.15	8.34	-	2.92	10.70	28.63	**1.43**
Median:	32.93	1.77	26.01	4.61	0.67	0.27	1.70	9.61	0.47	5.35	-	-	9.21	33.21	1.25
Traffic: 7 sequences															
Mean:	19.83	25.07	12.83	1.80	7.75	0.58	1.11	0.73	1.35	2.34	**0.1-**	1.67	2.91	3.02	0.71
Median:	19.55	23.79	11.45	0.00	0.49	0.00	1.40	0.73	0.19	0.19	0.-	-	0.00	0.18	0.36
Articulated: 2 sequences															
Mean:	16.85	7.25	21.38	2.71	21.08	**1.42**	5.41	6.91	4.26	8.51	15.-	6.38	5.60	44.89	2.13
Median:	28.66	7.25	21.38	2.71	21.08	0.00	5.41	6.91	4.26	8.51	15.-	-	5.60	44.89	2.13
All: 35 sequences															
Mean:	28.66	9.73	22.94	8.23	6.69	2.45	5.16	8.49	2.11	7.66	-	-	8.85	26.18	**1.32**
Median:	28.26	2.33	22.03	1.76	0.67	0.20	1.58	8.36	0.37	5.60	-	-	3.19	31.74	1.17

4 Experimental Results

We have carried out experiments on real image sequences from the Hopkins 155 database [4]. It includes motion sequences of 2 and 3 objects, of various degrees of classification difficulty and is corrupted by tracking noise, but without any missing entries or outliers. Typicall parameter settings for these experiments were: M=10-40 K-means clusters at the first or last frame of the sequence, reject clusters of ≤ 7 points, and K=50-100 mixture samples for the final merge (where necessary). Our results for 2 and 3 motions and the whole database are presented and compared with other state-of-the-art and baseline methods in Tables 1–3.

Our approach (Multiple Six Point Consistency - MSPC) outperforms every other method in the literature overall, in 2 and 3 motions and for all sequences combined. We achieve an overall classification error of 0.37% for two motions, less than 1/2 than the best reported result (SSC); an overall error of 1.32% for three motions, about 2/3 of the best reported result (SC); and an overall error of 0.59% for the whole database, less than 1/2 than the best reported result (SC). We also come first for the checkerboard sequences constituting the majority of the data, with almost 1/2 the classification errors reported by the SC method. For the articulated and traffic sequences (which are problematic for most methods) we perform well, coming a very close second to the best performing SSC or NNMF.

From the cumulative distributions in Fig. 2 we see that our method outperforms all others (where available) with only the SSC being slightly better (between 0.5-1% error) for 20-30% of the sequences. However, SSC soon degrades quite rapidly for the remaining 5-20% of the data with an error differential between 15-35% relative to MSPC. Furthermore, our method degrades gracefully from 2 to 3 motions as we do not have misclassification errors greater than 5% for any of the sequences, unlike SSC which produces a few errors between 10-20% and 40-50%. This is better illustrated in the histograms in Fig. 3.

Table 3. All motion results (italics are approximated from Tables 1 and 2)

	GPCA	LSA	RANSAC	MSL	ALC	SSC	SCC	SPC	SC	LP	NNMF	NSI	LLMC	CCS	MSPC
Checkerboard: 104 sequences															
Mean:	*12.55*	*3.37*	*11.33*	*5.94*	2.47	*1.58*	2.88	6.05	1.17	*4.49*	-	*3.54*	5.95	*19.43*	**0.66**
Median:	-	-	-	-	0.31	-	-	5.27	0.00	-	-	-	-	-	0.25
Traffic: 38 sequences															
Mean:	*4.80*	*9.04*	*4.44*	*2.15*	2.77	*0.12*	0.71	0.31	0.98	*0.70*	*0.1-*	*1.68*	1.22	*4.85*	0.20
Median:	-	-	-	-	1.10	-	-	0.00	0.00	-	-	-	-	-	0.00
Articulated: 13 sequences															
Mean:	*5.02*	*4.58*	*9.42*	*6.53*	13.71	*0.74*	*4.23*	2.91	2.10	*4.74*	10.76	7.79	6.07	*21.78*	1.13
Median:	-	-	-	-	3.46	-	-	0.00	0.00	-	-	-	-	-	0.00
All: 155 sequences															
Mean:	10.34	4.94	9.76	5.03	3.56	1.24	*2.46*	4.38	1.20	*3.43*	-	-	*4.8*	*15.32*	**0.59**
Median:	2.54	0.90	3.21	0.00	0.50	0.00	-	1.95	0.00	-	-	-	-	-	0.00

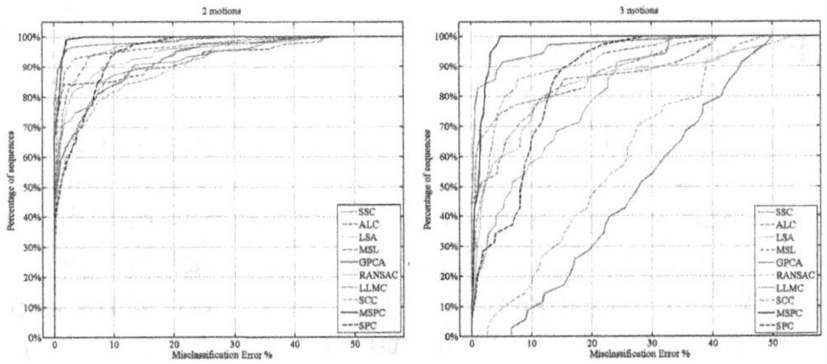

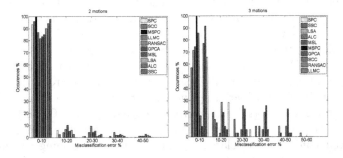

Fig. 2. Cumulative distributions of the errors per sequence for two and three motions

Fig. 3. Histograms of the errors per sequence for two and three motions

5 Conclusion

We have presented a method for segmenting moving objects using the geometry
of 6 points to infer motion consistency. Our evaluations on the Hopkins 155
database have shown superior results than current state-of-the-art methods, both

in terms of overall performance and in terms of maximum errors. The method finds initial cluster seeds in the spatial domain, and then classifies points as belonging to the cluster that minimizes a motion consistency score. The score is based on a geometric matching error measured in the image, implicitly describing how consistent the motion trajectories of 6 points are relative to a rigid 3D motion. Finally, the resulting clusters are merged by agglomerative clustering using a similarity criterion.

References

1. Black, M.J., Jepson, A.D.: Estimating Optical Flow in Segmentated Images Using Variable-Order Parametric Models With Local Deformations. PAMI 18, 972–986 (1996)
2. Shi, J., Malik, J.: Normalized cuts and image segmentation. PAMI 22, 888–905 (2000)
3. Tomasi, C., Kanade, T.: Shape from motion from image streams under orthography: A factorization method. IJCV 9, 137–154 (1992)
4. Tron, P., Vidal, R.: A Benchmark for the Comparison of 3-D Motion Segmentation Algorithms. In: CVPR (2007)
5. Elhamifar, E., Vidal, R.: Sparse Subspace Clustering. In: CVPR (2009)
6. Lauer, F., Schnorr, C.: Spectral clustering of linear subspaces for motion segmentation. In: ICCV (2009)
7. Chen, G., Lerman, G.: Motion Segmentation by SCC on the Hopkins 155 Database. In: ICCV (2009)
8. Hu, H., Gu, Q., Deng, L., Zhou, J.: Multiframe motion segmentation via penalized map estimation and linear programming. In: BMVC (2009)
9. da Silva, N.M.P., Costeira, J.: The normalized subspace inclusion: Robust clustering of motion subspaces. In: ICCV (2009)
10. Cheriyadat, A.M., Radke, R.J.: Non-negative matrix factorization of partial track data for motion segmentation. In: ICCV (2009)
11. Sugaya, Y., Kanatani, K.: Geometric Structure of Degeneracy for Multi-body Motion Segentation. In: SMVC (2004)
12. Roweis, S., Saul, L.: Think globally, fit locally: unsupervised learning of low dimensional manifolds. J. Mach. Learn. Res. 4, 119–155 (2003)
13. Goh, A., Vidal, R.: Segmenting motions of different types by unsupervised manifold clustering. In: CVPR, pp. 1–6 (2007)
14. Rao, S.R., Tron, R., Vidal, E., Ma, Y.: Motion Segmentation via Robust Subspace Separation in the Presence of Outlying, Incomplete, or Corrupted Trajectories. In: CVPR (2008)
15. Vidal, R., Tron, R., Hartley, R.: Multiframe Motion Segmentation with Missing Data Using PowerFactorization and GPCA. IJCV 79, 85–105 (2008)
16. Nordberg, K., Zografos, V.: Multibody motion segmentation using the geometry of 6 points in 2d images. In: ICPR (2010)
17. Quan, L.: Invariants of Six Points and Projective Reconstruction From Three Uncalibrated Images. PAMI 17, 34–46 (1996)
18. Carlsson, S.: Duality of Reconstruction and Positioning from Projective Views. In: Workshop on Representations of Visual Scenes (1995)
19. Weinshall, D., Werman, M., Shashua, A.: Duality of multi-point and multi-frame Geometry: Fundamental Shape Matrices and Tensors. In: ECCV (1996)

20. Torr, P.H.S., Zisserman, A.: Robust parameterization and computation of the tri-
    focal tensor. IVC 15, 591–605 (1997)
21. Nordberg, K.: Single-view matching constraints. In: Bebis, G., Boyle, R., Parvin,
    B., Koracin, D., Paragios, N., Tanveer, S.-M., Ju, T., Liu, Z., Coquillart, S., Cruz-
    Neira, C., Müller, T., Malzbender, T. (eds.) ISVC 2007, Part II. LNCS, vol. 4842,
    pp. 397–406. Springer, Heidelberg (2007)
22. Leadbetter, M.R., Lindgreen, G., Rootzn, H.: Extremes and related properties of
    random sequences and processes. New York (1983)

Systematic Evaluation of Spatio-Temporal Features on Comparative Video Challenges

Julian Stöttinger[1,3], Bogdan Tudor Goras[2], Thomas Pöntiz[3],
Allan Hanbury[4], Nicu Sebe[5], and Theo Gevers[6]

[1] CVL, Institute for Computer-Aided automation, TU Vienna
[2] Faculty of Electronics, Telecommuniction and Informatics, Tech. University of Iasi
[3] CogVis Ltd., Vienna
[4] IR Facility, Vienna
[5] Dept. of Information Eng. and Computer Science, University of Trento
[6] Faculty of Science, University of Amsterdam

Abstract. In the last decade, we observed a great interest in evaluation of local visual features in the domain of images. The aim is to provide researchers guidance when selecting the best approaches for new applications and data-sets. Most of the state-of-the-art features have been extended to the temporal domain to allow for video retrieval and categorization using similar techniques to those used for images. However, there is no comprehensive evaluation of these. We provide the first comparative evaluation based on isolated and well defined alterations of video data. We select the three most promising approaches, namely the Harris3D, Hessian3D, and Gabor detectors and the HOG/HOF, SURF3D, and HOG3D descriptors. For the evaluation of the detectors, we measure their repeatability on the challenges treating the videos as 3D volumes. To evaluate the robustness of spatio-temporal descriptors, we propose a principled classification pipeline where the increasingly altered videos build a set of queries. This allows for an in-depth analysis of local detectors and descriptors and their combinations.

1 Introduction

The bag-of-words approach, has been successfully adapted to the use of visual vocabularies describing images [1]. One central question for this approach is the choice of the right visual features. For set of local features the aim is to describe visual data successfully in a discriminative and robust way. Additionally, the data to be processed should be reduced as much as possible and should lead to a robust representation of the video. Video features based on local 3D patches are a popular representation for videos in tasks in retrieval, recognition and categorization (e.g. [2,3,4,5]). The most promising approaches for spatio-temporal features are corner detectors [6], blob detectors [7], periodic spatio-temporal features [8], volumetric features [9], and spatio-temporal regions of high entropy [10].

Recent work [11] points out that throughout the literature many experiments are not comparable. As such, the justification of specific properties of detectors and descriptors advocated in the literature is often insufficient. For example, results are frequently presented for different data-sets such as the KTH data-set [8,12,13,4,5,7,14], the Weizmann data-set [15] or the aerobic actions data-set [10]. Nevertheless, in that evaluation

R. Koch et al. (Eds.): ACCV 2010 Workshops, Part I, LNCS 6468, pp. 349–358, 2011.
© Springer-Verlag Berlin Heidelberg 2011

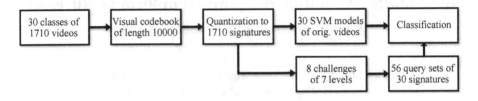

Fig. 1. Experimental setup to test the description's robustness against visual alterations

paper, combinations of detectors and descriptors are only measured on their final classification accuracy on the mentioned data-sets. A principled evaluation of every step of a matching framework, as is successfully done in "2D" images (e.g. [16]), is missing for "3D" video matching so far.

Therefore, we propose a new way for the evaluation of video retrieval approaches: We divide the evaluation of detectors and descriptors into two independent tasks. For detection, we use a repeatability measurement in 3D similar to [7]. For the descriptions we propose a pipeline to identify the robustness of local spatio-temporal descriptions in a principled way. These two tasks are measured by their performance under alterations of the visual input data. Therefore, we use a publicly available dedicated on-line data-set[1] providing 30 classes of videos [17]. Every video undergoes 8 types of transformations denoted as *challenges*. Each challenge is applied at 7 levels of increasing impact on the video leading to 1710 videos in total (compare Fig. 1). We use the original videos as ground-truth and observe to what extent the features change under the challenges. Example frames can be found in Fig. 2.

(a) HD movie   (b)   surveillance   (c) TV show
                      video

Fig. 2. Example videos and example transformations

We follow [11] and use the best performing approaches Harris3D, Hessian3D and the Gabor detector and HOG/HOF, SURF3D (also referred to as *extended SURF*) and HOG3D for our evaluation on videos. We use the same parameters and the same implementations.

The paper is organized as follows. The chosen features are described in more detail in Section 2. The experimental setup is described in Section 3. Results are given in Section 4. Section 5 gives a critical discussion and conclusions.

---

[1] www.feeval.org

2   Spatio-Temporal Features

An extension of the Harris corner detector [18] is the **Harris3D** detector [6]. The authors compute a spatio-temporal second-moment structure tensor at each video point using independent spatial and temporal scale values σ, τ, a separable Gaussian smoothing function G, and space-time gradients L. Extending the scale space to the temporal domain, we add the temporal variance τ^2 to get $L_{\mathbf{x},\sigma^2,\tau^2} = G_{\mathbf{x},\sigma^2,\tau^2} * f_{\mathbf{x}}^t$ and use the image data of the corresponding video frame f^t. The spatio-temporal Gaussian kernel is defined as

$$G_{\mathbf{x},\sigma^2,\tau^2} = \frac{1}{2\pi\sigma^4\tau^2}e^{-\frac{x^2+y^2}{2\sigma^2}-\frac{t^2}{2\tau^2}} \tag{1}$$

It is separable and thus can be calculated for each dimension on its own and in parallel. This extension gives then the structure tensor M for every location and scale. The final locations are extracted by applying $H = \det(M) - k \cdot \operatorname{trace}^2(M)$ and extracting the positive maxima of the corner function H. Points are extracted at multiple scales based on a regular sampling of the scale parameters s, t as suggested by the authors. We use the original implementation[3] and its settings $k = 0.0005$, $s^2 = 4, 8, 16, 32, 64, 128$, $t^2 = 2, 4$ with a detection threshold of 10^{-9}.

The **Hessian3D** detector [7] is the spatio-temporal extension of the Hessian blob detector [19]. The saliency of a location is given by the determinant of the 3D Hessian matrix. For efficiency, box-filter operations are applied on an integral video structure on multiple scales. Each octave is divided into 5 scales, with a ratio between subsequent scales in the range [1.2; 1.5] for the inner 3 scales. A non-maximum suppression algorithm selects the common extrema over space, time and scales: (x,y, t,s, τ). It is defined by the structure tensor Γ

$$\Gamma = \begin{pmatrix} L_x^2 & L_xL_y & L_xL_t \\ L_xL_y & L_y^2 & L_yL_t \\ L_xL_t & L_yL_t & L_t^2 \end{pmatrix} \tag{2}$$

where the *strength* S of an interest point is given by its tensor determinant $S = |det(\Gamma)|$. We use the authors' implementation[4] with the suggested parameters.

The **Gabor** detector is a set of spatial Gaussian convolutions and temporal Gabor filters [8]. The Gabor filters give a local measurement focusing not only on local changes in the temporal domain, but prioritize repeated events of a fixed frequency. The function gives $R_{\mathbf{x}t\sigma\tau\omega} = (f_{\mathbf{x}\sigma}^t * G_{\mathbf{x}\sigma} * H_{t\tau\omega}^{ev})^2 + (f_{\mathbf{x}\sigma}^t * G_{\mathbf{x}\sigma} * H_{t\tau\omega}^{od})^2$ where the 2D Gaussian smoothing is only applied in the spatial domain, whereas the two filters H^{ev} and H^{ov} are applied in the temporal domain only. H^{ev} and H^{ov} are the quadrature pair of 1D Gabor filters. The set of functions is available on-line as a toolbox[2]. As suggested and used in previous evaluations, we chose $\sigma = 3$ and $\tau = 4$.

To describe the detected patches by local motion and appearance, [4] compute histograms of spatial gradients and optical flow accumulated in space-time neighborhoods of detected interest points referred to as **HOG/HOF**. HOG results in a descriptor of length 72, HOF in a descriptor of length 90. For proper performance they are simply

---

[2] vision.ucsd.edu/~pdollar/toolbox/doc/index.html

concatenated. The descriptor size is defined by $D_x(\sigma) = D_y(\sigma) = 18\sigma$, $D_t(\tau) = 8\tau$. The approach is inspired by the SIFT descriptor. In the experiment, the grid parameters $n_x, n_y = 3$, $n_t = 2$ as suggested in [4]. The binaries are available online[3].

Willems et al. [7] proposed the **SURF3D** (ESURF) descriptor which extends the image SURF descriptor to videos. An image patch is represented by a 288 dimensional vector of weighted sums of uniformly sampled responses of Haar-wavelets. The binaries are also available[4]. 3D patches are divided into $n_x \times n_y \times n_t$ cells. The size of the 3D patch is given by $D_x(\sigma) = D_y(\sigma) = 3\sigma$, $D_t(\tau) = 3\tau$. For the feature descriptor, each cell is represented by a vector of weighted sums $v = (\sum d_x, \sum d_y, \sum d_t)$ of uniformly sampled responses of the Haar-wavelets d_x, d_y, d_t along the three axes.

For the third descriptor in the evaluation we use the **HOG3D** [13]. This is based on histograms of 3D gradient orientations efficiently computed using an integral video representation. It leads to a descriptor of length 960.

3   Experimental Setup

In this section, the experimental set-up used throughout the evaluation is described . Section 3.1 presents an overview of the evaluation data-set used. In Section 3.2, the methodology for the detector evaluation is given. The pipeline and the parameters of the classification task for the descriptor evaluation is described in detail in Section 3.3.

3.1   Video Data-Set and Features

Our experiments aim to quantify the robustness of the state-of-the-art spatio-temporal features described in the previous section. We challenge the robustness of these approaches on the FeEval data-set [17][1], which consists of 1710 videos of about 20 seconds each. Starting with 30 short clips from HDTV shows, Hollywood movies of a *full HD* resolution of 1920×1080, and surveillance videos, the full FeEval dataset is created as follows: (1) Every video undergoes 8 types of systematic alterations denoted as challenges. The challenges are noise, increasing lightness, decreasing lightness (darkness), median filtering, compression, scale and rotation, and reduction in frames per second. (2) Each challenge is applied at 7 levels of increasing impact, and encoded by a parameter (see Fig. 2). The parameters and the challenge abbreviations used throughout the experiments are given in Tbl. 1. This leads to about 34 Gigabytes (GB) of H.264 compressed video material.

3.2   Detector Evaluation

To evaluate the robustness of the three detectors Harris3D, Hessian3D, and Gabor, we measure their robustness or *repeatability* for each altered video with respect to its corresponding original video. Each of the 30 original video is regarded as a *boolean* 3D volume V_{o_i}, $i = \overline{1,30}$, sized according to the frame resolution and the total number of frames. $V_{o_i} = 1$ if a voxel is being detected by a feature or 0 otherwise. Every of the

---

[3] www.irisa.fr/vista/Equipe/People/Laptev/download.html#stip

[4] homes.psat.kuleuven.be/~gwillems/research/Hes-STIP/

Table 1. Video transformations for each of the 30 videos

Transformation	Abbreviation	Range
Gaussian blur: σ in pixels	blur	3 - 21
H.264 compression	compr	60 - 0
Noise in %	noise	5 - 35
Median Filter: σ in pixels	median	2 - 8
Increasing lightness in %	lighten	+30% - +90%
Decreasing lightness in %	darken	-30% - -90%
Frames per Second	fps	20 - 3
Scale + Rotation in degrees	scalerot	90% & $10°$ - 30% & $70°$

m detected feature $\xi_{c,1..m}$ in an altered video is defining a cuboid in space. Per repeatability test, we map the cuboid $\xi_{c,j}$ to V_o to get its position and expanion in the original video's volume V_o denoted as $\xi'_{c,j}$. This is done by applying its homography matrix Ω to $V_o \leftarrow \Omega * V_c$ For the challenge of scale and rotation, we use the provided "2D" matrices defined by the parameters given in Tbl. 1, as the alteration is per frame only and does not affect the temporal configuration. For the challenge of decreasing frames per second, we regard it as a simple scaling in the temporal direction and apply it on the t expansion of V_c only. Overlap ϱ of feature j is then defined by

$$\varrho = \frac{V_o \cap \xi'_{c,j}}{v(\xi_{t,i})} \tag{3}$$

where $v(\xi_{t,i})$ is the volume of the transformed feature's cuboid. The final repeatability score of a video is defined by the number of matched features divided by the total number of features in the challenge video.

3.3 Descriptor Evaluation

We want to test the ability of state-of-the art spatio-temporal descriptors to what extent they maintain their robustness under alteration of their input videos. We aim to test their performance in a large scale video classification experiment where the training data consists of 30 original videos forming 30 classes of challenges. For the three descriptors HOG/HOF, SURF3D and HOG3D and the combination with the detectors we carry out the following set-up:

We form a visual codebook of 10000 words by clustering all the features of the data-set with the *kshift* [20][5] algorithm. In contrast to many other clustering implementations, the data-set can be larger than the memory. For every cluster center, it is only necessary to have the *next* feature in the memory, not the whole data-set. It is feasible to cluster 45 GB of 960 dimensional features within 20 hours using 2 X5560@2.8GHz processors (4 cores each). A video's signature is built by quantizing its features to the codebook by the cluster center with the nearest Euclidean distance. For the training set, we use the 30 original videos with their normalized signatures of a length of 10000 each as ground truth classes. For every class, we train a linear one-against-all SVM model equally weighting every class. For this setup, the model is similar to a nearest neighbor

---
[5] www.cogvis.at

classification. We are using the well known LibSVM library[6] with default parameters. For the 8 challenges with 7 levels, we build 56 test sets of equal size to be evaluated. The experimental question is then until which alteration the description is still able to discriminate against the other videos and under which circumstances it fails. When an altered video is successfully classified as its original video, the description is regarded as robust to the alteration. In this context, the classification performance according to the alterations gives then the descriptor robustness in the challenge.

4   Results

Starting with the repeatability experiments in the following Section 4.1 we are able to evaluate the robustness of the detections of state-of-the-art spatio-temporal features. In Section 4.2 the three descriptors are evaluated in a classification experiment.

4.1   Detector Evaluation

Regarding the overall repeatability performance the Hessian3D detector outperforms the Harris3D detector, whereas the Gabor detector shows to be significantly less robust. The mean results on varying ϱ are given in Fig. 3. The single-scale Gabor detector is not much affected by the change of the overlap criterium, as the large number of small features tends to be matched almost perfectly or not at all. This is of course different for the multi-scale approaches Harris3D and Hessian3D, where different sizes of features are matched.

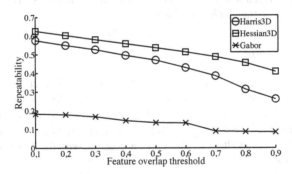

Fig. 3. Mean repeatability results for the whole data-set over varying overlap ϱ

Hessian3D has the best mean repeatability and performs best throughout the experiments. However, it provides a richer representation as its coverage is almost 10 times larger than Harris3D, thus making the probability for a geometrical match higher. Still, Harris3D performs comparably similar, which coincides closely to the evaluation of their 2D counterparts in [16] As we observe in Fig. 4(a), Harris3D and Hessian3D are almost equally robust to increasing blur. This also holds for increasing compression shown in Fig. 4(b). The two detectors are very robust to increasing compression,

---

[6] www.csie.ntu.edu.tw/~cjlin/libsvm

showing similar results on 2D images [16]. This is an important observation, since the spatio-temporal structure tensor has more degrees of freedom and a much bigger dataset than it has been done for 2D repeatability. In contrast to 2D detectors, the Harris3D and Hessian3D show to be very sensitive to change of lightness (see Fig. 4(e) and 4(f)). The number of features decreases rapidly with the decrease of contrast. This is the only challenge where the Gabor detector outperforms the other approaches in robustness at level 7. The decrease of frames per second (see Fig. 4(g)) can be seen as scaling in the temporal domain. As the approaches are not scale invariant, they perform worse than their 2D counterparts. Hessian3D regarding the most scales of the approaches evaluated remains rather stable until level 3, which is the reduction from 25fps to 13fps. Therefore the standard sampling rate of 2 for the Hessian3D approach can be easily set to 4 without a significant loss in performance, disregarding 50% of the data right away. For scale and rotation, Gabor and Harris perform poorly compared to the Hessian3D which is able to maintain a repeatability rate of 0,41 for a video scaled by a factor of 0.3 and rotated by 70 degrees. Harris3d and Gabor are very sensitive to noise, Hessian3D remains stable showing a repeatability of 0,62 with 35% of noise in the video. For increasing median filtering, Harris3D is equally robust as the Hessian3D.

Following these results, the following for noisy video data is proposed: Gaussian blur degrades the detections severely therefore it should not be used in pre-processing videos. Hessian3D on noise performs more robust than on blurred data. Gabor detections are neither reliable on noisy or blurred data. When using the Harris3D detector, it is recommended to use the median filter to remove the noise in advance.

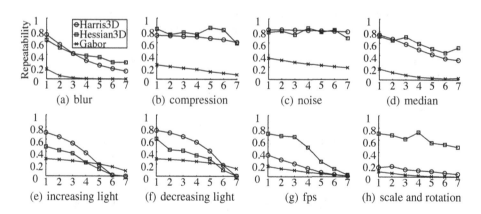

Fig. 4. Mean repeatability ($\varrho = 0.6$) of 30 videos per challenge. Legend is found in (a).

4.2 Descriptor Evaluation

Summary results are shown in Tbl. 2 Results per challenge are shown in Fig. 5. In Fig. 6 results of the experiments using the HOG3D descriptor are given. The combination of Harris3D and HOG3D outperforms other approaches.

As already argued in the previous section, Gaussian blur decreases the representation of the videos significantly. As seen in Fig. 5(a), the classification accuracy goes

Table 2. Overview experimental results descriptor evaluation

	Classification accuracy			Mean precision			Mean recall		
	Harris3D	Hessian3D	Gabor	Harris3D	Hessian3D	Gabor	Harris3D	Hessian3D	Gabor
HOG/HOF	23,57	-	-	19,40	-	-	23,57	-	-
SURF3D	-	39,52	-	-	40,46	-	-	44,80	-
HOG3D	**49,76**	37,96	34,75	**42,40**	38,80	28,15	**49,76**	42,20	35,30

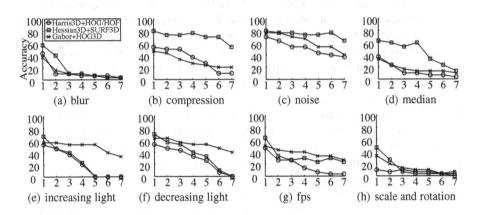

(a) blur   (b) compression   (c) noise   (d) median

(e) increasing light   (f) decreasing light   (g) fps   (h) scale and rotation

Fig. 5. Classification accuracy with increasing alterations of the query images with suggested descriptor and detector combinations. Legend is found in (a).

towards the prior probability of 3%. This is different for the HOG3D descriptor. For all detectors, there is a significant gain in classification performance, especially for the Harris3D+HOG3D raising to a mean accuracy of 54,76%.

Similar behavior is observed for change of lightness: For HOG/HOF and SURF3D, the classification accuracy goes down rapidly, whereas the HOG3D descriptor provides a stable description on data of varying contrast. Gabor+HOG3D outperforms these approaches (see Fig. 5(e) and 6(e)). When combining the detectors with HOG3D, we observe a correlation with the repeatability experiments of changing lightness. With a more stable descriptor, the more repeatable representation influences the classification performance. This does not hold for the fps challenge (see Fig. 5(g) and 6(g)). There is no correlation between detector robustness and classification performance. This suggests that none of the descriptors is scale invariant to a satisfying extent. We deduce that for performance reasons, detectors can be applied on a reduced data-set but the local description has to be performed on full resolution.

Descriptors revealed to be more robust to increasing noise than the local detectors. Worst performing Harris3D+HOG/HOF reaches a mean accuracy of 51,43%. Hessian3D + SURF3D remains almost stable throughout the challenge (see Fig. 5(c)). HOG3D shows to be more robust than HOG/HOF (see Fig. 6(c)), but decreases the performance for the Hessian3D. It is shown that SURF3D is more robust to noise than HOG3D in this context. Regarding noise reduction using the median filter (see Fig. 5(d) and 6(d)) performance decreases more than for the noise challenge. HOG/HOF and HOG3d are sensitive to the filtering, SURF3D performs best coherent to the repeatability rate of its detector. Increasing compression does not affect the description

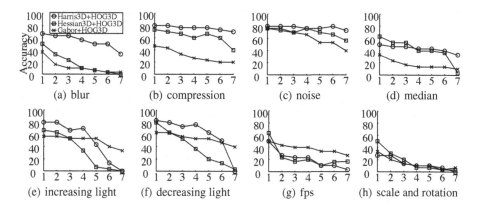

Fig. 6. Classification accuracy with increasing alterations of the query images with detectors and HOG3D descriptor. Legend is found in (a).

performance of the HOG3D and the SURF3D descriptor. Even strong JPEG artifacts are described in a stable and discriminative way (see Fig. 5(b) and 6(b)). For level 7 of the challenge, the data is compressed up to 10% of the original file size.

To sum up the evaluation, we interpret the results categorizing them to simple votes according to the challenges. '-' denotes sensitivity, '+' robustness to the challenge. '+/-' refers to undecided decision or room for improvements in the algorithmic details of the approach. Our final suggestions are given in Tbl. 3.

Table 3. Final suggestions based on the evaluation

	Detector Robustness			Descriptor Robustness		
	Harris3D	Hessian3D	Gabor	HOG/HOF	SURF3D	HOG3D
Gaussian blur	+/-	+/-	-	-	-	+/-
H.264 compression	+	+	-	-	+	+
Noise	-	+	-	+/-	+	+
Median Filter	+	+	-	-	+/-	+/-
Increasing lightness	+/-	+/-	+/-	-	-	+
Decreasing lightness	+/-	+/-	+/-	-	-	+
Frames per Second	-	+	-	+/-	+/-	+/-
Scale & Rotation	-	+	-	+/-	+/-	+/-

5 Conclusion

In this work, we perform the first principled evaluation of spatio-temporal features using comparative challenges inspired by prior evaluation of local 2D image features. For detector robustness, we experienced comparable results for spatio-temporal features with their image counterparts. Generally, it showed to be worse to reduce noise in input data than to let the features take care of it on their own. For change of lightness, both the Harris3D and the Hessian3D are more sensitive than their 2D counterparts. Description is most stable using the HOG3D descriptor, outperformed by the SURF3D descriptor in the challenges of compression, noise and median filtering. The high dimensionality of the HOG3D descriptor of 960 compared to 288 of the SURF3D descriptor is a drawback in terms of the complexity of all succeeding operations and should be considered when choosing the most appropriate descriptor.

References

1. Cula, O.G., Dana, K.J.: Compact representation of bidirectional texture functions. IEEE Computer Society Conference on Computer Vision and Pattern Recognition 1, 1041 (2001)
2. Duchenne, O., Laptev, I., Sivic, J., Bach, F., Ponce, J.: Automatic annotation of human actions in video. In: ICCV (2009)
3. Junejo, I., Dexter, E., Laptev, I., Pérez, P.: View-independent action recognition from temporal self-similarities. PAMI (2009)
4. Laptev, I., Marszalek, M., Schmid, C., Rozenfeld, B.: Learning realistic human actions from movies. In: CVPR, pp. 1–8 (2008)
5. Schüldt, C., Laptev, I., Caputo, B.: Recognizing human actions: a local SVM approach. In: ICPR (2004)
6. Laptev, I., Lindeberg, T.: Space-time interest points. In: ICCV (2003)
7. Willems, G., Tuytelaars, T., Gool, L.: An efficient dense and scale-invariant spatio-temporal interest point detector. In: Forsyth, D., Torr, P., Zisserman, A. (eds.) ECCV 2008, Part II. LNCS, vol. 5303, pp. 650–663. Springer, Heidelberg (2008)
8. Dollár, P., Rabaud, V., Cottrell, G., Belongie, S.: Behavior recognition via sparse spatio-temporal features. In: VS-PETS, pp. 65–72 (2005)
9. Ke, Q., Kanade, T.: Quasiconvex optimization for robust geometric reconstruction. In: ICCV, pp. 986–993 (2005)
10. Oikonomopoulos, A., Patras, I., Pantic, M.: Kernel-based recognition of human actions using spatiotemporal salient points. In: CVPR, p. 151 (2006)
11. Wang, H., Ullah, M., Kläser, A., Laptev, I., Schmid, C.: Evaluation of local spatio-temporal features for action recognition. In: BMVC (2009)
12. Jhuang, H., Serre, T., Wolf, L., Poggio, T.: A biologically inspired system for action recognition. In: ICCV, pp. 1–8 (2007)
13. Kläser, A., Marszałek, M., Schmid, C.: A spatio-temporal descriptor based on 3d-gradients. In: BMVC, pp. 995–1004 (2008)
14. Wong, S.F., Cipolla, R.: Extracting spatiotemporal interest points using global information. In: ICCV, pp. 1–8 (2007)
15. Gorelick, L., Blank, M., Shechtman, E., Irani, M., Basri, R.: Actions as space-time shapes. PAMI 29, 2247–2253 (2007)
16. Mikolajczyk, K., Tuytelaars, T., Schmid, C., Zisserman, A., Matas, J., Schaffalitzky, F., Kadir, T., Gool, L.V.: A comparison of affine region detectors. IJCV 65, 43–72 (2005)
17. Stöttinger, J., Zambanini, S., Khan, R., Hanbury, A.: Feeval - a dataset for evaluation of spatio-temporal local features. In: ICPR (2010)
18. Harris, C., Stephens, M.: A combined corner and edge detection. In: AVC, pp. 147–151 (1988)
19. Lindeberg, T.: Feature detection with automatic scale selection. IJCV 30, 79–116 (1998)
20. Pönitz, T., Donner, R., Stöttinger, J., Hanbury, A.: Efficient and distinct large scale bags of words. In: AAPR (2010)

Two-Probabilistic Latent Semantic Model for Image Annotation and Retrieval

Nattachai Watcharapinchai[1], Supavadee Aramvith[1], and Supakorn Siddhichai[2]

[1] Department of Electrical Engineering, Chulalongkorn University, Thailand
Nattachai.W@student.chula.ac.th, Supavadee.A@chula.ac.th
[2] National Electronics and Computer Technology Center (NECTEC),
Thailand Science Park, Thailand
yai@nectec.or.th

Abstract. A novel latent variable modeling technique for image annotation and retrieval is proposed. This model is useful for annotating the images with relevant semantic meanings as well as for retrieving images which satisfy the users query with specific text or image. The framework of two-step latent variable is proposed to support multi-functionality of the retrieval and annotation system. Furthermore, the existing and the proposed image annotation models are compared in terms of their annotating performance. Images from standard databases are used in the comparison in order to identify the best model for automatic image annotation, using precision-recall measurement. Local features, or visual words, of each image in the database are extracted using Scale-Invariant Feature Transform (SIFT) and clustering techniques. Each image is then represented by Bag-of-Features (BoF) which is a histogram of visual words. Semantic meanings can then be related to each BoF using latent variable for annotation purposes. Subsequently, for image retrieval, each image query is also related to semantic meanings. Finally, image retrieval results are obtained by matching semantic meanings of the query with those of the images in the database using a second latent variable.

1 Introduction

Due to the fact that the digital multimedia recording and storage devices become common. The number of digital multimedia, such as digital image and digital video, is also increased considerably. Therefore, to efficiently access the image/video collection requires a system to handle search and organization of this information. Such system is called image/video retrieval system. The ideal retrieval system should be designed to support intuitively search for the user, and requires minimal amount of human interaction and to be is applicable to large collections. In practical, the prevalent approach to image retrieval falls into two main categories: text-based and content-based image retrieval. In text-based image retrieval system, images are firstly annotated with text, and the traditional text retrieval techniques can be used to perform image retrieval. The main advantages of text-based retrieval are its simplicity and its conveniently direct adoption of mature textual information retrieval techniques. In addition, it

R. Koch et al. (Eds.): ACCV 2010 Workshops, Part I, LNCS 6468, pp. 359–369, 2011.

is easy to use text to express textual characteristic related to images. In general, there are two strategies to associate image with text. One strategy is to annotate image by human. These annotations provided by people usually are close to the actual semantic of images. However, this strategy suffers two drawbacks. First, it is very tedious and time-consuming to annotate image manually, especially when the size of image collection is huge. Although manual annotation involves a substantial amount of work, and often results in considerable cost, there is a system that utilizes a collaborative system approach called LabelMe [1] which takes advantage of its member as annotators. Secondly, these annotations are usually subjective because different people may give different descriptions to the same image. Another strategy automatically annotates image with terms, or words. The main advantage of this strategy is the complete automatic process of image annotation without human interference. Recently, automatic image annotation techniques are proposed to address the semantic gap problem. Automatic image annotation is the process by which a computer system automatically assigns keywords to a digital image. The primary purpose of a practical Content-Based Image Retrieval (CBIR) system is to discover images relating to a given concept in the absence of reliable metadata. In contrary to CBIR, annotations can facilitate image search through the use of semantic meaning such as text. This methodology assures the good performance of image retrieval that if the results of mapping between images and words are reasonable, text-based image retrieval system can be semantically more meaningful than search using CBIR. However, this technique is still in its infancy and is not sophisticated enough to extract perfect semantic concepts. Moreover, many experiments show that current image annotation techniques still have poor performance in the context of image retrieval because of the irrelevant keywords associated with images often lessen image retrieval performance.

2   Literature Reviews and Related Works

Different models and machine learning techniques are developed to learn about the correlation between low-level features and textual words from the examples of annotated image and then apply such correlation to predict words for new images. In this section we review some pioneer works about automatic image annotation which divided into two major categories namely, generative model and discriminative model. The basic idea of generative model [2,3,4,5,6,7] is to construct a model from joint probability of image features and words, and then use Bayes rule and marginalization of probability to estimate the conditional probability of words given by image features. But in the discriminative model, the model is directly being constructed the conditional probability of words given image features.

Co-occurrence Model [15]: is to count words and image features matrix, and use its matrix to predict annotated image words for images. In [8], Duygulu et al. proposed the improved co-occurrence model by utilizing machine translation models. In this method, it considers image annotation as a process of translation from visual feature to texts and collects the co-occurrence information by estimation of translation probability.

Relevance Model: some researchers used relevance language model which has been successfully applied to automatic image annotation. The essential idea is to firstly find annotated images which are similar between images and then use the words shared by the annotations of the similar images to annotate to an unlabeled image. There are two subcategories in relevance model, namely discrete variable, and continuous variable. In discrete variable, it is the basic idea of cross-media relevance model. This model is to improve the co-occurrence model described in [10], Jeon et al. assumes a one-to-one correspondence between blobs and words in images. Images are considered as a set of words and blogs, which are assuming independent. The conditional probability of word given a training image is estimated by the count of word in this image smoothed by the average count of this word in training set. These posterior distributions allow the estimation of the probability of a potential caption (set of words) and unseen blobs as an expectation over all training images. Multiple-Bernoulli relevance model [12] is based on Cross-media relevance model but it is different in the word distribution hypothesis. Cross-media relevance assumes that annotation words for any given image follow a multinomial distribution, while this model uses Bernoulli process to generate words. While the cross-media relevance model is to counting word in given training set that is discrete random variable techniques, but same authors considered blogs correspond to word which called Continuous-space relevance Model described in [11]. Annotation quality of the Cross-media relevance model is very sensitive to clustering errors, and depends on being able to *priori* select the right cluster granularity. Too many clusters will results in extreme sparse of the space, while too few will lead us to confuse different objects in the images. Continuous relevance model does not rely on clustering and consequently does not suffer from the granularity issues. Currently P. Huang et al [13] proposed combining three co-occurrence models including translation model, Cross-media relevance model and Multiple Bernoulli relevance model. They showed the comparison performance between individual model and combining models. The combining model gives the better performance for image annotation.

Latent Semantic Analysis Model: another way of capturing co-occurrence information is to introduce latent variables to associate visual features with words. Standard latent semantic analysis (LSA) and probabilistic latent semantic analysis (pLSA), are applied to automatic image annotation [2,3,4]. A significant step forward in this approach was proposed by Hofmann [2], who presented the probability LSA model, also known as the aspect model, as an alternative to LSI, which has used for text retrieval research. By this approach, Money et al [3,4] have extended the experiments to bigger collection of 8000 images and achieved encouraging results to using pLSA model. Some works extended pLSA model to annotation images. Blei and Jordan [5] extended the aspect model as the latent Dirichlet

Allocation (LDA) model and proposed a correlation LDA (CORR-LDA) model. This model assumes that a Dirichlet distribution can be used to generate a mixture of latent factors. This mixture of latent factors is then used to generate words and regions. Fei-Fei and Perona [6] modified LDA model and have extended the experiments to learn natural scene categories. Their algorithm provides a principle approach to learn relevant intermediate representation of scenes automatically and without supervision to what humans would to do. The aspect model approach is extended by Zhang et al [9] who proposed the aspect model as Gaussian mixture distribution. They assumed a latent aspect as the connection between the visual feature and the annotation words to explicitly exploit the synergy among the modalities. In [14], Pham et al. studied the effect of Latent Semantic Analysis on two different tasks namely image retrieval task and automatic annotation task. This result ensures that LSA model when combining to image retrieval system can improve automatic annotation image. Because the previous models are designed for an individual task, for instance, the co-occurrence model is useful for automatic annotation but it is not suitable for image retrieval task. While the latent semantic analysis is designed for image retrieval task but it is not suitable for automatic image annotation. So the model is not suitable for image retrieval. Therefore, to design a novel image retrieval system, the generative mode should be adopted as the expert system, i.e., a joint probability among low-level image features, semantic word feature vectors and image document terms. Our model can automatically annotate new images which will be added into that system and can also retrieve images by users query with text or image. So, in this work, we propose a new model, called the two latent aspects PLSA model, which can support multifunctionalities, including of image retrieval function and automatically image annotation for image retrieval system.

3   Two-Probabilistic Latent Analysis Model

In this section, we propose a novel model based on PLSA for image annotation and retrieval. We will call this new model, Two-Probabilistic Semantic Analysis model (Two-pLSA), because we use two hidden random variables, the first latent aspect is used for representing that the images on a corpus relate their word, and the second latent aspect is used for representing visual features of each image relating with their word. The graphical model is shown in Fig. 1. First, we define the following notations:

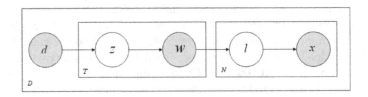

Fig. 1. Two-Probabilistic Latent Analysis Model (Two-pLSA Model)

- Images are represented by a observed random discrete variable D that can take the value $d_i, i \in \{1, ...D\}$, where is the number of document in training data set.
- Words on each image can be represented by observed random variable W that can take the value $w_j, j \in \{1, , T\}$ where T is the size of a word vocabulary including of several words that are used for labeling an image in image annotation task, and are used for searching images by texts in image retrieval task.
- Bag-of-Features, being basis unit in our approach, are presented by a observed visual random variable X that can take the value $x_n, n \in \{1, , N\}$, where N is the size of visual vocabulary, constructed by K-Mean algorithm, and are used for image annotation and retrieval.
- The first latent variables L to which is refered as a visual latent aspect can take the values $l_m, m \in \{1, , M\}$, where M is the number of visual aspects.
- The second hidden variable Z to which is refered as a word latent aspect can take the values $z_k, k \in \{1, , K\}$, where K is the number of word aspects.

Thus, the joint probability $P(d_i, w_j, x_n)$ of all observed d_i, w_j and x_n is given by the marginalization over all the possible values z_k and l_m: $P(d_i, w_j, x_n) = \sum_{k=1}^{K} \sum_{m=1}^{M} P(d_i, z_k, w_j, l_m, x_n)$. By the probabilistic graphical model in Fig. 1, the joint probability of all random variables including of observation and non-observation can expressed in $P(d_i, z_k, w_j, l_m, x_n) = P(d_i)P(w_j \mid z_k)P(z_k \mid d_i)P(x_n \mid l_m)P(l_m \mid w_n)$. Therefore, the joint probability distribution of all observed variable can rewrite as $P(d_i, w_j, x_n) = P(d_i) \sum_{k=1}^{K} P(w_j \mid z_k)P(z_k \mid d_i) \sum_{m=1}^{M} P(x_n \mid l_m)P(l_m \mid w_j)$. where, the first term descripes each image d_i as a mixture of word latent aspects, defined by the multinomial distribution $P(z \mid d_i)$. Each word latent aspect z_k is defined by the multinomial distribution $P(w \mid z_k)$ which gives the probability of each word w_j given by each word aspect z_k. Moreover, it also indicates that each w_n as a mixture of visual latent aspects, defined by the multinomial distribution $P(l \mid w_j)$, where each visual latent aspect is defined by the multinomial distribution $P(x \mid l_m)$ which gives the probability of each visual word x_n

3.1 Learning Parameters Using EM Algorithm

Our model consists of four conditional probabilities, $P(w_j \mid z_k)$, $P(z_k \mid d_i)$, $P(x_n \mid l_m)$, and $P(l_m \mid w_j)$ which are assumed as multinomial distribution. Their parameters are estimated by the Expectation Maximization Algorithm. For word vocabulary of T different words, $P(w \mid z)$ is a T-by-K table that stores the parameter of word latent aspects K being multinomial distribution. And the K-by-D table stores the parameters of the D multinomial distribution $P(z \mid d_i)$ that describes the training document d_i. Moreover, for visual vocabulary of N different visual words, $P(x \mid l)$ is a N-by-L table that stores the parameter of visual word latent aspects L, which still is multinomial distribution. On the contrary, the L-by-T table is relative between visual word and word, as it stored the parameter of T multinomial distribution $P(l \mid w_d)$ that describes the training words of word vocabulary. In order to learning these parameters, in this work, we use EM algorithm including of 2 steps: E-step complete posterior probabilities of two dimension latent aspects and M-step four parameters are updated based on expectation of the posterior probabilities of E-step. For E-step, the probability of two latent aspects depending on all observation can simply applies Bayes rule that can rewrite as

$$P(z_k, l_m \mid d_i, w_j, x_n) = \frac{P(w_j \mid z_k)P(z_k \mid d_i)P(x_n \mid l_m)P(l_m \mid w_n)}{\sum_{k=1}^{K} P(w_j \mid z_k)P(z_k \mid d_i)\sum_{m=1}^{M} P(x_n \mid l_m)P(l_m \mid w_j)} \tag{1}$$

For the M-step, we have to maximize the expected complete data log-likelihood $E(L^c)$ by Eq. (2).

$$\begin{aligned} E(L^c) = \sum_{k=1}^{K}\sum_{m=1}^{M}\sum_{i=1}^{D}\sum_{j=1}^{T}\sum_{n=1}^{N}(n(d_i, w_j, x_n)\ln(P(w_j \mid z_k)P(z_k \mid d_i))) \\ + \sum_{k=1}^{K}\sum_{m=1}^{M}\sum_{i=1}^{D}\sum_{j=1}^{T}\sum_{n=1}^{N}(n(d_i, w_j, x_n)\ln(P(x_n \mid l_m)P(l_m \mid w_j))) \end{aligned} \tag{2}$$

where $n(d_i, w_j, x_n)$ is the count of element x_n correspond to word w_j in document d_i. After applying the Lagrange multipliers, we can obtain the M-step re-estimation equations:

$$P(z_k \mid d_i) = \frac{\sum_{j=1}^{T} n(d_i, w_j)P(z_k \mid d_i, w_j)}{n(d_i)}, \tag{3}$$

$$P(w_j \mid z_k) = \frac{\sum_{i=1}^{D} n(d_i, w_j)P(z_k \mid d_i, w_j)}{\sum_{j=1}^{T}\sum_{i=1}^{D} n(d_i, w_j)P(z_k \mid d_i, w_j)}, \tag{4}$$

$$P(l_m \mid w_j) = \frac{\sum_{n=1}^{N} n(w_j, x_n)P(l_m \mid w_j, x_n)}{n(w_j)}, \tag{5}$$

$$P(x_n \mid l_m) = \frac{\sum_{j=1}^{T} n(w_j, x_n)P(l_m \mid w_j, x_n)}{\sum_{n=1}^{N}\sum_{j=1}^{T} n(w_j, x_n)P(l_m \mid w_j, x_n)}, \tag{6}$$

where $n(d_i, w_j) = \sum_{n=1}^{N} n(d_i, w_j, x_n)$ and $n(w_j, x_n) = \sum_{i=1}^{D} n(d_i, w_j, x_n)$.

3.2   Annotating an Unlabeled Image

Given a new BoF extracted from a new image and the previously estimated $P(l_m \mid w_j)$. By this process, firstly, the new BoF is matched with $P(x \mid l_m)$, being the parameter of each cluster in latent variable l_m, to estimate the similarity measurement between BoF and visual latent variable. And secondly, in the cluster l_m, the probability of words $P(l_m \mid w_j)$ from learning process is used for selecting the set of word by the criterion in Eq. (7).

$$P(w_j \mid x_1, x_2, ..., x_n) \approx \sum_{m=1}^{M} [\prod_{n=1}^{N} P(x_n \mid l_m)^{BoF_{new}(x_n)}]P(l_m \mid w_j) \tag{7}$$

The parameters $P(x_n \mid l_m)$, and $P(l_m \mid w_j)$ are estimated by the learning process. These parameters are the conditional probability, grouping words into a latent variable l_m to estimate the probability of the words. Based on Bayes rule, the probabilities of words are ranked by increasing of their probability to annotate a new image.

3.3   Querying by Words

Given a query words, which may be 1 or more, the Bag-of-Word (BoW) can compute from counting users word in our word vocabulary. This BoW is clustered into a latent variable z_k. And the a set of retrieved image at the clustered latent variable z_k is ranked by their conditional probability the latent variable z_k given d_i which each d_i is estimated from learning process by Eq. (8).

$$P(d_i \mid w_1, w_2, ..., w_T) \approx \sum_{k=1}^{K} [\prod_{j=1}^{T} P(w_j \mid z_k)^{BoW_{new}(w_j)}] P(z_k \mid d_i) \qquad (8)$$

4   Performance Measurement

The performance measure is used in this work is the mean average precision (mAP)[16]. The ability of mAP is to summarize the performance in a meaningful way. To compute mAP, the average precision (AP) of a query q is firstly defined as the sum of the precision of correctly retrieved words at rank i , divided by the total number of relevant images $rel(q)$ for this query when $AP(q) = \sum_{i \in relevant} precision(i)/rel(q)$. Afterward, the average precision measure of a query is sensitive to the entire ranking of images in term of words. Thus the mAP of entire image M_q is calculated by Eq. (9) to summarize the performance of annotation system.

$$mAP = \frac{\sum_{M_q} AP(q)}{M_q}.$$ (9)

Another performance measurement used in this paper is the processing time which measures the time from annotation of testing dataset per an image.

5   Simulation Results

5.1   Dataset and Simulation Condition

In this paper, we used the PASCAL 2008 database [17] which contains a total number of 4,340 annotated images, and 20 words. An image contains at least one word. The data from training/validation are divided into training and testing process by the ratio of 50%. Table 1 summarizes the number of words and images. In Table 1, the PASCAL 2008 dataset is an unbalance dataset, which unequally contains the number of word. The word Person is the highest number of both image and word than the others. In this simulation, the PASCAL

Table 1. Statistics of PASCAL 2008 Image dataset

Words	#image	#word	Words	#image	#word
Aeroplane	236	316	Diningtable	105	110
Bicycle	192	269	Dog	388	477
Bird	305	476	Horse	198	285
Boat	207	336	Motorbike	204	272
Bottle	243	457	Person	2002	4168
Bus	100	129	Pottedplant	180	361
Car	466	840	Sheep	64	145
Cat	328	378	Sofa	134	151
Chair	351	623	Train	151	166
Cow	74	130	Tvmonitor	215	274

2008 dataset is divided into 2 separated sets; 50% for training to construct the models, and the remaining for testing purpose. SIFT feature of training data set is extracted from images by Hessian Affine detection. Then, these SIFT features are used for constructing visual vocabulary or BoF prototype by K-mean, and Images are represented into BoF prototype as new features. BoF of entire training dataset are used for input feature to construct the model namely Naive Bayes, CMRM, pLSA and two-pLSA. To measure the performance among these models, the testing dataset is used for image annotation by mAP and processing time and mAP in each word to measure the efficiency of the models.

5.2   Image Annotation: mAP Performance and Processing Time

In this experiment, we investigate the annotating algorithms by varying the number of visual words which are constructed from K-mean algorithm ranging from 100, 300, 600 and 1000. For pLSA, the number of latent variable is equal to 5 variables, and for two-pLSA model, the number of latent variables K and M are equal to 10 and 100 variables, respectively. So the comparison results are shown in Table 2. From Table 2, in the case of $N = 1000$, the Naive Bayes model obtains the lowest mAP which equals to 0.438. Thus, this method is not suitable for annotation. When investigating the CMRM, its mAP performance is better than that Naive Bayes model. For our model,two-pLSA model, its mAP value is still less than CMRM and pLSA, but more than Naive Bayes Model.

Table 2. Mean Average Precision Performance of Image Annotation

$Models$	$N = 100$	$N = 300$	$N = 600$	$N = 1000$
Naive Bayes	0.369	0.391	0.420	0.438
CMRM	0.489	0.491	0.492	0.492
pLSA($K = 10$)	0.486	0.488	0.491	0.493
Two-pLSA($K = 10, M = 100$)	0.420	0.440	0.459	0.470

Table 3. Processing time per an image Performance of Image Annotation

Models	$N = 100$	$N = 300$	$N = 600$	$N = 1000$
Naive Bayes	≤ 0.1	0.4	1.3	1.8
CMRM	10.0	30.0	55.0	80.2
pLSA($K = 10$)	12	24	54	101
Two-pLSA($K = 10, M = 100$)	0.5	2.3	4.0	7.2

When increasing the number of latent variables K of pLSA, its performance is slowly increased by the number of latent variable. This implies that the number of latent variable is not a necessary factor to increase the mAP performance for an imbalance dataset. But, the mAP of two-pLSA is increased by raising the number of visual word, we obtain its mAP that equals to 0.470, where $N = 1000$. Moreover, we compare the processing time per image of the annotation of each algorithm, as shown in Table 3. For every algorithm, when increasing the number of visual words, the processing time is increased according to their number. The processing time of Naive Bayes Model is the lowest. The processing time of two-pLSA model is 7.2 msec per an image, which is lower than pLSA and CMRM respectively.For the results of processing time, although the mAP of two-pLSA is less than that of CMRM and pLSA, but the two-pLSA model can annotate an image faster than that of CMRM and pLSA.

5.3   Text-Based Image Retrieval: mAP and Processing Time

In this experiment, the performance of image retrieval by searching with text are evaluated by varying the number of visual word as 100,300, 600, and 1000 respectively. the algorithms of text-based image retrieval are performed namely CMRM, and two-pLSA model. In the experimental results, it indicates that our proposed model can obtain better performance in both mAP and processing image which are shown in Table 4 and 5. CMRM model obtains mAP value being around 0.11 and its processing time increases when the number of visual words are increased. The processing time of pLSA and two-pLSA are a constant time, 350 and 50 msec respectively. Their constant times occur from the set of captions, which are estimated by annotation process, stored in the file format, and the estimated captions are used them for the text-based image retrieval. The mAP of our model is about 0.55, and obtains a constant time being 50 msec

Table 4. Mean Average Precision on text-based image retrieval

Models	$N = 100$	$N = 300$	$N = 600$	$N = 1000$
CMRM	0.11	0.12	0.11	0.11
pLSA($K = 10$)	0.14	0.14	0.16	0.16
Two-pLSA($K = 10, M = 100$)	0.55	0.54	0.55	0.55

Table 5. Processing time (msec) on text-based image retrieval

Models	$N = 100$	$N = 300$	$N = 600$	$N = 1000$
CMRM	200	450	800	11500
pLSA($K = 10$)	350	350	350	350
Two-pLSA($K = 10, M = 100$)	50	50	50	50

which is faster than CMRM, and pLSA Model, So that its mAP is better than both.

6   Conclusions

In this paper, we investigate the image annotation and retrieval namely Naive Bayes model, Cross Media relevance model, and pLSA model which are compared to our proposed model, two-pLSA. From the annotation results, pLSA model achieves the best efficient mAP performance closed to our proposed model while our proposed model has faster processing time. Moreover, in text-based image retrieval, our propose model obtains better performance than CMRM in term of mAP and processing time, 0.55 for mAP and 50 msec per a given search. This indicates that our model can support multi-functionality of the retrieval and annotation system.

Acknowledgement

This research is support by Thailand Graduate Institute of Science and Technology (TGIST), National Science and Technology Development Agency (NSTDA), Thailand.

References

1. Russell, B.C., Torralba, A.: LabelMe: a database and web-based tool for image annotation. Intl. J. Computer Vision 77, 157–173 (2008)
2. Hofmann, T.: Unsupervised Learning by Probabilistic Latent Semantic Analysis. Machine Learning 41(2), 177–196 (2001)
3. Quelhas, P., Monay, F., Odobez, J.-M., Gatica-Perez, D., Tuytelaars, T.: A Thousand Words in a Scene. IEEE Trans. Pattern Analysis and Machine Intelligence 29(9), 1575–1589 (2007)
4. Monay, F., Gatica-Perez, D.: Modeling Semantic Aspects for Cross-Media Image Indexing. IEEE Trans. Pattern Analysis and Machine Intelligence 29(10), 1802–1817 (2007)
5. Blei, D., Jordan, M.: Modeling Annotated Data. In: Proc. Intl. Conf. Research and Development in Information Retrieval (2003)

6. Fei-Fei, L., Perona, P.: A Bayesian Hierarchical Model for learning Natural Scene Categories. In: Intl. IEEE Conf. Computer Vision and Pattern Recogntion, vol. 2, pp. 20–25 (2005)
7. Weber, M., Welling, M., Perona, P.: Unsupervised Learning of Models for recognition. In: Vernon, D. (ed.) ECCV 2000. LNCS, vol. 1842, pp. 18–32. Springer, Heidelberg (2000)
8. Duygula, P., Barnard, K., De Freitas, N., Forsyth, D.: Object Recognition as Machine Translation: Learning a lexicon for a fixed image vocabulary. In: Heyden, A., Sparr, G., Nielsen, M., Johansen, P. (eds.) ECCV 2002. LNCS, vol. 2353, pp. 97–112. Springer, Heidelberg (2002)
9. Zhang, R., Zhang, Z., Li, M., Ma, W.-Y., Zhang, H.-J.: A probabilistic semantic model for image annotation and multi-model image retrieval. Multimedia Systems 12, 27–33 (2006)
10. Jeon, J., Lavrenko, V., Manmatha, R.: Automatic Image Annotation and Retrieval using Cross-Media Relevance Models. In: Proc. Intl. Conf. Research and Development in Information Retrieval, SIGIR (2003)
11. Lavrenko, V., Manmatha, R., Jeon, J.: A Model for Learning the Semantics of Pictures. In: Proc. of Advances in Neural Information Processing Systems (2003)
12. Feng, S.L., Manmatha, R., Lavrenko, V.: Multiple Bernoulli Relevance Models for Image and Video Annotation. In: Intl. Conf. Computer Vision and Recognition, vol. 2, pp. II–2002–II–1009 (2004)
13. Huang, P., Bu, J., Chen, C., Liu, K., Qiu, G.: Improve Image Annotation by combining Multiple Models. In: Intl. IEEE Conf. Signal-Image Technologies and Internet-based system (2008)
14. Pham, T.-T., Maillot, N.E., Lim, J.-H., Chevallet, J.-P.: Latent Semantic Fusion Model for Image Retrieval and Annotation. In: Proc. ACM Conf. Information and Knowledge Management, pp. 439–444 (2007)
15. Mori, Y., Takahashi, H., Oka, R.: Image-to-word transformation based on dividing and vector quantizing images with words. In: Proc. of Intl. Workshop on Multimedia Intelligent Storage and Retrieval Management (1999)
16. Smeaton, A.F., Over, P., Kraaij, W.: Evaluation campaigns and TRECVid. In: MIR 2006. ACM Press, New York (2006)
17. Everingham, M., Van-Gool, L., Williams, C.K.I., Winn, J., Zisserman, A.: The PASCAL Visual Classes Challenge, VOC 2008 Results (2008)

Using Conditional Random Field for Crowd Behavior Analysis

Saira Saleem Pathan, Ayoub Al-Hamadi, and Bernd Michaelis

Institute for Electronics, Signal Processing and Communications (IESK)
Otto-von-Guericke-University Magdeburg, Germany
{Saira.Pathan,Ayoub.Al-Hamadi}@ovgu.de

Abstract. The governing behaviors of individuals in crowded places offer unique and difficult challenges. In this paper, a novel framework is proposed to investigate the crowd behaviors and to localize the anomalous behaviors. Novelty of the proposed approach can be revealed in three aspects. First, we introduce block-clips by sectioning video segments into non-overlapping patches to marginalize the arbitrarily complicated dense flow field. Second, flow field is treated as a 2d distribution of samples in block-clips, which is parameterized by using mixtures of Gaussian keeping the generality intact. The parameters of each Gaussian model, particularly mean values are transformed into a sequence of Gaussian mean densities for each block-clip namely a sequence of latent-words. A bank of Conditional Random Field model is employed, one for each block-clip, which is learned from the sequence of latent-words and classifies each block-clip as normal and abnormal. Experiments are conducted on two challenging benchmark datasets PETS 2009 and University of Minnesota and results show that our method achieves higher accuracy in behavior detection and can effectively localize specific and overall anomalies. Besides, a comparative analysis is presented with similar approaches which demonstrates the dominating performance of our approach.

1 Introduction

Crowd behavior analysis is an attractive research for computer vision with challenging issues due to jumble of objects, dynamics and self-organizing behaviors of crowds [1]. Some earlier approaches [2] [3] attempted to detect and track objects across the intervals to investigate object's activities. Performance of these approaches is not promising in the crowded scenes due to heavy occlusions, cluttering and varying proximity of objects. Therefore, to address the challenging aspects of crowd behavior analysis new methodologies are being devised by exploiting crowd-specific sociological studies [4].

The objectives of crowd behavior analysis are very diversified, for example, from sparse-level (i.e. scene level) [5] crowd density analysis to coarse level (i.e. subject level) [6] analysis. In recent attempts, Andrade et al. [7] built a generative model (i.e. ergodic HMM) at a sparse level for normal motion patterns and events of low probability from the defined threshold are equated as anomalous

R. Koch et al. (Eds.): ACCV 2010 Workshops, Part I, LNCS 6468, pp. 370–379, 2011.

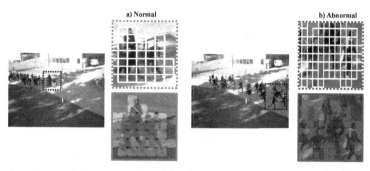

Fig. 1. (a). An example scene showing normal behavior whereas the detection results(bottom) are marked with green. (b). shows a sample frame for abnormal event and the red squares depict the abnormal behavior.

behaviors. In similar aspect, Kratz et al. [8] proposed spatio-temporal gradients in cuboid where the dominant patterns of inactivity are detected by modeling their statistics on coupled HMM in dense crowds. In contrast, Mehran et al. [9] combined the social force model with the optical flow based particle advection and simulate the normal forces of particles implicitly to detect the deviations. Another work is presented by Chan et al. [10] to holistically model the crowd flow in the scene using the dynamic texture model where Support Vector Machines (SVM) is used as classifier along with other classification techniques to detect the crowd events.

Unlike the above reviewed approaches, Albio et al. [11] maintain the probabilities of optical flow at corner points and constitutes histogram to detect the abnormalities on PETS 2009 dataset. In the same context, Benabbas et al. [12] build online probabilistic models of both density and orientation of flow patterns to detect the crowd activities. It can be notice that a dense field of flow is observed in crowded scenes, which needs to be marginalized prior to model for effective analysis of underlying crowd behaviors. However the suggested unsupervised approaches [11] [12] offer very specific solutions to this problem. In contrast, for generative modeling approaches such as HMM and LDA, stringent conditional independence over expanded space of observed flow field is an essential requirement for more tractable joint distributions. Consequently, it is challenging when modeling the dense and correlated flow patterns which indicate the likely situations.

In this paper, the problem of modeling and learning the crowd behaviors (i.e. normal and abnormal) is addressed as shown in Fig. 1. A novel top-to-down framework is constructed in which we begin by finding foreground regions. Later, the video sequence is sectioned into video segments to form the block-clips. The introduction of our block-clips thereby allows us to form the prototypes of dense flow vectors which are extended over an interval in video segments containing significantly correlated and uncorrelated flow field. There on, without losing generality, the computed flow vectors inside each block-clip are treated as a 2d distribution of observed data and are assigned to specific linearly superimposed

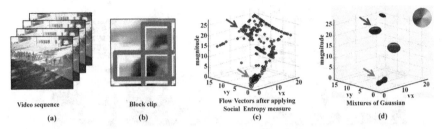

Video sequence   Block clip   Flow Vectors after applying   Mixtures of Gaussian
                                            Social Entropy measure
   (a)   (b)   (c)   (d)

Fig. 2. a) Presents the sectioning of video sequence into video segments. b) $M \times N$ block-clips are formed in each video-segment. c) demonstrates the observed flow field where green points highlight the normal behavior and red points indicate the abnormal behaviors (Also, arrows indicate the respective crowd behaviors). d) shows the resulting mixtures of Gaussian fitted over the point cloud as shown in (c), whereas colors of the mixtures of Gaussian show the respective orientation of the flow field (please zoom-in for the better visibility).

components of the Gaussian mixture model. Expectation Maximization (EM) is used to find the maximum likelihood estimators in the flow vectors. The parameter values obtained by the Gaussian models are interpreted as a sequence of latent words of flow patterns for each block-clip which we term as latent-words for simplicity. A bank of Conditional Random Field (CRF) models are constructed, one for each block-clip to model the sequence of latent-words with corresponding label sequence to characterize the crowd behavior at the specific and global level in an unconstrained environment.

The main contributions of the paper are twofold: 1) first, we characterize both the recurrent and non-recurrent behaviors of the crowd using a bank of CRF models unlike the work proposed by Andrade et al. [7] using ergodic HMM for severely limited environments. Typically, crowd data contains dense and correlated flow features, which are highly difficult to model directly using HMM. In the above referenced literatures [7] [9], parametric and clustering approaches are employed to form the analytical representation of flow field keeping the generality of data intact but still severely limited to address the issues of generality. Second contribution of our paper is the introduction of block-clips and performing the parametric modeling of the flow vectors in each block-clip by fitting mixtures of Gaussian. As a results, we obtained a sequence of latent-words with corresponding label which is modeled with CRF to the analyse the underlying crowd behaviors. Our proposed approach demonstrates promising results and outperforms when compared with other approaches [10] [9].

The rest of the paper is structured as: section 2 explains the proposed methodology in detail. Section 3 presents the results of our proposed approach, and concluding remarks are sketched in section 4.

2   The Methodology

Abrupt and independent activities in crowded scenes result in incoherency, which defines the overall self-organizing mechanism of the underlying crowded scenes.

The proposed framework is staged in several phases to model and characterize the crowd behaviors. Our proposed approach begins by extracting the region of interest (ROI) through segmentation, thus limiting the information redundancy and computational time unlike the holistic approach used in [9] [5]. In parallel, a grid of two by two is mapped over the ROI in each frame and optical flow is computed. Then, we introduce the concept of block-clips in which at first video sequence is fragmented into equal sized segments. We divide each video segment into non-overlapping blocks which result in spatio-temporal block-clips. Flow vectors inside block clips are modeled using the mixtures of Gaussian resulting in a sequence of latent-words. It acts as a sequential data for each of our block-clips with corresponding label sequence for learning of CRF parameters during the training and the crowd behaviors are inferred on test samples. The details are presented in the following:

2.1 Pre-processing

In pre-processing, we build an initial background model which is generated by using Gaussian Mixture Model (GMM). Foreground is extracted robustly with background subtraction, whereas background model is updated through MDI [13] for each time step (currently, the problem shadows is not addressed).

2.2 Creating Block-Clips

Each frame is sectioned into $N \times M$ blocks of size (i.e. $size = 16$), selected after conducting empirical studies over the dataset (i.e. PETS 2009). In parallel, a grid of two by two is placed over the detected ROI which we refer as points of interest (POI) and the optical flow is computed associated with these $POIs$ as presented in Fig. 2(a and b). The observed dense flow field is transformed into a flow vector $f = (v_x, v_y)$. Where v_x and v_y represent the velocities along the horizontal and vertical axis of the motion field.

As, it is observed that in crowded scenes the occupancy regions in each frame are equally important and provide distinctive attributes. So, we begin by marginalizing the video sequence into equally sized segments (i.e. video segments). The selection of video segment s_k size (i.e. K) depends upon the dataset and the frame rate of the video sequence. After this, we obtained the block clips $c_{l,k,t}$ inside each video segments as follows:

$$\mathbf{V} = [s_1, ..., s_K], \quad s_k = \left\{ c_{(1,1,1)}, ..., c_{(L,K,T)} \right\} \quad and \quad c_{(l,k,t)} = (f_1, ..., f_P)$$

where $\mathbf{V}$ is the video sequence, s_k are the segmented clips of the sequence which contains L block-clips in K-th segment at time t. Each block-clip keeps P cloud of flow vectors which serves as a fundamental information for the analysis of crowd behaviors as shown in Fig. 2(c).

2.3 Mixture Model

We can define our 2d flow vectors (i.e. $f = (v_x, v_y)$) in each block-clip as random variables which are extended over certain frames (i.e. K) in the corresponding

374   S.S. Pathan, A. Al-Hamadi, and B. Michaelis

video segment. Since, the flow points present in the block-clip can be significantly different and correlated, therefore, it is required to glean the information by applying the parametric approximation. Now, our objective is to learn the parameters and model the mixtures of Gaussian for this 2d distribution. The principle objective of using Gaussian mixture model is that it provides a theoretically straightforward way to model our data and forms a comprehensive representation of the flow vectors field inside each block-clip which we as latent-words in respective video segment. These latent-words in each block-clip are used to train and test CRF model for crowd behavior analysis.

Given our 2d distribution of flow vectors in each block-clip, a fairly informal way to initialize the model randomly followed by EM, an elegant optimization function for finding the maximum likelihood solution for our distributions. Particularly, parameters of the distribution are estimated to transform the observations (i.e. flow vectors in each block-clip) into G Gaussian models (i.e. $G = 3$ which is found empirically and optimal for our experiments) as illustrated in Fig. 2(d). The Gaussian mixture distribution can be written as:

$$p(x) = \sum_{g=1}^{G} \pi_g \mathcal{N}(x \mid \mu_g, \Sigma_g)$$

where the G represents the number of Gaussian models, π_g is the weight, μ is the mean and covariance $\sum$ are the parameters of each component of Gaussian model. The μ_g of the mixtures of Gaussian contains parameters for each dimension of the sample flow vectors (i.e. f). We compute the mean density for these mixtures of Gaussian, thus form a sequence of latent-words for each block-clip, which is to be processed by CRF whereas the length of the latent-word sequence (i.e. Seq or $\bar{x}$) is directly proportional to G. We can write as:

$$\mu_g = (\mu_{g_{vx}}, \mu_{g_{vy}}), \quad d_{\mu_g} = \sqrt{\mu_{g_{vx}}^2 + \mu_{g_{vy}}^2}, \quad and \quad Seq = \bar{x} = \{d_{\mu_1}, .., d_{\mu_G}\}$$

2.4   Conditional Random Field and Crowd Behavior Detection

Conditional Random Field is a discriminative modeling technique for labeling the sequential data and a special case of log linear model. CRF provides a probabilistic framework to specify the probability of particular label sequence given the observation sequence, a nice description on CRF is provided by Wallach et al. [14]. Specifically, $\bar{x}$ is our input sequence (i.e. $\bar{x} = x_1...x_w$) of w latent-words and $\bar{y}$ is the corresponding label sequence (i.e. $\bar{y} = y_1...y_w$) of respective behaviors. Here, we assume that both sequences $\bar{x}$ and $\bar{y}$ have the same length. As defined by Lafferty et al. [15], the probability of label sequence given observation sequence can be:

$$p(\bar{y} \mid \bar{x}; \theta) = \frac{1}{Z(\bar{x}, \theta)} exp \sum_i \theta_i F_i(\bar{x}, \bar{y}) \tag{1}$$

The numerator $F_i(\bar{x}, \bar{y})$ is the feature function which represents the paired mapping $F_i : X \times Y \longrightarrow \Re$ of the data space X and the label space Y at different level of granularity. Therefore, the feature function can be arbitrarily correlated and can be defined as follows:

$$F_i(\bar{x}, \bar{y}) = \sum_j f_i(y_{j-1}, y_j, \bar{x}, j) \tag{2}$$

where f_i is the low level feature function which is influenced by the subset of the above entities such as previous label y_{j-1}, current label y_j, observation sequence $\bar{x}$, and current position j.

The denominator in Eq.1 is the partition function commonly termed as a normalization factor which ranges over all the label sequence, but we assume that the feature-function can depend on at most two labels. So, instead of enumerating all possible $\bar{y}$, this assumption allows us to enumerate the possible $\bar{y}$ efficiently. The formulation of Z is as follows:

$$Z(\bar{x}, \theta) = \sum_{\bar{y}} exp \sum_i \theta_i F_i(\bar{x}, \bar{y}) \tag{3}$$

Training CRF. We perform training using stochastic gradient methods based on gradient of conditional likelihood function for nonlinear optimization. The goal of learning task is to compute parameter θ (i.e. weights) values of our model and learns the Conditional log-likelihood (CLL) of the training sequences. Our objective is to maximize CLL, we have used stochastic gradient ascent method for training. The formulation is defined as follows:

$$\frac{\partial}{\partial \theta_i \log p(y \,|x\, ; \theta)} = F_i(x, y) - \frac{\partial}{\partial \theta_i} log Z(x, \theta) \tag{4}$$

In the above equation, for each θ_i, the partial derivative of CLL is evaluated for single training sequences (i.e. one wight for each feature-function). Precisely, the partial derivative with respect to θ_i is the, i-th value of the feature function for its true label y, minus the averaged feature-function values for all possible labels $\bar{y}$. So, above equation can be rewritten as:

$$\frac{\partial}{\partial \theta_i} \log p(y \,|x\, ; \theta) = F_i(x, y) - E_{\bar{y} \approx p(\bar{y}|x\, ; \theta)} [F_i(x, \bar{y}] \tag{5}$$

In practice, the function $\log(\theta)$ does not maximize in a closed form solution. Therefore, we invoke BFGS (Broyden Fletcher Goldfarb Shanno) as an optimization routine to estimate curvature numerically from the first derivative of CLL avoiding the requirement of exact Hessian inverse computation [16] with stochastic gradient ascent.

Inferencing CRF. Given the test sequence of latent-words for each block-clip $\bar{x}$ and the learned parameter values of θ from the training data, the corresponding label for the sequence is obtained as:

$$\bar{y}^* = \text{argmax}_{\bar{y}} p(\bar{y} \,|\bar{x}\, ; \theta) = \text{argmax}_{\bar{y}} \sum_i \theta_i F_i(\bar{x}, \bar{y}) \tag{6}$$

Using the definition of feature function in Eq.2, we get:

$$\bar{y}^* = \text{argmax}_{\bar{y}} \sum_i \theta_i \sum_j f_i(y_{j-1}, y_j, \bar{x}, j) \qquad (7)$$

Each label sequence is aggrandize from $< start, end >$ states of labels (i.e. y_0 to $y_n + 1$), so, for efficient computation an alternative choice is to employ matrices. For this, g_j is a $q \times q$ matrix where q is the cardinality of the set vectors in the label sequence $\bar{y}$ and is defined over each pair of labels y_{j-1} and y_j as follows:

$$g_j(y_{j-1}, y_j \,|\bar{x}) = exp(\sum_i \theta_i f_i(y_{j-1}, y_j, \bar{x}, j)) \qquad (8)$$

For each j, we will get different g_j function which depends on weight θ, test observation sequence $\bar{x}$ and the position j. The sequence probability of the label $\bar{y}$ given observation sequence $\bar{x}$ can be rewritten in compact manner as follows:

$$p(\bar{y}\,|\bar{x}\,;\theta) = \frac{1}{Z(\bar{x}, \theta)} \prod_j g_j(y_{j-1}, y_j \,|\bar{x}) \qquad (9)$$

$$Z(\bar{x}, \theta) = \prod_j g_j(y_{j-1}, y_j) \qquad (10)$$

Our main contention in obtaining the local sequence of latent-words in each block-clip is that by interpreting the motion flow field globally (i.e. at a video segment level), it is difficult to reveal the required level of detail, which can differentiate the coherent and incoherent dynamics. Therefore, in the above sections, we are able to acquire the intrinsic flow patterns in a compact manner that faithfully characterizes the behavior of the crowd dynamics. The mixtures of Gaussian are invoked to parameterize dense flow vectors into a sequence of latent-words, which are modeled with CRFs to characterize the normal and abnormal behaviors in the crowds.

3   Experiments and Discussion

The proposed approach is tested on publicly available benchmark datasets from PETS 2009 [17] and University of Minnesota (UMN) [18]. Ideally, the normal situation is represented by the usual walk of large number of people. In contrast, the abnormal situations such as running, panic, and dispersion are observed when individuals or group of individuals deviate from the normal behavior. There is a major distinction between these two datasets. For example, in PETS 2009, the abnormality begins gradually unlike UMN dataset making PETS more challenging due to the transitions from normal to abnormal situations.

We conducted our test experiments on PETS S3 dataset, comprises of 16 outdoor sequences containing different crowd activities such as walking, running and dispersion. To evaluate the diversity of proposed approach, we conduct experiments on UMN crowd dataset which contains 11 videos of different scenarios

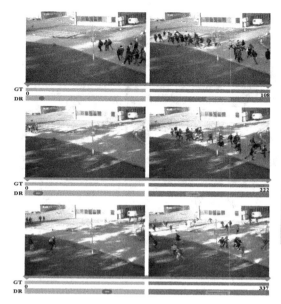

Table 1. Confusion Matrix

Events	Normal	Abnormal
Normal	97.3	2.7
Abnormal	3.1	96.9

Table 2. Comparative Analysis

Methods	Results(%)
Our Methods	97.1
Mehran et al. [9]	96
Chan et al. [10]	81

Fig. 3. Presents the quantitative analysis on PETS 2009. The left frames indicate absolute normal (green) behavior and the right frames depict absolute abnormal (red) behaviors along the time-line (gray) in each row.

showing normal and escape cases. The training is performed on PETS S0 and S1 scenarios. However, the UMN sequences are tested without special training.

We have demonstrated a qualitative analysis of the test sequences. Fig. 3 presents ground truth (GT) and detection result (DR) in each row for normal and abnormal situations in the sequences. The left and right frame in each row depict the normal and abnormal behavior of the crowds. Thin color bars (i.e. green and red) in each row show the GT, whereas thick bars indicate the detection results. Colors of the bars define crowd behaviors and timings of the occurrences. Meanwhile, incorrect localization of the crowd behaviors are marked with respective colors of false detections in Fig. 3.

Fig. 4(left) demonstrates the detection results of crowd behaviors on PETS 2009. Normal behaviors are marked as green patched and abnormal behaviors are highlighted with red patches. Results in Fig. 4(right) demonstrate the detection of crowd behaviors for normal and panic situations in UMN crowd sequences. Results show that the proposed approach is capable of detecting the governing dynamics of crowd and able to capture the transition's period (normal to abnormal) successfully. Table. 1 shows the confusion matrix of the probabilities of normal and abnormal behaviors for each class in crowds. The diagonal elements in the confusion matrices represent the percentage probability of each class in the group. Misclassification between the classes is shown by the non-diagonal elements which are observed due to the prominent motion field at the objects leg's parts as compared to body and head.

(a)   (b)

Fig. 4. a) Shows detection results on PETS 2009 and b) presents behaviors detection on UMN datasets. Left frames show the normal behavior detection indicated by green patches and right frames depict the abnormal behaviors marked with red patches.

To analyze the performance of our proposed approach in detecting the crowd dynamics effectively, we have made a comparative analysis from two recent proposed techniques [9] [10]. In the first approach, the computed social forces are model with LDA, whereas in the second approach, SVM is used to classify different categories of crowd behaviors (Note. we consider the results of specific categories which define the abnormal crowd behaviors). As it can be seen in Table. 2, performance of our method is promising and achieves higher detection rate for the detection of the crowd behaviors in specific and overall manner when compared with related approaches.

4   Conclusion

We propose a novel approach for detecting crowd behaviors by modeling computed sequence of latent-words using Conditional Random Field. We define block-clips as non-overlapping spatio-temporal patches and parameterize the flow vectors in each block-clip with mixtures of Gaussian to obtain a sequence of latent-words. Block-clip specific sequence of latent-words allows an effective representation of features. CRF is learned from these sequences to characterize the behaviors. The presented results demonstrate promising performance and outperform when compared with the related work.

Acknowledgment

This work is supported by Transregional Collaborative Research Centre SFB/ TRR 62 Companion-Technology for Cognitive Technical Systems funded by DFG and Forschungspraemie (BMBF-Froederung, FKZ: 03FPB00213).

References

1. Zhan, B., Monekosso, D., Remagnino, P., Velastin, S., Xu, L.: Crowd analysis: a survey. Machine Vision Application 19, 345–357 (2008)
2. Boghossian, A., Velastin, A.: Motion-based machine vision techniques for the management of large crowds. In: Proceedings of IEEE International Conference on Electronics, Circuits and Systems, vol. 2, pp. 961–964 (2002)
3. Maurin, B., Masoud, O., Papanikolopoulos, N.: Monitoring crowded traffic scenes. In: IEEE International Conference on Intelligent Transportation Systems, pp. 19–24 (2002)
4. Johansson, A.: Constant-net-time headway as a key mechanism behind pedestrian flow dynamics. Physical Review E 80, 26120-1 (2009)
5. Wu, S., Moore, B.E., Shah, M.: Chaotic invariants of lagrangian particle trajectories for anomaly detection in crowded scenes. In: IEEE Conference on Computer Vision and Pattern Recognition, pp. 2054–2060 (2010)
6. Ge, W., Collins, R.: Marked point processes for crowd counting. In: IEEE Computer Society Conference on Computer Vision and Pattern Recognition, pp. 2913–2920 (2009)
7. Andrade, E.L., Scott, B., Fisher, R.B.: Hidden markov models for optical flow analysis in crowds. In: Proceedings of the International Conference on Pattern Recognition, pp. 460–463. IEEE Computer Society, Los Alamitos (2006)
8. Kratz, L., Nishino, K.: Anomaly detection in extremely crowded scenes using spatio-temporal motion pattern models. In: IEEE Conference on Computer Vision and Pattern Recognition, pp. 1446–1453 (2009)
9. Mehran, R., Oyama, A., Shah, M.: Abnormal crowd behavior detection using social force model. IEEE Computer Society Conference on Computer Vision and Pattern Recognition, 935–942 (2009)
10. Chan, A.B., Morrow, M., Vasconcelos, N.: Analysis of crowded scenes using holistic properties. In: Performance Evaluation of Tracking and Surveillance workshop at CVPR 2009, pp. 101–108 (2009)
11. Albiol, A., Silla, M., Albiol, A., Mossi, J.: Video analysis using corner motion statistics. In: Performance Evaluation of Tracking and Surveillance workshop at CVPR 2009, pp. 31–37 (2009)
12. Benabbas, Y., Ihaddadene, N., Djeraba, C.: Global analysis of motion vectors for event detection in crowd scenes. In: Performance Evaluation of Tracking and Surveillance Workshop at CVPR 2009, pp. 109–116 (2009)
13. Al-Hamadi, A., Michaelis, B.: An intelligent paradigm for multi-objects tracking in crowded environment. Journal of Digital Information Management 4, 183–190 (2006)
14. Wallach, H.M.: Conditional random fields: An introduction. Technical Report MS-CIS-04-21, University of Pennsylvania (2004)
15. Lafferty, J., McCallum, A., Pereira, F.: Conditional random fields: Probabilistic models for segmenting and labeling sequence data. In: Proc. 18th International Conf. on Machine Learning, pp. 282–289 (2001)
16. Sha, F., Pereira, F.: Shallow parsing with conditional random fields (2003)
17. Ferryman, J., Shahrokni, A. (PETS (2009), http://www.cvg.rdg.ac.uk/PETS2009
18. UMN: (Detection of unusual crowd activity), http://mha.cs.umn.edu

Understanding Interactions and Guiding Visual Surveillance by Tracking Attention

Ian Reid, Ben Benfold, Alonso Patron, and Eric Sommerlade

Department of Engineering Science
University of Oxford
OX1 3PJ
Oxford, UK

Abstract. The central tenet of this paper is that by determining where people are looking, other tasks involved with understanding and interrogating a scene are simplified. To this end we describe a fully automatic method to determine a person's attention based on real-time visual tracking of their head and a coarse classification of their head pose. We estimate the head pose, or coarse gaze, using randomised ferns with decision branches based on both histograms of gradient orientations and colour based features. We use the coarse gaze for three applications to demonstrate its value: (i) we show how by building static and temporally varying maps of areas where people look we are able to identify interesting regions; (ii) we show how by determining the gaze of people in the scene we can more effectively control a multi-camera surveillance system to acquire faces for identification; (iii) we show how by identifying where people are looking we can more effectively classify human interactions.

1 Introduction

This paper summarises work in the Active Vision Group in which we have developed methods for estimating the gaze of individuals from low resolution video, and a variety of applications based on that idea. In particular, we observe that by determining where people are looking, other tasks involved with understanding and interrogating a scene can often be simplified. Humans are remarkably good at inferring where others are looking, even from brief views, or from very small image fragments in video, and it is clearly an important cue: consider, for instance, a car driver approaching a pedestrian about to cross the road; if the pedestrian has not been seen to look towards the driver then more caution may be exercised.

While there is a substantial body of work on finding gaze for human computer interaction, this is a relatively controlled scenario in which the face is close to the camera and more often than not facing it. In contrast, there has been little work on finding people's gaze automatically using passive visual sensing in the case of visual surveillance, for instance, where the face may occupy only a very small fraction of the image. To our knowledge the earliest work on trying to find gaze from low resolution images is [1] (and a more advanced version of the

R. Koch et al. (Eds.): ACCV 2010 Workshops, Part I, LNCS 6468, pp. 380–389, 2011.

same work [2]). At a similar time [3] investigated estimation of coarse eye-line. Subsequently [4] proposed using a combination of segmentation and randomised tree classification, and refined the tree decisions in [5]. [6] proposed a different technique based on template matching and SVM classification and showed this to be superior to [1]. We compare the methods of [6] and [5] in section 2.

In this paper we draw together three pieces of work, bonded by the common theme of using estimates of coarse gaze. In the first application, first published in [5] we show how by monitoring the gaze of two or more individuals we can determine areas that are of collective interest. The second application [7] combines gaze estimation with our work on active surveillance. Active cameras cooperate to obtain facial images of all individuals in a scene; this is achieved by monitoring the head pose (the coarse gaze direction) and choosing pan, tilt and zoom values for each camera to maximise the information gained by looking at each person. The final application, first published in [8] shows how using head pose information can aid in the discovery and recognition of interactions in video. Knowledge of head pose yields a degree of viewpoint invariance since descriptions about the local vicinity of a person can be made with reference to their gaze, and furthermore can be used to hypothesize and verify local interactions, under the assumption that people interacting usually look at each other. We discuss these applications briefly in sections 3, 4 and 5 (though the reader is referred to the respective publications for full details). Prior to this, in the following section we introduce our method for estimating the gaze from low resolution video.

2   Coarse Gaze Estimation

Here we briefly describe our method for tracking people and estimating their coarse gaze direction. More details appear in [5].

The first stage requires each pedestrian in a scene to be tracked, with the purpose of providing stable head images for the following pose estimation step. We track only the heads of pedestrians rather than their entire bodies for two reasons. The first is that security cameras are generally positioned sufficiently high to allow pedestrian's faces to be seen, so their heads are rarely obscured. The second is that the offset between the centre of a pedestrian's body and their head changes as they walk, so tracking the head directly provides more accurately positioned head images. The general approach we adopt is one of tracking by repeated detection. We use Dalal and Trigg's HOG detection algorithm [9] trained on cropped head images to provide detections. In each frame of video, sparse optical flow measurements from KLT feature tracking [10] are used to predict the head location and the head detections provide absolute observations which are fused with the predictions using a Kalman filter. The resulting location estimates provide stable head images which are used for gaze estimation (see figure 1). The 2D image locations are converted into a 3D location estimates by assuming a mean human height of 1.7 metres using the camera calibration with a ground plane assumption. This knowledge of the ground plane also permits us to restrict the set of scales over which the head detector must be evaluated (e.g. large head will not appear near the horizon). This restriction in scale, together with our CUDA-based implementation of HOG [11] mean that we achieve real-time tracking on full 480p Standard Definition video.

We address finding the coarse gaze of a person as a multi-class classification problem. We make use of so-called randomised ferns, a simplified version of the randomised decision trees that have previously been successfully applied to tasks in object detection and classification from small image patches [12]. A single fern comprises a fixed set of binary decisions, which are typically chosen randomly (with the assumption that by considering a large enough set of such decisions, sufficient informative ones will be included). Interrogation of an image patch using such a fern results in a binary string (each bit is the outcome of one decision), which can be considered as an index into a set of histograms. Each histogram encodes the relative frequency with which the binary string is selected by patches of each class. These histograms are quickly and easily trained by presenting many examples to the fern and accumulating the relative frequencies. Likewise, at query time, the fern is applied to the query image patch to yield an index, and the likelihood read directly from the indexed histogram. The classification accuracy can be improved by combining the output from several ferns (i.e. a *forest*).

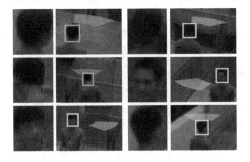

Fig. 1. Sample head images with the corresponding gaze estimations

To estimate the head pose we discretise the gaze direction by dividing the full 360° range into a fixed set of 8 classes, and learn a forest of ferns classifier from a large corpus of pre-labelled heads of 16x16 pixels. The choice of what sort of decisions to use within the ferns is very important; decision outcomes must be able to recognise general properties of each direction class irrespective of the large variations in appearance between people. We make use of decisions based on two different feature types, both of which compare values from different image locations against one another, rather than against a fixed threshold. This yields a degree of robustness to brightness variations and colour tints.

The first decision type is based on normalised histograms of gradient orientations. A fern decision is then a simple comparison between the sizes of a pair of histograms bins. The second type of decision is based on Colour Triplet Comparisons (CTCs). Each CTC decision samples colours from pixels at three different locations within the tracked head region and makes a binary decision based on whether the first and second colours are more similar than the second and third colours in RGB colourspace.

Since few other authors have considered the question of finding gaze from such low resolution data, direct comparisons with the state-of-the-art are difficult. Orozco *et al.* [6] describe a competing method for low resolution gaze, based on templates and Support Vector Machines, in which heads are localised using a combination of HOG detection and background subtraction (note that unlike that work, we do not assume that background subtraction is possible). Their work showed remarkable improvement relative to early work in the field [2]. We tested our method against theirs (using our own implementation of their algorithm) while varying the nature of the training and testing data sets. Table 1 summarises our findings. Each entry in the table gives the percentage of the test data that was classified to the correct class. Where the testing and training sets are the same, ninety percent of the data was used for testing and the remaining ten percent for testing. In all cases, both our method and Orozco's perform significantly better than random (which would be 12.5% for an 8 class problem), but it is noticeable that the performance figure closest to that cited in [6] is in the biased scenario of testing and training on the same data. The "Photo dataset" comprises a set of still image of people harvested from the web, manually classified, and scaled to the standard size of 16x16. "Hand-cropped video" is data taken from a variety of video sources, observing many individuals, but with many exemplars from the same individuals. "i-LIDS" data are taken from the standard UK Home Office dataset (from UK Home Office Scientific Development Branch, i-LIDS: Image Library for Intelligent Detection Systems).

Table 1. Each cell gives the testing performance as a percentage of correctly classified testing examples, based on training from one of the three datasets, for either the method of [5] or [6]

	Test data					
	Photo dataset		Hand-cropped Video		Tracked i-LIDS	
	[5]	[6]	[5]	[6]	[5]	[6]
Photo dataset	44	36	30	26	26	23
Hand-cropped Video	49	36	86	82	29	32
Tracked i-LIDS	23	23	30	29	63	50

(Training data label on left of table)

3 Measuring Attention Using Coarse Gaze

In [5] we combined the tracking and head-pose to make a fully automatic system for measuring the amount of attention received by different areas of a scene. Over the course of an entire video, the locations and gaze directions of the pedestrians are projected onto the ground plane and used to build up an *attention map* representing the amount of attention received by each square metre of the

(a)   (b)

Fig. 2. (a)A frame showing the gaze direction estimates and the paths along which pedestrians were tracked. The lower images show the resulting attention map and the result of projecting it onto a video frame, identifying the shop window as a popular subject of attention. The blue lines on the attention map show the edges of the road. (b) Sequence showing how the attention map can be used to highlight transient areas of interest. The left column shows video frames with annotated gaze directions, the middle column shows the corresponding attention maps and the third column shows the video frame modulated with the projected attention map, under the assumption that the subject of interest is between 0 and 2 metres above the ground.

ground. The projected attention density is reduced linearly with the distance from the pedestrian to correct for the increasing field of view width.

In one experiment we analyse a video of a busy town centre street with up to thirty pedestrians visible at a time over twenty-two minutes. The results from tracking approximately 2200 people are shown in figure 2(a). A second application is to use the estimated gaze to identify a transient source of interest. In particular here we are interested in finding areas of the image that are simultaneously viewed by two individuals. The resulting intersection, shown in figure 2(b) identifies the subject of attention. Further experiments and details can be found in [5] .

4   Using Gaze Estimates for Camera Control

Face recognition in surveillance situations usually requires high resolution face images to be captured from remote active cameras. Since the recognition accuracy is typically a function of the face direction – with frontal faces more likely to lead to reliable recognition – we propose a system which optimises the capturing of such images by using coarse gaze estimates from a static camera. This work builds on ideas we have been pursuing for some time based on the idea that a set of pan-tilt-zoom cameras can achieve cooperative surveillance of an scene by choosing their PTZ values to make the most informative sensor readings [13].

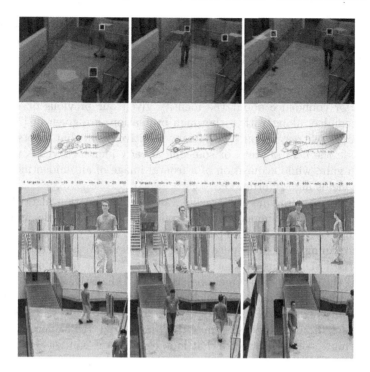

Fig. 3. Sample images showing the operation of the live system at three different times (left to right). The top row shows images from the static camera. The tracked heads are annotated with boxes and their gaze direction represented by a circular section. The second row shows a schematic of the camera control method. The active cameras are depicted as cones. The objective function for pan, tilt and zoom settings for each camera is marked in red, with darker representing higher expected information gain. Each ring shows variation of expected information gain with pan value, while the concentric rings represents varying zoom. Active targets are circled, and the target with the most information gain for the left camera is green. The trajectories of all targets from the last 30 seconds are drawn. The third row shows the images recorded by the left active camera, and the last row contains images from the right camera.

More formally, we define a probabilistic model of the scene, which naturally en- codes uncertainty (for example, distributions over the locations of all targets, distributions over the identities of each target, the possibility that new targets have appeared, etc). The uncertainty in this model can be quantified formally as the *entropy*; our goal then becomes to choose actions that lead to greatest expected decrease in entropy, or in other words, the greatest expected gain in mutual information.

In the task at hand, we seek to choose the PTZ parameters for a pair of cameras which are most likely to result in acquisition of face images suitable for identification (i.e. close to frontal views). To that end we develop an abstract model of a face identification system in terms of its true/false positive/negative rates. While we omit the details (and refer the reader to [7]), the key notion is that the identification performance of the system is a function of the gaze of the individual, and that the longer we observe an individual, the more likely we are to make a positive identification. The expected gain in information is then a function of the field of view of each camera, the observed gaze direction, and our prior confidence about the identity given our previous observations. Detailed formulae, along with a thorough evaluation, are given in [7]. In this overview paper, we restrict ourselves to the intuitive result that a person facing away, outside the field of view, or someone already identified yields very little information gain, while acquisition of a frontal image of an unidentified person potentially yields high information.

A supervisor camera runs the tracking and gaze estimation algorithm from section 2, and two active cameras are controlled according to the the information theoretic objective function. The parameters of the active cameras are pan, tilt and zoom, meaning the search space has six dimensions. We conduct a full search over discretised 6D parameter space, in which we evaluate the information gain by summing the gain for each target observed. This exhaustive search runs at two frames per second per camera. Sample frames from the live system are shown in figure 3. While the behaviour of the system is difficult to convey in still images, the main result from this live system is that, as desired, the cameras follow the targets, trying to keep the targets that look towards the camera centred in the field of view.

5   Understanding Human Interactions via Attention

In this work, published first in [8], we make use of coarse gaze classification in order to simplify the problem of recognising human interactions in TV shows. More specifically, we are interested in classifying video as containing one, or none of the interactions: handshake, hug, kiss, and high-five, in spite of background clutter, camera movement, shot changes, arbitrary lighting, etc. We make use of coarse gaze in two novel ways. The first is to achieve a weak view invariance in a local descriptor of the appearance and motion in the vicinity of a person; this is because we reference the descriptor relative to the facial orientation. The second is based on the assumption that people generally face each other while interacting. We use this to learn a structured SVM [14] that learns the spatial relations between actors for each action class, and is trained to obtain the best *joint* classification of a group of people in frame. Using structured learning improves the retrieval results over those obtained when classifying each person independently.

Unlike the work described in the two preceding sections, here we use a slightly different classification scheme and classifier for obtaining the coarse gaze. This is because the heads/faces in TV shows generally occupy a larger fraction of the image than in surveillance data, and also because we are content to have an even coarser classification into one of five classes: profile left and right, inclined left and right, and facing away (note that in TV shows it is very rare for a face to

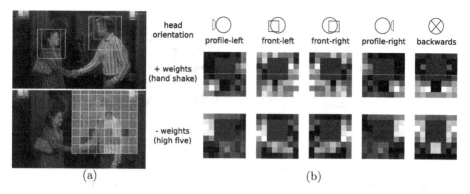

Fig. 4. (a) Upper body detections and estimated discrete head orientation are shown above, while the below shows the grid indicating dominant gradients per cell and significant motion (red cells) for a hand shake. (b) Motion weights outside upper body detection (blue square) learnt by a linear SVM classifier trained to discriminate between hand shakes and high fives. Higher weights are indicated by lighter areas. As expected, the more important motion regions are in lower locations for hand shakes and in higher ones for high fives. These also follow the direction of the face.

be absolutely frontal in a scene). We train a one-vs-rest linear SVM classifier to distinguish these five gaze classes.

Interaction recognition proceeds as follows: using a combination of upper-body and head detections together with local feature tracking we track each person in a scene [15]. For each person track, and for each frame of video we then classify the gaze and create a local descriptor of motion, as illustrated in figure 4(a) (details are given in [8]). The descriptor is then used as a data vector for training a linear SVM classifier for interaction class. An illustrative example of the results that we obtain, figure 4(b), shows the motion regions (outside the upper body detection) learnt by a linear SVM classifier trained to discriminate between hand shakes and high fives. As expected, important motion regions are correlated with the head orientation and occur in lower locations for hand shakes and higher ones for high fives.

Each of these simple one-vs-all SVM interaction classifiers returns a score, and we then aim to improve the classification; to that end we learn relative locations of people given both their head orientation and an interaction label, in a structured learning framework similar to the one described in [16]. The goal is simultaneously to estimate the best joint classification for a set of detections in a video frame rather than classifying each detection independently. We first define a set of spatial neighbourhood relations, and then learn the weights for these spatial relations using the SVM^{struct} package [17]. The weights learned this way using this method can be seen in figure 5. Examples of the retrieval results using the system are shown in figure 6. The reader is referred to [8] for the detail of the structured SVM implementation and a more thorough evaluation.

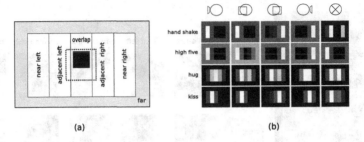

Fig. 5. (a) Spatial relations used in our structured learning method. The black square at the centre represents the head location inside an upper body detection. (b) Weights learnt for each interaction class and head orientation combination. Lighter intensity indicates a higher weight.

Fig. 6. Highest ranked true positives (left) and false positives (right) for each interaction obtained using the automatic method. The red squares indicates negative videos.

6   Conclusions

In this paper we have reviewed our previous work on and related to the problem of estimating the head pose, or coarse gaze, of people in video. We described a method for finding coarse gaze in real-time that includes real-time tracking, and real-time classification and showed that it performs at the current state of the art. We then briefly described three applications that demonstrate both the efficacy of the method, and also the value of being able to estimate gaze from video.

The accuracy with which we can obtain the gaze is a limiting factor and falls well below what humans can achieve with the same stimuli. We hope to investigate methods for improving the accuracy, such as better use of local motion information, and also methods for reducing the training effort by moving to unsupervised learning techniques. While randomised ferns are certainly fast and expedient, we do not propose them as the only solution and other classification techniques, or possibly even regression techniques, may well yield better results.

We believe that there is a wide variety of other applications yet to be explored. Driver (or pedestrian) assistance is one clear application alluded to in the introduction, in which gaze can be used as a powerful cue in automated braking situations. A further intriguing possibility would be to monitor patterns of gaze within a crowd to detect anomalous behaviour.

Acknowledgements. This research was funded by the EU project HERMES (IST-027110), an EPSRC Platform Grant, the Mexican Council of Science and Technology (CONACYT), and Oxford Risk.

References

1. Robertson, N., Reid, I., Brady, J.: What are you looking at? gaze estimation in medium-scale images. In: Proc. HAREM Workshop (in assoc. with BMVC) (2005)
2. Robertson, N., Reid, I.D.: Estimating gaze direction from low-resolution faces in video. In: Leonardis, A., Bischof, H., Pinz, A. (eds.) ECCV 2006. LNCS, vol. 3952, pp. 402–415. Springer, Heidelberg (2006)
3. Ono, Y., Okabe, T., Sato, Y.: Gaze estimation from low resolution images. In: Chang, L.-W., Lie, W.-N. (eds.) PSIVT 2006. LNCS, vol. 4319, pp. 178–188. Springer, Heidelberg (2006)
4. Benfold, B., Reid, I.: Colour invariant head pose classification in low resolution video. In: Proceedings of the 19th British Machine Vision Conference (2008)
5. Benfold, B., Reid, I.: Guiding visual surveillance by tracking human attention. In: Proc. BMVC (2009)
6. Orozco, J., Gong, S., Xiang, T.: Head pose classification in crowded scenes. In: Proc. BMVC (2009)
7. Sommerlade, E., Benfold, B., Reid, I.: Gaze directed camera control for face image acquisition. under review. In: Intl Conf. on Robotics and Automation (2011)
8. Patron, A., Marszalek, M., Zisserman, A., Reid, I.: High five: Recognising human interactions in tv shows. In: Proc. BMVC (2010)
9. Dalal, N., Triggs, B.: Histograms of oriented gradients for human detection. In: IEEE CVPR, vol. 2, pp. 886–893 (2005)
10. Lucas, B.D., Kanade, T.: An iterative image registration technique with an application to stereo vision. In: IJCAI, pp. 674–679 (1981)
11. Prisacariu, V., Reid, I.: fastHOG - a real-time GPU implementation of HOG. Technical Report 2310/09, Dept of Engineering Science, Oxford University (2009)
12. Lepetit, V., Lagger, P., Fua, P.: Randomized trees for real-time keypoint recognition. In: CVPR, vol. 2, pp. 775–781. IEEE Computer Society, Los Alamitos (2005)
13. Sommerlade, E., Reid, I.: Probabilistic surveillance with multiple active cameras. In: Proc. IEEE Int'l Conf. on Robotics and Automation, pp. 440–445 (2010)
14. Tsochantaridis, I., Hofman, T., Joachims, T., Altun, Y.: Support vector machine learning for interdependent and structured output spaces. In: Proc. ICML (2004)
15. Ferrari, V., Marin-Jimenez, M., Zisserman, A.: Pose search: retrieving people using their pose. In: Proc. IEEE CVPR (2009)
16. Desai, C., Ramanan, D., Fowlkes, C.: Discriminative models for multi-class object layout. In: Proc. IEEE ICCV (2009)
17. Joachims, T., Finley, T., Yu, C.: Cutting plane training of structural svms. Machine Learning 77, 27–59 (2009)

Algorithm for Discriminating Aggregate Gaze Points: Comparison with Salient Regions-Of-Interest

Thomas J. Grindinger[1], Vidya N. Murali[1], Stephen Tetreault[2],
Andrew T. Duchowski[1], Stan T. Birchfield[1], and Pilar Orero[3]

[1] Clemson University
[2] Rhode Island College
[3] Universitat Autònoma de Barcelona

Abstract. A novel method for distinguishing classes of viewers from their aggregated eye movements is described. The probabilistic framework accumulates uniformly sampled gaze as Gaussian point spread functions (heatmaps), and measures the distance of unclassified scanpaths to a previously classified set (or sets). A similarity measure is then computed over the scanpath durations. The approach is used to compare human observers's gaze over video to regions of interest (ROIs) automatically predicted by a computational saliency model. Results show consistent discrimination between human and artificial ROIs, regardless of either of two differing instructions given to human observers (free or tasked viewing).

1 Introduction

A compelling means of analysis of human visual perception is drawn from the collection of eye movements over dynamic media, *i.e.*, video. The video stream can either be a scene captured by a forward-facing camera worn during the performance of some natural task [1], or of film presented to the viewer [2]. Analysis of the former leads to improved understanding of how humans function in the world, and in particular, how vision is used in concordance with basic motor actions such as walking or reaching [3]. Analysis of the latter leads to better understanding of how artistic media is perceived, and in turn, how its design and production can be altered to affect its perception.

Analysis of eye movements over dynamic media has largely been performed manually, *e.g.*, by hand-coding saccadic events as they occur in relation to events present in the media such as scene cuts [4]. What is needed, and what this paper addresses, is an automatic means of classification of disparate viewing patterns, or *scanpaths*—defined as the temporal sequence of gaze or fixation coordinates cast over the stimulus.

This paper contributes a means of classification of scanpaths accumulated over temporal event samples. Event samples happen to coincide with video frames in this instance, but the technique can assume any sampling rate and is thus also applicable to still imagery presented for extended viewing durations [5]. Applications of the approach include gaze-based discrimination between classes of human viewers (*e.g.*, experts from novices—eye movements are known to be task-dependent [6]), or discrimination between human gaze and artificially predicted regions of interest, or ROIs. The paper focuses on the latter, in a manner differing from previous work with images [7], distinguishing between *perceptually salient* and *computationally salient* gaze coordinates.

R. Koch et al. (Eds.): ACCV 2010 Workshops, Part I, LNCS 6468, pp. 390–399, 2011.
© Springer-Verlag Berlin Heidelberg 2011

2 Background

Scanpath comparison can be classified as either *content-* or *data-driven*. The former is largely based on regions of interest, or ROIs, identified *a priori* in the stimulus and subsequently by associating those regions with fixations, leading to analysis of image regions or elements fixated by the viewer. The latter approach, in contrast, is made on scanpaths directly, independent of whatever was presented as the stimulus. An important advantage of the latter is that it obviates the need for establishing a reference frame within which the ROI stipulation must take place.

Consider two recent approaches to the scanpath comparison problem. The vector-based similarity measure is content-driven, as it relies on the quantization of the stimulus frame into an arbitrarily-sized 5×5 grid which serves as the method's source of ROI labeling [8]. A label is added to the scanpath stream whenever a fixation is present within a grid cell. In contrast, the revisited string-editing approach is data-driven, as it operates directly on scanpaths [9]. String (ROI) labels are determined by overlapping fixation clusters. Both approaches consider fixation durations and are therefore potentially suitable for analysis of gaze collected over dynamic media, however, their means of scanpath aggregation are derived from pairwise vector or string comparisons. For groups of viewers, considerable additional organization is required.

As an alternative to string-editing approaches, heatmaps have become a common tool for visualization of eye tracking data [10,11]. To our knowledge, to date they have not been successfully used for quantitative classification of aggregate eye movements.

Perhaps most similar to the present work are two previous efforts of calculation of the "average scanpath" [12] and of the computation of the scanpath distance via the Earth Mover's Distance [13]. The former was based on string-based multiple sequence alignment, although the derivative notion of variance (distance from the average) was omitted. The latter relied on the conceptualization of a scanpath composed of "piles of earth", with a comparison scanpath represented by "holes". The minimum amount of energy required to move earth from piles to holes gave the scanpath similarity.

The present paper extends a framework for multiple scanpath comparison and classification [5]. Although the previous approach was inspired by dynamic media, it was only implemented over still images viewed for very short durations. In this paper the analysis framework is applied to dynamic media for which it was originally conceived, namely video sequences. The resultant procedure may be conceptualized as a measure of deviation, over time, of one or more scanpaths of unknown classification from a set of scanpaths of known classification. This is similar to a prior effort based on machine learning, which was also intended to act as a classifier, although its aim was to classify content (*i.e.*, image regions) [14], whereas the present approach is directed at classification of the data (*i.e.*, scanpaths).

3 Classification Framework

Following Airola *et al.*'s nomenclature [15], let D be a probability distribution over the sample space $\mathcal{Z} = \mathcal{X} \times \mathcal{Y}$, with input space $\mathcal{X}$ and output space $\mathcal{Y} = \{-1, 1\}$, where $y \in \mathcal{Y}$ denotes the labeling of the input $x \in \mathcal{X}$ as a non-class ($x_- \in X_-$) or class member ($x_+ \in X_+$), respectively. We define a classifier as a function $C_Z(x)$ that outputs a set of threshold-based decisions $Z = \{z_1, \ldots, z_m\} \in \mathcal{Z}^m$ where $z_i = (x_i, y_i)$, for the training set of m training examples $X = \{x_1, \ldots, x_m\} \in \mathcal{X}^m$.

There are three steps to building and evaluating the real-valued prediction function C_Z produced by a learning algorithm developed with fixed training set Z. First, similarity scores are extracted from X. Second, a discrimination threshold h is computed from the similarity scores assigning the positive class X_+ to x if $C_Z(x) > h$ and the negative class X_- otherwise. Third, classifier reliability is gauged by the *conditional expected AUC*, or AUC, the area under Receiver Operating Characteristic (ROC) curve, $A(C_Z) = E_{x_+ \sim D_+, x_- \sim D_-}[H(C_Z(x_+) - C_Z(x_-))]$ where $H(a)$ is the Heaviside step function, which returns 1 when $a > 0$, $1/2$ when $a = 0$, and 0 when $a < 0$. In practice, because the probability distribution D cannot be accessed directly, the AUC estimate $\hat{A}$ is calculated *e.g.*, via cross-validation, or by the Wilcoxon-Mann-Whitney statistic:

$$\hat{A}(S, C_Z) = \frac{1}{|S_+||S_-|} \sum_{x_i \in S_+} \sum_{x_j \in S_-} H(C_{\overline{\{i\}}}(x_i) - C_{\overline{\{j\}}}(x_j))$$

where $S_+ \subset S$ and $S_- \subset S$ are the positive and negative examples of the set S, and $C_{\overline{\{i\}}}(x_i)$ is the classifier trained without the ith training example.

Along with AUC, classifier accuracy is reported by evaluating $C_Z(w)$ on test data $w \in W$, assumed to be disjoint from X. Accuracy is defined as the ratio of correctly classified examples of W (true positives and true negatives) to all classified examples.

Accuracy and AUC measures can be seen to correspond to two different metrics of interest. The former is related to the quality of the learning algorithm, *i.e.*, how well on average C_Z generalizes to new test and training data. The latter addresses how well $C_Z(x)$ generalizes to future test examples once learned from the given training set. In the present context, the latter is more of interest as it provides a better indication of the discriminability of the given training data set against the test set or sets, *i.e.*, does a given scanpath class differ from another class or classes of scanpath sets.

3.1 Extracting Similarity Scores

The classifier's similarity measure computes a scanpath's deviation from a probabilistic model of one (or more) class(es) of scanpaths classified *a priori*. Scanpath classes can be operationalized arbitrarily, *e.g.*, based on some characterization of viewers. The classifier functions over dynamic stimuli, *i.e.*, video, which may be considered as a collection of static stimuli, *i.e.*, frames. Scanpath similarity metrics developed for static stimuli can thus be applied on a frame-by-frame basis and aggregated in some way (*e.g.*, averaged). The trouble with prior vector- or string-based approaches is their reliance on pairwise comparisons for aggregation. This leads to rather complicated bookkeeping requirements for pairwise organization, *e.g.*, labeling each pair as *local*, *repetitive*, *idiosyncratic*, or *global* based on the dyadic permutations of viewer and stimulus [7].

Presently, each frame is composed of a sampled set of gaze points (or fixations), sampled from as many sets as there are scanpath classes, with each set composed of scanpaths collected from multiple viewers. A per-frame similarity measure is then derived and averaged over the duration of the video sequence to compute the total similarity of an unclassified scanpath to the one or more sets of classified scanpaths.

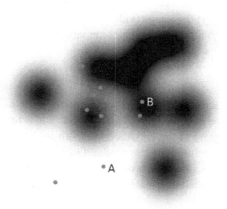

Fig. 1. Heatmap of a classified scanpath set S at a discrete timestamp. As yet unclassified scanpaths' (gray circles not used in heatmap generation) similarities are calculated as the average Gaussian similarity, *e.g.*, $d(A, S) < d(B, S)$ in this example.

With video acting as the temporal reference frame, a scanpath $s(t)$ is parametrized by the frame timestamp t, such that $s(t) = \{(i(t), j(t)) \mid t \in [t - w, t + w]\}$ for some window w, with $w = 0$ identifying a single frame, yielding the scanpath's 0^+ gaze points over a video frame at t^1. This *event-driven* model, effectively samples a scanpath at a single point in time, and affords notational interchangeability between a gaze point, fixation, and scanpath, when considered on a per-event, or in this case per-frame, basis. A set of scanpaths $S(t) = \{s_1(t), s_2(t), \ldots, s_m(t)\}$ is similarly parametrized to define the combined set of gaze points over frame t from the scanpath set collected from m viewers. Over each frame, multiple sets are represented, e.g., S_+ member and S_- non-member sets (in the experiment described below, three such sets are established).

Modeling a classified scanpath s by a normally distributed point spread function $f(s) = 1/\sqrt{2\pi\sigma^2} \exp\left(-s^2/2\sigma^2\right)$ produces the well-known *heatmap* scanpath visualization (on a per-frame basis; see Fig. 1), typically visualized with the Gaussian kernel's support truncated beyond 2σ for computational efficiency [16]. Extending kernel support also defines the scanpath's first moment $\mu_s = \int_{-\infty}^{\infty} sf(s)ds$ so that the (Gaussian) similarity of an unclassified scanpath s' to s is estimated by its deviation

$$g(s', \mu_s) = \frac{1}{\sqrt{2\pi\sigma^2}} \exp\left(-\frac{(s' - \mu_s)^2}{2\sigma^2}\right)$$

with frame timestamp t made implicit and σ set to the expected eye tracker error, as illustrated in Fig. 1. In practice, the above model is necessarily discrete and s is

---

[1] With a 50 Hz eye tracking sampling rate and a common video refresh rate of 30 Hz, it is assumed that a scanpath will yield at most two gaze point samples per frame; alternatively, if operationalized by a sequence of fixations, a scanpath will yield a single fixation coordinate per frame (or none if the frame happened to sample an inter-fixation saccade).

understood to be two-dimensional, $s(t) = (i(t), j(t))$, $s(t) \in \mathbb{R}^2$, with t denoting the frame timestamp and (i, j) the image (video frame) coordinates.

The similarity of s' to a set of classified scanpaths S (at t) is defined as

$$d(s', S) = \frac{1}{|S|} \sum_{s \in S} g(s', \mu_s)$$

where the weighting factor $1/|S|$ is used for similarity score normalization. The measure $d(s', S)$ is averaged over the entire video sequence to estimate the mean similarity of an unclassified scanpath to the classified scanpath set, $\bar{d}(s', S) = 1/T \sum_t d(s', S)$ with $t \in T$, the sequence duration. The resultant mean similarity lies between 0 and 1, but tends to fall near 0. Its value, however, is not as important as the probability that the score lies within the expected distribution of scores for a specific class.

3.2 Computing the Classification Threshold

Gaussian similarity scores serve as input to the classification mechanism that estimates an optimal discrimination threshold for scanpaths of unknown classification. An unclassified scanpath is accepted by the classifier if its similarity score is higher than the computed threshold.

The ROC curve plots the true positive response against the false positive response of the threshold at each threshold level and provides two convenient facilities. First, it facilitates the choice of an optimal threshold, by selecting the level at which the threshold is closest to $(0, 1)$, where the ratio of false positives to true positives is balanced. Second, AUC indicates the classifier's discriminative capability. Ideally, AUC should equal unity (1), while a completely random classifier yields AUC close to 0.5. AUC represents the probability of an arbitrarily-chosen class member obtaining a similarity score greater than some arbitrarily-chosen non-class member.

3.3 Estimating Classifier Performance via Cross-Validation

A typical strategy used for estimating the performance, or reliability, of a classifier, when operating in a small sample setting, is cross-validation[2]. Specifically, leave-pair-out cross-validation, or LPOCV, is adopted since the intent is to estimate the conditional AUC as an indicator of the classifier's performance while avoiding the pitfalls associated with pooling and averaging of LOOCV (leave-one-out cross-validation) [15].

Cross-validation is performed by repeatedly partitioning the data set into two non-overlapping parts: a training set and a hold-out set. For each partitioning, the hold-out set is used for testing while the remainder is used for training. Accuracy is computed as the percentage of hold-out sets successfully classified. For each partitioning, LPOCV leaves out at a time from the training set each possible positive-negative pair of training examples. With LPOCV, AUC is estimated as

$$\hat{A}(X, C_Z) = \frac{1}{|X_+||X_-|} \sum_{s_i \in X_+} \sum_{s_j \in X_-} H(C_{\overline{\{i,j\}}}(s_i) - C_{\overline{\{i,j\}}}(s_j))$$

---

[2] Scanpath data sets generally number in the tens, whereas classifiers tend to operate on data sets numbering in the thousands.

(a) Seq. A, chosen for its mis-placed pair of modern sneakers (b) Seq. B, chosen for its unfamiliarity (c) Seq. C, chosen for its large number of prominent faces

Fig. 2. Frames from stimulus sequences. Seqs. A and C were excerpts from Sofia Coppola's *Marie Antoinette* © 2006, Columbia Pictures and Sony Intl., obtained with permission for research purposes by the Universitat Autònoma de Barcelona. Seq. B shows the mouse vasculature in the spinal cord at $0.6 \times 0.6 \times 2$ μm resolution with blood vessels stained black, as obtained by a knife-edge microscope (courtesy of Texas A&M).

where $X_+ \subset X$ and $X_- \subset X$ are the positive and negative examples of the training set X, $C_{\overline{\{i,j\}}}(s_i)$ is the classifier trained without the ith and j^{th} training examples, and $H(a)$ is the Heaviside step function. Because AUC estimate $\hat{A}(X, C_Z)$ is equivalent to the Wilcoxon-Mann-Whitney U statistic, AUC > 0.7 is generally considered a statistically significant indicator of discriminability, although a test of significance should be performed by computing the standardized value under assumption of normality of class distributions.

The training data generally consists of multiple classes, very often two, but possibly more. The current approach generates multiple classifiers, each trained to a single class, with all other classes acting as non-class training data. Generally, when there are more than two classes, a "one-to-many" comparison may be carried out first, with all non-class training data pooled into the negative class set. Should the classifier AUC be significant, "one-to-one" comparisons can then be performed, in a manner analogous to ad-hoc pairwise t-tests following ANOVA.

4   Empirical Evaluation

The classifier was applied to scanpaths drawn from three classes: two from human observers distinguished by differing tasks, and the third from a bottom-up saliency model (simulating artificial observers), developed by Itti *et al.* [17]. The model is part of iLab's Neuromorphic Visual C++ Toolkit and is freely available online[3]. At the model's core is a neuromorphic simulation that predicts elements of a visual scene that are likely to attract the attention of human observers. This has wide applications in machine vision, *e.g.*, automated target detection in natural scenes, smart image compression, *etc.* The model was compared to human scanpaths captured over video sequences.

---

[3] http://ilab.usc.edu/bu/, last accessed Aug., 2010.

Stimulus. Stimuli consisted of three video sequences, named A, B, and C, shown to human observers in Latin square counterbalanced order, with approximately each third of the viewers seeing the sequences in order $\{A, B, C\}$, $\{B, C, A\}$, or $\{C, A, B\}$. Seq. A contained a misplaced modern pair of sneakers in an 18th century setting, while a modern popular song played in the background. Seq. C was from the same feature film, with scenes containing a large number of human faces. Seq. B was composed of CT-like scans of the mouse vasculature in the spinal cord. Select frames from the clips are shown in Fig. 2.

Apparatus. Eye movements were captured by a Tobii ET-1750 eye tracker, a 17 inch (1280×1024) flat panel with built-in eye tracking optics. The eye tracker is binocular, sampling at 50 Hz with $0.5°$ accuracy.

Participants. Twenty-seven college students volunteered in the study (seven male, twenty female). Participants' ages ranged from 18 to 21 years old.

Procedures. Participants sat in front of the eye tracker at about 60 cm distance. Following 9-point calibration, subjects were asked to naturally watch the first of two viewings of each of the three sequences (amounting to "free viewing"). They then received viewing instructions prior to the second viewing of the same sequence. For seq. A, they were asked to look for anything unusual (they were meant to notice the sneakers). For seq. B, they were asked to focus on the vascular stains (they were meant to avoid the aberrant artifacts at the top and sides of the frames). For seq. C, they were asked to avoid looking at faces (they were meant to simulate autism, since autistic viewers have been shown to exhibit reduced face gaze [18]).

Artificial gaze points over video were generated by the iLab Neuromorphic Toolkit. The toolkit contains a program called *ezvision* that can be executed on static images to produce a primary point of focus that is expected to match the visual attention of a human viewing the scene, followed by other salient points in the scene that are connected by a trajectory depending on the exposure time stipulated. However, the model also operates in video mode by extracting images from the video at the video frame rate. This causes the algorithm to be forced to find a salient point within the frame within the frame's exposure duration. For a typical video, this means the algorithm has only 33 ms to arrive at a salient viewpoint in the frame.

To compare the model's prediction with gaze points captured from human observers *ezvision* was run in video mode with the timestep set to 33 ms for the faces and shoes video, and 40 ms for the mouse video. Itti's algorithm is able to produce predictions with small amounts of noise added to the predictions [17]. This helped simulate results for 27 hypothetical users, by running *ezvision* on each video 27 times with random noise added to the predictions made each time.

5   Results

Classifier AUC and accuracy shows significantly consistent discriminability (AUC > 0.7) between perceptual (top-down) and computational (bottom-up) saliency (see Tab. 1).

Table 1. Results composed of classifier accuracy (ACC) and area under ROC curve (AUC) for one-to-many and one-to-one comparisons of two classes of viewers ("free viewing" and tasked) vs. the computational model for each of the three video stimuli

| | One-to-many Cross-Validation | | | One-to-one Cross-Validation | | | | | |
| | Perceptual (pooled) vs. computational saliency | | | Perceptual "free viewing" vs. computational saliency | | | Perceptual tasked vs. computational saliency | | |
	A	B	C	A	B	C	A	B	C
ACC	1.000	1.000	0.997	1.000	0.999	0.999	0.999	1.000	1.000
AUC	1.000	1.000	1.000	1.000	1.000	1.000	1.000	1.000	1.000

Consistency refers to the evaluation of the Heaviside step function $H(a)$, where the classifier correctly discriminates between human and artificial scanpath classes in all of the $m \times (m-2)$ cross-validation partitionings, over all frames of each of the three video stimuli. The classifier is not as consistent in distinguishing between the two human scanpath classes, able only to distinguish between them in two of the three cases (Seq. B and C; these results are discussed at length elsewhere [19]).

Human observers tend to exhibit extreme preferential behavior over Seq. C, *i.e.*, when free viewing, heatmap visualization (see Fig. 3) suggests most viewers fixate faces, particularly in "close shots". Tasked viewers, in contrast, who were told to avoid faces, did so, but without apparent agreement on scene elements. Both strategies employ top-down directives that are apparently different from the strategy employed by the computational saliency model. The model fails to match human scanpaths over Seq. B even though it seems well suited to this stimulus (high contrast elements and sudden onset stimulus). Visualization suggests that both the model's and free viewers' gaze fell atop the sudden onset aberrant artifacts at the video frame edges. However, once humans were tasked to avoid these artifacts, they did so, whereas the model was not privy to this top-down goal-directed information. In either case, insufficient gaze overlap was detected over the length of this short video clip to diminish classifier output below unity. Seq. A yields similarly consistent discriminability results. Verbal instructions had little

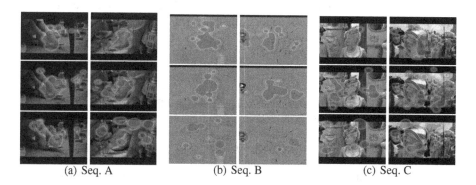

| (a) Seq. A | (b) Seq. B | (c) Seq. C |

Fig. 3. Heatmap visualizations of two excerpted video frames viewed freely (top row), with task (middle row) or by the saliency model (bottom row)

impact on perturbing human gaze (tasked scanpaths were not discriminable from free viewers' scanpaths by the classifier). Seq. A appears sufficiently complex to foil the saliency model from accurately predicting features selected by human observers.

6   Discussion

The saliency model works well on simple videos/images of traffic signals, or on tracks of single or multiple persons moving against fairly non-complex backgrounds, or in interactive visual environments [20]. However, for complex video segments with multiple objects of interest in the foreground and background and with rapid motion between the frames such as the *Marie Antoinette* videos, the bottom-up saliency model's gaze selection differs from that of natural viewing by humans. Two hypothetical parameters describe the extent of success/failure of the model: (1) the complexity of a single frame in the video, and (2) the amount of motion (apparent or real) between frames. When the two are low (simple images with small motion between frames), the model is likely to match human gaze points. However, when the complexity of the image and/or inter-frame motion increase(s), results diverge. The model could probably be used to describe the human visual system's tropism to salient points in a video, but only under fairly simple conditions. Once video complexity increases, bottom-up saliency can be clearly distinguished from tasked as well as natural viewing.

Given sufficiently clear instructions (*e.g.*, avoid looking at faces), the tropism of the human visual system, driven by top-down cognitive processes, differs from free viewing such that it can generally be distinguished by the classifier. The saliency model is, in contrast, task-independent and models bottom-up processes. Although it is possible to modify the relative feature weights in the construction of the saliency map with supervised learning to achieve some degree of specialization, it is at present unlikely that such specialization is sufficient to adequately model top-down visual processes.

7   Conclusion

A classification algorithm was developed to distinguish scanpaths collected over dynamic media. The algorithm successfully discriminated between perceptual and computational saliency over video sequences, illustrating the disparity between top-down visual processes and their bottom-up computational models.

References

1. Land, M.F., Tatler, B.W.: Looking and Acting: Vision and Eye Movements in Natural Behavior. Oxford University Press, New York (2009)
2. Smith, T.J., Henderson, J.M.: Edit Blindness: The Relationship Between Attention and Global Change Blindness in Dynamic Scenes. Journal of Eye Movement Research 2, 1–17 (2008)
3. Franchak, J.M., Kretch, K.S., Soska, K.C., Babcock, J.S., Adolph, K.E.: Head-Mounted Eye-Tracking of Infants' Natural Interactions: A New Method. In: ETRA 2010: Proceedings of the 2010 Symposium on Eye Tracking Research & Applications, pp. 21–27. ACM, New York (2010)

4. d'Ydewalle, G., Desmet, G., Van Rensbergen, J.: Film perception: The processing of film cuts. In: Underwood, G.D.M. (ed.) Eye guidance in reading and scene perception, pp. 357–367. Elsevier Science Ltd., Oxford (1998)

5. Grindinger, T., Duchowski, A.T., Sawyer, M.: Group-Wise Similarity and Classification of Aggregate Scanpaths. In: ETRA 2010: Proceedings of the 2010 Symposium on Eye Tracking Research & Applications, pp. 101–104. ACM, New York (2010)

6. Yarbus, A.L.: Eye Movements and Vision. Plenum Press, New York (1967)

7. Privitera, C.M., Stark, L.W.: Algorithms for Defining Visual Regions-of-Interest: Comparison with Eye Fixations. IEEE Transactions on Pattern Analysis and Machine Intelligence (PAMI) 22, 970–982 (2000)

8. Jarodzka, H., Holmqvist, K., Nyström, M.: A Vector-Based, Multidimensional Scanpath Similarity Measure. In: ETRA 2010: Proceedings of the 2010 Symposium on Eye Tracking Research & Applications, pp. 211–218. ACM, New York (2010)

9. Duchowski, A.T., Driver, J., Jolaoso, S., Ramey, B.N., Tan, W., Robbins, A.: Scanpath Comparison Revisited. In: ETRA 2010: Proceedings of the 2010 Symposium on Eye-Tracking Research & Applications, pp. 219–226. ACM, New York (2010)

10. Pomplun, M., Ritter, H., Velichkovsky, B.: Disambiguating Complex Visual Information: Towards Communication of Personal Views of a Scene. Perception 25, 931–948 (1996)

11. Wooding, D.S.: Fixation Maps: Quantifying Eye-Movement Traces. In: ETRA 2002: Proceedings of the 2002 Symposium on Eye Tracking Research & Applications, pp. 31–36. ACM, New York (2002)

12. Hembrooke, H., Feusner, M., Gay, G.: Averaging Scan Patterns and What They Can Tell Us. In: ETRA 2006: Proceedings of the 2006 Symposium on Eye Tracking Research & Applications, p. 41. ACM, New York (2006)

13. Dempere-Marco, L., Hu, X.P., Ellis, S.M., Hansell, D.M., Yang, G.Z.: Analysis of Visual Search Patterns With EMD Metric in Normalized Anatomical Space. IEEE Transactions on Medical Imaging 25, 1011–1021 (2006)

14. Torstling, A.: The Mean Gaze Path: Information Reduction and Non-Intrusive Attention Detection for Eye Tracking. Master's thesis, The Royal Institute of Technology, Stockholm, Sweden, Techreport XR-EE-SB 2007:008 (2007)

15. Airola, A., Pahikkala, T., Waegeman, W., De Baets, B., Salakoski, T.: A Comparison of AUC Estimators in Small-Sample Studies. In: Proceedings of the 3rd International workshop on Machine Learning in Systems Biology, pp. 15–23 (2009)

16. Paris, S., Durand, F.: A Fast Approximation of the Bilateral Filter using a Signal Processing Approach. Technical Report MIT-CSAIL-TR-2006-073, Massachusetts Institute of Technology (2006)

17. Itti, L., Koch, C., Niebur, E.: A Model of Saliency-Based Visual Attention for Rapid Scene Analysis. IEEE Transactions on Pattern Analysis and Machine Intelligence (PAMI) 20, 1254–1259 (1998)

18. Leigh, R.J., Zee, D.S.: The Neurology of Eye Movements, 2nd edn. Contemporary Neurology Series. F. A. Davis Company, Philadelphia (1991)

19. Grindinger, T.J.: Event-Driven Similarity and Classification of Scanpaths. PhD thesis, Clemson University, Clemson, SC (2010)

20. Peters, R.J., Itti, L.: Computational Mechanisms for Gaze Direction in Interactive Visual Environments. In: ETRA 2006: Proceedings of the 2006 Symposium on Eye Tracking Research & Applications, pp. 27–32. ACM, New York (2006)

Gaze Estimation Using Regression Analysis and AAMs Parameters Selected Based on Information Criterion

Manabu Takatani[1], Yasuo Ariki[2], and Tetsuya Takiguchi[2]

[1] Graduate School of Engineering, Kobe University
takatani_m@me.cs.scitec.kobe-u.ac.jp
[2] Organization of Advanced Science and Technology, Kobe University
{ariki,takigu}@kobe-u.ac.jp
http://www.me.cs.scitec.kobe-u.ac.jp/

Abstract. One of the most crucial techniques associated with Computer Vision is technology that deals with the automatic estimation of gaze orientation. In this paper, a method is proposed to estimate horizontal gaze orientation from a monocular camera image using the parameters of Active Appearance Models (AAM) selected based on several model selection methods. The proposed method can estimate horizontal gaze orientation more precisely than the conventional method (Ishikawa's method) because of the following two unique points: simultaneous estimation of horizontal head pose and gaze orientation, and the most suitable model formula for regression selected based on each model selection method. The validity of the proposed method was confirmed by experimental results.

1 Introduction

The human gaze is thought to be effective for understanding or measuring the degree of his / her interest or attention because the information from the gaze is the most vital for humans to understand their environment. Thus, estimating gaze orientation automatically is expected to be applied not only to robot vision, artificial intelligence, and human interaction but also to the analysis of image and video content, and analysis and retrieval based on human percipient models [1,2].

In order to estimate gaze orientation, two main types of methods have been proposed. One approach employs a special device (such as an infrared camera) as proposed by Ohno [3]. This approach can estimate gaze orientation with a high degree of accuracy. The other approach processes monocular camera images. The advantage of this approach is that gaze orientation can be estimated inexpensively because only a monocular camera is required. From this view point, we employ the latter approach in this study.

Many methods have been proposed for estimating gaze orientation from a monocular image. For instance, Yamazoe proposed the use of the Lukas-Kanade's feature tracking method [4] and 3D-eyeball model[5]. Gaze can be estimated stably by this method, even if the subject is not included in the training data.

R. Koch et al. (Eds.): ACCV 2010 Workshops, Part I, LNCS 6468, pp. 400–409, 2011.
© Springer-Verlag Berlin Heidelberg 2011

However, the precision is not so accurate because the gaze orientation is estimated after the head pose estimation.

Ishikawa proposed the use of 3D AAM (Active Appearance Models) to extract the coordinates of the feature points, and the gaze orientation was estimated by the 3D eyeball model. The gaze orientation can be computed more precisely than Yamazoe's method due to the improvement of head pose estimation error using this method. However, the positioning error of the feature points causes the gaze estimation error because gaze orientation is computed using the coordinates of the feature points relative to the eye [6].

Thus, there were few methods that address the relationship between head pose and gaze orientation simultaneously. This is the reason why we propose a method in this paper to estimate them simultaneously using regression-based AAM parameters.

Moreover, the feature parameters extracted by AAM contain unessential information for the estimation. To select the essential feature parameters, we employ the model selection method (e.g. AIC[10], MDL[11], BIC[12]).

The rest of this paper is organized as follows. In Section 2, the method to estimate horizontal gaze orientation is proposed. Experimental results are presented in Section 3, followed by concluding remarks in Section 4.

2   Proposed Method

In this section, the method to estimate horizontal gaze orientation is proposed. Fig. 1 shows a processing flow of the proposed method. First, the facial area in the test image is detected using AdaBoost based on Haar-like features for stable AAM performance. Next, the feature parameters are extracted by AAM on this facial area. Finally, the head pose and the horizontal gaze orientation are simultaneously estimated using a regression model that is selected based on AIC (Akaike Information Criterion), MDL (Minimum Description Length), and BIC (Bayes Information Criterion).

2.1   Facial Area Search

The performance of AAM feature extraction depends on the initial search points. To make AAM search performance more stable, the facial area in an image is roughly computed using AdaBoost based on Haar-like features proposed by Viola [7]. Haar-like features for face detection are based on the difference between the sums of the pixels within two rectangular regions of the same size and shape that are adjacent to one another horizontally or vertically.

Since the total number of Haar-like features is far larger than the number of pixels on the image, simple and efficient classifiers can be constructed by selecting a small number of important features using AdaBoost from a huge library of potential features.

Actually, we employed "haarcascase_frontalface" in OpenCV library for searching facial area.

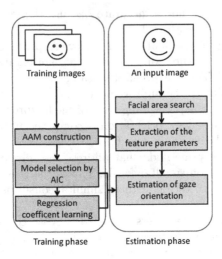

Fig. 1. Processing flow of the proposed method

2.2   Active Appearance Models

Cootes proposed AAM to represent shape and texture variations of an object with a low dimensional parameter vector **c** [8]. Vector **c** can represent various facial images with arbitrary orientation of face and gaze using the training images that contain varying faces and gazes.

Since AAM is constructed statistically from training images, some elements of vector **c** represent the information related to the variance in face and gaze orientation. Therefore, this parameter vector **c** is employed as the feature parameter for the estimation of gaze orientation because parameter vector **c** is thought to be linearly associated with the displacement of the feature points caused by changes in head pose and gaze orientation.

In the AAM framework, shape vector **s** and texture vector **g** of the face are represented as shown in Eq. (1) and Eq. (2), respectively. In particular, shape vector **s** indicates the coordinates of the feature points, and texture vector **g** indicates the gray-level of the image within the shape,

$$\mathbf{s}(\mathbf{c}) = \bar{\mathbf{s}} + \mathbf{P_s}\mathbf{W_s^{-1}}\mathbf{Q_s}\mathbf{c} \tag{1}$$

$$\mathbf{g}(\mathbf{c}) = \bar{\mathbf{g}} + \mathbf{P_g}\mathbf{Q_g}\mathbf{c} \tag{2}$$

where $\bar{\mathbf{s}}$ and $\bar{\mathbf{g}}$ are the mean shape and mean texture of training images, respectively. $\mathbf{P_s}$ and $\mathbf{P_g}$ are a set of orthogonal bases of shape and texture variation, respectively. $\mathbf{Q_s}$ and $\mathbf{Q_g}$ are eigen matrices (including the eigenvectors). $\mathbf{W_s}$ is a diagonal weight matrix for each shape parameter, allowing for the difference in units between the shape and texture models. **c** is a vector of parameters controlling both the shape and gray-levels of the model.

Fig. 2. 43 feature points for construction of AAM

In this paper, AAM is constructed using 43 shape points as shown in Fig. 2.
On the test image **I**, the goal of the AAM search is to minimize the error $\mathbf{e}(\mathbf{p}, \mathbf{c})$
as shown in Eq. (3) with respect to parameter vector $\mathbf{c}$ and pose parameter
vector $\mathbf{p}$.

$$\mathbf{e}(\mathbf{p}, \mathbf{c}) = \| \mathbf{g}(\mathbf{c}) - \mathbf{I}(\mathbf{W}(\mathbf{p})) \| \tag{3}$$

where $\mathbf{W}$ denotes the Affine warp function, $\mathbf{p}$ denotes the pose parameter vector
for Affine warp (translation, scale, rotation), and $\mathbf{I}(\mathbf{W}(\mathbf{p}))$ indicates the Affine-
transformed image controlled by the pose parameter $\mathbf{p}$ on the test image $\mathbf{I}$. $\mathbf{g}(\mathbf{c})$
is given in Eq. (2).

Thus, we can extract the most optimized parameter vector $\mathbf{c}$ as feature parameters from the test image.

2.3   Regression Analysis and Model Selection Method

The head pose and gaze orientation are estimated by regression analysis using
the feature parameters extracted by AAM. In order to estimate horizontal face
orientation ϕ, Cootes proposed a face rotation model [9]. In this paper, we propose a nobel method for simultaneous estimation of horizontal face orientation ϕ
and horizontal relative gaze orientation θ based on the relationship between the
displacement of feature points and rotation angles ϕ or θ by expanding Cootes's
method.

In the proposed method, the regression formula to estimate horizontal facial
orientation ϕ and horizontal gaze orientation θ can be represented by Eq. (4).

$$\mathbf{y} = \mathbf{a_0} + \mathbf{Ac} \tag{4}$$

where $\mathbf{y} = (\phi, \acute{\theta})^T \in \mathbf{R}^{2 \times 1}$ is the vector of the objective variable. $\acute{\theta}$ is the
total orientation of facial orientation ϕ and gaze orientation θ ($\acute{\theta} = \phi + \theta$),
which means the horizontal gaze orientation relative to the image plane. $\mathbf{a_0} =$
$\left(a_{1,0}, a_{2,0} \right)^T \in \mathbf{R}^{2 \times 1}$ is the constant vector of regression. $\mathbf{c} = \left(c_1 \ldots c_d \right)^T \in$
$\mathbf{R}^{d \times 1}$ is the parameter vector (explanatory variable) as given in Eq. (1) and
Eq. (2). $\mathbf{A} \in \mathbf{R}^{2 \times d}$ is the matrix of the regression coefficients as given in Eq. (5).

$$\mathbf{A} = \begin{pmatrix} a_{1,1} \cdots a_{1,d} \\ a_{2,1} \cdots a_{2,d} \end{pmatrix} \tag{5}$$

where d is the dimension of parameter vector $\mathbf{c}$.

Some components of parameter vector **c** are thought to be unessential when estimating horizontal facial orientation and horizontal gaze orientation because they sometimes cause over estimation when learning the regression coefficient matrix **A**.

To solve this problem and improve the precision of this method, in this paper, the model selection method is employed to select the most suitable formula. In Eq. (4), for example, $\acute{\theta}$ can be represented as in Eq. (6).

$$\acute{\theta} = a_{2,0} + \sum_{i=1}^{d} a_{2,i} c_i + \epsilon(0, \sigma^2) \tag{6}$$

where estimation error ϵ is assumed to have Gaussian distribution with mean 0 and variance σ^2. We want to select only the essential components of parameter vector **c**, but there are many combinations of the components. Therefore, we make S_k denote a set of the components among the following $2^d - 1$ sets.

$$S_1 = \{c_1\}$$
$$S_2 = \{c_2\}$$
$$S_3 = \{c_1, c_2\}$$
$$\vdots$$
$$S_{2^d-1} = \{c_1, \ldots, c_d\}$$

Then, the regression formula for S_k given in Eq. (6) can be represented as follows.

$$\acute{\theta}_k = a_{2,0} + \sum_{i \in S_k} a_{2,i} c_i + \epsilon(0, \sigma_k^2) \tag{7}$$

After learning the regression coefficients using the least squared method among $k = 2^d - 1$, the least scored model is selected as the most suitable model.

Akaike Information Criterion[10] (AIC) is one of model selection methods in regression analysis, which indicates generalization capability of regression formular using training data.

The maximum log-scaled likelihood $l(\Theta_k; X)$ and the degrees of freedom of the model are evaluated by AIC as shown in Eq. (8).

$$\text{AIC}_k = -2l(\Theta_k; X) + 2 \dim(\Theta_k) \tag{8}$$

$$l(\Theta_k; X) = -\frac{n}{2}(1 + \log(2\pi\sigma_k^2)) \tag{9}$$

where n denotes the number of training images. Θ_k denotes the model parameters given in Eq. (7), and maximum log-scaled likelihood $l(\Theta_k; X)$ is assumed to be given in Eq. (9). The lower the AIC score, the better the evaluation of the model. This means that AIC gives an answer with a trade-off between the complexity of the model and the variance σ_k^2 of the fitting error ϵ to the training image set X as given in Eq. (9).

In a similar way, Minimum Description Length[11] (MDL), and Bayesian Information Criterion[12] (BIC) are respectively defined as shown in Eq. 10, and Eq. 11.

$$\text{MDL}_k = -l(\Theta_k; X) + \frac{\dim(\Theta_k) \ln n}{2} \tag{10}$$

$$\text{BIC}_k = -2l(\Theta_k; X) + \dim(\Theta_k) \ln n \tag{11}$$

Thus, matrix $\mathbf{A}$ and vector $\mathbf{a_0}$ are trained in the above mentioned methods. Horizontal gaze orientation relative to the image plane can be estimated as shown in Eq. (12) using the parameter vector $\mathbf{c}$ of the test image.

$$k = \arg\min_k \begin{cases} \text{AIC}_k \\ \text{MDL}_k \\ \text{BIC}_k \end{cases} \tag{12}$$

3 Experiment

To confirm the validity of our method in estimating horizontal gaze orientation, we conducted the following experiment.

3.1 Experimental Conditions

Since there was no open dataset with variation of face and gaze orientation, we prepared a dataset by asking each subject to look at each of the markers on a wall in turn. The markers were placed horizontally on the wall at every 5 degrees. The variations of head pose and gaze orientation were in the horizontal direction and ranged from approximately -20 degrees to +20 degrees relative to the front. The dataset contained 4 subjects, 63 training images and 252 test images with 640 x 480 pixels for each subject, as shown in Table 1.

Table 1. Overview of our dataset for each subject ("Training" means the number of training images, and "Test" means the number of test images)

Face [deg]	Gaze [deg]	Training	Test
0(Frontal)	± 20, ± 15, ± 10, ± 5, 0	9	36
5	+20, +15, ± 10, ± 5, 0	7	28
-5	-20, -15, ± 10, ± 5, 0	7	28
+10	+20, +15, ± 10, ± 5, 0	7	28
-10	-20, -15, ± 10, ± 5, 0	7	28
+15	-20, -15, ± 10, ± 5, 0	7	28
-15	-20, -15, ± 10, ± 5, 0	7	28
+20	-20, -15, -10, ± 5, 0	6	24
-20	-20, -15, -10, ± 5, 0	6	24
In total		63	252

In this experiment, we used AAM parameters with up to 95% cumulative contribution ratio. In fact, the number of dimensions were about 10-20.

AAM construction and regression analysis were performed for each subject. The proposed method was evaluated by comparing it with the method proposed by Ishikawa et al (a conventional method). The horizontal gaze orientation estimation method was evaluated by means of absolute error degree (MAE).

Moreover, we conducted this experiment with the purpose of showing the validity and the contribution of the two unique points of the proposed method.

At first, we compared the proposed method with the Ishikawa et al.'s method. In our method, we didn't use any model selection methods. On the other hand, we gave true coordinates of AAM shape points in Ishikawa's method because it required a lot of time for us to implement 3D AAM. Through this comparison, the validity of our method can be confirmed.

Next, we compared the difference between simultaneous method and sequential method. In the sequential method, the angle θ of the gaze in relation to the face is described in Eq. (4). After regression analysis, total horizontal gaze orientation $\acute{\theta}$ is computed as $\phi + \theta$. Through this comparison, the validity of "simultaneous" estimation can be confirmed.

Finally, we compared AIC with other methods of model selection. MDL (Minimum Description Length) [11] and BIC (Bayesian Information Criterion) [12] are well-known methods for selecting the model in recent years. Thus, the validity of "model selection by AIC" can be confirmed by comparing it with MDL and BIC.

3.2   Results

Fig. 3 shows the experimental results. We can confirm the validity of simultaneous method. The graph shows the average estimation error [deg]. Though the difference among these methods seems small, a significant difference is confirmed with significance level of 95%.

This graph shows that our approach contributes the improvement of the gaze estimation error from 4.2 [deg] to 2.7 [deg]. I think there is no critical difference

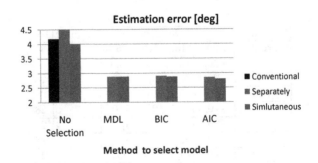

Fig. 3. Experimental results (Mean estimation error)

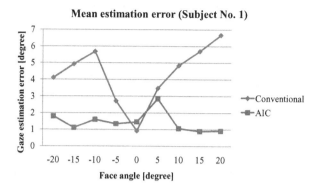

Fig. 4. Experimental results (estimation error) of subject No. 1 in each face angle

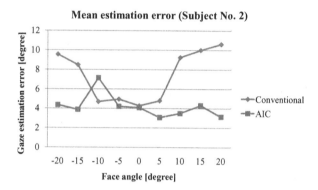

Fig. 5. Experimental results (estimation error) of subject No. 2 in each face angle

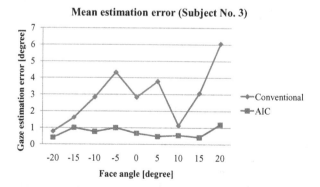

Fig. 6. Experimental results (estimation error) of subject No. 3 in each face angle

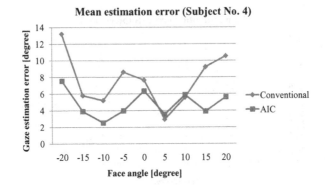

Fig. 7. Experimental results (estimation error) of subject No. 4 in each face angle

among three model selection methods. But it is important to reduce the dimension of the feature vector using model selection method.

Next, We analyzed the performance of our approach (e.g. AIC) in each face direction. Fig. 4, Fig. 5, Fig. 6, and Fig. 7 respectively show that the mean estimation error in each face angle and each subject. From these graph, we can confirm the face angle robustness of the proposed method.

From these experimental results, the validity of the proposed method to estimate horizontal gaze orientation from monocular images was confirmed. Also, it was confirmed that the model selection by AIC contributes the most to reducing the degree of error.

4   Conclusion

In this paper, the nobel method was proposed in which horizontal gaze orientation is estimated by using model selection method to select the necessary parameters from to the AAM parameters and then carrying out regression analysis on those paramters. This method contributes to the improvement of horizontal gaze estimation error from 4.2 [deg] to 2.7 [deg].

In near future research, we will address the problem of AAM adaptation to an unseen subject for a wide range of gaze estimation applications.

References

1. Miyahara, M., Aoki, M., Takiguchi, T., Ariki, Y.: Tagging Video Contents with Positive/Negative Interest. In: The 14th International Multimedia Modeling Conference, pp. 210–219 (2008)
2. Pang, D., Kimura, A., Takeuchi, T., Yamato, J.: A Stochastic Model Of Selective Visual Attention With A Dynamic Bayesian Network. In: IEEE International Conference on Multimedia and Expo., pp. 1073–1076 (2008)
3. Ohno, T., Mukawa, N., Yoshikawa, A.: FreeGaze: gaze tracking systems for everyday gaze interaction. In: Proceedings of the symposium on Eye tracking research & applications, pp. 125–132 (2002)

4. Lucas, B., Kanade, T.: An interactive image registration technique with an application to stereo vision. In: Proc Int'l Joint Conference on Atrificial Intelligence, pp. 674–679 (2005)
5. Yamazoe, H., Utsumi, A., Yonezawa, T., Abe, S.: Remote Gaze Estimation with a Single Camera Based on Facial-Feature Tracking without Special Calibration Actions. In: Proceedings of the symposium on Eye Tracking Research & Applications Symposium, pp. 245–250 (2008)
6. Ishikawa, T., Baker, S., Matthews, I., Kanade, T.: Passive Driver Gaze tracking with Active Appearance Models. In: Proc. 11th World Congress in Intelligent Transport Systems (2004)
7. Viola, P., Jones, M.J.: Robust Real-Time Face Detection. In: International Journal of Computer Vision, vol. 2, pp. 137–154 (2004)
8. Cootes, T.F., Edwards, G.J., Taylor, C.J.: Active Appearance Models. In: European Conference on Computer Vision, pp. 484–498 (1998)
9. Cootes, T.F., Walker, K., Taylor, C.J.: view-based Acitve Appearance Models. In: Forth IEEE Conference on Automatic Face and Gesture Recognition, pp. 227–232 (2000)
10. Hirotugu, A.: A new look at the statistical model identification. In: IEEE Transactions on Automatic Control, vol. 19, pp. 716–723 (1974)
11. Rissanen, J.: Infomation and Complexity in Statistical Modeling. Springer, Heidelberg (2007)
12. McQuarrie, A.D.R., Tsai, C.L.: Regression and Time Series Model Selection. World Scientific, Singapore (1998)

Estimating Human Body and Head Orientation Change to Detect Visual Attention Direction

Ovgu Ozturk, Toshihiko Yamasaki, and Kiyoharu Aizawa

Fac. of Eng. Building 2, 7-3-1 Hongo, Bunkyo-ku, Tokyo 113-8656, Japan

Abstract. This paper presents a method to estimate human body and head orientation change around yaw axis from low-resolution data. Body orientation is calculated by using Shape Context algorithm to match the outline of upper body with predefined shape templates within the ranges of 22.5 degrees. Then, motion flow vectors of SIFT features around head region are utilized to estimate the change in head orientation. Body orientation change and head orientation change can be added to the initial orientation to compute the new visual focus of attention of the person. Experimental results are presented to prove the effectiveness of the proposed method. Successful estimations, which are supported by a user study, were obtained from low-resolution data under various head pose articulations.

1 Introduction

Visual focus of attention analysis of humans has recently attracted remarkable interest from many researchers. To evaluate a given scene, detection or tracking of humans are not enough any more. More semantic understanding of human motions in the scene is required. In this respect, estimating gaze direction of humans in a given scene has become an important problem to detect focus of attention and social interactions in the scene.

In case of low-resolution data, gaze direction estimation becomes a challenging problem. It is very difficult to detect and locate facial features of a human from a low-resolution data, such as when head region size has resolution in the ranges between 20x20-pixels and 40x50-pixels. Especially, locating eye balls accurately to detect gaze direction becomes impossible. In such cases, a person's body and head orientations [1] can provide a hint about where the person is looking at in the scene. Hence, visual focus of attention and intention of the person can be understood, and social interactions in the scene can be interpreted.

In this work, we present an algorithm to estimate body and head orientation change of humans walking in an environment from low-resolution data. Estimated orientation change can be added to the initial orientation to calculate gaze direction in the next step. Our work was inspired by the work of Ozturk et al. [2] for body orientation calculation and builds a new head orientation change estimation algorithm on the top of it. To determine the body orientation, outline of human head-shoulder region is extracted as an edge contour. First of all, various appearances of the upper human body are studied and divided into

R. Koch et al. (Eds.): ACCV 2010 Workshops, Part I, LNCS 6468, pp. 410–419, 2011.

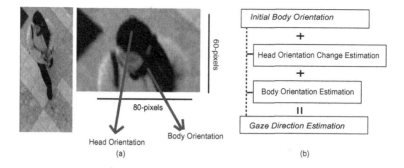

Fig. 1. (a)Human body and head orientation estimation.(b)Overflow diagram.

7 clusters within the ranges of 22.5 ($\pi/8$) degrees. A representative outline of the upper body for each cluster is constructed by studying edge-contours. Then, human body orientation for a given case is estimated by matching the outline of head-shoulder region seen from the camera to the corresponding orientation class. Shape Context [3] is utilized in our framework. After detecting body orientation in a coarse level, head orientation change is estimated, which is later added to the body orientation to estimate gaze direction. Head orientation change is calculated by examining local motion changes in motion flow vectors of SIFT [4] features around head region. SIFT features are chosen because they can be tracked robustly under complex motions compared to other features such as KLT [5], SURF [6], corner-based features, etc.

In our work, the environment is monitored by a single camera mounted sufficiently high above to provide a top-view of the scene. Human appearance resolution in the captured data varies from 70x90 to 100x150 pixels with the head region in the ranges of 20x20 to 40x50 pixels.

Related Work Until now, there have been many researches inspecting human body and/or head orientation problem from various aspects. The earlier researches used multi-camera systems or a combination of camera and sensors [7,8,9,10,11] to detect body orientation or gaze direction. In [7,8,9,11], Voit et al. deal with the problem of head pose estimation in low-resolution images by using multiple-cameras to analyze dynamic meeting scenarios. Glas et al. [12] study the orientation estimation problem by combining video data and laser scanner data. Their work extracts the position of the arms and the head from a top-view appearance and finds the orientation of the human body in the scene. By using a laser scanner system, head orientation can not be detected, where head shape is extracted as a circle.

A big group of researchers have dealt with the problem of detecting facial features [13,14,15,16,17,18] by using high-resolution images and established the geometric localization. Then, they used relative positions of facial features to estimate the gaze orientation. Chutorian and Trivedi [19] give a comprehensive recent survey of head pose estimation. Most of the work in this topic utilizes medium or high-resolution images, captured from a close distance.

A few researches have attempted to solve head orientation estimation with low-resolution data [20,21,22], but they assumed different problem settings such as side-view camera. In [20], Shimizu and Poggio study body orientation estimation of walking people captured from a side-view camera. Body orientation of a person is divided into 16 groups with ranges of 22.5 degrees. They use Haar wavelet responses of each image to generate feature vectors and classify them with Support Vector Machines (SVM). They do not consider the head orientation of the person. Gandhi and Trivedi follow an approach similar to us [21]. They use Histogram of Gradients (HOG) as feature descriptors and SVM for classification. However, they group the orientation of a person very roughly without distinguishing between body and head. They form an orientation set of 45 degrees range for in-plane rotation only. Benfold and Reid [22] proposes a head pose classification algorithm by segmenting the head into skin and non-skin pixels, they try to estimate the head pose in low-resolution images within the ranges of 45 degrees. The work of Zhao et al. [23] is very close to our approach and their result supports the effectiveness of SIFT features. They utilize SIFT feature tracking between two head images to estimate the motion change of the head in 3D with an additional dimension coming from the range image. They work with high-resolution images.

The contribution of the proposed work is that it introduces a framework to estimate *both* body and head orientation change of humans from low-resolution data employing only a single camera. The proposed algorithms can classify body orientation within 22.5 degrees and estimate the head orientation change within five degrees error range. Also, another advantage is that unlike most of the other methods it avoids the necessity of distinguishing between skin or non-skin pixels.

2   System Overview

Figure 1 shows overview of the proposed idea with sample images. To estimate body and head orientation, hence visual focus of attention, the key idea is based on two main processes. The first one is calculating the body orientation; the second one is calculating the change in the head orientation. In other words, starting with an initial orientation of the body, the orientation of the head is calculated at some intervals by adding the orientation change during that interval to the previous orientation. Initially, it is assumed that head orientation is in accordance with the body orientation, which is correct for general cases.

We calculate the initial body orientation of the person by using the Shape Context [3] method. Then, at some intervals we calculate the orientation change of the head by analyzing motion flows of SIFT[4] features around head region. In our work, we propose and show that examining the motion changes of local features around head region for short intervals can help us to describe the motion of the head. Simultaneously, body orientation is also checked to detect major body orientation changes, which is also added while calculating the new gaze direction. In addition, motion flow vector of the center of mass is utilized to find the direction of the global motion and its effects on local motions. In our

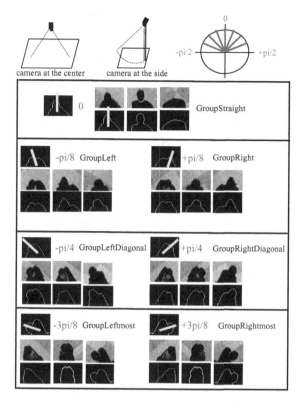

Fig. 2. Body orientation categories calculated from head-shoulder region

dataset, human appearance resolution in the captured data varies from 70x90 to 100x150 pixels with the head region in the ranges of 20x20 to 40x50 pixels.

3   Estimating Body Orientation

Detection of the body orientation correctly is significant at two points. First, body orientation determines the initial state of the orientation calculation which serves as a reference to the successive gaze direction estimation. Second, body orientation estimation is repeated at certain intervals to calculate the change in body orientation, which is later added to the total orientation change.

In our work, to match the appearance of the upper body to the corresponding body orientation, first, the outer contour of the head-shoulder region is extracted by detecting edges on the boundary. The outer contour composed of edge points forms the representative shape of the upper body appearance. Canny edge detection algorithm is used to detect the edges. Figure 2 shows various appearances of upper body and the representation of an example upper body image in terms of edges on the boundary. In our experiments, camera monitors the scene from a very high place. In Figure 2, two types of placement of a camera are shown. We use the placement with the camera at a side. If the camera was placed in a central position on the ceiling, the number of categories would be doubled to include the other half. There are seven groups (-3pi/8, -pi/4, -pi/8, 0, pi/8, pi/4, 3pi/8) in the set, which is used to categorize all possible cases of head region. Shape Context matching algorithm is used to find the correspondence between the input image and the orientation category.

During body orientation estimation step, when an image of the upper body of the person is given, the edge map of the image is compared to the edge maps in the category set in Figure 2. The best three results from the output of Shape Context matching are chosen to vote for the estimated orientation and the winner of the majority voting is selected as the orientation. This helps to improve the matching results by combining various possibilities when the edge contour is insufficient or includes noise. Global motion flow vector is used to define the direction of the body along the estimated orientation. Direction information is important during the estimation of the gaze direction later.

4   Head Orientation Change Estimation

A head has more flexibility in terms of motion types (yaw, pitch, roll) and head orientation of a person can change easily and frequently depending on the person's intent. Our aim is to estimate the head orientation change around yaw axis (assuming that the person is turning his/her head right or left). The orientation change of the head combined with the orientation of the body and global motion, can be used to find the object that the person is paying attention.

To track the orientation change, our key idea is to make use of the motion change in SIFT features around head region. Figure 4 gives the starting point of our estimation method. Considering the ideal case, top-view of a human head can be represented by a circle. SIFT features coming from head region (face, hair, etc.) are represented by the points traveling along the circumference whenever the person rotates his/her head around yaw axis. When $\widehat{AMC}$ represents the orientation change of the person head, the feature A moves to C. Orientation change of the motion flow of the feature is represented by $\widehat{ZAC}$. Table 1 gives

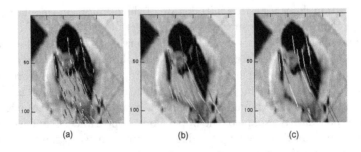

(a)   (b)   (c)

Fig. 3. SIFT flow vectors for interval lengths of: (a)1-frame (b)3-frames (c)5-frames

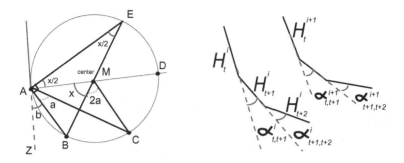

Fig. 4. Angle relations diagram

Table 1. Orientation change of a center point affected by the motion of a point moving around on a circle

Motion of feature point	Motion change: $\widehat{ZAC}$	Orientation change: $\widehat{AMC}$
From A to B	goes to b	goes to x
From A to C	goes to (b+a)	goes to (x+2a)
From A to D	goes to pi/2	goes to pi

some orientation changes of the head and motion flow of A. Considering $\widehat{ZAM} = \pi/2$, $\widehat{BAM} + \widehat{ABM} + \widehat{AMB} = \pi$, $(2*(\pi/2 - b) + x = \pi)$, it can be derived that $x = 2b$. This shows that orientation change of the head can be estimated as two times of the orientation change of the motion flow of the feature. This is the ideal case and it is assumed that previous motion flow vector of A is tangent to the circular motion. This case inspires us to approximate the effect of orientation change of head on the orientation change of motion flow of the feature as 1/2 times. Experimental results show close estimations by using this idea.

SIFT features around head region are tracked at some intervals and their motion flow vectors are constructed as in Figure 4. Motion flow vector of the ith SIFT feature at time t is represented by H_t^i. $\alpha_{t,t+1}^i$ is the angle showing the orientation difference between two motion flow vectors of the ith feature between time t and $t+1$. In the same way, center of mass of the person's body is also tracked and motion flow vector is constructed and represented by C_t. $\beta_{t,t+1}$ is the angle showing the orientation difference between two motion flow vectors of center of mass. The average orientation change of SIFT features around head region between time t and $t+1$ is estimated by the equation 1. The effect of global motion on each feature is suppressed by subtracting $\beta_{t,t+1}$ from each $\alpha_{t,t+1}$. k is the threshold value to eliminate noisy data, n is the number of valid features.

$$\Delta O_{t,t+1} = \frac{\sum_{i=1}^{n}(\alpha_{t,t+1}^i - \beta_{t,t+1})}{n}, \; if |\alpha_{t,t+1}^i| \geq k, |\beta_{t,t+1}| \geq k \qquad (1)$$

$$\Delta VOC_n^{n+N} = 2 \times \sum_{t=n}^{n+N} \Delta O_{t,t+1} + \Delta B_n^{n+N} \qquad (2)$$

To obtain useful information, the tracking interval between t and $t+1$ is important. Figure 3 shows the resultant motion flow vectors, when the interval length is one, three and five. When the interval length is one, motion flow vectors are too short to represent a robust orientation change. When the interval length is five, it becomes harder to keep the continuous tracking of SIFT features, and only a few vectors are obtained. In our method three is chosen to be the interval length, which empirically gives the best results.

N is the length of the duration of observation step. We set observation length to nine or multiples of nine, which gives the best results for various cases. Head orientation change is estimated by the following equation. For example, if N is three, it corresponds to nine frames of video sequence observation, since the interval length between t and $t+1$ is three. ΔB_n^{n+N} symbolizes the orientation change coming from body orientation change. As a result, equation 2 is used to estimate the overall orientation change in the visual attention direction of the person.

5   Experimental Results

Data used in the experiments was captured from a market place in an airport. People are ordinary customers wandering in the area. Camera is placed 12m above the ground. Data was captured in HD mode, with 1440x1080 pixels resolution. Size of the human appearance varies from 70x90 to 100x150 pixels with the head region in the ranges of 20x20 to 40x50 pixels. To validate the experimental results a user study was conducted. 17 people participated in the user study to evaluate the head orientation change.

In the experiments, interval length to construct the motion flows and observation length to evaluate the orientation change are important. In our experiments, interval length is three and observation length is 18 frames. Figure 5 shows head orientation change results of a person walking in the market place. He is turning his head towards left, changing his direction of attention from one place to another. Head region size(including face and hair) is about 25x45 pixels. During 18 frames of motion, the user study gives the orientation change as 27 with the standard deviation of 4.5. Our algorithm calculates the orientation change as 32.

In Figure 6, a woman is wandering in the environment. She is paying attention to an object on her left, and then she switches her attention to another object far in front of her on the right. She turns her head towards right and continues to walk in that direction. Between frames 330 and 342, the user study says that head orientation change is 41.5 with standard deviation of 8. Our algorithm calculates the orientation change as 48 between frames 330-342. It gives 70.5 orientation change between frames 330-348 when the 22.5 body orientation change between frames 342-348 is also added.

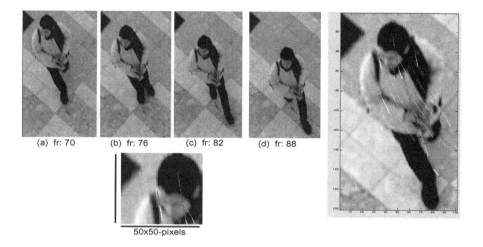

Fig. 5. Experimental results 1

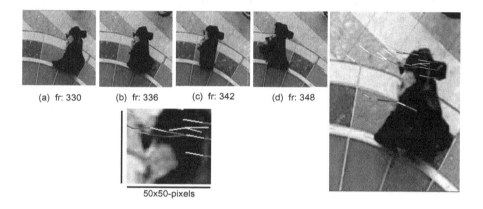

Fig. 6. Experimental results 2

Figure 7 shows one of the exceptional cases. The man turns his head away from the camera and the face of the man is not seen most of the time. Enough number of SIFT features can not be obtained. Hence, only body orientation estimation part works for this case. Body orientation is estimated as $-3\pi/8$, $-\pi/4,-3\pi/8,3\pi/8$, respectively from (a) to (d).

In our work, the cases where people wear bags or hats which occlude head-shoulder region are eliminated. In some other cases, when face is not seen most of the time, SIFT features cannot be tracked. Dark hair region does not provide distinctive image features to track. These situations are challenging for all of the researchers and should be studied further.

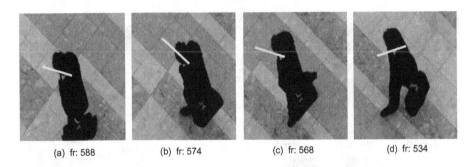

(a) fr: 588   (b) fr: 574   (c) fr: 568   (d) fr: 534

Fig. 7. An example of body orientation change of a walking person

6   Conclusions

In this work, orientation change estimation methods for human body and head orientation from low-resolution images, have been presented to detect gaze direction. With the proposed methods gaze direction change of a walking human can be calculated and new visual focus of attention can be determined. We have assumed a top-view single camera set-up which resulted in low-resolution data with a wide range of human appearances. Shape Context matching of head-shoulder region is used to find the corresponding body orientation. Then, orientation changes in motion flow vectors of SIFT features around head region are utilized to estimate head orientation change. Experimental results on real-world data show the successful estimation of gaze direction change within the error range of five degrees.

References

1. Stiefelhagen, R., Finke, M., Yang, J., Waibel, A.: From gaze to focus of attention. In: Huijsmans, D.P., Smeulders, A.W.M. (eds.) VISUAL 1999. LNCS, vol. 1614, pp. 761–768. Springer, Heidelberg (1999)
2. Ozturk, O., Yamasaki, T., Aizawa, K.: Tracking of humans and estimation of body/head orientation from top-view single camera for visual focus of attention analysis. In: THEMIS 2009 Workshop held within ICCV (2009)
3. Belongie, S., Malik, J., Puzicha, J.: Shape matching and object recognition using shape contexts. IEEE Trans. on Pattern Analysis and Machine Intelligence 24, 509–522 (2002)
4. Lowe, D.: Distinctive image features from scale-invariant keypoints. Intl. Journal of Computer Vision 60, 91–110 (2004)
5. Lucas, B.D., Kanade, T.: An iterative image registration technique with an application to stereo vision. IJCAI (1981)
6. Bay, H., Ess, A., Tuytelaars, T., Gool, L.V.: Surf: Speeded-up robust features. Computer Vision and Image Understanding 110, 346–359 (2008)
7. Voit, M., Nickel, K., Stiefelhagen, R.: Multi-view head pose estimation using neural networks. In: Canadian Conf. on Computer and Robot Vision, pp. 347–352 (2005)

8. Voit, M., Steifelhagen, R.: A system for probabilistic joint 3d head tracking and pose estimation in low-resolution, multi-view environments. In: Fritz, M., Schiele, B., Piater, J.H. (eds.) ICVS 2009. LNCS, vol. 5815, pp. 415–424. Springer, Heidelberg (2009)

9. Voit, M., Steifelhagen, R.: Visual focus of attention in dynamic meeting scenarios. In: Popescu-Belis, A., Stiefelhagen, R. (eds.) MLMI 2008. LNCS, vol. 5237, pp. 1–13. Springer, Heidelberg (2008)

10. Ba, S., Odobez, J.: Multi-person visual focus of attention from head pose and meeting contextual cues. IEEE Transactions on Pattern Analysis and Machine Intelligence 99 (2010)

11. Murphy-Chutorian, E., Trivedi, M.: Hybrid head orientation and position estimation (hyhope): a system and evaluation for driver support. In: Proc. IEEE Intelligent Vehicles Symp. (2008)

12. Glas, D., Miyashita, T., Ishiguro, H., Hagita, N.: Laser tracking of human body motion using adaptive shape modeling. In: Proc.IEEE/RSJ Conf. on Intelligent Robots and Systems, pp. 602–608 (2007)

13. Brown, L., Tian, Y.: Comparative study of coarse head pose estimation. MOTION (2002)

14. Matsumoto, Y., Ogasawara, T., Zelinsky, A.: Behavior recognition based on head pose and gaze direction measurement. In: Proc. IEEE/RSJ Conf. on Intelligent Robots and Systems, pp. 2127–2132 (2000)

15. Wu, Y., Toyama, K.: Wide-range person and illumination-insensitive head orientation estimation. In: Proc. of the Fourth Int'l Conf. on Automatic Face and Gesture Recognition, pp. 183–188 (2000)

16. Sherrah, J., Gong, S., Ong, E.: Face distributions in similarity space under varying head pose. IVC 19, 807–819 (2001)

17. Otsuka, K., Takemae, Y., Yamato, J., Murase, H.: Probabilistic inference of gaze patterns and structure of multiparty conversations from head directions and utterances. In: Washio, T., Sakurai, A., Nakajima, K., Takeda, H., Tojo, S., Yokoo, M. (eds.) JSAI Workshop 2006. LNCS (LNAI), vol. 4012, pp. 353–364. Springer, Heidelberg (2006)

18. Ma, Y., Konishi, Y., Kinoshita, K., Lao, S., Kawade, M.: Sparse bayesian regression for head pose estimation. In: Proc. 18th Intfl Conf. Pattern Recognition, pp. 507–510 (2006)

19. Chutorian, E.M., Trivedi, M.: Head pose estimation in computer vision: a survey. IEEE Transactions on Pattern Analysis and Machine Intelligence 31 (2009)

20. Shimizu, H., Poggio, T.: Direction estimation of pedestrians from multiple still images. In: IEEE Intelligent Vehicle Symposium (2004)

21. Gandhi, T., Trivedi, M.: Image based estimation of pedestrian orientation for improving path prediction. In: IEEE Intelligent Vehicles Symposium, pp. 506–511 (2008)

22. Benfold, B., Reid, I.: Colour invariant head pose classification in low resolution video. In: BMVC (2008)

23. Zhao, G., Chen, L., Song, J., Chen, G.: Large head movement tracking using sift-based registration. In: Proc. ACM Intfl Conf. Multimedia, pp. 807–810 (2007)

Can Saliency Map Models Predict Human Egocentric Visual Attention?

Kentaro Yamada[1], Yusuke Sugano[1], Takahiro Okabe[1]
Yoichi Sato[1], Akihiro Sugimoto[2], and Kazuo Hiraki[3]

[1] The University of Tokyo, Tokyo, Japan, 153-8505
{yamada,sugano,takahiro,ysato}@iis.u-tokyo.ac.jp
[2] National Institute of Informatics, Tokyo, Japan, 101-8430
sugimoto@nii.ac.jp
[3] The University of Tokyo, Tokyo, Japan, 153-8902
khiraki@idea.c.u-tokyo.ac.jp

Abstract. The validity of using conventional saliency map models to predict human attention was investigated for video captured with an egocentric camera. Since conventional visual saliency models do not take into account visual motion caused by camera motion, high visual saliency may be erroneously assigned to regions that are not actually visually salient. To evaluate the validity of using saliency map models for egocentric vision, an experiment was carried out to examine the correlation between visual saliency maps and measured gaze points for egocentric vision. The results show that conventional saliency map models can predict visually salient regions better than chance for egocentric vision and that the accuracy decreases significantly with an increase in visual motion induced by egomotion, which is presumably compensated for in the human visual system. This latter finding indicates that a visual saliency model is needed that can better predict human visual attention from egocentric videos.

1 Introduction

Our visual focus of attention is an important clue for inferring our internal state and therefore can be used effectively for developing human-centric media such as interactive advertising, intelligent transportation systems, and attentive user interfaces. Since our visual focus of attention is closely related to our gaze, many gaze sensing techniques based on various approaches have been developed. However, it is still a difficult task to measure our gaze in unconstrained settings.

An alternative way of estimating the visual focus of attention is to use a visual saliency map model. Inspired by psychological studies of visual attention [1], Koch and Ullman proposed the concept of the saliency map model [2]. Itti et al. subsequently proposed a computational model [3] for predicting which image locations attract more human attention. Since then, many types of saliency map models have been proposed [4,5,6,7,8,9]. The models have been applied not only to static images but also to video clips by incorporating low-level dynamic image

R. Koch et al. (Eds.): ACCV 2010 Workshops, Part I, LNCS 6468, pp. 420–429, 2011.

features such as motion and flicker [4]. Studies based on actual gaze measurement [10,11,12] have demonstrated that such saliency maps match distributions of actual human attention well. However, those studies considered only recorded images and videos. The saliency maps were computed from images shown to human subjects, and their effectiveness was evaluated against the gaze coordinates on the display. While such visual saliency map models can be used for certain applications such as image editing, they lack an important aspect: consideration of the visual motion caused by motion of the observer, i.e., visual motion seen in a static scene captured by a moving camera.

Egocentric vision refers to a research field analyzing dynamic scenes seen from egocentric perspectives, e.g., taken from a head-mounted camera. Egocentric perspective cameras are well suited for monitoring daily ego activities. Accurate prediction of visual attention in egocentric vision would prove useful in various fields, including health care, education, entertainment, and human-resource management. However, the mechanism of visual attention naturally differs significantly in egocentric perspectives. For instance, visual stimuli caused by egomotion are compensated for in egocentric vision, but such a mechanism is not considered in conventional saliency map models. Since conventional models have not been examined for egocentric videos, whether they are valid for such videos is unclear. We have investigated the validity of using conventional saliency map models for egocentric vision. Egocentric videos were captured using a head-mounted camera, and gaze measurements were made using a wearable gaze recorder. The performances of several saliency models and features were quantitatively determined and compared, and the characteristics of human attention in egocentric vision were discussed. To the best of our knowledge, this is the first experimental evaluation of the performance of saliency map models for egocentric vision.

2   Related Work

In this section, we first introduce background theory on visual saliency and briefly review previous work on computational saliency map models.

Due to a person's limited capacity to process incoming information, the amount of information to be processed at a time must be limited. That is why a mechanism of attention is needed to efficiently select and focus on an important subset of the available information [13]. The same holds true for the human visual system; visual attention is necessary to enable a person to handle the large amount of information received through the eyes.

A key to understanding the mechanism of visual attention is feature integration theory [1]. The human visual system first divides incoming images into simple visual features [14]. Since natural objects usually have two or more features, after processing each simple feature separately, the visual system reintegrates the incoming image information. Treisman et al. concluded from their studies that the human mechanism of visual attention includes integration of such visual cues. On the basis of this theory, Koch and Ullman proposed the concept of a visual

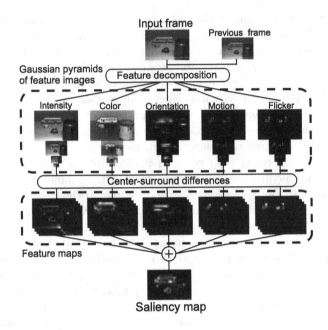

Fig. 1. Procedure for computing saliency maps for videos

saliency map: a two-dimensional topographic map that encodes saliency values for a scene [2]. Those values are generated by integrating simple visual features, and they represent how strongly the region attracts a person's attention.

Itti et al. [3] proposed and developed a fully bottom-up computational saliency map model. They introduced procedures for extracting simple visual features from images; the saliency values are computed through procedures for imitating visual receptive fields. The input is static images, and the output is saliency maps corresponding to the input images. The model was later extended by adding two dynamic features, motion and flicker, so that it can deal with dynamic scenes [4].

Other approaches to saliency map modeling have been proposed. For instance, in the recently introduced graph-based approach [7,8,9], graph representations of input images are generated by defining dissimilarity functions and distance functions between nodes. Saliency values are computed through steady-state analysis of the graphs. Studies using this approach focused mainly on the procedures for computing the saliency values from simple image features rather than on validating the efficiency of the image features used in the models.

3   Procedure for Computing Saliency Maps for Videos

In this study, we used two representative saliency map models to evaluate the validity of using saliency map models for egocentric vision. One is Itti et al.'s model [4] which is based on the center-surround mechanism, and the other is

Harel et al.'s graph-based model [7]. We first introduce the computational procedure of Itti et al.'s model, and then explain Harel et al.'s model.

Figure 1 illustrates the procedure which consists of three main stages. In the first stage, feature decomposition generates Gaussian pyramids of *feature images* from an input frame. In the second stage, "center-surround" mechanism generates *feature maps* from feature images; i.e., saliency maps are computed from each feature. In the third stage, the feature maps are normalized and integrated into a single saliency map.

In the first stage, the input image is decomposed into five types of visual feature images using simple linear filters. The features are typically intensity, color and orientation as static features, and motion and flicker as dynamic features. The intensity feature image is obtained as the average of the red, green, and blue channels of the input images. Itti et al. used two difference images generated by sets of two color channels, i.e., red-green and blue-yellow, for the color feature images. In contrast, we use the Derrington-Krauskopf-Lennie (DKL) color space [15] as color features instead of these difference images. The DKL color space is defined physiologically by three channels used for color processing in the retina and thalamus. Orientation feature images are computed from the intensity image using four oriented ($0°, 45°, 90°, 135°$) Gabor filters.

Two input frames are required for obtaining flicker and motion feature images. The flicker feature image is computed from the absolute difference between the intensity feature images in the current and previous frames. The motion feature images are obtained from the spatially shifted differences between every four orientation feature images of the current and previous frames. As a result, 12 feature images are obtained: one for intensity, two for color, four for orientation, one for flicker, and four for motion. Next, nine spatial scales (scale zero = 1:1 to scale eight = 1:256) are created using dyadic Gaussian pyramids [16] for each feature image.

In the next stage, feature maps are computed from these Gaussian pyramids using the center-surround mechanism. We made six sets of two different sizes of Gaussian pyramids. Six feature maps were computed from each feature image using across-scale image subtraction, which is obtained by interpolation to the finer scale and point-wise subtraction.

In the last stage, the final saliency map is obtained by combining the 72 normalized feature maps (six for intensity, 12 for color, and 24 for orientation, six for flicker, 24 for motion). The normalization is performed by globally multiplying each feature map by $(M - \bar{m})$, where M is the map's global maximum and $\bar{m}$ is the average of its other local maxima. This normalization process suppresses the feature maps with more peaks and thus enhances the feature maps with fewer peaks.

Harel et al.'s model [7] follows the graph-based approach in the second and the third stages. The feature maps and final saliency map are generated by computing the equilibrium distributions of Markov chain graphs. For the second stage, they defined a dissimilarity function and a distance function between nodes and multiplied them together to obtain the weight of each node. In the

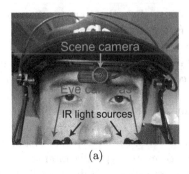

(a)   (b)

Fig. 2. (a) EMR-9 [17], mobile eye tracking system developed by NAC Image Technology. EMR-9 has two eye cameras and two IR light sources to measure gaze movements at 240 [Hz]. It captures egocentric video at 30 [fps] using head-mounted scene camera. Horizontal view angle of scene camera is 121°, and resolution of recorded video is 640 × 480. (b) Example video frame captured the scene camera during experiment.

last stage, they obtained the weight of each node by multiplying the value of the location on the feature maps by the distance function.

4   Experiment

As summarized above, conventional saliency map models use simple, low-level image features as sources to compute saliency maps. They are designed to compute visual saliency for recorded images and videos, but no consideration is given to dealing with visual motion induced by camera motion. To evaluate the validity of using conventional models for egocentric vision, we conducted an experiment.

4.1   Experimental Procedure

To enable us to evaluate the validity of saliency map models for egocentric vision, we designed an experiment that would enable us to determine the correlation between the actual gaze points and the saliency maps for videos captured with a head-mounted camera.

We used the EMR-9 mobile eye tracking system developed by NAC Image Technology [17] to determine the gaze points and to capture egocentric videos. As shown in Figure 2(a), the EMR-9 has two eye cameras and two IR light sources for measuring gaze movement at 240 [Hz]. The scene camera attached to the head captures egocentric video at 30 [fps]. The horizontal view angle of the scene camera was 121°, and the resolution of the recorded video was 640 × 480.

We used the saliency map models of Itti et al. [4] and Harel et al. [7] as baseline models. The experiment was conducted in a room. Four human subjects (one at a time) sat on a chair while another person walked randomly around the room. The subjects were asked to look around the room by moving their head freely for one minute. Figure 2(b) shows an example video frame captured by the scene

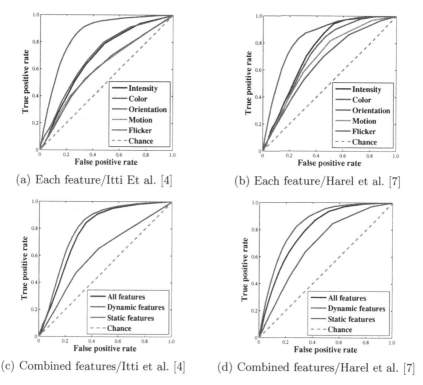

(a) Each feature/Itti Et al. [4]   (b) Each feature/Harel et al. [7]

(c) Combined features/Itti et al. [4]   (d) Combined features/Harel et al. [7]

Fig. 3. ROC curves for each feature ((a) Itti et al. [4], (b) Harel et al. [7]) and for static, dynamic, and all features ((c) Itti et al. [4], (d) Harel et al. [7]). Curves were calculated by changing saliency threshold values from minimum to maximum. Horizontal axis indicates false positive rate, i.e., rate of pixels above threshold. Vertical axis indicates true positive rate, i.e., rate of gaze points for which saliency value of corresponding point on saliency map was higher than threshold.

camera. We obtained about 12,000 gaze points for each subject after removing errors caused by eye blinks.

Human attention is affected by performing a task, but the high-level mechanism of attention cannot be treated efficiently with conventional saliency map models. Since the purpose of our study was to examine the validity of saliency map models for egocentric vision, and thus, we did not assign a task to the subjects.

4.2   Results

To examine how each feature contributes to the accuracy of estimating attention, we compared the correlation between each *feature saliency map*, computed using only one feature, and the actual gaze points. The curves in Figure 3 are the average receiver operating characteristic (ROC) curves, which were calculated

Fig. 4. Examples of gaze trajectory of subject facing moving object (walking person). Images are overlaid with motion feature saliency maps. Crosses show gaze points.

by changing the saliency threshold values from minimum to maximum. The horizontal axis indicates the false positive rate, i.e., the rate of pixels on the map above a threshold. The vertical axis indicates the true positive rate, i.e., the rate of gaze points for which the saliency value of the corresponding point on the saliency map was higher than the threshold.

Figures 3 (a) and (b) compare the feature saliency maps explained in Section **3**. Figure 3 (a) shows the results of using Itti et al.'s model [4], and Figure 3 (b) shows the results of using Harel et al.'s model [7]. Figures 3 (c) and (d) compare the static, dynamic, and standard saliency maps. The static maps were computed using only the static features (intensity, color, and orientation), and the dynamic maps were computed using only dynamic features (motion and flicker). The standard maps were computed using all the features. Figure 3 (c) shows the results of using Itti et al.'s model [4], and Figure 3 (d) shows the results of using Harel et al.'s model [7]. The areas under the curves (AUC) of these three curves, a measure of prediction performance, are shown in Table 1. These results indicate that these saliency map models can predict human egocentric visual attention better than chance. However, with both models, the dynamic features did not contribute to performance. In fact, they even reduced accuracy.

Table 1. AUC of combined saliency maps for two models

	Static features	Dynamic features	All features
Itti et al. [4]	0.803	0.615	0.778
Harel et al. [7]	0.838	0.690	0.793

4.3   Discussion

Our experimental results show that the performance of the dynamic features, motion and flicker, significantly degrades prediction performance for egocentric vision. However, during the experiment, we observed situations in which dynamic visual stimuli attracted the subject's attention. Figure 4 shows examples of the gaze trajectory when the subject was facing a moving object (walking

Fig. 5. Example of the scene in which object quickly changed its color (laptop monitor). Images are overlaid with flicker feature saliency maps. Crosses show gaze points.

person). Figure 5 shows an example scene in which an object quickly changed its color (laptop monitor). In these cases, the subject paid attention to the dynamic changes of the visual features; however, large saliency values are given to the other locations which did not dynamically change. Hence, previously proposed features could not capture dynamic visual stimuli appropriately in our experimental situation.

Unlike the case with recorded images and videos, the case in which we are interested includes the effects of egomotion. While human beings have the ability

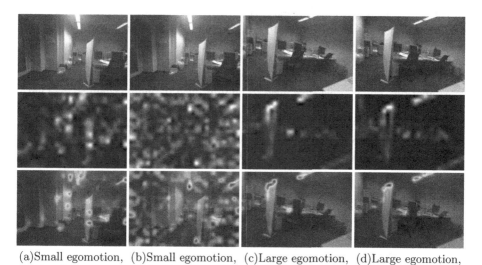

(a)Small egomotion, (b)Small egomotion, (c)Large egomotion, (d)Large egomotion,
motion feature map   flicker feature map   motion feature map   flicker feature map

Fig. 6. Examples of dynamic feature saliency maps with and without effect of egomotion. (a) and (b) show video frames with small egomotion, and (c) and (d) show frames with large egomotion. (a) and (c) are motion feature saliency maps, and (b) and (d) are flicker feature saliency maps. Images in the top row are input images, those in middle row are feature saliency maps, and those in bottom row are input images overlaid with feature saliency maps.

to compensate for egomotion [18], conventional saliency map models do not have a mechanism for such compensation, so high saliency values appear in dynamic feature saliency maps regardless of whether they are caused by egomotion.

Figure 6 shows example dynamic feature saliency maps with and without the effect of egomotion. Figure 6 (a) and (b) shows video frames with small egomotion, and (c) and (d) show ones with large egomotion. Figure 6 (a) and (c) are motion feature saliency maps, and (b) and (d) are flicker feature saliency maps. The images in the top row are input images, those in the middle row are feature saliency maps, and those in the bottom row are input images overlaid with feature saliency maps. As shown in Figures 6 (a) and (b), many peaks appear within dynamic feature saliency maps when the egomotion is small. Since they are suppressed by the normalization in the last combining step, explained in Section 3, these peaks do not substantially affect the final saliency map. In contrast, as shown in Figures 6 (c) and (d), large saliency values are given to the locations with large disparity and to the edges of large intensity difference caused by large egomotion. These feature saliency maps can greatly affect the final saliency map. This indicates that, to model dynamic visual stimuli efficiently, it is necessary to compensate for large egomotion.

5   Conclusion and Future Work

We have investigated the validity of using saliency maps computed from videos captured from an egocentric perspective by experimentally examining the correlation between saliency maps and gaze points. The results show that saliency map models can predict human egocentric visual attention better than chance; however, the dynamic features decreased their performance for egocentric vision because these models cannot model the way a person compensates for the effects of egomotion. The models thus need to be improved to enable them to deal with egocentric videos.

We plan to conduct more experiments under various conditions, e.g., in outdoor scenes and with walking subjects. We also plan to develop a motion compensation mechanism so that the dynamic feature maps work better for egocentric vision.

References

1. Treisman, A., Gelade, G.: A feature-integration theory of attention. Cognitive Psychology 12, 97–136 (1980)
2. Koch, C., Ullman, S.: Shifts in selective visual attention: towards the underlying neural circuitry. Human neurobiology 4, 219–227 (1985)
3. Itti, L., Koch, C., Niebur, E.: A model of saliency-based visual attention for rapid scene analysis. IEEE Transactions on Pattern Analysis and Machine Intelligence 20, 1254–1259 (1998)
4. Itti, L., Dhavale, N., Pighin, F., et al.: Realistic avatar eye and head animation using a neurobiological model of visual attention. In: SPIE 48th AnnualInternational Symposiumon Optical Science and Technology, vol. 5200, pp. 64–78 (2003)

5. Avraham, T., Lindenbaum, M.: Esaliency (extended saliency): Meaningful attention using stochastic image modeling. IEEE Transactions on Pattern Analysis and Machine Intelligence 32, 693–708 (2010)
6. Cerf, M., Harel, J., Einhäuser, W., Koch, C.: Predicting human gaze using low-level saliency combined with face detection. Advances in Neural Information Processing Systems 20, 241–248 (2008)
7. Harel, J., Koch, C., Perona, P.: Graph-based visual saliency. Advances in Neural Information Processing Systems 19, 545–552 (2006)
8. Costa, L.: Visual saliency and atention as random walks on complex networks. ArXiv Physics e-prints (2006)
9. Wang, W., Wang, Y., Huang, Q., Gao, W.: Measuring visual saliency by site entropy rate. In: IEEE Computer Vision and Pattern Recognition, pp. 2368–2375 (2010)
10. Foulsham, T., Underwood, G.: What can saliency models predict about eye movements? Spatial and sequential aspects of fixations during encoding and recognition. Journal of Vision 8, 1–17 (2008)
11. Itti, L.: Quantitative modelling of perceptual salience at human eye position. Visual Cognition 14, 959–984 (2006)
12. Parkhurst, D., Law, K., Niebur, E.: Modeling the role of salience in the allocation of overt visual attention. Vision Research 42, 107–123 (2002)
13. Ward, L.M.: Attention. Scholarpedia 3, 1538 (2008)
14. Broadbent, D.: Perception and communication. Pergamon Press, Oxford (1958)
15. Derrington, A., Krauskopf, J., Lennie, P.: Chromatic mechanisms in lateral geniculate nucleus of macaque. The Journal of Physiology 357, 241–265 (1984)
16. Greenspan, H., Belongie, S., Goodman, R., Perona, P., Rakshit, S., Anderson, C.: Overcomplete steerable pyramid filters and rotation invariance. In: IEEE Computer Vision and Pattern Recognition, pp. 222–228 (1994)
17. nac Image Technology Inc.: EMR-9 (2008),
    http://www.nacinc.com/products/Eye-Tracking-Products/EMR-9/
18. Howard, I.: The optokinetic system. The Vestibulo-ocular Reflex and Vertigo, 163–184 (1993)

An Empirical Framework to Control Human Attention by Robot

Mohammed Moshiul Hoque[1], Tomami Onuki[1], Emi Tsuburaya[1],
Yoshinori Kobayashi[1], Yoshinori Kuno[1], Takayuki Sato[2], and Sachiko Kodama[2]

[1] Graduate School of Science and Engineering, Saitama University, 255
Shimo-Okubo, Sakura-ku, Saitama-shi, Saitama 338-8570, Japan
{moshiul,t.onuki,e.tsuburaya,yosinori,kuno}@cv.ics.saitama-u.ac.jp
[2] Graduate School of Informatics and Engineering, The University of
Electro-Communications, Chofu, Tokyo, Japan
jirowaruq@feel.ocn.ne.jp, kodama@hc.uec.ac.jp

Abstract. Human attention control simply means that the shifting of
one's attention from one direction to another. To shift someone's
attention, gaining attention and meeting gaze are two most important
pre-requisites. If a person would like to communicate with another, the
person's gaze should meet the receiver's gaze, and they should make eye
contact. However, it is difficult to set up eye contact when the two people are not facing each other in non-linguistic way. Therefore, the sender
should perform some actions to capture the receiver's attention so that
they can meet face-to-face and establish eye contact. In this paper, we focus on what is the best action for a robot to attract human attention and
how human and robot display gazing behavior each other for eye contact.
In our system, the robot may direct its gaze toward a particular direction
after making eye contact and the human will read the robot's gaze. As a
result, s/he will shift his/her attention to the direction indicated by the
robot gaze. Experimental results show that the robot's head motions can
attract human attention, and the robot's blinking when their gaze meet
can make the human feel that s/he makes eye contact with the robot.

1 Introduction

Attention attraction or control is a fundamental skill in human social interaction and cognition. People direct their gaze at each other to signal that their
attention directed at the other [1]. Therefore, we may define attention control
as a means of gaze control in which one can shift/control someone's gaze from
one direction to the direction of his/her interest. Control one's attention plays
a critical role in a wide range of social behaviors: it sets the stage for learning,
facilitates communication, and supports inferences about other people's current
and future activities, both in overt and covert. A major challenge is to develop
robots that can behave like and interact with humans. In order to fully understand what humans do or intend to do, robots should be capable of detecting
and understanding many of the communicative cues used by humans.

R. Koch et al. (Eds.): ACCV 2010 Workshops, Part I, LNCS 6468, pp. 430–439, 2011.

In the case of human-robot interaction, the perception of robots is important because it facilitates a bi-directional flow of information: robots must understand what is being conveyed by human's attention behavior, as well as direct their own verbal and non-verbal behavior to humans in an appropriate manner. Current robot researchers developed some robotic systems for eye-contact [2] and joint attention that uses several social cues (for example, gaze [3,4], head and gaze [5], reference term and pointing [6,7]. Most of the previous work assumed that the human faces to the robot when their interaction starts. However, this assumption may not be practical in natural human-robot communication. Therefore, our major concern is how the robot can make eye contact with a human if s/he is not facing to the robot, in other words, if the robot cannot capture his/her eyes or whole face due to the spatial arrangements of the person and the robot.

Humans may use appropriate actions depending on the situation. Although there might be various situations, basic situations can be classified into the following four cases . Case 1 (Fig. 1 (a)): A human subject (H) is not paying his/her attention to any particular thing, and is just looking around by moving his/her head and gaze without any intention. Robot (R) tries to communicate with H. Case 2 (Fig. 1 (b)): While H is concentrating on some object, R tries to contact with H. Case 3 (Fig. 1 (c)): While H1 and H2 are communicating each other, R tries to communicate with one of them (say, H1). Case 4 (Fig. 1 (d)): H is facing the back and R cannot capture the face of H.

In any case, H is not looking at the robot. Therefore, if R would like to start communicating with H, R should perform some action to make H turn his/her face toward R and to gain his/her attention. However, what kind of action is appropriate or effective to attract human attention is solely depends on the social situation. Humans can easily understand the social situation of others by using the social cues and perform an appropriate action based on the situation [8]. Generally, humans use verbal and non-verbal means to attract others' attention. For example, one can attract another's attention simply by calling his/her name if s/he knows it, or using hand movements, head movements, eye movements and reference term. Therefore, robots also need to perform an effective action according to the social situation. However, to the best of our knowledge, there have been no systems that address these issues: how robots can attract human attention and what actions can be effective in a particular situation before establishing eye contact. Thus, we have developed robot heads that can move and on which the eyes generated by CG are projected. Then, we have performed experiments using human participants to examine these issues.

2   Robot Behavior and Architecture

2.1   Robot Behaviors

The purpose of our research is to develop a robot system that can control human attention. We would like to realize a robot that can pick up a particular person and make him/her turn his/her gaze in the direction where the robot would like him/her look. To do so, the robot should first attract the person's attention

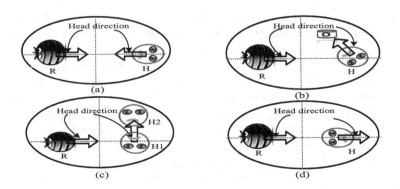

Fig. 1. Possible cases. H is not looking at R in any case.

by some action. The appropriate action may depend on the person's current situation as mentioned in the introduction. Thus, the robot should recognize the current situation to determine its action. In this paper, however, we assume that a target person is in Case 1 situation. Situation recognition is left for future work.

We consider only non-verbal actions since voice or sound may tend to attract attention of others than that of the target person. Although non-verbal actions in general may also gain others' attention, we may be able to design such actions that only the target person may feel that they are intended to him/her. A person's eyes are attracted by the objects in her environment. That is, Eyes are attracted by moving objects and tend to pursue them [9]. In this paper, we consider head and eye movements as robot's non-verbal actions.

Fig.2 illustrates our robot action design to control human attention in Case 1. In Case 1, the person is looking around without any particular attention target (Step 1). If the robot makes some action such as moving its head (Step 2), s/he may notice it and turn his/her head toward the robot (Step 3). The important thing for the robot is to turn its head toward the person at this time. Now the person and the robot face each other. Then, the robot performs some action, blinking in the current implementation, to show the awareness (Step 4). The person will feel that s/he has made eye contact with the robot, and mutual gaze is expected to start. Thus, the robot looks in the direction where it wants him/her to pay attention (Step 5), he/she will look in the direction (Step 6).

We have confirmed that the above scenario can work by developing robot heads and performing experiments using human participants with them.

2.2   Hardware Configuration

We have developed two identical robot heads for human-robot interaction experiments. Fig. 3 shows an overview of our robot head. Each head consists of a spherical 3D mask, an LED projector (Bit AD-MP15A), and a pan-tilt unit (PTU-D46). The 3D mask and projector are mounted on the pan-tilt unit. The projector projects CG generated eyes on the mask as proposed in [10]. Thus,

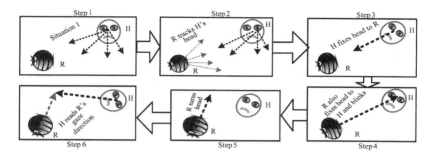

Fig. 2. How to attract and shift participant's attention in Case 1

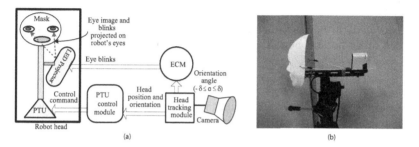

Fig. 3. System overview: (a)Robot configuration consists of three software modules, one USB camera and a robot head (b) Prototype of robot head with a pan-tilt unit PTU-D46, an LED projector and a 3D mask

the robot head can show non-verbal behaviors by its head movement and eye movement. A USB camera (Logicool Inc. Qcam) is installed to track a human head. In the current implementation, the camera is put on a tripod placed at an appropriate position for observing the human.

2.3   Software Configuration

The system has three main software modules: the head detection and tracking module (HDTM), eye-contact module (ECM), and the pan-tilt unit control module. The last module controls the head movement based on the output of the first module. We use FaceAPI [11] to detect and track a human face. It also computes the head direction.

The robot head continuously moves to track the target person's face and to compute its direction. If the face direction is toward the robot, the robot considers that the person looks at it. The robot should show gaze awareness to complete eye contact. In the current implementation, we use eye blinking to show gaze awareness. Since the eyes are CG images, the robot can easily blink the eyes in response to the human's gazing at it. Fig. 4 shows some screenshots of eye behaviors of the robot.

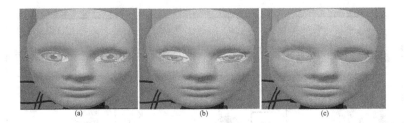

Fig. 4. Eye behaviors of the robot: (a) Fully opened eyes b) Partially opened eyes (c) Closed eyes

3 Experiments

We performed three experiments using human participants to examine if our proposed robot actions can attract and gain human attention.

3.1 Experiment 1: To Attracts Human Attention

In Case 1, the person is looking around without any particular attention target. Although existence of robots may attract his/her attention to some extent, some robot movements, head motion in the current implementation, may help attract more his/her attention. Although this might be apparent, we performed the first experiments to confirm this hypothesis.

Experimental procedure: As mentioned in the previous section, we prepared two identical robot heads. In this experiment, we programmed each to show different behaviors. One is programmed as a static robot (RS). RS does not move at all, just stays as in the initial condition. The second is a moving robot (RM). It is initially static as RS. After some time, however, it starts head movements about 3 seconds, and then becomes static again.

We prepared the experimental settings as shown in Fig. 5.

We hanged five pictures on the wall at the same hight (a bit above the eye level of participants sitting on the chair). We placed two robot heads to the left of the leftmost picture and to the right of the rightmost picture. One of the robot heads worked as RS and the other as RM. The roles of the left and right robot

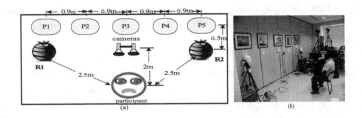

Fig. 5. Experimental settings: (a) Five pictures are hanged on the frontal wall above the eye level. Positions of human, RS and RM are fixed (b) Experimental scene.

heads were exchanged randomly so that the number of participants experienced each case could be almost the same.

We used 12 Japanese students at Saitama University as participants. We instructed them just to look around the pictures. Each experimental session for a participant lasted 60 seconds. We divided the session into two parts (i) static period (first 30 seconds): RS and RM were static, (ii) moving period (last 30 seconds): RM moved several times (3 seconds/motions) depending on the participants head locations and then stopped while RS remained static.

We videotaped all sessions to analyze human behaviors.

Experimental Results: Fig. 6 shows the experimental results. Fig. 6 (a) shows the average numbers of participants' gazing behaviors toward each robot in the static and moving periods. The average number toward RM is significantly greater than that toward RS (t-test, $p < 0.05$). However, participants sometimes ignored the RM's motions due to mental inertia. We measured participants response from their head direction only. Though most of times participants response well by shifting their head and eyes together, sometimes response via eyes only. Fig. 6(b) shows the response rate of each participant, that is, the number of times that the participant turns toward the robot divided by the number of times of the robot movements in each RM session. As it is observed, the proportion of participants responses varied from 50% to 86%. Participants shift their gaze overall 69% of times after getting motion signal from RM. Although sometimes participants ignore the RM's motions, most of the times moving the head is effective to attract human attention than the static head.

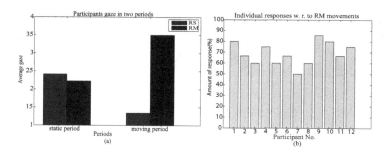

Fig. 6. Attention attraction experiment: (a) Average numbers of participant's gaze actions at the robot during static and moving periods (b) Rate of each participant's response against actions of RM

3.2 Experiment 2: Eye-Contact Experiment

The first experiment has confirmed that robot head motion can attract human gaze. However, if the robot does not look toward the human direction at the time of human's gazing at it, the human will not think that the robot really wants to contact with him/her. Thus, we have developed robot heads as described in the previous section to detect and track the person's face and to move its head toward the person synchronously when s/he turns his/her head toward it. We designed the second experiment to verify the effectiveness of this robot head action.

Experimental procedure: In this experiment, the participants and the experimental settings were the same as in the first experiments except in the behaviors of two robots. We programmed the robot heads in two ways. (i) Gaze Avoiding Robot, RA: if participants look toward the robot, the robot directs gaze in another direction, i.e., they never meet face-to-face, and (ii) Gaze Tracking Robot, RT: the robot tracks the participant's face. If s/he looks at the robot in response to its motions, the robot also directs its gaze (head) to the human, which ensures their face-to-face orientation.

As in the first experiment, participants were asked to look around the pictures in front of them. RT placed in the right position starts tracking action when the participant directs his/her gaze toward P1 (the leftmost picture). This means that RT and the participant are not in face-to-face initially. If the participant notices the robot action, s/he will direct his/her gaze toward RT. RT also reorients its head orientation toward the participant so that they can meet face-to-face. Then, the robot performs eye blinking action to show gaze awareness. Some snapshots in the experimental scene are depicted in Fig. 7. After making eye contact, the robot turns its head toward a particular direction.

(a)   (b)

Fig. 7. Eye contact experiment scenes

After the session, we asked the participants to rate each robot by answering the following two questions: (i) "Did you feel that the robot try to attract your attention?" and (ii) "Did you feel that you made eye contact with the robot after observing the robot's gaze behavior?" on a seven-point Likert scale. We made two predictions in designing these questions: *Prediction 1:* participants feel that RT is trying to attract their attention much more, and *Prediction 2:* participants feel that they made eye contact with RT.

Experimental Results: Fig. 8 (a) shows the average scores for the first question. Comparing the 12 resultant pairs with the Wilcoxon rank sum test gives a p-value of 0.000012 for the first question results. Participants rated RA and RT with an average accuracy of 13.2% (SD=13.2%) and 84.5% (SD=22.9%). The Wilcoxon rank sum test for the second question results gives also a p-value of 0.000019. Fig. 8 (b) illustrates this result. The Scheffe test shows that there are significant differences between the two robots in attracting participant's attention as well as in making them to feel eye contact ($p < 0.05$).

We have proposed and developed the robot head to detect and track a target person and to turn its face toward the target person when s/he notices the robot and turns toward the robot. The experimental results confirm that the proposed robot can effectively attract a target person's attention and establish a communication channel between him/her and the robot.

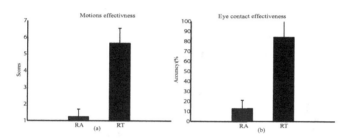

Fig. 8. Eye contact experiment: (a) Participants rating of different actions of robot in terms of effectiveness to attract their attention (b) Comparison of how participants perceived eye contact for the gaze behaviors of RA and RT

3.3   Experiment 3

In this paper, we have mainly considered head movement. However, gaze direction, which can be indicated by the pupil position in the whole eye, is also known as an important communicative non-verbal behavior. Since our robot head uses CG images to display the eyes, it can easily change its gaze direction. Fig. 9 shows some photographs in which the robot is changing its gaze direction.

In the third experiment, we examined preliminarily effect of gaze movement, i.e, eye movement. We programmed the robot head in three motion modes: when the robot detects a human face, (i) it turns only head toward the face,(ii) it turns only eyes toward the face, and (iii) it first turns eyes toward the face, then head motion follows to turn toward the face. Fig. 10 (a) shows the experimental

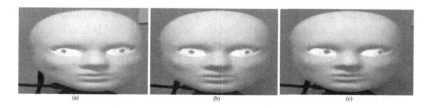

Fig. 9. Some snapshots of robot's gaze changing behavior: (a) Gaze at right (b) Gaze forward (c) Gaze at left

setting. A participant is asked to sit down on the chair and to look at the robot. However, the robot head does not face to the participant. The robot shows three motion modes in random order. After the session, we ask participants, "Which robot motion made you feel most that the robot really looked at you?". We used 54 undergraduate students at Saitama University as participants.

Fig. 10 (b) shows the result of the experiments. Psychological studies [12] show that usually the eyes move first and the head follows when something attracts a person. Thus, we had expected the third would be most supported by the participants. However, the experimental result show that the second mode gained the most votes. This might be because we used the verb "look" in the question. Actually, the CG eye images were well designed and made feel people that the robot looked at them. Although we need to further study the combination of head and eye movements, this experimental result confirms that the eye movement is an effective means in attention control.

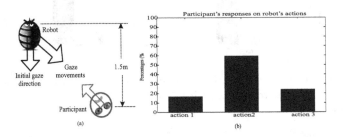

Fig. 10. Eye motions experiment: (a) Participants observed robot's gaze behavior from the fixed position (b) Proportion of participants choices on three actions of the robot

4   Discussion and Conclusion

The main goal of our work is to develop a robot that can shift a particular human's attention from one direction to another. If the human and the robot have already started communication, this turns to be a mutual gaze problem. In this case, since the communication channel has been already established, the robot can control the human's attention easily. The human usually looks in the direction where the robot looks. The robot can further control his/her attention if it uses pointing gestures. In this study, however, we consider the necessary robot actions that bring the person and the robot to such a situation. That is, the robot first gains a little attention of the person, then holds his/her attention completely to establish their communication channel. We have hypothesized that the robot head motion is effective for the first part and that making eye contact by turning its head when the person turns his/her head toward it can achieve the second part. Then, we have developed two identical robot head systems that can be programmed to show the proposed actions and others. Finally, we have verified the hypotheses through experiments using human participants with the robot heads programmed to show various actions.

We have considered the case where the person does not pay attention to a particular target in this paper. If s/he is paying attention to a particular object or talking with another person, the robot needs to use some other actions than non-verbal behaviors, such as voice and patting. There might be other useful actions even in the case treated in this paper. These are left for our future work.

Acknowledgements. This work was supported by JST, CREST.

References

1. Goffman, E.: Behavior in public place. The Free Press, Glencoe (1963)
2. Miyauchi, D., Nakamura, A., Kuno, Y.: Bidirectional eye contact for human-robot communication. IEICE Transactions on Information and Systems, 2509–2516 (2005)
3. Yucel, Z., Salah, A.A., Mericli, C., Mericli, T.: Joint visual attention modeling for naturally interacting robotic agents. In: 14th International Symposium on Computer and Information Sciences, North Cyprus, vol. 14, pp. 242–247 (2009)
4. Mutlu, B., Shiwa, T., Kanda, T., Ishiguro, H., Hagita, N.: Footing in human-robot conversations: how robots might shape participant roles using gaze cues. In: 4th ACM/IEEE International Conference on Human Robot Interaction, La Jolla, California, USA, pp. 61–68 (2009)
5. Nagi, Y., Hosoda, K., Morita, A., Asada, M.: A constructive model for the development of joint attention. J. of Connection Science 15, 211–229 (2003)
6. Hanafiah, Z.M., Yamazaki, C., Nakamura, A., Kuno, Y.: Understanding inexplicit utterances using vision for helper robots. In: 17th International Conferenve on Pattern Recognition, Cambridge, UK, vol. 4, pp. 925–928 (2004)
7. Sugiyama, O., Kanda, T., Imai, M., Ishiguro, H., Hagita, N.: Human-like conversation with gestures and verbal cues based on three-layer attention drawing mode. J. of Connection Science 18, 379–402 (2006)
8. Cranach, M.: The role of orienting behavior in human interaction. J. of Behavior and Environment, 217–237 (1971)
9. Argyle, M., Cook, M.: Gaze and mutual gaze. Cambridge University Press, London (1976)
10. Delaunay, F., Greeff, J.D., Belpaeme, T.: Towards retro-projected robot faces: an alternative to mechatronic and android faces. In: 18th IEEE International Symposium on Robot and Human Interactive Communication, Toyama, Japan, pp. 306–311 (2009)
11. FaceAPI: Face tracking for oem product development, http://www.facepi.com
12. Khan, A.Z., Blohm, G., McPeek, R.M., Lefvre, P.: Differential influence of attention on gaze and head movements. J. of Neurophysiology, 198–206 (2009)

Improvement and Evaluation of Real-Time Tone Mapping for High Dynamic Range Images Using Gaze Information

Takuya Yamauchi, Toshiaki Mikami, Osama Ouda,
Toshiya Nakaguchi, and Norimichi Tsumura

Department of Information and Image Sciences, Chiba University 1-33 Yayoi-cho,
Inage-ku, Chiba 2638522, Japan

Abstract. Using gaze information in designing tone-mapping operators has many potentials over traditional global tone-mapping operators. In this paper, we evaluate a recently proposed real-time tone mapping operator based on gaze information and show that it is highly dependent on the input scene. We propose an important modification to the evaluated method to relief this dependency and to enhance the appearance of the resultant images using smaller processing area. Experimental results show that our method outperforms the evaluated technique.

1 Introduction

With recent advances in systems and techniques that can capture high dynamic range (HDR) images, the need for smart techniques that can handle such images is gaining increasing attention due to their importance for a variety of applications such as digital photography and realistic rendering [6]. Since most of the current display devices have a limited dynamic range which can not accommodate the wide range of intensities of the HDR images, different tone mapping operators have been proposed to scale down the wide range of intensities of HDR images to the narrow range of intensities of such displays. Most of these operators are based on human visual models and traditional photography. A thorough survey of many tone mapping operators for HDR images can be found in [1].

The vast majority of tone-mapping techniques treat all parts in the HDR images equally. However, since people are more interested in image parts they are looking at, some researchers have proposed to use gaze information in tone-mapping. It has been shown that image quality is highly influenced by quality of the gazing area [3]. Therefore, integrating gaze information in tone-mapping techniques is expected to improve the performance achieved using conventional global tone-mapping methods. Rahardja et al. [4] proposed a dynamic approach for displaying HDR images on low-dynamic-range displays that adapt itself interactively based on the user's view. More recently, We [2] implemented a real-time tone-mapping system for HDR images, which is based on the global tone mapping operator proposed by Reinhard et al. [5], taking into account information of the gazing area and showed the superiority of our approach over the conventional

R. Koch et al. (Eds.): ACCV 2010 Workshops, Part I, LNCS 6468, pp. 440–449, 2011.

global tone-mapping operator. However, We did not investigate the impact of variations in content and size of input scenes on the performance of Our system.

In this paper, our previous method in [2] is evaluated using different gazing areas and different scenes. We show that the appearance of the displayed image is highly dependent on the content of the input image. Moreover, based on this evaluation, we propose an important modification to our method in order to relief this dependency and to get more realistic appearance of the displayed image using smaller processing area.

The rest of this paper is organized as follows. In Section 2, we give a brief overview of the global tone mapping operator proposed by Reinhard et al. [5]. This operator is the basis of the gaze information-based modification is presented. In Section 5, we illustrate the effectiveness of our method compared to the other two techniques using several experiments. Section 6 concludes the paper.

2   Reinhard's Global Tone Mapping Operator

Reinhard et al. [5] proposed a tone mapping method for HDR images by simulating Ansel Adams' zone system in traditional photography. The zone system, illustrated in Fig.1, predicts how the scene intensities would map to a set of print zone. A print zone is defined as a region of the scene luminance. There are eleven zones, ranging from pure black (zone 0) to pure white(zone X). The middle gray is the subjective middle brightness region of the scene, which is mapped to print zone V.

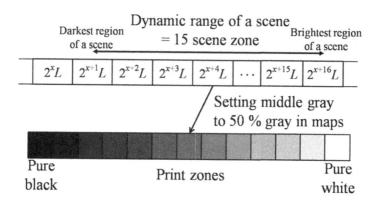

Fig. 1. Mapping from dynamic range of a real scene to print (or display) zones. Scene dynamic range is separated to the scene zones decided by coefficient x and luminance L (reproduced from [5]).

Based on the framework of the zone system, Reinhard et al. assumed the log-average of luminance values as an approximation to the middle gray. This quantity is computed according to the following equation:

$$\bar{L}_{HDR} = exp\left(\frac{1}{N}\sum_{x,y} log\left(\delta + L_{HDR}\left(x,y\right)\right)\right) \tag{1}$$

$$x, y \in V_{Global}$$

where $L_{HDR}(x,y)$ is the luminance of an HDR image for pixel (x,y), N is the total number of pixels in the HDR image, and δ is a constant that prevent a log value to a minus. V_{Global} denotes all pixels of the HDR image.

Accordingly, the luminance of an HDR image is scaled using the following formula:

$$L\left(x,y\right) = \frac{a}{\bar{L}_{HDR}}L_{HDR}\left(x,y\right) \tag{2}$$

where $L(x,y)$ is the scaled luminance of an HDR image, and a is a constant $(\in [0,1])$ specified by the user to allow him to map the log-average to different values of a. The displayed luminance on display is given by:

$$L_{scale}\left(x,y\right) = \frac{L\left(x,y\right)}{1 + L\left(x,y\right)} \tag{3}$$

This global operator expands low luminance levels and compresses high luminance levels.

3   Gazing Area Based Tone Mapping Operator

We proposed a real-time tone mapping system using gaze information [2]. Figure 2 shows the flow diagram of this system.

First, the eyes of the observer are captured by the eye-camera system, and the coordinate of the gazing point on the display is computed from the captured image. Then, the gazing area is determined by the adjacent region of gazing point. Then, based on the global method described in the previous Section,

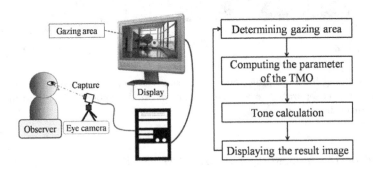

Fig. 2. The real-time gaze information based tone mapping approach presented in [2]

The middle gray is computed by the log-average of the luminance values within the gazing area. The quantity is computed by:

$$P_{HDR}(t) = exp\left(\frac{1}{N'}\sum_{x,y} log\left(\delta + L_{HDR}\left(x,y,t\right)\right)\right) \qquad (4)$$

$$x,y \in V_{Eye}$$

where L_{HDR} , and δ are as in Eq.(1), N' is the total number of pixels in the gazing area, The gazing area is set to the region corresponding to 4 visual degree in [2], t is the frame number, and V_{Eye} denotes all pixels within the gazing area of a still HDR image. V_{Eye} varies with the change of the gazing point while viewing the HDR image.

If only the current frame t is used to compute the parameter $P_{HDR}(t)$, the luminance of the displayed still image would unexpectedly change by cascade. For addressing this problem, we suggested to calculate $P_{HDR}(t)$ using several previous frames as follows:

$$\bar{P}_{HDR}\left(t\right) = P_{HDR}\left(t\right) \times W\left(t\right) + P_{HDR}\left(t-1\right) \times W\left(t-1\right) + \\ ... + P_{HDR}\left(t-n\right) \times W\left(t-n\right) \qquad (5)$$

where W is a weighting function, and n is the number of the considered previous frames. In our previous paper [2], W was a linear function and n was set to 9. Now, the scaling luminance of a still HDR image is computed by the following Equation:

$$L\left(x,y,t\right) = \frac{a}{\bar{P}_{HDR}\left(t\right)}L_{HDR}\left(x,y\right) \qquad (6)$$

where a is constant as described in Eq.(1). After this calculation, the mapping luminance is given by Eq.(3). Finally, the displayed luminance L_d is given by following normalization:

$$L_d\left(x,y,t\right) = \frac{L_{scale}\left(x,y,t\right) - L_{min}}{L_{max} - L_{min}} \qquad (7)$$

4   Proposed Method

In this section, we describe an extension of our gazing area based tone mapping operator. The middle gray of our previous operator runs from very low luminance through very high luminance. It allows dynamic changes of image appearance, but it causes unrealistic appearance. On the other hand, Reinhard's global operator assumes middle gray very well. Therefore, we introduce an important modification to our previous method by imposing a simple constraint on the way of considering the middle gray value. First, we compute the middle gray the same way as in Eq. 4:

(a) Proposed operator   (b) Previous operator   (c) Reinhard's operator

Fig. 3. Result images of each method when looking at the sun (high luminance area)

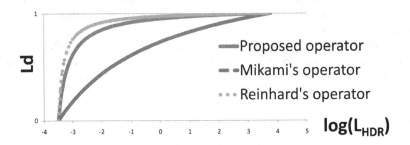

Fig. 4. Input - Output for each algorithm in Fig.3

$$P'_{HDR}(t) = exp\left(\frac{1}{N'}\sum_{x,y} log\left(\delta + L_{HDR}\left(x,y,t\right)\right)\right) \qquad (8)$$

$$x, y \in V_{Eye}$$

Then, according to the following condition, the considered middle gray is either the one calculated using Reinhard's *global* method [5] or the one calculated using our previous gaze information based approach [2]:

$$P_{HDR}(t) = \begin{cases} P'_{HDR}(t) & P'_{HDR}(t) < \bar{L}_{HDR} \times \sigma \\ \bar{L}_{HDR} \times \sigma & otherwise \end{cases} \qquad (9)$$

where L_{HDR} is the middle gray of Reinhard's global tone mapping operator (see Eq.1), and σ is a constant that should be > 1.0. In this paper, σ is set to 2.0 in this study. After this condition is applied, the displayed luminance L_d is computed using equations 5 - 7 in the same way.

Eq.9 aims to restrict the sliding range of middle gray based on global middle gray. Fig.3 and fig.5 shows tone mapped examples for each algorithm. When looking at a high luminance region, our method(Fig.3(a)) works like Reinhard's global operator(Fig.3(c)). Our previous operator(Fig.3(b)) acts like linear scale mapping due to too high middle gray(Fig.4) ,which causes an unrealistic apperance. When looking at a low luminance region, our proposed method improves apperance of dark area like our previous method(Fig.5).

(a) Proposed operator   (b) Previous operator   (c) Reinhard's operator

Fig. 5. Result images of each method when looking at the rock (low luminance area)

5 Experiments and Results

5.1 Experimental Conditions

All experiments were performed in a dark surround. The total number and age range of observers are given in Table 1. We chose five HDR images (Fig. 7) Their images include indoor scenes, outdoor scenes, a scene with people. The long-dimensions of images were approximately 800 pixels. The tone-mapped images were displayed on a 21.5-inch LCD Display(1920×1080 pixels). The images were presented on a 20% gray background. Color normal observers sat on three times the image height from the display.

Gaze information was acquired by using the commercial eye tracking system (EMR-NL8B, nac Image Technology Inc.). The main components of this system include an eyeball photography camera which houses an infra-red LED illuminator and an eye camera, a controller, a LED power supply box, a signal conversion box, and a chin level. Eye movement is detected by calculating the distance between the image of pupil and the reflected image of an infra-red LED illuminator on the cornea.

Table 1. Statistics of each experiment

	Numbers of observers	Age range
Experiment 1	10	21-24
Experiment 2	10	21-24
Experiment 3	10	21-23

5.2 Experiment 1: Checking Suitable Processing Area on Previous Operator

We decided 4 visual degrees for the processing area in computing the middle gray, but the suitable processing area is considered to be dependent on the ratio by size of processing area to image size, image content or display size. Therefore, we conducted the subjective experiment aimed to check the suitable

St. George   People

Falls   Cathedral   Memorial

Fig. 6. Experimental images courtesy of High Dynamic Range Imaging [Reinhard, Ward et al]

processing area on our previous method for each image. We conducted the paired-comparison. This method is a frequently-used technique for generating interval scales of algorithm performance, which are derived using Thurstone's law of comparative judgement [7]. For each pair, observers make a dichotomous choice to which image was preferred in a comprehensive manner. From these comparison data, An interval value is computed based on Z-value.

Figure 8 is the preference scores of experiment 1. The result shows that the suitable processing area is scene dependant and we cannot fix the processing area as 4 visual degree. Two visual degree suited for the scene "People" but not suited for three scenes. 4 visual degree is good for the scene "Falls". 10 visual degree is good for "St.George".

5.3   Experiment 2: Checking Processing Area for Proposed Method

We conducted the suitable processing area on proposed method in the same way as Experiment 1. Figure 9 shows the results of experiment 2. It shows that the suitable processing area is scene dependant too, but 10 visual degree was not chosen. So we can fix the narrower degree than the previous method. Thanks to using narrower visual degree, Proposed method can map the tone more adaptively than the previous method.

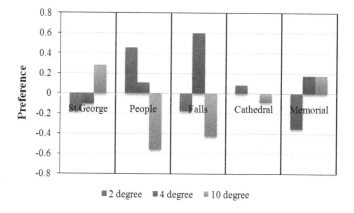

Fig. 7. Preference scores of our previous method in [2]

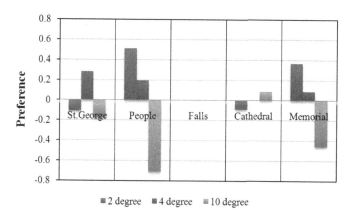

Fig. 8. Preference scores of proposed method

5.4   Preference Comparison of Algorithm Performances

In this experiment, we conducted two comparison to verify the advantage of our proposed method. In the first comparison we confront proposed method fixed 2 visual degree with the previous method fixed 4 visual degree. In the second comparison we confronts proposed method fixed 2 visual degree with the previous method suited the best degree for each scene. We used the same analysis method as in [2], which is one of subjective ratings. Observer answered the subjective score from -3 to 3. Plus score means that observers preffered the tone-mapped images. From these comparison data, we got algorithm scores.

Figure 9 is a result of the first comparison, and Figure 10 is a result of the second comparison. They show that the proposed method is better than the previous method in image preference.

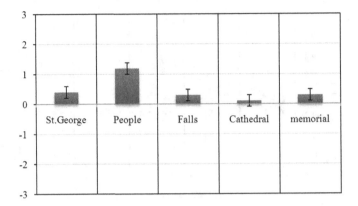

Fig. 9. Proposed method fixed 2 visual degree vs. Previous method fixed 4 visual degree

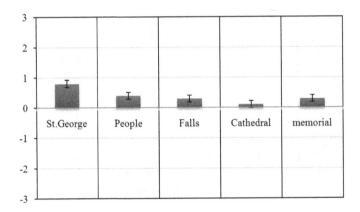

Fig. 10. Proposed method fixed 2 visual degree and the previous method suited the best visual degree. 2 visual degree is suited for "People" and "Cathedral", 4 visual degree is suited for "Falls" and "Memorial", 10 visual degree is suited for "St.George".

6   Discussion

In this paper, we evaluated our real-time tone mapping technique based on gaze information. The experimental result shows that the size of processing area assumed by previous method is not particularly suited, the value is highly scene dependant. Based on this evaluation, we proposed a modification in order to reduce scene dependency. Modification is simple, we add a binary selection when computing the middle gray. It effected not only to reduce scene dependency but also to achieve better appearance, which implies that it is very important to assume the middle gray too. Our proposed method can use smaller processing area than our previous method, so we can map the tone more adaptively.

We impose an upper limit on computing the middle gray in this paper, so we will investigate the influence of imposing a lower limit. We focused on real-time tone mapping for still HDR images, it is interesting to extend our work to cover HDR videos in the future.

References

1. Devlin, K.: A review of tone reproduction techniques. Technical Rep CSTR-02-005 Department of Computer Science University of Bristol (2002)
2. Mikami, T., Hirai, K., Nakaguchi, T., Tsumura, N.: Real-time tone-mapping of high dynamic range image using gazing area information. In: Proc. International Conference on Computer and Information (2010)
3. Miyata, K., Saito, N., Tsumura, N., Haneishi, H., Miyake, Y.: Eye movement analysis and its application to evaluation of image quality IS&T/SID's. In: 5th Color Imaging Conference Color Science Systems and Applications, pp. 116–119 (2009)
4. Rahardja, S., Farbiz, F., Manders, C., Zhiyong, H., Ling, J.N., Khan, I.R., Ping, O.E., Peng, S.: Eye HDR: gaze-adaptive system for displaying high-dynamic-range images. In: ACM SIGGRAPH ASIA 2009 Art Gallery & Emerging Technologies, pp. 68–68 (2009)
5. Reinhard, E., Stark, M., Shirley, P., Ferwerda, J.: Photographic tone reproduction for digital images. ACM Transactions on Graphics 21(3), 267–276 (2002)
6. Reinhard, E., Ward, G., Pattanaik, S., Debevec, P.: High Dynamic Range Imaging, p. 520. Morgan Kaufmann, San Francisco (2006)
7. Thurstone, L.: A law of comparative judgment. Psychological Review (8) (1927)

Evaluation of the Impetuses of Scan Path in Real Scene Searching

Chen Chi [1], Laiyun Qing [1], Jun Miao [2], and Xilin Chen [2]

[1] Graduate University of Chinese Academy of Science, Beijing 100049, China
[2] Key Laboratory of Intelligent Information Processing, Institute of Computing Technology, Chinese Academy of Sciences, Beijing 100190, China
{chichen,lyqing,jmiao,xlchen}@jdl.ac.cn

Abstract. The modern computer vision systems usually scan the image over positions and scales to detect a predefined object, whereas the human vision system performs this task in a more intuitive and efficient manner by selecting only a few regions to fixate on. A comprehensive understanding of human search will benefit computer vision systems in search modeling. In this paper, we investigate the contributions of the sources that affect human eye scan path while observers perform a search task in real scenes. The examined sources include saliency, task guidance, and oculomotor bias. Both their influence on each consecutive pair fixations and on the entire scan path are evaluated. The experimental results suggest that the influences of task guidance and oculomotor bias are comparable, and that of saliency is rather low. They also show that we could use these sources to predict not only where humans look in the image but also the order of their visiting.

1 Introduction

The recognition and localization of objects in complex visual scenes is still a challenge problem for computer vision systems. Most modern computer vision algorithms [1,2] scan the image over a range of positions and scales. However, humans perform this task in a more intuitive and efficient manner by selecting only a few regions to focus on (Figure 1 shows the scan paths of two observers on one stimulus [3]). A comprehensive understanding of human search will benefit computer vision systems in search modeling.

Many studies have been engaging in exploring the mechanisms underlying the human eye movement [4,5,6]. Several sources were considered as its impetuses. Itti et al. considered the role of top-down and bottom-up in visual searching and combined them to speed up the search [7]. Malcolm et al. investigated how the visual system combines multiple types of top-down information to facilitate search and their results indicated either a specific target template or scene context can facilitate search [8]. Considering that there are some biases in oculomotor behaviors (e.g., saccades in horizontal directions are more frequent), Tatler et al. showed that incorporating an understanding of oculomotor behavioral biases into models of eye guidance is likely to significantly improve the prediction of

R. Koch et al. (Eds.): ACCV 2010 Workshops, Part I, LNCS 6468, pp. 450–459, 2011.

Fig. 1. The scan paths of two observers on one stimulus

fixations [9]. Kollmorgen et al. quantified the influences of several sources in a set of classification tasks [10].

Although there are abundant works in understanding the human visual processing, the experimental stimuli and tasks they used were only designed for their purposes that were far from real applications. e.g. searching a red circle in the background consists of messy green triangles. The main purpose of our work is to evaluate the relative importance of the saliency, task guidance and oculomotor biases as impetuses of scan path generation on data collected by Ehinger et al. They recorded observers' eye movements while they searched pedestrians in outdoor scenes   [3]. The task is a nature human behavior and the scenes are the challenges for some computer systems, so the conclusion we got can be see as a valuable consultant for human eye movement prediction designing and can be extended to real applications. We concentrate on the sequential properties of fixations. Specifically, the location of nth fixation is depended on a "guide map" generated based on the location of the n-1th fixation. The contributions of the three sources, saliency, task guidance, oculomotor bias, are examined within each fixation. In addition, computational scan paths are produced to evaluate the predicting power of the model with all of these sources.

The paper is organized as follows. The three kinds of sources and the evaluation framework are introduced In Section 2. Section 3 evaluates the performances of the sources as the impetuses of scan path. Computational scan paths are produced to evaluate the predicting power of the model as well in this section. Section 4 concludes the paper.

2   The Impetus of Scan Path Generation

2.1   Database

The contributions of saliency, task guidance and oculomotor bias are evaluated on the data collected by Ehinger et al. that is available online. The eye movements were collect as 14 observers searched for pedestrians in 912 scenes. Half the images are target present and half are target absent. The image contents include parks, streets and buildings which are close to our daily lives and real applications. Observers were asked to search pedestrians in each trail as quickly as possible and respond whether pedestrians appeared. More details about the

452   C. Chi et al.

database can be found in [3].We excluded the data of which the responding correct rates were lower than 70%. The data with wrong bounding boxes were also excluded.

2.2 Saliency Map

It is widely accepted that in the early stage of visual processing, attention mechanisms bias observer towards selecting stimuli based on their saliency. Saliency is an important bottom-up factor and is regarded as the impetus for the selection of fixation points, especially in free viewing.

In this work, the saliency maps were generated by using the toolbox developed by Itti et al. which computes the color, intensity, orientation features then combines the results of center-surround difference on these features to generate a saliency map [11]. Examples are shown in Figure 2. The second column is the saliency maps. The lighter regions are regarded as salient regions.

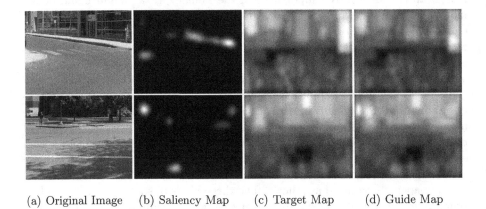

(a) Original Image   (b) Saliency Map   (c) Target Map   (d) Guide Map

Fig. 2. Examples of the saliency maps, the target maps and their combinations. The lighter regions have higher probabilities to be fixated on.

It could be observed that saliency finds the regions that differ from their surroundings. People who appear on the road, grass, horizon and other homogeneous scenes will be found efficiently with saliency.

2.3 Task Guidance

Task is the high-level information on eye movement guidance. Many studies have already proved that in a search task saccades would be directed to the image regions similar to the target, e.g., an item has a rectangular shape, a circle on the top, and some complexional regions would catch more fixations in a searching people task.

In this work we employ a "target map" to implement the guidance of the task. A target map is a matrix of which each value measures how likely the target locates at that point. The more similar an item to the target, the more likely it will be chosen as the next fixation. The target map finds out all the target-like items including both target and distracters. We used the output of detector developed by Dalal and Trigger [2] which were publicized with the eye data by Ehinger et al.. Some examples are shown in the third column of Figure 2.

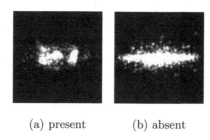

(a) present   (b) absent

Fig. 3. The preferred saccade amplitude modes learned from the validation sets for the pedestrians present and absent. Each saccade start point was aligned to the figure center and the saccade end point was plotted on the figure according to the saccade length and angle. All the points were convolved by a Gaussian.

2.4   Oculomotor Bias

It has been noticed that there are some biases in oculomotor behaviors, e.g., saccades in horizontal directions are more frequent. Oculomotor biases guide the eye movement mainly depend on spatial constrains and properties of human oculomotor system rather than any outside information. Many studies observed that people tend to view the image with preferred saccade lengths and angles. It could reflect the ways that people deal with information in their daily lives which are "trained" for several years.

We counted the first six saccades for each scan path over images and observers in the validation set consisting of 120 images half with pedestrians and half without to learn the saccade amplitude that people preferred. Figure 3 shows the preferred range of saccade which is irrespective of the image contents. For each case (with pedestrian or without pedestrian), we aligned all the beginning points of saccades to the center. Then each end point was plotted on the figure according to the relative position to the corresponding beginning point (which is controlled by saccade length and angle). All saccades were put on one figure. The final figure was found by convolving a Gaussian over all the end points.

It can be found that in both the target present case and target absent case people tend to saccade horizontally and the center is preferred in Figure 3. Lack of the target attraction, the ranges of saccades are more uniformly in the horizontal direction in the target absent case.

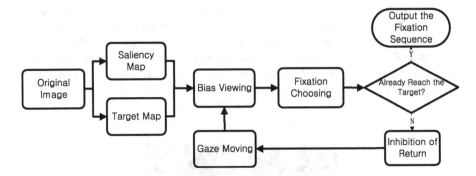

Fig. 4. The flow of fixation sequence generation. The saliency map and target map are combined at the first stage. Within each round of fixation choosing, the combined map is adjusted by oculomotor biases then the points with the highest value pop-out. An Inhibition of Return mechanism also takes part in.

2.5   Scan Path Generation Flow

In visual search task, eye movement employs not only parallel processing but also serial fixation choosing [12]. Thus we modeled the generation process of scan path by sequentially choosing fixations as shown in Figure 4. As stated in [13], within each fixation all the points in the view compete for being biased as next fixation. A guide map(Gui) which is the combination of target map(Tar) and saliency map(Sal) can be regarded as the basis of the competition among all the points. The combination was described as follows:

$$Gui(x,y) = Sal(x,y)^{\gamma_1} \cdot Tar(x,y)^{\gamma_2} \tag{1}$$

The two exponents (γ_1, γ_2) are constants confined in range$[0,1]$, their value($\gamma_1 = 0.95, \gamma_2 = 0.05$) were selected by various tests on the validation set consisting of 120 images to make the guide map best predicted the fixation distribution.Two examples of combined results are shown in the last column in Figure 2.

Within each fixation, the guide map will be adjusted by oculomotor bias as Equation 2, where OB represents the oculomotor bias map.After that the new fixation with the highest value pops-out and a new round of fixation selection starts. This iteration operates until the selected fixation land on the target.

$$\widetilde{Gui}(x,y) = Gui(x,y) \cdot OB(x,y) \tag{2}$$

3   The Performances of the Sources on Scan Path Generation

We compared the several combinations of the target map, saliency map and oculomotor bias with recorded human fixations to evaluate the contributions of the sources. The combinations were achieved by integrating or removing the related modules in Figure 4. The ROC curves were employed to measure the predicting power of the attended sources by comparing the corresponding guide map to the following fixation.

This process was done for each fixation number and the results were averaged over the observers and images. Then a series of ROC curves corresponding to each fixation number were attained. Because the observers used 3.5 fixations to land on the target averagely, for each combination the AUC was computed and averaged over the first four fixation numbers.

Computational scan paths were produced to evaluate the performance of the model with all the sources on whole scan path generation. The Edit Distance was introduced to evaluate the performance.

3.1 Single and Combined Sources

We measured the performance for each source at first. The results are shown in Table 1. We use T, S, and O to represent Target map, Saliency map and Preferred saccade amplitude respectively for short.

Table 1. The mean AUC for the first four fixations for each source

Impetus	Present	Absent	Impetus	Present	Absent
T	0.8115	0.7529	$T+O$	0.8178	0.8540
S	0.6282	0.5792	$S+O$	0.7939	0.8283
O	0.7886	0.8469	$T+S+O$	0.8201	0.8568

We attain the following observations.

(1)All the sources outperform the chance (AUC=0.5) ,which indicates that these sources can be seen as the impetuses of human eye movement. (2)Comparing with the target map and oculomotor bias, saliency map performs worst. It suggests that in searching task, the contribution of saliency is limited. Saliency is important but not crucial. (3) Without the target, eye movements are largely controlled by oculomotor bias whereas when targets are present, the eye movements highly connect to the target.

We also examined the performance of combined sources. Shown in the righter columns of Table 1.

In conjunction with the previous observations, Table 1 suggests that the combinations with oculomotor bias perform better than single source . As shown in the righter columns of Table 1 the performance of the model with all the sources is highest in both the target present and target absent. Figure 5 shows ROC curves.

In Figure 5,the performance goes down from the fourth fixation in the target present case. It coincides with the fact that observers used 3.5 fixations to find the pedestrians averagely. After reaching the pedestrians, the fixations are less controlled by the the sources. However,in the target absent case, there are not obvious differences among the fixation numbers. It suggests that the observers stayed at the "searching" condition for a longer time.

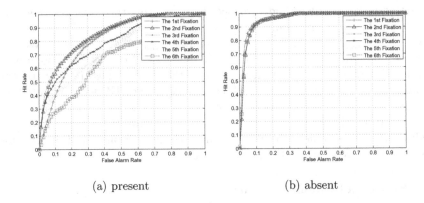

(a) present   (b) absent

Fig. 5. ROC curves for each fixation number. For each case, all the three sources take part in . The mean AUC for the first four fixations is 0.8201 for target present and 0.8568 for target absent.

3.2   Whole Scan Path

As discussed in the last section, the model with all the sources performs best. In this section we evaluate the model combining all the source on whole scan path generation. Edit Distance [14] was used to measure the similarity between the computational scan paths and recorded scan paths.

Edit Distance is a useful measurement for describing the similarity of two strings of characters. Different string manipulations (Insert, Delete, Replace and None) are appointed to different costs. We used Edit Distance to plan operation serial which costs least to transform one scan path into another.

Before comparing, the scan paths were discretized into strings of chars based on the location of each fixation. The images were meshed and the fixations in the same grid were tagged identically. The mesh size is 3 degree visual angels. The fixations closed to each other were regarded as the same points even they located in different grids. Then the scan paths were translated into the stringed tags of its fixations.

The computational scan paths were generated for each image and compared with the recorded scan paths of the 14 observers. Because the standard Edit Distance calculation returns the total costs, we averaged it to the cost of each manipulation. Similarity is defined as the reciprocal of the normalized Edit Distance and averaged over the observers and images. The final result is 0.3138 for pedestrians present and 0.2198 for pedestrians absent by averaging the value over all the images.

For comparison we examined the inter-observer consistency on the data as the upper bound. For each stimulus, we used scan paths of all-except-one observers to predict scan path of the excluded observer. Then the values were averaged over the observers and images. As for the lower bound, we generated a set of scan paths by randomly selecting several points as fixations. The consistency

(a) present   (b) absent

Fig. 6. The artificial fixation sequences generated by our model for the two cases. The red dots wrapped by yellow circles denote the fixation of collected human eye data and the green dots wrapped by blue circles denote the artificial fixations generated by our model. Two time adjacent points are connected by red or blue edges. The red rectangle in (a) denotes the ground truths (pedestrians).

458   C. Chi et al.

Table 2. The performance of the artificial model

Impetus	Present	Absent
Human consistency	0.3616	0.2850
Computational scan paths	0.3138	0.2198
Random	0.1919	0.1919

among the random paths was also computed. The results are shown in Table 2. Figure 6 show some examples of the computational scan paths.

In Table 2,the human consistency is higher than the random paths in both the two cases. It indicates that the scan paths of observers' visiting are similar. The consistency in target present case outperforms the target absent case shows that with the target, the ways people visit the images are more alike.

As for the performance of the combined sources, there are two observations. First, it outperforms the random scan paths. This indicates that these sources could be used in modeling the scan paths of human eye movement. In addition, in target present case, it achieves 86% of human consistency and in the target absent case, the number is 76%. The difference of the two numbers suggests that in the target absent cases, there maybe some other sources controlling the scan paths e.g. context, semantic interpretation of the objects beside the sources we considered.

4   Conclusions and Future Work

In this work, we mainly evaluated the performances of the target map, saliency map and oculomotor bias as the impetuses of scan path generation. Several combinations were considered. We found that in target present case, the task guidance performs best and the oculomotor bias takes the second place whereas in the target absent case, the oculomotor bias plays best and the task guidance takes the second place. In both the two cases, saliency performs somewhat lower. We also evaluated their performances on the whole scan path generation. The result shows that these sources could be used in modeling the human eye movements.

However, the visual processing is a complicated system that a lot of factors may contribute to it. Although we proved that the three sources can be seen as the impetuses, they are still imperfect. Many other sources should be introduced to make the model performs better such as context, the past experiences of the observers, people habits and so on.

A significant work in the future is to use the analyses to real applications to improve efficiency and accuracy. Many studies such as image understanding, image and video compression will be benefited from the eye movement in searching.

Acknowledgements

This research is partially sponsored by National Basic Research Program of China (No. 2009CB320902), Natural Science Foundation of China (Nos.60702031,

60970087, 61070116 and 61070149), Hi-Tech Research and Development Program of China (No. 2006AA01Z122) and Beijing Natural Science Foundation (Nos. 4072023 and 4102013).

References

1. Viola, P., Jones, M.J.: Robust real time object detection. In: Workshop on Statistical and Computational Theories of Vision (2001)
2. Dalal, N., Triggs, B.: Histograms of oriented gradients for human detection. In: CVPR, pp. 886–893 (2005)
3. Ehinger, K., Hidalgo-Sotelo, B., Torralba, A., Oliva, A.: Modelling search for people in 900 scenes: A combined source model of eye guidance. Visual Cognition 17, 945–978 (2009)
4. Yarbus, A.: Eye movements and vision. Plenum press, New York (1967)
5. Rayner, K.: Eye movements in reading and information processing. Psychological Bulletin 85, 618–660 (1978)
6. Robinson, D.: The mechanics of human saccadic eye movement. The Journal of Physiology 174, 245 (1964)
7. Navalpakkam, V., Itti, L.: An integrated model of top-down and bottom-up attention for optimizing detection speed. In: CVPR, pp. 2049–2056 (2006)
8. Malcolm, G., Henderson, J.: Combining top-down processes to guide eye movements during real-world scene search. Journal of Vision 10 (2010)
9. Tatler, B., Vincent, B.: The prominence of behavioural biases in eye guidance. Visual Cognition, 1–26 (2009)
10. Kollmorgen, S., Nortmann, N., Schrder, S., Knig, P.: Influence of Low-Level Stimulus Features, Task Dependent Factors, and Spatial Biases on Overt Visual Attention (2010)
11. Itti, L., Koch, C., Niebur, E.: A model of saliency-based visual attention for rapid scene analysis. IEEE Transactions on pattern analysis and machine intelligence 20, 1254–1259 (1998)
12. Maioli, C., Benaglio, I., Siri, S., Sosta, K., Cappa, S.: The integration of parallel and serial processing mechanisms in visual search: Evidence from eye movement recording. European Journal of Neuroscience 13, 364–372 (2001)
13. Findlay, J., Gilchrist, I.: Eye guidance and visual search. Eye guidance in reading and scene perception, 295–312 (1998)
14. Levenstein, A.: Binary codes capable of correcting deletions, insertions and reversals. In: Soviet Physics-Doklandy, vol. 10 (1966)

Author Index